D0753251

EXCITED STATES
IN
ORGANIC CHEMISTRY

EXCITED STATES
IN
ORGANIC CHEMISTRY

J. A. BARLTROP

Brasenose College, Oxford

and

J. D. COYLE

Department of Physical Sciences,
The Polytechnic, Wolverhampton

JOHN WILEY & SONS

London · New York · Sydney · Toronto

Copyright © 1975, by John Wiley & Sons, Ltd.

All rights reserved.

No part of this book may be reproduced by any means, nor
transmitted, nor translated into a machine language without the
written permission of the publisher.

Library of Congress Cataloging in Publication Data:

Barltrop, J. A.
Excited states in organic chemistry.

'A Wiley-Interscience publication'
1. Photochemistry.   2. Chemistry, Physical organic.
I. Coyle, John D., joint author.   II. Title.

QD715.B37   547'.3   74-22400
ISBN 0 471 04995 6

Printed in Great Britain by J. W. Arrowsmith Ltd.,
Winterstoke Road, Bristol

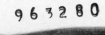
963280

Preface

This book is about the electronically excited states of organic molecules, and it treats all aspects of their behaviour—their creation, the radiative and non-radiative pathways by which they are deactivated, and their chemical reactions, i.e. photochemical reactions.

The early chapters are concerned with the theoretical foundations of the subject—with the production of excited states and with their time-independent properties and their time-dependent behaviour. Here, an attempt has been made to get beneath the mathematical formalism of quantum mechanics and electronic spectroscopy and to display the underlying concepts in simple (and perhaps simplistic) terms which may be readily understood and assimilated. Whether or not a particular photochemical reaction is observed depends on the (often complex) relationships between its rate constant and the rate constants for alternative energy-dissipating processes, so that kinetics are given a fundamental place in our treatment. Group theory, another powerful tool for excavations in this area, is treated in 'black box' fashion in an appendix.

In the later chapters, in which photochemical reactions are examined in detail, we have made a classification by chromophore rather than by reaction type. This latter classification has been widely used by other authors, but we think that it can be misleading. We believe that it is more important to the development of new photochemical synthetic techniques and to the interpretation and rationalization of photochemical reactions to have a synoptic view of, say, the (n, π^*) states of the carbonyl group or of the (π, π^*) states of the C$=$C chromophore, rather than to see the (often formal) analogies between the two systems. Hence we present our treatment by chromophores.

To whom is this book addressed? We hope that postgraduate students will find it useful, for it deals with the subject in greater depth than do introductory texts which have appeared in recent years, and the references to review articles and to the original literature may permit it to serve as a reference work for practising photochemists. Equally, we hope that the book will be of use to undergraduates, of whatever seniority, taking courses in photochemistry. A particular lecture course may have its own emphasis, whether towards photophysical processes or towards descriptive organic photochemistry, and we have attempted to give an up-to-date, reasoned and balanced account of all that can vi

happen when a photon of ultraviolet or visible radiation is absorbed by an organic molecule.

We are indebted to Drs Howard Carless, John Nelson, David Phillips, Graham Richards, Christopher Samuel and Peter Wagner for their helpful and constructive criticism of the manuscript.

August 1974

JOHN BARLTROP, Oxford
JOHN COYLE, Wolverhampton

Contents

xii

Chapter 1

Introduction and Basic Principles

The field of photochemistry covers all processes which involve chemical change brought about by the action of visible or ultraviolet radiation, and these processes generally involve the direct participation of an electronically excited state of a molecule. Many life processes involve photochemical reactions, such as those of photosynthesis and vision,[1] and this reflects the fact that the major source of energy on earth is the sun's radiation. Photographic processes have been in use for well over a century, and these too are based on the employment of visible radiation to produce chemical change in a system.[2] In this book is described a rather different range of photochemical reactions, namely those brought about in relatively simple organic compounds by the deliberate use of visible or ultraviolet radiation as an energy source. The qualitative study of such reactions began long ago, and more detailed study of simple gas phase processes followed, but it is only since about 1950 that intensive and systematic study of liquid and solid phase photochemical processes has emerged.

With the growth of such investigations the horizons of chemistry have widened considerably. In principle, the ground electronic state of any compound can give rise to a number of different excited electronic states, each with its own characteristic properties and electron distribution, and each might have a chemistry as varied as that of the ground state. In practice, the range of observed chemical reactions of excited states is restricted by the existence of (very) rapid physical processes by which one excited state of the molecule is converted to another state of lower energy, but this still allows for an enormous amount of 'new' chemistry. The importance of photophysical processes in a complete description of photochemistry is reflected in the space allotted to them in this text, and the combined study of physical and chemical processes is leading to a greater understanding of the nature and properties of molecules in different electronic states.

Photochemical reactions have made a considerable impact in synthetic chemistry, both in research laboratories and in commercial processes. Compounds can be made by a photochemical route which are difficult, if not impossible, to prepare by a thermal method, and others can be made more readily or at lower cost. Looking to the future it seems certain that a much wider range of useful applications of photochemical reactions in synthesis will be developed, and linked with this will be the continuing search for precise mechanistic information and development of the theoretical basis.

1.1   THERMAL CHEMISTRY AND PHOTOCHEMISTRY

Thermal and photochemical reactions are simply different aspects of chemistry, and on the whole the same basic theoretical considerations and descriptive models can be used in both areas. The rationalization of observed chemical change in terms of electron distribution in molecules and electron re-organization during the course of a reaction step can be applied to all chemical processes, as can such secondary considerations as the effects of sterically bulky groups on the rate of reaction, or the rationalization of the stereochemical course of concerted reactions on the basis of orbital interaction. One of the major causes of difference between thermal chemistry and photochemistry lies in the differences in electron distribution in ground and excited electronic states of a molecule, which can lead to major alteration of chemical behaviour. Similarly, although thermodynamic considerations of the feasibility of reaction apply throughout chemistry, it is in this area that the cause of a second major difference between thermal chemistry and photochemistry is found. Since an electronically excited state of a molecule has a higher (often a much higher) internal energy than the ground state, there exists a much greater choice of reaction product for the excited state on thermodynamic grounds. The comparison is between a reaction $A \rightarrow B$ and a reaction $A \xrightarrow{h\nu} A^* \rightarrow B$, and there will be many systems for which $A^* \rightarrow B$ is thermodynamically favourable where the corresponding reaction $A \rightarrow B$ is not.† In particular, an excited species can give rise to high energy products such as radicals, biradicals or strained ring compounds which are not readily formed (if they can be formed at all) from the ground state.

In a photochemical reaction, thermal equilibrium between excited state, intermediate(s) and product(s) is rarely achieved, because of the magnitude of some of the energy changes involved and because of the high rate constants for many of the individual steps. It is a kinetic model of the system which is therefore often of great value in mechanistic interpretation.

For a photochemical reaction to be readily observable, the rate constant for the initial photochemical step involving the excited state must be high (typically 10^6–10^9 s^{-1}). This is because the excited state is short-lived, decaying to the ground state very rapidly, and an efficient photochemical reaction must compete successfully with these very rapid photophysical processes. The decay processes may be radiative (fluorescence or phosphorescence) or non-radiative (internal conversion or intersystem crossing). It may be that there are many photochemical reactions as yet unobserved because their quantum yields are very low as a result of such competition.

---

† In this book a superscript asterisk (e.g., M*) denotes an electronically excited species. A double asterisk (M**) implies an upper excited state, and where relevant a numerical superscript before the symbol (^{3}M* or ^{1}M*) denotes spin multiplicity as triplet or singlet respectively.

The rate constants for primary photochemical processes, like the rate constants for individual thermal reaction steps, vary with temperature, and the empirical variation can be expressed in the Arrhenius form

$$k = A \exp(-E_a/RT)$$

The activation energies for excited state reactions are generally small, often less than 30 kJ mol^{-1} (7 kcal mol^{-1}), and this is a corollary of the fact that only fast photochemical reactions are detectable.

1.2   ELECTRONIC STRUCTURE OF MOLECULES

Molecular orbitals probably afford the clearest understanding of the electronic structure of molecules and of the changes in electronic structure brought about by the absorption of electromagnetic radiation.[3,4] The molecular orbitals are formulated as linear combinations of the valence shell atomic orbitals—it is assumed that the inner electrons remain in their original atomic orbitals. For example, the interaction of two identical atomic orbitals ϕ_A and ϕ_B gives rise to two molecular orbitals of the form (equation 1.1).

$$\psi_1 = \phi_A + \phi_B$$
$$\psi_2 = \phi_A - \phi_B$$

$$(1.1)$$

One molecular orbital is bonding (i.e. more stable than the initial atomic orbitals), and the other is antibonding (i.e. of higher energy than the initial atomic orbitals). The situation is depicted in Figure 1.1.

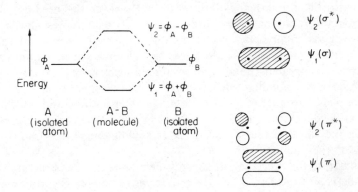

Figure 1.1. Interaction of two identical atomic orbitals

The form of the molecular orbitals is important and is also depicted in Figure 1.1. Those orbitals which are completely symmetrical about the internuclear axis are designated σ (sigma) if bonding or σ^* (sigma star) if antibonding, and these can arise if ϕ_A and ϕ_B are 's'-orbitals, for example. Molecular orbitals derived by mixing two parallel 'p'-orbitals are called π (pi) and π^* (pi star).

4

If the atomic orbitals are each singly occupied, or if one is doubly occupied and the other is vacant, the electrons in the molecular system both occupy the low-energy bonding molecular orbital. This leads to a gain in stability over the isolated atoms, and it is the basis of the molecular orbital description of electron-pair covalent bonding.

Molecular orbitals can encompass more than two atomic centres, and this leads to electron delocalization. The π-molecular orbitals of buta-1,3-diene are an example of this, and they are obtained by taking linear combinations of the four C(2p) orbitals. Their form is shown in Figure 1.2.

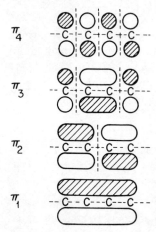

Figure 1.2. The π-molecular orbitals of buta-1,3-diene. The wavefunction has opposite signs in regions of the orbitals which are cross-hatched or left blank. Nodes, where the wavefunction changes sign, are shown as dotted lines

In a similar way to this, the σ-framework of an organic molecule consists of molecular orbitals embracing all the atoms.† However, this need not normally be considered, and for most purposes an adequate representation of the σ-framework is obtained by making use only of two-centre molecular orbitals. The framework thus consists of localized two-electron covalent bonds.

In certain compounds, notably those containing elements of Groups V, VI or VII, there are non-bonding valence shell electrons (designated n) which, as their name implies, are not involved in bonding relationships and which can be regarded as being localized on their atomic nuclei. The energy of such electrons is much the same as that of electrons occupying the corresponding atomic orbitals on the isolated atom.

To illustrate the preceding ideas, consider the electronic structure of formaldehyde ($H_2C=O$). This is described in terms of pairs of electrons occupying three localized σ-bonding molecular orbitals ($C-H$, $C-H$, $C-O$), one localized π-bonding molecular orbital ($C-O$), and two non-degenerate, non-bonding orbitals on oxygen (a p-orbital and a sp hybrid orbital).[5] The 'core' 1s electrons on carbon and oxygen are ignored. The shapes of these orbitals and the antibonding orbitals are shown in Figure 1.3.

† Note that the $\sigma-\pi$ approximation is adopted here, which implies a lack of interaction between the σ- and π-orbitals because of their different symmetry. This approximation is adequate for most discussions of molecular structure.

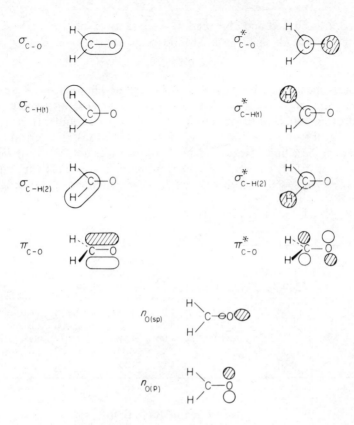

Figure 1.3. Molecular orbitals of formaldehyde

1.3   ELECTROMAGNETIC RADIATION

Electromagnetic radiation, of which visible light and ultraviolet radiation are examples, can be envisaged in terms of an oscillating electric field and an oscillating magnetic field operating in planes which are perpendicular to each other and to the direction of propagation. The time-variable strength of each field at a given point is described by a sinusoidal function. In a beam of normal radiation the orientation of the fields with respect to the surroundings is random, but plane polarized radiation, in which this orientation is restricted to a particular plane, can be produced using certain ordered arrays of ions in crystals or of molecules in a matrix to absorb and transmit selectively radiation with a particular direction of polarization. A property of such plane polarized light is that the direction of polarization is changed by passage through an ordered or a random array of chiral molecules.

For some purposes it is more convenient to use a particle description of electromagnetic radiation, since radiation of a given frequency is quantized and is emitted, transmitted and absorbed in discrete units (photons) whose energy (E) is directly related (equation 1.2) to the frequency (v).

$$E = hv \qquad (1.2)$$

The units most commonly employed by organic photochemists are s^{-1} for frequency, nm or Å (1 nm = 10 Å) for wavelength ($\lambda = c/v$, where c is the speed of propagation of the radiation), cm^{-1} for wavenumber ($\bar{v} = \lambda^{-1}$), and $kJ\,mol^{-1}$, $kcal\,mol^{-1}$ or eV for energy. Table 1.1 shows the numerical relationship between these units. Note that cm^{-1} is sometimes used as a unit of energy, but this is not strictly correct, since the wavenumber associated with radiation, though correlated with the energy, is simply the number of wavelengths per cm.

Table 1.1. Units used in photochemistry

	Energy			Wave-length nm	Wave-number cm^{-1}	
	$kJ\,mol^{-1}$	$kcal\,mol^{-1}$	eV			
$100\,kJ\,mol^{-1}$	100	23·9	1·04	1200	8 360	near infrared
$100\,kcal\,mol^{-1}$	418	100	4·34	286	35 000	near ultraviolet
1 eV	96·5	23·1	1·00	1240	8 070	near infrared
100 nm	1200	286	12·4	100	100 000	far ultraviolet
$10\,000\,cm^{-1}$	120	28·6	1·24	1000	10 000	near infrared

1.3.1   Absorption of Radiation

When a photon passes close to a molecule there is an interaction between the electric field associated with the molecule and that associated with the radiation. This perturbation may result in no permanent change in the molecule, but it is possible for a 'reaction' to occur in which the photon is absorbed by the molecule. The photon ceases to exist and its energy is transferred to the molecule, whose electronic structure changes. This change is visualized in simple molecular orbital terms as a change in the occupation pattern of a set of orbitals which is the same set in the excited state as in the ground state. This is the one-electron excitation approximation, and the approximation holds good for visualizing most absorption processes, although in some instances (e.g. for Rydberg transitions in alkenes see Chapter 2, p. 30) it is necessary to consider orbitals not normally envisaged for a description of the ground state. Formaldehyde offers a simple example (Figure 1.4 : note that only the C—O orbitals are shown). The absorption of a photon corresponding to radiation of wavelength around 280 nm produces an electronically excited state of the carbonyl group in which there is only one electron in the non-bonding orbital of higher energy, and one electron in the anti-bonding π^* orbital.

Such a transition is referred to as an $n \rightarrow \pi^*$ (n to pi star) transition, and the excited state as an (n, π^*) excited state of formaldehyde. Other types of transition

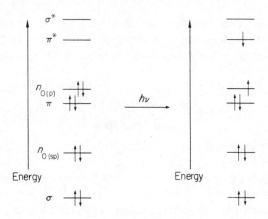

Figure 1.4. $n \rightarrow \pi^*$ Electronic excitation of formaldehyde

are possible with different wavelengths of radiation or with different classes of compound. Those most commonly encountered in organic compounds are $n \rightarrow \pi^*, \pi \rightarrow \pi^*, n \rightarrow \sigma^*$, and $\sigma \rightarrow \sigma^*$. The energy of the photon will not often match exactly the energy difference between the lowest vibrational levels of the ground and excited states of the absorbing molecule, so that in most cases the state initially produced is an upper vibrational/rotational state of the excited electronic state.

Under normal circumstances the bulk absorption characteristics of a compound (vapour, liquid, solid or solution) can be represented[6] by equation (1.3).

$$I = I_0 \, 10^{-\varepsilon cl}$$

or

$$\log(I_0/I) = \varepsilon cl \qquad (1.3)$$

I_0 is the intensity of the incident monochromatic radiation, I is the intensity of transmitted radiation, c is the concentration (or partial pressure or density) of the sample, l is the path-length of the radiation through the sample, and ε is a constant which is characteristic of the particular compound and the particular wavelength of radiation. ε is known as the extinction coefficient, or more specifically as the decadic molar extinction coefficient if c is in molarity units, l is in cm, and logarithms are to base 10.

This empirical law (the Beer–Lambert law) is valid except when very high intensities of radiation are employed (e.g., when using lasers) and a significant proportion of the molecules in a given region are in the excited state rather than the ground state at any one time.

An alternative measure of absorption intensity, which can be related more readily to theoretical principles,[7] is the oscillator strength, f, given by equation (1.4).

$$f = 4 \cdot 315 \times 10^{-9} \int \varepsilon \, . \, dv \qquad (1.4)$$

The major difference between oscillator strength and extinction coefficient is that the former is a measure of the integrated intensity of absorption over a whole band, whereas ε is a measure of the intensity of absorption for a single wavelength.

1.4   EXCITED STATES

1.4.1   Spin Multiplicity

An electronically excited state contains two unpaired electrons in different orbitals, and these can be of the same (parallel) spin or of different (opposed) spin. Such states are triplet and singlet states respectively, and the two are distinct species, with different physical and chemical properties. A triplet state has a lower energy than the corresponding singlet state because of the repulsive nature of the spin–spin interaction between electrons of the same spin (Hund's rule of maximum multiplicity for atomic electronic structure has the same basis). The magnitude of the difference in energy varies according to the degree of spatial interaction (overlap) between the orbitals involved. For orbitals which occupy substantially different regions of space [as in (n, π^*) states of carbonyl compounds] orbital overlap is small and the singlet–triplet energy difference (splitting) is relatively small. For orbitals which occupy similar regions of space [as in (π, π^*) states of alkenes] the difference is much larger.

The excited states initially produced by absorption of a photon are almost always singlet states. This is because practically all molecules encountered in organic chemistry have a singlet ground state (i.e. are fully electron-paired), and the selection rules for absorption strongly favour conservation of spin during the absorption process. Singlet $\rightarrow$ triplet absorption bands in the absorption spectra of some compounds can be observed with sensitive spectrophotometers, and these bands can often be enhanced by the presence of a paramagnetic species such as molecular oxygen, but they are, in general, very much weaker than singlet $\rightarrow$ singlet bands.

1.4.2   State Diagrams

Orbital energy level diagrams have been used in this chapter to show the electronic structure of a particular state of a molecule. A different type of diagram, a state energy level diagram, can be used to represent diagrammatically the various electronic states of a molecule. The singlet and triplet states are ranged in order of increasing energy (Figure 1.5) and numbered in the same order $S_0, S_1, S_2 \ldots$ and $T_1, T_2 \ldots$ respectively. For clarity the states of different multiplicity are separated horizontally, and normally only the first few (lowest energy) states are of interest. Each of the energy levels is the lowest energy point of a complete potential energy surface (energy versus all the variable parameters of the molecule), and the horizontal axis in a simple state diagram has no significance.

It is possible to represent on such a diagram all the physical processes involved in the interconversion of the states (Figure 1.6). These are absorption of a photon

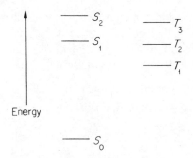

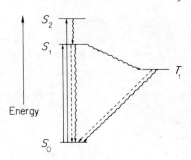

Figure 1.5. A simple state diagram

Figure 1.6. A simple Jablonski diagram

by the ground state to produce an excited singlet state ($\rightarrow$), radiative decay processes ($\dashrightarrow$), namely fluorescence (spin-allowed) and phosphorescence (spin-forbidden), and non-radiative decay processes ($\rightsquigarrow$), namely internal conversion (spin-allowed) and intersystem crossing (spin-forbidden). Such a diagram is known as a *Jablonski* diagram.

Other pathways open to each excited state are chemical reaction and energy transfer. The latter is a bimolecular process, often referred to as *quenching*, in which the excited state is converted to ground state in a non-radiative manner whilst the second molecule (the 'quencher' Q) is excited to a higher energy state. All the processes can be encompassed in a flow diagram (equation 1.5).

$$S_0 \xrightarrow{h\nu} S_1 \longrightarrow T_1 \qquad (1.5)$$

S_1: S_0; $S_0 + h\nu'$; $+Q \rightarrow S_0 + Q^*$; chemical reaction

T_1: S_0; $S_0 + h\nu''$; $+Q \rightarrow S_0 + Q^*$; chemical products

In the chemical reaction the process may be a simple concerted formation of products from reagents, or it may occur through one or more intermediates which may also be able to revert to ground state starting material (equation 1.6). The nature of the secondary chemical processes will depend on the system in question.

$$S_1 \text{ or } T_1 \rightarrow \text{intermediate} \nearrow S_0 \searrow \text{product} \qquad (1.6)$$

Such diagrams bring out the point that the overall efficiency of any particular process is governed by the relative magnitude of a number of individual rate constants. A useful parameter in quantitative photochemistry to which these rate constant ratios can be related is the quantum yield (ϕ). This is a measure of the efficiency of photon usage (cf. the chemical yield, which is a measure of the efficiency of reagent usage), and quantum yield is defined in equation (1.7). This definition can be modified to give expressions for the quantum yield of light emission or for the quantum yield of disappearance of starting material.

$$\phi_{product} = \frac{\text{number of moles (molecules) of product formed}}{\text{number of einstein (photons) of radiation absorbed}}$$

$$= \frac{\text{rate of formation of product}}{\text{intensity of absorbed radiation}} \tag{1.7}$$

REFERENCES

1. J. B. Thomas, *Primary Photoprocesses in Biology*, North-Holland, Amsterdam (1965).
2. T. H. James (ed.), *The Theory of the Photographic Process*, 3rd edition, Macmillan, New York (1966).
3. A. Streitwieser, *Molecular Orbital Theory for Organic Chemists*, Wiley, London (1961), chapter 1.
4. H. H. Jaffé and M. Orchin, *Theory and Applications of Ultraviolet Spectroscopy*, Wiley, London (1962), chapter 3.
5. For an alternative molecular orbital description of formaldehyde, see reference 4, p. 105.
6. Reference 4, p. 8.
7. R. S. Mulliken, *J. Chem. Phys.*, 7, 14 (1939).

Chapter 2

Excited States. Production and Time-independent Properties

There are many ways of introducing energy into molecules in order to produce electronically excited states, but by far the most important involves absorption of light (usually visible or ultraviolet). This is therefore treated first; consideration of other methods of generating excited states is deferred to p. 36.

2.1 FACTORS AFFECTING INTENSITIES OF ABSORPTION SPECTRA

Analysis of absorption and emission spectra and of their variation as a function of experimental conditions provides a powerful tool for exploring the nature of excited states. An understanding of the factors determining the form and intensities of such spectra is essential, and this requires a consideration of the probabilities associated with particular transitions. The following section presents a non-mathematical theoretical treatment of absorption and emission phenomena—the latter will be studied in more detail in Chapter 3.

2.1.1 Absorption and Emission of Light

The absorption of light can be regarded as an exercise in time-dependent perturbation theory. A molecule in an initial stationary state described by the wavefunction Ψ_i is subject to the Schrödinger equation $H_0\Psi_i = E\Psi_i$. If the system is perturbed by exposing it to light, the sinusoidal oscillating electric vector of the light wave induces oscillating forces on the charged particles of the molecule. Thus the static Hamiltonian operator H_0 no longer prescribes the energy of the system, and it has to be replaced by $(H_0 + H')$, where H' is the perturbation operator which takes into account the effect of the radiation field.

The eigenfunctions of $(H_0 + H')$ will be different from the initial wavefunction Ψ_i, and they will also be functions of time. Thus:

$$(H_0 + H')\Psi(x, t) = E\Psi(x, t)$$

These new wavefunctions can be expanded in terms of the wavefunctions of the unperturbed system:

$$\Psi(x, t) = \sum a_k(t)\Psi_k$$

where the coefficients $a_k(t)$, being functions of time, bring in the required time-dependence.

Thus the effect of the perturbation can be thought of as a time-dependent mixing of the initial wavefunction of the system with all the other possible wavefunctions. In other words, the initial state evolves with time into other states, and if the perturbation is suddenly removed at a time t there will be a finite probability that the system will be found in some final state (Ψ_f) other than the initial one. This probability is given by the square of the corresponding coefficient $a_f(t)$ in the above expansion. Mathematical analysis[1] leads to the expression in equation (2.1).

$$[a_f(t)]^2 = \frac{8\pi^3}{3h^2}\langle\Psi_i|\boldsymbol{\mu}|\Psi_f\rangle^2\rho(v_{if})t \qquad (2.1)$$

where $\rho(v_{if})$ is the radiation density (energy per unit volume) at the frequency v_{if} corresponding to the transition, t is the time of irradiation, and $\langle\Psi_i|\boldsymbol{\mu}|\Psi_f\rangle$, which is merely a shorthand way of writing the integral $\int\Psi_i . \boldsymbol{\mu} . \Psi_f \, d\tau$, is an extremely important quantity known as the *transition moment* ($\boldsymbol{\mu}$ is the dipole moment operator $e\sum r_j$, where e is the electronic charge and r_j is the distance of the jth electron).

This shows, then, that the probability of a given transition is proportional to the square of the transition moment. This quantity, which, in principle, can be calculated quantum mechanically, is also obtainable from absorption spectra through its relation (equation 2.2) to the oscillator strength f (see Chapter 1, p. 7).

$$f = \frac{8\pi^2 v_{if} m_e \langle\Psi_i|\boldsymbol{\mu}|\Psi_f\rangle^2}{3he^2} \qquad (2.2)$$

(m_e is the electronic mass.)

If the final state has higher energy than the initial state, the energy deficit must be made up from the radiation field and equation (2.1) gives the probability of *absorption* of a photon. However, equation (2.1) applies equally to the converse situation in which the initial state has higher energy than the final state. This implies that, on irradiation with light of the frequency corresponding to the transition, an excited species may be induced to revert to the ground state with the emission of a photon. This phenomenon, known as *stimulated emission*, provides the basis for laser action (see p. 41).

Absorption and emission of radiation are often discussed in terms of the Einstein coefficients which give the transition rate for absorption and emission per unit radiation density and time. The coefficient of (stimulated) absorption

B_{lu} (where the subscripts u and l refer to upper and lower states) is given by

$$B_{lu} = \frac{[a_u(t)]^2}{\rho(v_{lu})t} = \frac{8\pi^3 \langle \Psi_l|\mathbf{\mu}|\Psi_u\rangle^2}{3h^2} \quad \text{from equation (2.1)}$$

Similarly, the Einstein coefficient of stimulated emission is B_{ul}, and from the preceding discussion,

$$B_{ul} = B_{lu}$$

This being so, if there were no other effect, irradiation of atoms and molecules at the transition frequency would produce an equal population of the upper and lower states, for the probabilities of absorption and stimulated emission are the same. This is well known not to be the case.

In order to account for thermal equilibrium in a radiation field, Einstein found it necessary to assume that there was another emission process with a rate *independent* of the radiation density and given by his coefficient of *spontaneous* emission A_{ul}. Hence the overall rate of emission is equal to

$$A_{ul} + B_{ul}\rho(v_{ul})$$

It can be shown that A_{ul} and B_{ul} are related by the expression in equation (2.3).

$$A_{ul} = \frac{8\pi h v^3}{c^3} \cdot B_{ul} = \frac{64\pi^4 v^3}{3hc^3} \cdot \langle \Psi_u|\mathbf{\mu}|\Psi_l\rangle^2 \tag{2.3}$$

This relation codifies the important points that the rates of absorption and emission (spontaneous and stimulated) depend on the square of the transition moment, but that the rate of spontaneous emission depends also on the *cube* of the transition frequency. Thus upper excited states may be expected for this reason, if for no other, to be very short-lived.

2.1.2 **Radiative Lifetimes**

If no other decay process exists, a population of electronically excited species will decay radiatively to the ground state. Spontaneous emission, being random, obeys first-order kinetics with a rate constant equal to the Einstein coefficient A_{ul} which is the number of times per second that an excited state emits a photon.

$$\frac{d[M^*]}{dt} = -A_{ul}[M^*]$$

$$\therefore \quad [M^*] = M_0^* \, e^{-A_{ul}t} \tag{2.4}$$

We can therefore define a *radiative lifetime*

$$\tau_0 = \frac{1}{A_{ul}} \quad \text{(units, seconds per transition)}$$

which is the time for the population to diminish to $1/e$ of its initial concentration, assuming that *no radiationless processes are occurring*.

Since A_{ul} is directly related to B_{ul} and the oscillator strength f (via equations 2.2 and 2.3), it follows that when the absorption leading to a particular excited state is intense (i.e., the transition is allowed and f is large), the radiative lifetime of the excited state is short. Conversely, for a forbidden transition (e.g., singlet $\rightarrow$ triplet) the corresponding radiative lifetime is long. In other words, if the absorption of radiation is forbidden/allowed, the emission is also forbidden/allowed.

An order-of-magnitude estimate of the radiative lifetime may be obtained from the relation (2.5) for absorption bands in the near ultraviolet.

$$\tau_0 \sim 10^{-4}/\varepsilon_{max} \tag{2.5}$$

Thus, an allowed transition with $\varepsilon \sim 10^5 \, l \, mol^{-1} \, cm^{-1}$ will give rise to an excited state having an approximate radiative lifetime of 10^{-9} s (1 ns). Alternatively, for forbidden $S \rightarrow T$ transitions $\varepsilon \sim 10^{-1}-10^{-4} \, l \, mol^{-1} \, cm^{-1}$ and τ_0 can be of the order of seconds. This enormous radiative lifetime is one of the reasons why triplet species play such an important role in photochemistry. For more precise estimates of radiative lifetimes, several more complex relations have been proposed.[2]

Actual Lifetimes

It is important to make a clear distinction between the *radiative* (τ_0) and *actual* lifetimes (τ). Consider a system (2.6) in which an excited species A* both fluoresces with a rate constant $k_f (\equiv A_{ul})$ and undergoes a reaction with rate constant k_r.

$$A \xrightarrow[hv]{I} A* \begin{array}{c} \overset{k_f}{\nearrow} A + hv' \\ \underset{k_r}{\searrow} \\ products \end{array} \tag{2.6}$$

The actual rate constant for the decay of A* is $(k_r + k_f)$, and the *actual* lifetime of A* (as measured by the decay of fluorescence) will be found to be:

$$\tau = \frac{1}{k_r + k_f}$$

Generalizing, if there are several modes of decay (radiative and non-radiative) each characterized by a first-order or pseudo-first-order rate constant k_i ($i = 1, 2, \ldots, n$), then the actual lifetime (i.e., decay time) will be:

$$\tau = \frac{1}{\sum_i k_i} \tag{2.7}$$

The two lifetimes are related by the expression:

$$\tau = \tau_0 \phi_f$$

where ϕ_f is the quantum yield of fluorescence.

2.1.3   The Intensities of Electronic Transitions

We have seen that the rates of absorption and spontaneous emission, and therefore the intensities, of electronic transitions are proportional to the square of the transition moment (T.M.) $\langle \Psi_i | \mu | \Psi_f \rangle$, where Ψ is the *total* wavefunction (nuclear and electronic) for the system. This integral cannot be evaluated, because we do not know the exact form of the wavefunctions of any molecule. In order to make progress, approximations must be introduced.

Born–Oppenheimer Approximation

Because of the mass difference, nuclear motion is very sluggish in comparison with electronic motion, so that the electrons may be thought of as moving in the potential field of the static nuclei. If the Schrödinger equation can be solved for a variety of nuclear configurations and the electronic potential energy is plotted as a function of these nuclear configurations, we obtain a potential energy surface. The wavefunctions obtained in this way differ from the true wavefunctions, but the differences are usually important only near degeneracies, where potential surfaces cross.

This, the Born–Oppenheimer approximation, is equivalent mathematically to factorizing the total wavefunction into a nuclear (vibrational) wavefunction θ and an electronic wavefunction ψ. Thus $\Psi = \theta . \psi$ and the transition moment is given by equation (2.8).†

$$\text{T.M.} = \int \theta_i \psi_i . \mu . \theta_f \psi_f \, d\tau \tag{2.8}$$

Since the operator μ operates only on the electrons, we can write this in the form of equation (2.9):

$$\text{T.M.} = \int \theta_i \theta_f \, d\tau_N . \int \psi_i . \mu . \psi_f \, d\tau_e \tag{2.9}$$

where subscripts N and e refer to nuclei and electrons. This expression is still too difficult, and further approximations must be made.

(a) It is assumed (i) that ψ can be represented as the product of one-electron wavefunctions (orbitals) ϕ (which may themselves be linear combinations of atomic orbitals) and that the orbitals are the same in both ground and excited states, and (ii) that only one electron is promoted during a transition. The transition moment then reduces to the expression (2.10), where ϕ_i and ϕ_f are the initial and final *orbitals* of the excited electron and μ operates only on these orbitals.

$$\text{T.M.} = \int \theta_i \theta_f \, d\tau_N . \int \phi_i . \mu . \phi_f \, d\tau_e \tag{2.10}$$

† Strictly, complex wavefunctions should be used, with T.M. $= \int \theta_i^* \psi_i^* . \mu . \theta_f \psi_f \, d\tau$, but in this discussion it will be assumed that orbital wavefunctions are real. Complex wavefunctions are likely to be important only in systems of high symmetry such as linear polyatomics.

(b) The final approximation is that the orbitals can be factorized into a product of space and spin orbitals (φ and S respectively),

$$\phi = \varphi . S$$

Since μ operates only on the space coordinates,

$$\text{T.M.} = \int \theta_i \theta_f \, d\tau_N . \int S_i S_f \, d\tau_s . \int \varphi_i . \mu . \varphi_f \, d\tau_e \tag{2.11}$$

The first term is the overlap integral of the wavefunctions for nuclear vibrations—it embodies a quantum mechanical formulation of the Franck–Condon principle (see p. 17). The second term is a spin overlap integral, and its value depends on the initial and final spin states of the promoted electron. The third term is called the *electronic transition moment*, and its value depends on the symmetries and amount of overlap of the initial and final spatial orbitals.

The previous discussion shows that if the Born–Oppenheimer approximation is valid, the large and insoluble Schrödinger equation, $H\Psi = E\Psi$, can be split into three simpler equations:

$$H\varphi = E\varphi$$

$$H\theta = E\theta$$

$$HS = ES$$

and the total molecular energy can be expressed as the sum of electronic, vibrational and spin energies:

$$E = E_e + E_v + E_s$$

The Born–Oppenheimer approximation is a good one for most purposes, but it is important to realize (i) that it breaks down near degeneracies, i.e. where potential energy surfaces cross, for in this region a small change in nuclear coordinates can take a molecule from one surface to the other and thus produce a large change in the electronic wavefunction, and (ii) that it fails to account for the non-zero intensities observed with forbidden transitions (see Vibronic coupling, p. 25).

2.1.4   Selection Rules

The transition moment in the approximate form given in equation (2.11) is seen to be the product of three separate integrals, and its value will be zero if any one of the component integrals is zero. When this happens, the transition has a zero probability of occurrence. Such a transition is said to be *forbidden*, in contrast to *allowed* transitions in which the transition moment is non-zero. Forbidden transitions are only totally forbidden within the framework of approximations implicit in the derivation of equation (2.11). More refined calculations show that forbidden transitions should have small intensities. Thus, they can usually still be observed, although with intensities much less than those of allowed transitions.

Using equation (2.11), the transition moment can be evaluated and hence the approximate oscillator strength can be predicted. This, though not difficult with the aid of modern computer programs, is often unnecessary, because it is sufficient to know whether the transition is allowed or forbidden, and selection rules permit us to determine this point without detailed calculation.

Selection rules derive from symmetry considerations, which often indicate whether, for a particular transition, the integrand of one of the component integrals is an *odd* (antisymmetric) function of the coordinates. In such a case, the integral will be zero and the transition will be forbidden.

To illustrate this point, consider Figures 2.1 and 2.2 exhibiting typical even and odd functions. For an odd (antisymmetric) function, it is clear that for each individual positive infinitesimal area $f(x)\,dx$ there is a corresponding negative contribution. Thus the integral $\int_{-\infty}^{+\infty} f(x)\,dx$, which is merely the sum of these infinitesimal contributions to the area under the curve, will be zero. For an even (symmetric) function, the integral will be non-zero. With this in mind, the three component integrals of the transition moment integral will be examined.

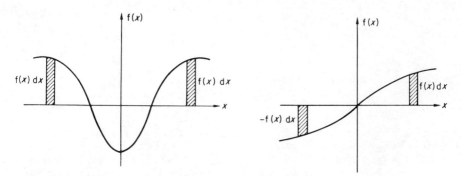

Figure 2.1. Symmetric (even) function of x **Figure 2.2.** Antisymmetric (odd) function of x

2.1.5 Vibrational Overlap Integral (Franck–Condon Principle)

In order to evaluate the integral $\int \theta_i \theta_f \, d\tau_N$, it is necessary to know the vibrational wavefunctions. For simplicity, consider a simple harmonic oscillator. For such a system, the potential energy for varying internuclear separations is described by the parabolic curve of Figure 2.3, the equilibrium separation (r_e) being given by the point (a) where the potential energy is a minimum. Solution of the appropriate Schrödinger equation leads to the result that the vibrational energy is quantized with values

$$E = h\nu(\nu + \tfrac{1}{2})$$

where v, the vibrational quantum number, must be integral $0, 1, 2, \ldots, n$. These values are represented by the equally spaced horizontal lines in Figure 2.3. (Note that the minimum vibrational energy is not zero but $\tfrac{1}{2}h\nu$, and that at room temperature in condensed phases most molecules will be in the $v'' = 0$ level). The oscillating curves represent the wavefunctions associated with each vibrational level. Since the square of the amplitude (θ) of the wavefunction gives the probability of a particular nuclear configuration, we can see that in the bottom level $(v'' = 0)$ the nuclei are most likely to be found at the equilibrium distance r_e, and that as v'' increases it becomes increasingly probable that the molecule will be found close to the configuration corresponding to the intersection of the horizontal lines and parabolic curve (the turning point of the vibration, where the total energy equals the potential energy, the kinetic energy is zero, and the nuclei are static).

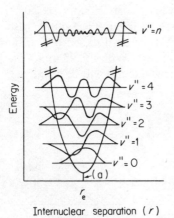

Figure 2.3. Vibrational wavefunctions and energy levels for a harmonic oscillator

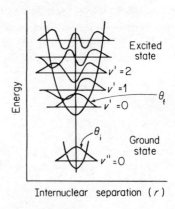

Figure 2.4. No change in geometry on excitation; the transition $v'' = 0$ to $v' = 0$ is the most probable

In Figure 2.4 are represented the potential energy curves for the ground and excited states of such an idealized molecule. In this case there is no change in geometry on excitation (r_e is the same for both curves). To evaluate the vibrational overlap integral, we multiply the individual values of the wavefunctions (θ_i and θ_f) for each internuclear separation and sum the infinitesimal contributions $\theta_i \theta_f \cdot dr$. It is clear from the Figure that the value of the vibrational overlap integral is positive for the transition $v'' = 0 \rightarrow v' = 0$ [a so-called $(0 \rightarrow 0)$ transition], and zero for the $0 \rightarrow 1$ transition because for each positive infinitesimal contribution to the overlap integral there is an equal negative contribution. The same is true for transitions to higher vibrational levels. Hence even though real molecules are not harmonic oscillators, when there is only a small change in geometry on excitation the $(0 \rightarrow 0)$ vibrational transition is expected to be the strongest; $(0 \rightarrow n)$ transitions will be much less probable and thus less intense.

Figure 2.5 represents the more common situation where excitation leads to stretching of a bond. This is to be expected wherever the excited electron is

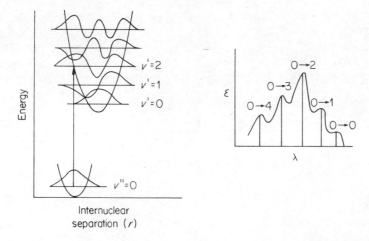

Figure 2.5. Change in geometry on excitation, leading to the vibrational fine structure in the spectrum shown

promoted into an antibonding orbital, thereby weakening the bond. In this particular example, inspection reveals that the $(0 \rightarrow 2)$ vibrational transition will be strongest.

Similar considerations apply to emission spectra, with the proviso that in condensed phases at room temperature the excited state will normally become thermally equilibrated before emission, which therefore takes place almost exclusively from the bottom vibrational level of the excited state.

The above discussion, based on the vibrational overlap integral, is a quantum mechanical formulation of the *Franck–Condon Principle* first expressed classically by Franck. The principle states that electronic transitions occur in an exceedingly brief interval of time so that no change in nuclear position or nuclear kinetic energy occurs during the transition. This implies that the transition may be represented by a *vertical* line connecting the two potential energy surfaces, and the most probable transition will be to that vibrational level with the same internuclear distance at the turning point of the oscillation, e.g. lines AY and ZB in Figure 2.6. A transition represented by line AX is extremely improbable because the molecule, in arriving at point X, would have suddenly acquired an excess kinetic energy given by XY.

It must be emphasized that real molecules are *anharmonic* oscillators, and for diatomics the potential energy diagram is approximated by a Morse curve (Figure 2.7), the higher vibrational levels of which become progressively closer together until they merge into a continuum at the dissociation limit. Furthermore, for any molecule containing more than two atoms the potential energy curve becomes a surface in many dimensions because of the very large number of possible vibrational modes. Hence corresponding to a particular electronic transition there will be a multiplicity of associated vibrational and rotational transitions so closely spaced that they overlap, giving rise to a smooth broad

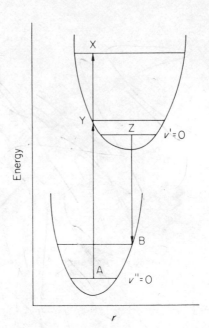

Figure 2.6. Franck–Condon Principle.
The most probable transitions are AY
and ZB

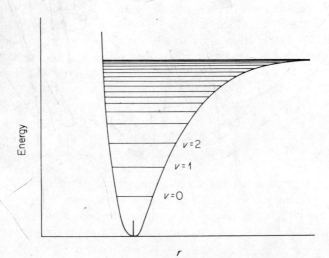

Figure 2.7. Energy diagram for a diatomic molecule (an-harmonic oscillator)

absorption. Therefore in the majority of cases little vibrational structure may be seen in a visible or ultraviolet absorption band. Even so, the shape of the absorption band is determined by the Franck–Condon principle, and its envelope is for this reason known as the Franck–Condon envelope.

In those cases where vibrational structure may be discerned,† important information about the shape of an excited molecule and its vibrational modes may be gleaned from an analysis of this *vibronic* (*vibra*tion + electr*onic*) structure.

2.1.6   **Spin**

The effect of electron spin upon transition intensities is given by the factor $\int S_i S_f \, d\tau_s$ in the transition moment expression. There are three common situations.

(a) *Singlet → singlet transitions.* The electron can have only two spin states (wavefunctions designated α and β). In a singlet → singlet transition no change occurs in the spin state of the promoted electron, and the spin overlap integral is $\int \alpha\alpha \, d\tau_s = \int \beta\beta \, d\tau_s = 1$ because the spin wavefunctions are assumed to be normalized. There are no spin restrictions on such transitions, which are therefore fully allowed.

(b) *Triplet → triplet transitions.* Since the transition occurs with no change in multiplicity, again the spin overlap integral $\int \alpha\alpha \, d\tau_s = 1$ and the transition is fully allowed. This fact is used in *flash photolysis* (see Chapter 5, p. 137), when triplet species are produced in high concentration by intersystem crossing after an intense burst of light and are examined by absorption spectroscopy.

(c) *Singlet → triplet transitions.* Since the promoted electron changes its spin state the spin overlap integral is $\int \alpha\beta \, d\tau_s = \int \beta\alpha \, d\tau_s = 0$, because the α and β spin wavefunctions are orthogonal. The transition moment is thus zero and the transition is strongly forbidden.

Spin-orbit Coupling

Singlet → triplet transitions are strongly forbidden; that they occur at all is due to spin–orbit coupling. Since the electron is charged and 'spinning', it is expected to have not only spin angular momentum but also a magnetic moment. In a $S → T$ transition, an electron inverts its spin, i.e. changes the direction of its magnetic moment. Clearly this calls for a magnetic interaction (consider trying to change the orientation of a bar magnet without touching it). The required magnetic interaction is provided by the magnetic field produced by the *orbital* motion of the charged electron. The magnetic moment of the *spinning* electron becomes coupled to the orbital magnetic field—hence the term spin–orbit coupling.

The effect may be seen more clearly by basing the co-ordinate system on the electron. Seen from the electron, there is a charged nucleus in orbit around it

---

† Vibrational fine structure is best observed in the gas phase or in solution in non-polar solvents where perturbations in the energy of the solute caused by intermolecular interactions are minimized.

which therefore generates a magnetic field which depends upon both the nuclear charge Z and the distance r between the electron and the nucleus. This simple treatment implies that although the total angular momentum (spin and orbital) is conserved, neither orbital nor spin angular momentum are individually conserved—that because of the spin–orbit coupling, angular momentum being continually switched between the spin and orbital modes. This means that one can no longer think of pure spin states (singlet or triplet), but intermediate situations must be contemplated in which a nominal singlet state has a certain degree of triplet character and vice-versa.

The spin–orbit interaction is treated quantum mechanically by introducing into the Hamiltonian operator a term H_{SO} for each electron of the form:

$$H_{SO} = k\zeta(\mathbf{L}.\mathbf{S})$$

where $\mathbf{L}$ is the orbital angular momentum operator, $\mathbf{S}$ is the spin angular momentum operator, and ζ is a factor depending on the nuclear field. ζ, and therefore H_{SO}, is proportional to Z/r^3, i.e. to Z^4 because of the reciprocal relation between Z and r. Perturbation theory shows that if Ψ_S^0 and Ψ_T^0 are the wavefunctions of 'pure' singlet and triplet states respectively, then the triplet state produced under spin–orbit coupling can be written[3] in the form in equation (2.12), where S_k is the kth singlet state, and E_T and E_S are the energies of the triplet and the perturbing singlet states respectively.

$$\Psi_T = \Psi_T^0 + \sum_k \frac{\langle \Psi_{S_k}^0 | H_{SO} | \Psi_T^0 \rangle}{(E_T - E_{S_k})} . \Psi_{S_k}^0 \qquad (2.12)$$

A similar expression can be written for the singlet state. Thus the effect of spin–orbit coupling is to mix a small amount of singlet character into the triplet states and vice-versa, so that 'pure' singlet and triplet states no longer exist. Singlet $\rightarrow$ triplet transitions can then be thought of as occurring between the pure singlet and pure triplet components of each hybrid state. Figure 2.8 represents the situation in diagrammatic form.

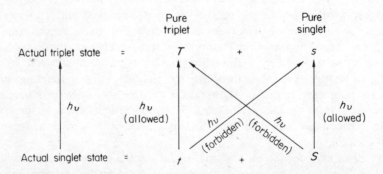

Figure 2.8. Diagrammatic resolution of actual states into their pure components. Forbidden $S \rightarrow T$ transitions appear as allowed transitions between the impurity components of the nominal states

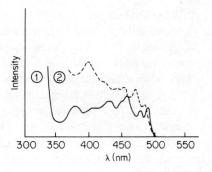

Figure 2.9. $S_0 \rightarrow T_1$ absorption spectra of 1,1-chloronaphthalene, and 2,1-iodonaphthalene (reproduced with permission from A. P. Marchetti and D. R. Kearns, *J. Amer. Chem. Soc.*, **89**, 768 (1967); copyright by the American Chemical Society)

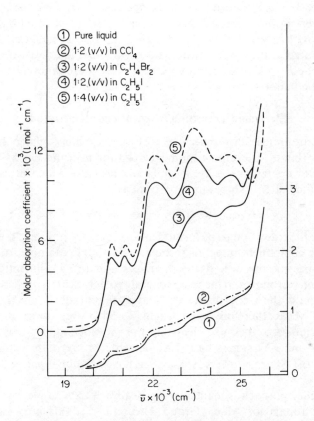

Figure 2.10. The effect of 'heavy-atom' solvents on the $S_0 \rightarrow T_1$ transition of 1-chloronaphthalene. Curves 1, 2 and 3 refer to the right-hand scale, curves 4 and 5 to the left-hand scale. (From S. P. McGlynn, T. Azumi and M. Kasha, *J. Chem. Phys.*, **40**, 507 (1964), reproduced by permission of the American Institute of Physics)

The probability of the $S \rightarrow T$ transition depends upon the energy gap between the states concerned and upon the size of matrix elements such as $\langle \Psi^0_{S_k}|H_{SO}|\Psi^0_T \rangle$ in equation (2.12). The latter quantity increases very rapidly with increasing atomic number, giving rise to the *heavy atom effect*. As explained earlier, H_{SO} depends on Z^4, and the matrix elements and therefore the probability of $S \rightarrow T$ transitions will show a similar dependence on the fourth power of the atomic number in atomic systems. In organic molecules this simple relation cannot be expected to hold, but we may still expect that $S \rightarrow T$ transitions will be a highly sensitive function of the presence of heavy atoms. There are both internal and external heavy atom effects depending upon whether the heavy atom is incorporated into the molecule itself or into the environment, but, in both cases, the presence of the heavy atom is manifested by an increase in the probability of singlet–triplet transitions, whether absorptive, emissive or radiationless (see Chapter 3, p. 82). An example of the operation of the internal heavy atom effect is given by the increased $S \rightarrow T$ absorption of 1-iodonaphthalene over that shown by 1-chloronaphthalene (Figure 2.9). The external heavy atom effect is demonstrated in Figure 2.10, where the use of ethyl iodide as solvent greatly intensifies the $S \rightarrow T$ absorption spectrum of 1-chloronaphthalene over that observed in ethanol.

2.1.7 Electronic Transition Moment and Polarization

The electronic transition moment (E.T.M.) is intimately related to the symmetries of orbitals. The dipole moment operator μ in the E.T.M. $\int \varphi_i \mu \varphi_f \, d\tau$, being a vector operator ($= e \sum r_i$), can be resolved along the Cartesian axes of space, and the E.T.M. can be similarly resolved:

$$\text{E.T.M.}_{\text{total}} = \text{E.T.M.}_x + \text{E.T.M.}_y + \text{E.T.M.}_z$$

In order for the transition to be forbidden on symmetry grounds, it is necessary for all three component integrals to be zero. A simple example from the field of atomic spectroscopy will clarify the issues. Figure 2.11 illustrates the symmetry of s and p orbitals and the vector operator μ. It is clear that whereas ψ_{1s} is an even function of the x-coordinate, ψ_{p_x} and μ are odd functions. The integrand of the E.T.M. is therefore even $\times$ odd $\times$ odd = even function,† and the E.T.M.$_x$ is non-zero. It is also clear that the wavefunction ψ_{p_x} has zero value along the y- and z-axes, making E.T.M.$_y$ and E.T.M.$_z$ zero. In such a situation, where only E.T.M.$_x$ is non-zero, we say that the transition is *polarized* along the x-axis.

This simple approach is inadequate to deal with the complex symmetries of molecular orbitals for which Group Theoretic methods are essential. The technique is described in the Appendix, but in essence it involves assigning symmetry symbols to $\psi_i, \psi_f, \mu_x, \mu_y,$ and μ_z, and hence determining the symmetry of the integrands of the components of the E.T.M. Unless the integrand is

† The rule is odd $\times$ odd = even $\times$ even = even; odd $\times$ even = odd. Note that the terms symmetric and antisymmetric are frequently used instead of even and odd respectively.

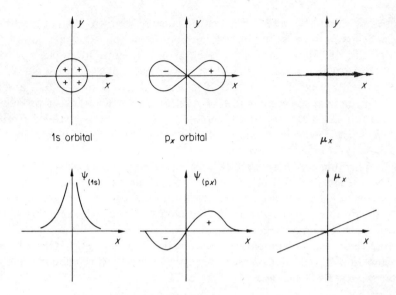

Figure 2.11. Symmetries of 1s and p_x orbitals and of μ_x

totally symmetric, the transition is forbidden. If the integrand of just one of the components of the E.T.M. (say E.T.M.$_x$) is totally symmetric, the transition will be polarized along the x-axis. For such a case, if we illuminate an oriented assembly of molecules (crystal or thin layer) with plane-polarized light (the electric vector of which oscillates in a particular plane), maximum absorption will occur when the plane of polarization of the incident light is parallel to the molecular x-axis. Rotation of the crystal will cause a decrease in absorption which falls to zero when the plane of polarization and the x-axis are perpendicular.

In a similar way, emission is frequently polarized along a particular molecular axis. Polarization measurements can give important information about the electronic transitions responsible for particular absorption bands.[4]

2.1.8   Vibronic Coupling and Forbidden Transitions

The previous discussion has shown that under the Born–Oppenheimer approximation the transition moment can be factorized (equation 2.11), so that if the electronic transition moment is zero, then the transition moment is zero and the transition will be forbidden. Therefore symmetry-forbidden transitions should have zero intensity and be unobservable; nonetheless, they are found to have small but finite intensities. Examples of such forbidden but observed transitions are the benzene absorption near 254 nm ($^1A_{1g} \rightarrow {}^1B_{2u}$, see Chapter 9), the $n \rightarrow \pi^*$ transitions of aliphatic aldehydes and ketones (see Chapter 7, p. 176), and the $d \rightarrow d$ transitions of the metal atom in centrosymmetric complexes. The question then arises as to why such symmetry-forbidden transitions should be observable, and the answer seems to be that the Born–Oppenheimer approximation is only an approximation (though a good one); i.e. that the vibrational (nuclear) and electronic motions are in fact not completely independent of each other but are weakly coupled. In other words, the motions of the electrons and nuclei affect each other to a small extent, and the Born–Oppenheimer representation of a vibronic wavefunction as a product as in equation (2.11) is not strictly correct. It is this coupling, called *vibronic coupling* (*vibr*ational–electr*onic*), which is responsible for the non-zero intensity of symmetry-forbidden transitions.

This phenomenon may be discussed at various levels of sophistication. Taking the benzene absorption $^1A_{1g} \rightarrow \, ^1B_{2u}$ as an example, the transition is forbidden only if the molecule is in the form of a regular hexagon. However, there are vibrations which distort the hexagon and reduce its symmetry so that the transition becomes weakly allowed.

More rigorously, in computing the transition moment, Born–Oppenheimer wavefunctions should not be used, but rather *vibronic* wavefunctions, and the value of integrals of the type

$$\langle \psi_i | \mathbf{\mu} | \Psi_f \rangle$$

should be assessed, where the Ψ terms are now vibronic wavefunctions. The value of the transition moment so obtained will not differ greatly from that estimated using Born–Oppenheimer wavefunctions, but for vibrations of the appropriate symmetries it will assume a small value when the Born–Oppenheimer transition moment is zero. Thus symmetry-forbidden transitions do occur, though with small intensities.

The symmetries of vibronic wavefunctions are dependent upon the symmetries of the component vibrations. This makes it possible, with the aid of Group Theory, to predict which vibrations are effective in making the transition 'slightly allowed' (see Appendix for a worked example). Vibronic coupling is also of importance in non-radiative transitions (see Chapter 3, p. 89).

2.1.9   Orbital Overlap

Suppose that the initial and final orbitals occupy such different regions of space that they have but little overlap. The overlap integral $\int \psi_i \psi_f \, d\tau$ is a measure of the amount of this spatial overlap, and it is evaluated by summing the infinitesimal contributions $\psi_i \psi_f \, d\tau$. The E.T.M. is obtained by multiplying each of these infinitesimals by the appropriate value of $\mathbf{\mu}$ before summing. Thus, if ψ_i and ψ_f are so oriented that large values of the wavefunction ψ_i occur where ψ_f is small and *vice versa*, then the product $\psi_i \psi_f$ will always be small, as will the triple product $\psi_i \mathbf{\mu} \psi_f$ and also the E.T.M. In such a situation, there will be a symmetry-allowed transition which is weak. An example is afforded by the $(n, \rightarrow \pi^*)$ transition of pyridine. Group Theory (see Appendix) shows that this transition is allowed, but inspection of Figure 2.12 shows that the n and π orbitals occupy quite different regions of space. The transition is thus forbidden on overlap grounds, and it is not surprising that the extinction coefficient of the band is only $400\,\mathrm{l\,mol^{-1}\,cm^{-1}}$. The $(n \rightarrow \pi^*)$ transition of formaldehyde $(\varepsilon_{max} \sim 20\,\mathrm{l\,mol^{-1}\,cm^{-1}})$ is forbidden on both symmetry and overlap grounds.

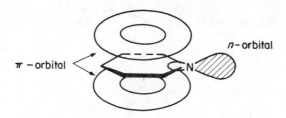

Figure 2.12. Limited overlap of π- and n-orbitals in pyridine

2.1.10 Oscillator Strengths and Forbidden Transitions

From the previous sections it is seen that a transition contravening a selection rule is forbidden, and the corresponding absorption band will be less intense than it would otherwise have been. The effect of such contraventions of the selection rules may be estimated very approximately as follows. If a completely allowed transition has an oscillator strength $F_A \sim 1$, then other transitions have oscillator strengths F given roughly by:

$$F = f_s \cdot f_0 \cdot f_{sym} \cdot F_A$$

where the f factors correct for the varying degrees of forbiddenness:

$f_s(\text{spin}) \sim 10^{-8}$ for aromatic hydrocarbons, $\sim 10^{-5}$ for second-row elements,

$f_0(\text{orbital overlap}) \sim 10^{-2}$ for $(n \rightarrow \pi^*)$ transitions of second-row elements,

$f_{sym}(\text{symmetry}) \sim 10^{-1}$–$10^{-3}$.

2.1.11 Other Factors Affecting the Intensities of Absorption Spectra

Hot Bands

When vibrational fine structure can be discerned, the lowest energy (longest wavelength) vibrational band is normally due to the $(0 \rightarrow 0)$ transition, because it is the $v'' = 0$ level of the ground state which is predominantly populated at room temperature. At higher temperature, or when the $(0 \rightarrow 0)$ transition is weak (symmetry forbidden) as with benzene, weak absorption bands can be discerned at longer wavelengths than the $(0 \rightarrow 0)$ transition. These are ascribed to transitions from levels higher than the lowest vibrational level of the ground state, because their intensity increases with temperature as the population of these higher levels increases. Such bands are known as *hot bands*.

28

External Perturbations[5]

Heavy atoms incorporated either into the substrate molecule or into the solvent enhance $S \rightarrow T$ absorption through a spin–orbit coupling effect. Similarly, an increase in the intensity of $S \rightarrow T$ transitions may be induced by observing the spectrum of the substance under a high pressure (20–100 atm) of xenon, nitric oxide or oxygen (see Figure 2.13). The xenon probably operates

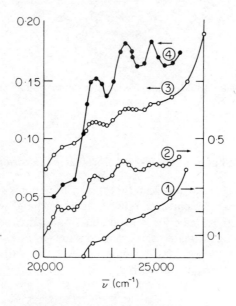

$\bar{\nu}$ (cm⁻¹)

Figure 2.13. Absorption spectrum of 1-chloronaphthalene: 1, pure; 2, with oxygen (30 atm); 3, with xenon (143 atm); 4, with ethyl iodide. Curves 1 and 2 refer to the right-hand scale, curves 3 and 4 to the left-hand scale (from A. Grabowska, *Spectrochim. Acta,* **19**, 307 (1963); reproduced by permission of Pergamon Press Ltd.)

through the external heavy atom effect, but a different mechanism must be responsible for the enhanced absorption observed with oxygen and nitric oxide. Evans,[6] who devised the oxygen perturbation technique, ascribed it to the paramagnetism of the oxygen molecule. It now seems,[7] however, that paramagnetic interactions are probably negligible and that the absorption enhancement is to be attributed to a mixing of states within contact charge–transfer complexes (see p. 35) involving oxygen or nitric oxide. These molecules also enhance non-radiative $S \rightarrow T$ transitions (intersystem crossing, see Chapter 4, p. 116).

The common types of molecular orbitals encountered in organic molecules were briefly described in Chapter 1. Electronic transitions between these orbitals give rise to different types of excited state, the nature of which can often be specified from an analysis of the absorption spectrum. Whether or not this initially produced excited species is the one responsible for a particular photochemical reaction depends on the magnitude of the rate constant for the chemical transformation relative to those for competing radiationless processes (internal conversion, intersystem crossing) which lead to alternative excited states. How this point is elucidated is discussed in Chapter 5. In this section the utilization of absorption spectra to identify the initial state is considered.

2.2.1   Nomenclature

Different systems of describing excited states are employed depending on which properties of the state are to be emphasized and on the amount of information available concerning the nature of the originating transition. Therefore, if symmetry considerations are dominant, group theoretical symbols will be adopted. For example, the ground state of benzene ($^1A_{1g}$) is excited by the 253·7 nm mercury line into its first electronically excited state, designated $^1B_{2u}$, and the transition would be symbolized $^1A_{1g} \rightarrow {}^1B_{2u}$.† Alternatively, the molecular orbitals involved in the transition may be specified. In the benzene example, an electron is promoted from a π to a π^* orbital; the transition is then ($\pi \rightarrow \pi^*$) and the state is symbolized (π, π^*). Since there are several π and π^* orbitals, clearly this nomenclature is less precise than that based upon group theory. In a similar manner, the long wavelength transition in aliphatic aldehydes and ketones can be designated ($n \rightarrow \pi^*$) and it gives rise to an (n, π^*) excited state. This scheme is due to Kasha. When all that is at issue is the ordering of the states, a simple enumerative scheme is adopted. The singlet states are characterized $S_0, S_1 \ldots S_n$, where S_0 is the ground state and S_1 etc. are the higher singlets. The corresponding triplets are then $T_1, T_2, \ldots, T_n$. (There is no T_0 state except for some paramagnetic molecules such as oxygen.) Other schemes are occasionally found in the literature, notably that of Platt[8] for the characterization of the states of aromatic molecules, but they will not be used in this book.

The commoner transitions encountered in organic molecules are: $\pi \rightarrow \pi^*$, $n \rightarrow \pi^*$, $n \rightarrow \sigma^*$, charge–transfer and Rydberg.

2.2.2 $\pi \rightarrow \pi^*$ Transitions

(a) Ethylene provides the simplest example of a ($\pi \rightarrow \pi^*$) transition. The energetic ordering of the levels (Figure 2.14) makes this the lowest energy (longest wavelength) transition.

---

† It is a convention among spectroscopists to write the upper state first regardless of the direction of the arrow, so that $^1B_{2u} \leftarrow {}^1A_{1g}$ represents absorption of radiation but $^1B_{2u} \rightarrow {}^1A_{1g}$ designates fluorescence. Photochemists, however, tend to write the initial state or orbital first, irrespective of the ordering of the energy levels.

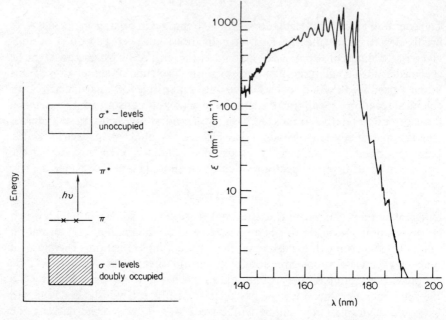

Figure 2.14. Energy levels of ethylene

Figure 2.15. Ultraviolet absorption spectrum of ethylene (from M. Zelikoff and K. Watanabe, *J. Opt. Soc. Amer.*, **43**, 756 (1953), reproduced by permission of the Optical Society of America)

The transition is allowed and gives rise to an intense absorption ($f \sim 0.3$) in the vacuum ultraviolet. The absorption is broad and extends from ~ 137.5 to ~ 200 nm with superimposed sharp lines (Figure 2.15) which seem[9] to be the first members of a *Rydberg* series and in which a π electron is promoted into an orbital probably of σ-type and embracing a central $C_2H_4^+$ ion.

The triplet ($\pi \rightarrow \pi^*$) transition is so weak ($\varepsilon \sim 10^{-4} \, l \, mol^{-1} \, cm^{-1}$) that a path length of 14 metres of liquid ethylene is required for its detection. It gives rise to a band with λ_{max} at 270 nm and extending as far as 325 nm.

(b) Conjugation shifts the ($\pi \rightarrow \pi^*$) transition to longer wavelengths (Figure 2.16). In molecular orbital terms, the phenomenon arises because the energy gap between the highest occupied molecular orbital (HOMO) and the lowest unoccupied molecular orbital (LUMO) becomes progressively smaller as the number of conjugated double bonds increases. Within the Hückel approximation the energy of the rth one-electron orbital adopts[10] the analytical form

$$E_r = \alpha + 2\beta \cos \frac{r\pi}{2n + 1} \tag{2.13}$$

where n is the number of conjugated double bonds, and α and β are the coulomb and resonance integrals for carbon. Hence the energy of the longest wavelength transition absorption band is

$$\Delta E = E_{n+1} - E_n = -4\beta \sin \frac{\pi}{4n+2} \tag{2.14}$$

This is a quantity which diminishes as n increases. Figure 2.17 shows how the interaction of two ethylenic double bonds in butadiene leads to a *bathochromic* (longer wavelength) displacement of the low energy transition of butadiene relative to that of ethylene.

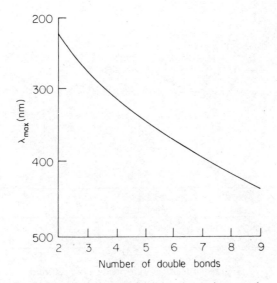

Figure 2.16. Ultraviolet/visible absorption maximum for the lowest energy transition in α,ω-dimethylpolyenes

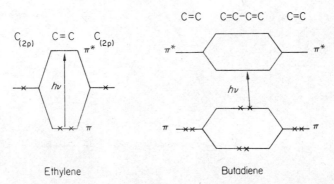

Figure 2.17. Orbital energies for ethylene and buta-1,3-diene

2.2.3   ***n → π* Transitions***

$(n \rightarrow \pi^*)$ Absorption is characteristic of molecules possessing chromophores with multiply bonded hetero-atoms (e.g. $C{=}O$, $C{=}N$, $C{=}S$, $N{=}N$, $N{=}O$). It is of low intensity ($f \sim 10^{-2}$–10^{-4}, $\varepsilon \sim 10$–$100\,l\,mol^{-1}\,cm^{-1}$) being symmetry and/or overlap forbidden, and it is normally the band occurring at longest wavelength. Acetone will serve as an example. The relevant orbitals are shown in Figure 2.18 and the ultraviolet spectrum in Figure 2.19. The $(n \rightarrow \pi^*)$

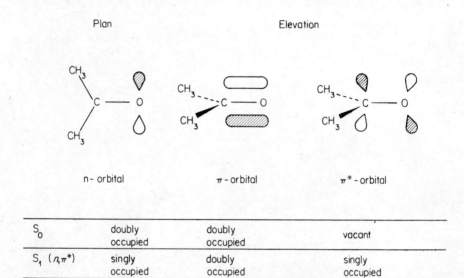

	Plan	Elevation	Elevation
	n- orbital	π- orbital	π^*- orbital
S_0	doubly occupied	doubly occupied	vacant
$S_1\ (n,\pi^*)$	singly occupied	doubly occupied	singly occupied

Figure 2.18. Molecular orbitals of acetone

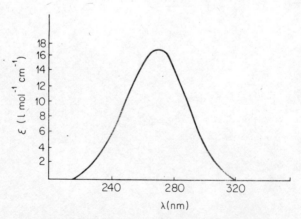

Figure 2.19. Absorption spectrum of acetone

transition has λ_{max} at ~ 280 nm. In the region of 190 nm is an intense band which may be due to $(n \rightarrow \sigma^*)^{11}$ or $(\sigma \rightarrow \pi^*)$ transitions,[12] and on the long wavelength tail of the $(n \rightarrow \pi^*)$ absorption are several weak bands due to the triplet $(n \rightarrow \pi^*)$ transition.

2.2.4 $n \rightarrow \sigma^*$ Transitions

The first absorption band of alkyl halides is due to the promotion of a non-bonding halogen p-electron into the σ^* antibonding C—X orbital.[13] The transition is partially forbidden, and the absorption is weak (see Chapter 11, p. 344). The promotion of an electron into the σ^* anti-bonding orbital largely neutralizes the bonding induced by the two electrons in the carbon–halogen σ-orbital, so that efficient dissociation is a consequence of irradiating alkyl halides in the near ultraviolet:

$$RX \xrightarrow{h\nu} R\cdot + X\cdot$$

Other compounds with hetero-atoms singly linked to carbon (R—OH, R—SH, R—NH$_2$ etc.) also show $(n \rightarrow \sigma^*)$ absorption, but this tends to occur at wavelengths shorter than 200 nm (e.g. for MeOH, $\lambda_{max} = 183$ nm, $\varepsilon \sim 500$ l mol^{-1} cm^{-1}). As might be expected, there is a correlation between the ionization potential of the compound (involving ionization of the n-electron) and λ_{max}—the lower the ionization potential, the greater λ_{max}.

2.2.5 Charge–Transfer (CT) Transitions

Mixtures of electron donors and electron acceptors in solution often exhibit a new absorption band which is shown by neither component separately and which is attributed to the presence in such mixtures of a donor–acceptor complex (DAC). Typical acceptors are picric acid and other polynitro aromatics, maleic anhydride, quinones and iodine, and typical donors include aromatic hydrocarbons and their derivatives, dienes and amines. The transition is referred to as a charge–transfer (CT) transition and in general is broad and structureless (see Figure 2.20). The nature of the bonding in DACs and the theory of CT absorption has been worked out by Mulliken.[14] Briefly, the ground state of a DAC is described by a wavefunction:

$$\psi_{(S_0)} = a\psi_{(DA)} + b\psi_{(D^+ A^-)} \tag{2.15}$$

in which $\psi_{(DA)}$ corresponds to a 'no-bond' structure, the components being held only by weak intermolecular forces such as hydrogen bonding and dipole–dipole interaction, and $\psi_{(D^+ A^-)}$ corresponds to the structure in which an electron has been totally transferred from the donor to the acceptor. The wavefunction for the excited states of the DAC is then given by:

$$\psi_{(S_1)} = a\psi_{(D^+ A^-)} - b\psi_{(DA)} \tag{2.16}$$

For the majority of DACs $b \ll a$, so that $\psi_{(S_0)} \sim \psi_{(DA)}$, and $\psi_{(S_1)} \sim \psi_{(D^+ A^-)}$.

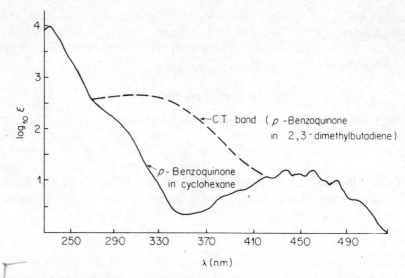

Figure 2.20. Absorption due to charge–transfer complex p-benzoquinone–2,3-dimethylbutadiene

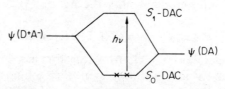

Figure 2.21. Stabilization of the ground state and destabilization of the excited state of a DAC, induced by mixing of states

Thus the spectroscopic transition corresponds approximately to the light-induced transfer of an electron from the donor to the acceptor, and hence the name charge–transfer transition. It should be noted that in such cases there is very little charge–transfer in the ground state, and it is perhaps inappropriate to call them charge–transfer complexes.

The energetics of DACs can be explored with the aid of energy diagrams. Figure 2.21 shows how mixing the wavefunctions of $\psi_{(DA)}$ and $\psi_{(D^+ A^-)}$ leads to stabilization of the ground state and destabilization of the excited state. In Figure 2.22 ΔH_0 and ΔH_1 are the energies of formation of the ground state of a DAC from D and A, and of the excited state from D^+ and A^-, A_A and I_D are the electron affinity of the acceptor and the ionization potential of the donor. It is then clear that the energy of the CT transition (ΔE_{CT}) is given by

$$\Delta E_{CT} = I_D - A_A - (\Delta H_1 - \Delta H_0)$$

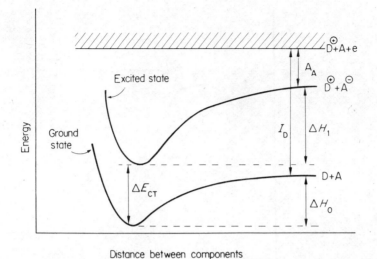

Figure 2.22. Energetics of donor–acceptor complexes

Thus, λ_{max} of the CT transition moves to longer wavelength as the donor and acceptor powers of the components increase. Extensive investigations[15] confirm the general validity of these principles.

2.2.6   Contact DACs and Contact Charge-Transfer Absorption

The phenomenon of contact CT absorption became known through the observations of Evans[16] that iodine in saturated hydrocarbons shows a new structureless absorption in a region where both components are transparent (Figure 2.23). The optical density of the band is proportional to both iodine and hydrocarbon concentrations, showing that the concentration of the DAC is extremely small. This point is borne out by a Benesi–Hildebrand[17] analysis which shows that the equilibrium constant for complex formation is zero. The data are consistent with the hypothesis that the transition occurs during a collision between donor and acceptor.

A number of other systems exhibiting contact charge–transfer absorptions are now known, notably oxygen in solution in saturated and aromatic hydrocarbons and in alcohols, ethers and even water. The basic difference between ordinary DACs and contact DACs seems to be twofold:

(i) the binding energy of contact DACs is very much smaller ($< kT$, where k is Boltzmann's constant), and

(ii) that, whereas ordinary DACs have definite structures, in contact DACs the two components are oriented randomly.

The theory of the spectra of contact CT absorption has been worked out by Murrell.[18]

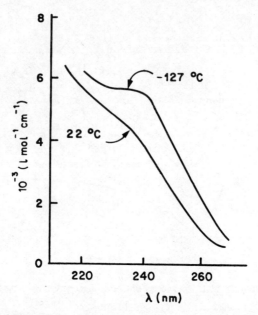

Figure 2.23. Absorption spectrum of iodine in methylcyclohexane. Contact charge–transfer band appears at ~ 240 nm in low temperature spectrum (from D. F. Evans, *J. Chem. Soc.*, 4229 (1957); reproduced by permission of the Chemical Society)

2.3 OTHER METHODS OF PRODUCING EXCITED STATES

2.3.1 Electrical Discharges

If a high potential is applied across an inert gas containing mercury vapour, the few stray electrons present are accelerated to energies sufficient to ionize the gas. Recombination of ions and electrons then gives rise to excited atoms of the inert gas. Direct excitation of the gas can also occur.

$$Ne + e^- \rightarrow Ne^+ + 2e^-$$
$$Ne^+ + e^- \rightarrow Ne^*$$

The excited atoms on collision with Hg atoms excite the latter (a phenomenon known as *energy transfer*, see Chapter 4), and these subsequently emit radiation:

$$Ne^* + Hg \rightarrow Ne + Hg^*$$
$$Hg^* \rightarrow Hg + h\nu$$

This is the basis for the functioning of mercury arc lamps so extensively used in photochemistry.

The spectral distribution of the emission of mercury arcs is markedly dependent on the pressure.

(a) Low pressure arcs, which contain Hg at pressures $\sim 10^{-3}$ mm, emit predominantly (95%) in the resonance line at 253·7 nm [Hg($^3P_1$) $\rightarrow$ Hg($^1S_0$)] and to a much lesser extent at 184·9 nm [Hg($^1P_1$) $\rightarrow$ Hg($^1S_0$)], see Figure 2.24.

(b) Medium pressure arcs are operated at Hg pressures in the region of 1 atmosphere with much higher currents, and this has the effect of raising the concentration of electrons, ions and excited species so that the metastable (relatively long-lived) Hg($^3P_1$) atoms, which would otherwise radiate at 253·7 nm, are promoted to many higher levels. Emission from these other states then gives rise to the multiplicity of lines characteristic of this type of arc. This rich and intense spectral output makes the medium pressure Hg arc an extremely important light source for photochemical investigations.† For physical measurements, where spectral purity is important, monochromators would be used to isolate particular mercury lines, but for preparative work, where it is often sufficient to use light embracing a range of wavelengths, filters are more appropriate. Compilations of filters are to be found in books by Calvert and Pitts[13] and by Bowen.[19]

(c) High pressure arcs, which run at pressures of up to several hundred atmospheres, are constructed with the electrodes close together thereby restricting the discharge to a small volume, and they constitute near point sources of brilliance comparable with that of the sun. The emission from such arcs is almost continuous. The continuum arises partly from pressure and temperature (Doppler effect) broadening of the numerous lines which consequently tend to overlap. In addition, excited Hg* atoms form excited dimers (excimers, see Chapter 4, p. 103) with Hg atoms in their ground state:

$$Hg^* + Hg \rightarrow Hg_2^* \rightarrow Hg + Hg + h\nu$$

Since the excimers are dissociated in their ground state, which therefore has no quantized levels, the potential diagram resembles that of Figure 2.25, and emission will clearly be continuous.

Other metals, such as Na, Zn, or Cd, which have appreciable vapour pressure at relatively low temperatures, may be used in place of Hg in discharge lamps with concomitant changes in the spectral output. In particular, Zn and Cd arc lamps have strong emission in the region 200–230 nm‡ in which the Hg arcs are notably deficient. High pressure arcs in inert gases, particularly xenon, are important sources of continuous radiation from < 200 nm to the infrared. It should be noted that whereas electrical discharges provide an important way of exciting stable entities such as atoms, they are useless for systems of greater

---

† The long lifetime of Hg($^3P_1$) atoms means that the 253·7 nm line is very narrow both in emission and in absorption (Uncertainty Principle). However, in medium or high pressure Hg arcs the emission is broadened by pressure and temperature (Doppler effect), and the centre of the line at 253·7 nm is virtually absent (line reversal) because light generated in the body of the discharge is absorbed by 'cool' Hg atoms near the walls. This means that light from medium or high pressure arcs is useless for exciting Hg atoms, as in Hg sensitization experiments (see Chapter 8, p. 241), for which low pressure Hg arcs must be used.

‡ Strong ultraviolet emission from Zn vapour occurs at 214, 307 and 330–334 nm, and from Cd at 229 and 326 nm.

38

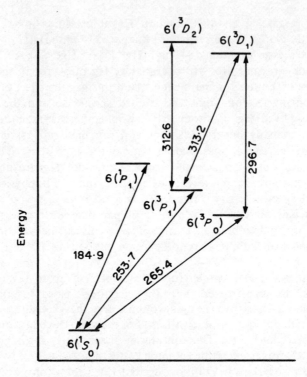

Figure 2.24. Simplified energy level diagram for mercury, showing a few of the states and the genesis of some of the more important mercury lines (in nm)

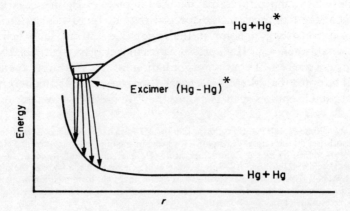

Figure 2.25. Continuous emission from mercury excimer

complexity. Diatomics are dissociated, and organic molecules are rapidly destroyed.

2.3.2   Ionizing Radiation

Media exposed to α, β or γ radiation become ionized, and the ensuing recombination reactions generate excited states, often in high yield. The phenomenon is exploited in pulse radiolysis (the ionizing radiation equivalent of flash photolysis) and in the scintillation detection and counting of such radiation. Here, an ionizing particle excites molecules of an organic solvent, energy transfer (see Chapter 4) to fluorescent solute molecules excites the latter, and the resulting emission is detected as a pulse by a photomultiplier.

2.3.3   Thermal Activation

Simple consideration of the Boltzmann distributed law:

$$\frac{N_u}{N_1} = e^{-\Delta E/RT}$$

where N_u and N_1 are the number of molecules in the upper and lower states and ΔE is their energy separation, reveals the virtual impossibility of producing electronically excited species by thermal methods except at extreme temperatures. If $\Delta E = 209$ kJ mol^{-1} (50 kcal mol^{-1}), then at 20 °C

$$\frac{N_u}{N_1} = e^{-209\,000/(8\cdot31)\times293)} = 2\cdot4 \times 10^{-38}$$

but at 500 °C

$$\frac{N_u}{N_1} = 6\cdot2 \times 10^{-3}.$$

Such temperatures may be achieved in flames, and even higher temperatures may be generated in shock tubes, but since these temperatures destroy organic molecules, thermal excitation is largely confined to atoms and diatomics.

2.3.4   Chemical Activation (Chemiluminescence)[20]

The excess energy of product molecules formed in chemical reactions is normally dissipated by collisions with the environment. Occasionally, however, in strongly exothermic reactions, the energy of the reaction is trapped as electronic energy, so that one of the products is formed in an electronically excited state. If this product is fluorescent, then radiation is emitted and luminescence is observed. Such chemically induced emission is known as *chemiluminescence*. The more efficient chemiluminscent systems fall into three groups:

(a) *Electron Transfer Reactions.*[21] The removal of an electron from the radical anion of a polynuclear aromatic (carbocylic and heterocyclic) or, more rarely, the addition of an electron to the radical cation often results in the production of the neutral substrate in an electronically excited state from which emission may occur. The electron transfers may be induced by oxidizing or reducing agents. Thus, oxidation of the radical anion of 9,10-diphenylanthracene (DPA) by 9,10-dichloro-9,10-diphenylanthracene ($DPACl_2$) induces an emission identical with that of DPA. A possible mechanism is:

$$DPA^{\cdot -} + DPACl_2 \rightarrow DPA + Cl^- + DPACl^{\cdot}$$
$$DPACl^{\cdot} + DPA^{\cdot -} \rightarrow DPA + Cl^- + DPA^* \qquad (2.17)$$
$$DPA^* \rightarrow DPA + h\nu$$

In molecular orbital terms, the process can be represented diagrammatically as in Figure 2.26.

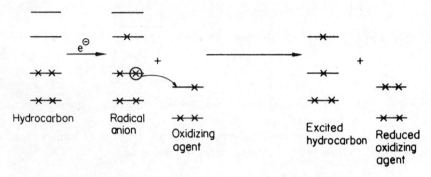

Hydrocarbon   Radical anion   Oxidizing agent   Excited hydrocarbon   Reduced oxidizing agent

Figure 2.26. Energy level scheme for electron transfer chemiluminescence

The required electron transfers may be induced purely electrically. Many substances, dissolved in an aprotic solvent such as dimethylformamide containing tetra-n-butylammonium perchlorate as electrolyte, luminesce strongly when a few volts a.c. are impressed upon electrodes immersed in the solution, particularly when a fluorescer such as rubrene is present which may be excited by energy transfer from the substrate excited state. It should be emphasized that obscurity surrounds many of the details of the mechanism by which chemical energy is converted into radiation.

(b) *Singlet Oxygen.* When hydrogen peroxide is oxidized by chlorine in alkaline solution, a red glow is observed; this emission is from excited oxygen. The lowest excited states of the oxygen molecule are singlets $^1\Delta_g$ and $^1\Sigma_g$ (contrast the ground state, which is a triplet). These excited states have energies of 92 (22) and 160 kJ mol^{-1} (38 kcal mol^{-1}) respectively, and they are sufficiently energetic to emit in the red. It seems that it is a dimer of the $^1\Delta_g$ state ($E = 185$ kJ mol^{-1} (44 kcal mol^{-1}) $\equiv 640$ nm) which is the actual emitter. Equally, the dimer ($^1\Delta_g + {}^1\Sigma_g$) emits at 478 nm. Since the dimer ($^1\Sigma_g)_2$ would have 320 kJ mol^{-1} (76 kcal mol^{-1}) available energy, Khan and Kasha[22] advanced a general theory of chemiluminescence which proposed that energy transfer to suitable fluorescent species might be responsible for the chemiluminescence of those systems in which singlet oxygen is generated.

(c) *Peroxide Decompositions.* The most efficient (brilliant) chemiluminescent reactions involve the decomposition of peroxides, and many of these processes have been shown to involve the concerted and highly exothermic decomposition of intermediate dioxetanes (McCapra mechanism[23]). Tetramethyldioxetane was found to be weakly chemiluminescent on heating and much more so in the presence of a fluorescer excited by energy transfer (equation 2.18).

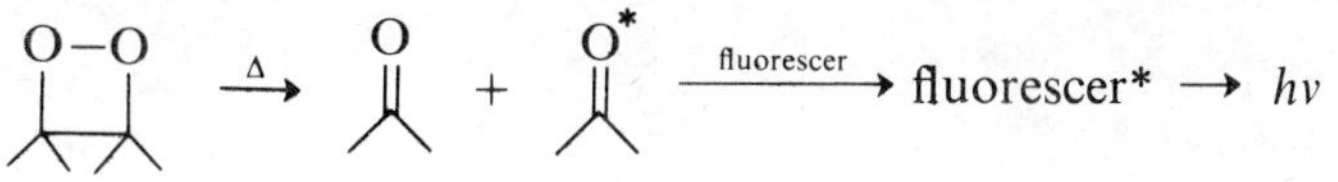

$$\text{(structure)} \xrightarrow{\Delta} \text{(structure)} + \text{(structure)}^* \xrightarrow{\text{fluorescer}} \text{fluorescer}^* \rightarrow h\nu \qquad (2.18)$$

Similarly, indolenyl peroxides glow with a beautiful green light on treatment with potassium t-butoxide in aprotic solvents, probably by the process (2.19).[24] The emission is identical with the fluorescence emission spectrum of the product.

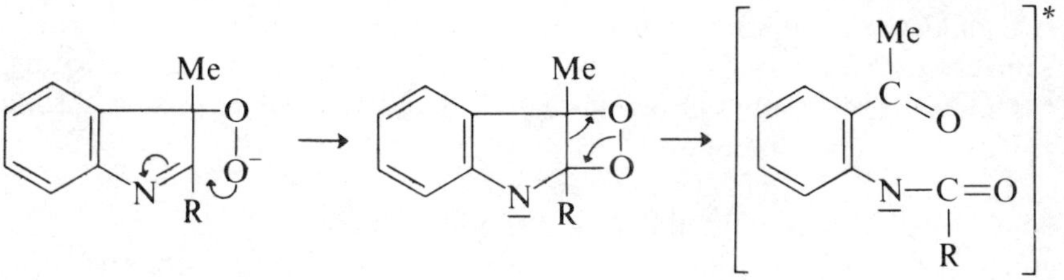

$$(2.19)$$

There are numerous other examples of chemiluminescent reactions which depend on intermediate dioxetane formation.

2.4 LASERS

It seems appropriate, at this point, to discuss lasers. A laser (*Light Amplification by Stimulated Emission of Radiation*), as is implied by its name, depends for its operation on stimulated emission of radiation—contrast conventional sources which function through spontaneous emission. Recalling (p. 12) that the rates of absorption and of stimulated emission for a given radiation density ρ are $n_l B_{lu}\rho$ and $n_u B_{ul}\rho$, where n_l and n_u are the number of molecules in the lower and upper states and B_{ul} and B_{lu} are the Einstein coefficients, it can be seen that if spontaneous emission could be suppressed in some way, a system exposed to radiation at the transition frequency would absorb radiation until $n_l = n_u$, i.e. until the ground state and excited state were equally populated. However, should a *population inversion* exist (i.e. if $n_l < n_u$), then light amplification can take place, for irradiation with light of the transition frequency will cause emission whose rate will be greater than the rate of absorption. This process will continue until $n_l = n_u$.

Lasers, which exploit this principle, comprise an emitting material set in a tuned optical cavity—a pair of parallel mirrors, one of which is partially transparent (Figure 2.27). The mirrors are separated by an integral number of

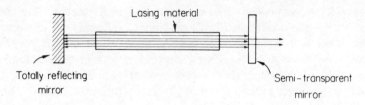

Figure 2.27. Schematic diagram of a laser

$\frac{1}{2}\lambda$ (where $E_u - E_l = hc/\lambda$). As a result, the light reflected from a mirror will be in phase with the incident wave (constructive interference). When a population inversion has been generated in the emitting material, *spontaneous* emission provides a few photons, which on collision with molecules of the excited lasing material stimulate it to emit in phase with the incident photons. Thus the wave builds up in intensity as it travels back and forth between the mirrors. A chain reaction is set up which destroys the population inversion in an exceedingly brief time interval (~ 1 μs), producing a burst of radiation which escapes through the partially transmitting mirror.

The characteristics of laser radiation are:

(i) Coherence—the emitted light waves are all in phase.

(ii) Very high monochromaticity—because the cavity is tuned to one particular frequency, light of other frequencies suffers destructive interference.

(iii) Accurate parallelism, because the cavity will not be tuned for off-axis radiation, which, in any case, would escape after only a few reflections. This lack of divergence of the beam permits it to be focussed to a spot of very small dimensions (of the order of a wavelength), thus generating very high radiation densities ($> 10^9$ W cm^{-2}) for short laser pulses.

(iv) Enormous brilliance of a laser pulse—a 'long' pulse of 1 μs containing 1 J of energy will have a power output of 1 MW, a figure which can be increased by orders of magnitude by shortening the laser pulse duration.

The population inversion essential to laser action cannot be achieved by direct radiative excitation to the lasing state, since this can at best cause n_u to equal n_l. Radiationless transitions or energy transfer must be employed in order to gain access to some excited state other than the one being populated by the energy input. For example, in the well-known ruby laser the lasing material is a cylindrical ruby rod, the Cr^{3+} ions of which are excited to the $^4T_{2g}$ state by an intense flash of light from a flash tube surrounding the rod. Rapid intersystem crossing populates the ^2E_g state from which the characteristic red laser emission occurs (Figure 2.28). Such a three-level system is inefficient— it can only work if more than 50% of the Cr^{3+} ions are excited to the ^2E_g state during the exciting flash. More efficient are the four-level systems (Figure 2.29) in which laser emission results from transition not to the ground state but to an unpopulated state C. If the state C, which is populated by the transition $B \rightarrow C$, is also rapidly depopulated by a fast conversion to the ground state, then it is possible to maintain a continuous population inversion of B with respect to C.

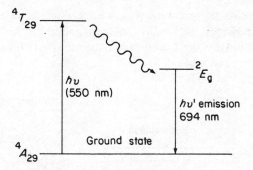

Figure 2.28. Energy levels for Cr^{3+} ions in ruby laser

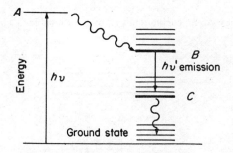

Figure 2.29. Energy level scheme for continuously operating laser

Such lasers will give a continuous light output. Some continuously operating gas lasers operate on similar principles. In the He–Ne laser, He is excited by an electrical discharge to the state A in Figure 2.29. Energy transfer affords Ne* (B in the Figure), which is the emitting species.

Recently it has been shown that solutions of many strongly-fluorescent organic dyes (2.20) can be induced to lase. The importance of these dye lasers[25] is that they can be tuned, i.e. the wavelength of their light output can be varied

Rhodamine 6G   Some lasing dyes   (2.20)

(often over as much as 50 nm). One type of apparatus, shown diagrammatically in Figure 2.30, replaces the totally reflecting plane mirror by a diffraction grating which returns to the cavity only light of a particular frequency determined by its angle of rotation.

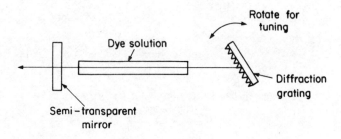

Figure 2.30. Schematic diagram of a tuneable dye laser

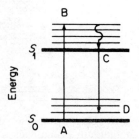

Figure 2.31. Energy level scheme for a dye laser

Now consider the energy level diagram for the lasing dyestuff (Figure 2.31). Pumping and vibrational relaxation promotes molecules of the dye to point C, the bottom vibrational level of S_1, from which they can radiatively collapse to the quasi-continuum of densely packed vibrational and rotational levels of S_0. If the cavity is tuned to a particular frequency corresponding, say, to the transition $C \rightarrow D$, then the laser will emit light of this frequency only. A moment's consideration shows that although only two electronic states are involved dye lasers are, in fact, four-level systems and as such may be made to operate continuously. Up to 70 % of the energy trapped in the S_1 state can be obtained in the form of laser emission at the prescribed frequency.

Laser Pulse Width[25b]

The simple laser (Figure 2.27) gives relatively long pulses (~ 1 μs) because spontaneous emission starts immediately upon excitation, but laser emission cannot begin until a population inversion has been achieved. The pulse width may be reduced to ~ 10 ns by the technique of Q-switching, which involves the interposition between the end of the laser rod and the totally reflecting mirror a shutter which can be opened with extreme rapidity (e.g. a Kerr cell, see Figure 2.32). If the laser rod is pumped with the shutter closed, lasing is impossible because light leaving the rod cannot get back to stimulate further emission (the amplification factor of the cavity is $\ll 1$). If the shutter is now opened, the amplification factor of the cavity suddenly rises to a value far beyond unity, and the stored energy is released as a giant pulse. A small ruby laser storing 1 J of energy, Q-switched to give a pulse of 10 ns duration, will now deliver a peak power of 100 MW.

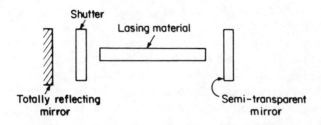

Figure 2.32. Schematic diagram of a Q-switched laser

Slight modification of the system to include the use of a dilute solution of a dye as the shutter results in a laser output consisting of a train of pulses ('mode-locking') having a half width of $1/\Delta v$, where Δv is the half-width of the fluorescent emission of the laser material (Figure 2.33). By these means, pulses with pulse widths as short as 0·4 ps (0·4 $\times$ 10^{-12} s) have been obtained with peak powers of 5 GW (to comprehend such short time intervals, remember that light requires 3 ps to travel 1 mm).

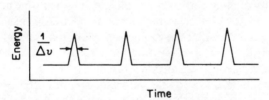

Figure 2.33. Pulse output of a mode-locked laser

2.4.1   **Laser Applications**[26]

Energy transfer, intersystem crossing and vibrational relaxation are extremely fast processes. Their rates can now be measured directly using picosecond flashes from a mode-locked laser (see Chapter 3, p. 75). Also the very narrow line-width of laser emission means that the output from a dye laser can be tuned to excite a specific vibrational level of a molecule in the gas-phase in order to study its decay characteristics.

Forbidden processes are characterized by very low extinction coefficients, so that only a small fraction of the incident light can be absorbed and used to effect these processes. The enormous intensities of laser pulses overcome this difficulty. For example, the photon scattering giving rise to the Raman effect is extremely weak—only about 10^{-6} of the incident light appears as Raman scattering. The very high power output per unit area of a focussed laser makes it an ideal light source and permits the use of very much smaller sample specimens. Further-more, at the focussed output of a giant pulse laser, novel phenomena are ob-served—the stimulated Raman effect and the inverse and hyper-Raman effects, which are excellently summarized by Long.[27] Another forbidden process is multiphoton absorption. This was first demonstrated[28] through the blue fluorescence ($\lambda = 425$ nm) of Eu^{2+} ions when subjected to the focussed red output of a pulsed ruby laser ($\lambda = 694 \cdot 3$ nm). The phenomenon is due to the 'simultaneous' absorption of two photons. There are now many organic examples of two-photon and even three-photon absorption. A recent applica-tion of this phenomenon is two-photon spectroscopy,[29] a simple procedure for recording the population of otherwise inaccessible ultraviolet vibronic levels using the visible light output from a tunable dye laser.

2.5   PROPERTIES OF EXCITED STATES

Excited species are subject to rate processes such as emission, quenching, etc. They also have a range of static properties which are independent of time and are not the consequence of the conversion of the species into another entity. It is with these latter properties that we are now concerned. Dynamic properties will be discussed in the next chapter.

2.5.1   Geometry of Excited Molecules

The impossibility of obtaining over a sufficient length of time a sufficiently high concentration of excited species to be examined by the usual techniques of structural analysis means that the geometries of excited molecules are extremely difficult to determine. The analysis of vibrational and rotational fine structure of high resolution absorption spectra obtained in the gas phase is one of the few methods available, and provides information about the vibrational modes and rotational constants. Another technique can be applied to transitions which are symmetry forbidden. If the molecule in its ground state has a vibration which transforms its shape towards that of the excited state, then the correspond-ing vibronic band appears strongly in the gas-phase spectrum. Identification of strong vibronic transitions then provides valuable clues to the structure of the excited state. Such techniques, which can only be applied to very simple mole-cules, show that excitation can lead to profound changes in geometry. The results can be qualitatively rationalized by simple molecular orbital considera-tions and supported by numerical calculations.

Acetylene[30]

The ultraviolet absorption band of acetylene at 250–210 nm corresponds to a singlet ($\pi \rightarrow \pi^*$) transition ($^1\Sigma_g^+ \rightarrow {}^1A_u$) to an excited state which has the transoid planar trigonal structure of Figure 2.34, in which the pair of electrons originally in one of the π orbitals now occupy sp^2 orbitals on the carbon atoms. The transition is symmetry forbidden and is further weakened ($f \sim 10^{-4}$) because the change in molecular architecture causes such a large relative displacement of the potential energy surfaces of the ground state and excited state that the vibrational overlap integral is very small.

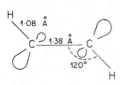

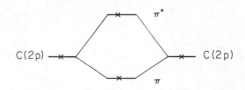

Figure 2.34. Geometry of excited acetylene (^1A_u)

Figure 2.35. In a double bond, the π^* level is more antibonding than the π level is bonding

Reference to Figure 2.35 illustrates that the antibonding induced by the electron in the π^*-orbital is greater than the bonding of the single electron in the π-orbital. In other words, stability may be obtained by uncoupling the π and π^*-bond and 'locating' the two electrons in C 2p orbitals, when further stability may be attained by rehybridization to sp^2. In valence bond terms this excited state could be crudely represented as a resonance hybrid:

$$H\dot{C}=\dot{C}H \leftrightarrow H\overset{+}{C}=\overset{-}{C}H \leftrightarrow H\overset{-}{C}=\overset{+}{C}H$$

It must be emphasized that the electronic transition from linear ground state acetylene gives rise (Franck–Condon principle) to a linear excited state. The relaxation to the more stable bent state occurs after excitation.

Ethylene[31]

The most stable structure for the lowest energy (π, π^*) state of ethylene appears from both experiment and calculation to be one in which the two CH_2 groups are joined by what is essentially a single bond and lie in perpendicular planes (Figure 2.36) with two electrons occupying p-orbitals on the carbon atoms. The data can be rationalized by noting that, as with excited acetylene, the promotion of an electron to a π^*-orbital leads to a system which becomes more stable by uncoupling the π-orbital to leave only a σ-bond between the carbon atoms. The two electrons, now in the C 2p orbitals, experience electrostatic repulsion, and this is readily minimized by rotation about the C—C single bond (Figure 2.37). Presumably the carbon atoms of the two orthogonal CH_2 groups rehybridize to a pyramidal structure in order to incorporate stabilizing 's' character into the orbitals of the two electrons. Similar considerations apply to the triplet (π, π^*) state of ethylene.

48

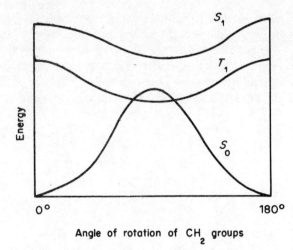

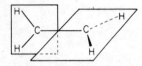

Figure 2.36. Most stable geometry for the (π, π^*) state of ethylene

Figure 2.37. Energy of the ethylene molecule as a function of the relative rotations of the relative rotations of the CH_2 groups

Again, the perpendicular state is not the Franck–Condon state obtained immediately upon excitation, and it is referred to as a *non-vertical* or *non-spectroscopic* state. The rotations consequent upon excitation of ethylene presumably also occur in substituted olefins, and they are thought to constitute the underlying basis for *cis–trans* photoisomerization (see Chapter 8, p. 230).

Formaldehyde

The important difference between the (n, π^*) state of formaldehyde and the (π, π^*) states of ethylene and acetylene is that in the former there are *two* electrons in the bonding π-orbital. The antibonding induced by the single electron in the π^*-orbital is inadequate to neutralize completely the bonding power of the doubly occupied π-orbital. However, it does weaken the bond and hence lengthen it to a value intermediate between those for C—O and C=O. Furthermore, the electron in the π^*-orbital increases the electron density at the carbon atom. As a result, this atom rehybridizes in order to confer some 's' character

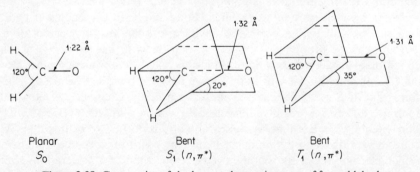

Figure 2.38. Geometries of the lowest electronic states of formaldehyde

upon the promoted electron. This implies loss of coplanarity (Figure 2.38). The experimental data[32] support these expectations.

It is to be expected that in more extended chromophores the geometrical changes consequent upon excitation will be much smaller. In the lowest (π, π^*) state of ethylene, 50% of the π-electrons are promoted. In naphthalene, for example, only 1 π-electron in 10 would be involved in the $(\pi \rightarrow \pi^*)$ transition.

2.5.2   Acid–Base Properties[33]

It has been known for some time that the fluorescence spectra of certain phenols and aromatic amines are pH-dependent. In Figure 2.39 curves 1 and 5 are the fluorescent emissions of 2-naphthoxide and 2-naphthol respectively. At intermediate pH (curves 2, 3 and 4) emission is observed from both 2-naphthol and 2-naphthoxide even though at these acid concentrations there is no significant amount of the anion present in the solution being scanned. Förster explained the phenomenon by postulating that proton exchanges are so rapid that an acid–base equilibrium is established (2.21) between the excited phenol (HA*) and its *excited* conjugate base (A⁻*):

$$HA \xrightarrow{h\nu} HA^* \rightleftarrows H^+ + A^-{}^*$$

$$\downarrow \qquad\qquad\qquad \downarrow \qquad\qquad\qquad (2.21)$$

$$HA + h\nu' \qquad\qquad A^- + h\nu''$$

The implication of Figure 2.39 is that the phenols in their S_1 states are much stronger acids than in their ground states.

Two methods have been employed to measure the pK_a values of the excited singlets of phenols and amines. Both assume proton equilibrium during the lifetime of the excited state (10^{-8} to 10^{-9} s).

(i) The pK_a is that pH at which the fluorescence of HA* drops to one-half of its intensity in solutions so strongly acid that only HA* emits. It is also the pH at which the intensity of the fluorescence of A⁻* rises to one-half of the value observed in solutions so alkaline that only A⁻* emits.

(ii) The second method depends on the Förster–Weller cycle (Figure 2.40). It is clear from the diagram that:

$$\Delta E_{HA} - \Delta E_{A^-} = \Delta H - \Delta H^* = \Delta G - \Delta G^*$$

assuming that the entropy of dissociation is the same in both ground state and excited state. It follows that

$$\ln\left(\frac{K^*}{K}\right) = \frac{\Delta E_{HA} - \Delta E_{A^-}}{RT} = \frac{h\,\Delta\nu}{kT} \qquad (2.22)$$

where K^* and K are the dissociation constants of HA* and HA, and $\Delta\nu$ is the difference in the frequency of absorption or emission of HA and A⁻ measured in the (O—O) bands. Hence K^* may be obtained. Similarly, the pK_a values of triplet states have been estimated by comparing the absorption and phosphorescent emission spectra of triplet species. Some values of pK_a for deprotonation are collected in Table 2.1.

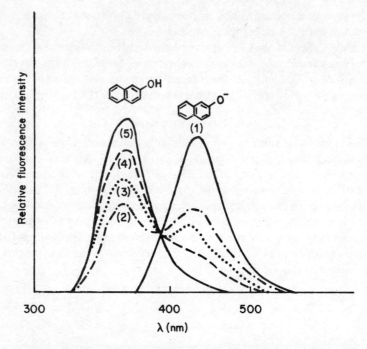

Figure 2.39. Fluorescence spectrum of 2-naphthol in solutions of different pH. (1) 0·02 M NaOH; (2) 0·02 M acetic acid + 0·02 M sodium acetate; (3) pH 5–6; (4) 0·004 M HClO$_4$; (5) 0·15 M HClO$_4$ (from A. Kearwell and F. Wilkinson, in G. M. Burnett and A. M. North (ed.), *Transfer and Storage of Energy by Molecules*, volume 1, (1969), John Wiley and Sons Ltd.)

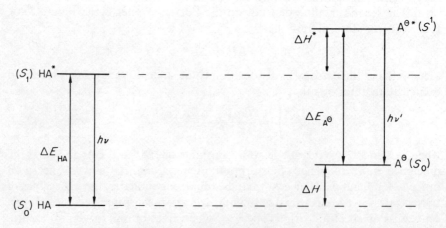

Figure 2.40. Förster–Weller cycle for deriving pK values for excited states

Table 2.1. pK_a values for deprotonation of ground and excited states

Molecule	p$K_a(S_0)$	p$K_a(S_1)^a$	p$K_a(T_1)^a$
p-Cresol	10·3	4·1–5·7	8·6
1-Naphthol	9·2	2·0	
2-Naphthol	9·5	2·5–3·4	7·7–8·1
2-Naphthylamine salts	4·1	−1·5–−2·9	3·1–3·3
1-Naphthoic Acid	3·7	10–12	3·8–4·6
Acridine salts	5·5	10·6	5·6
Indole	very large	12·3	

^{a} The values obtained depend upon the method used for their estimation.

Examination of the data in the Table reveals the following features:

(i) For phenols and the salts of aromatic amines, the lowest *singlet* excited state is much more acidic than the ground state by a factor of 10^5–10^6.

(ii) Conversely, aromatic acids and the conjugate acids of certain heterocyclic amines are much weaker acids in the singlet excited state than in the ground state.

(iii) The pK_a values of excited triplet states are comparable with those of the ground states. This phenomenon is perhaps most readily explained in valence bond terms. The wavefunctions for S_0 and S_1 states of phenol can be approximated by mixing the wavefunctions corresponding to the structures **I, II** and **III** (2.23).

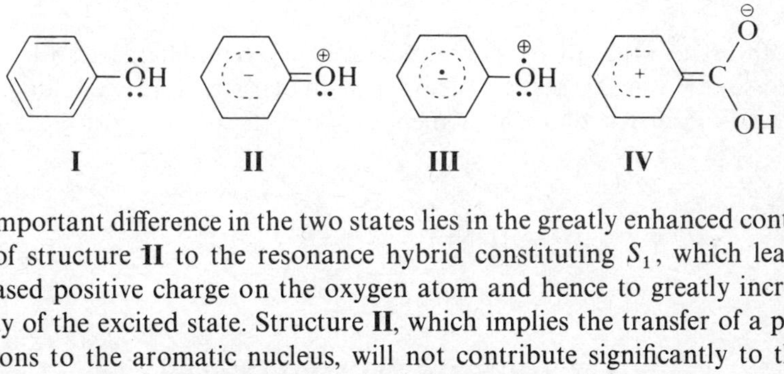

$$\text{(2.23)}$$

$$\quad \text{I} \qquad\qquad \text{II} \qquad\qquad \text{III} \qquad\qquad \text{IV}$$

The important difference in the two states lies in the greatly enhanced contribution of structure **II** to the resonance hybrid constituting S_1, which leads to increased positive charge on the oxygen atom and hence to greatly increased acidity of the excited state. Structure **II**, which implies the transfer of a pair of electrons to the aromatic nucleus, will not contribute significantly to the T_1 state because spin correlation tends to keep electrons with parallel spins apart (see discussion, p. 54). Hence the oxygen atom is less positively charged in the T_1 state than in the S_1 state, leading to the observed sequence p$K_a(S_1)$ < p$K_a(T_1)$ < p$K_a(S_0)$. A similar argument explains the increased acidity and diminished basicity of aromatic amines.

In aromatic carboxylic acids, excitation to S_1 leads to charge migration away from the ring, something which can be represented by an increased contribution of the charged structure **IV** (2.23) to the resonance hybrid constituting S_1. The effect will be to reduce the acidity of the S_1 states of aromatic carboxylic acids and greatly enhance their basicity.

2.5.3   Dipole Moments[34]

Because electronic transitions lead to redistribution of electrons, the dipole moments of a molecule in its ground state and excited state will, in general, be different. Estimates of the dipole moments of excited states may be obtained by analysing the effect of solvents on the absorption and fluorescence maxima,[35] and more directly by Czekalla's method.[36] This depends on the application of an electric field to a solution being irradiated with plane polarized light. The molecules in both their ground and excited states tend to align themselves with the field to an extent determined by their dipole moment, and this is manifested as a change in the degree of polarization (see Chapter 3, p. 76) of the fluorescent emission. Table 2.2 summarizes some of the data.

Table 2.2. Dipole moments of some singlet states

Compound	Dipole moment (in Debyes)	
	S_0	$S_1{}^a$
p-Nitroaniline	6	14
4-Amino-4′-nitrobiphenyl	6	18–23
2-Amino-7-nitrofluorene	7	19–25
4-Dimethylamino-4′-nitrostilbene	7·6	20·7–32
Formaldehyde	2·3	1·6

 ^{a} The values obtained depend on the method adopted for their measurement.

It can be seen that there is a considerable degree of charge–transfer in the excited states of the aromatic systems. The dipolar structure (2.24) for p-nitro-aniline has a calculated dipole moment of about 25 D.

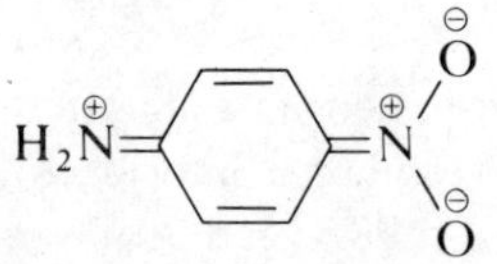

$$(2.24)$$

2.6   ENERGIES OF EXCITED STATES

The energy of a singlet or triplet excited state (E_S or E_T) is normally taken to be the energy difference between the $v = 0$ levels of the excited state and the corresponding ground state. It is therefore a quantity determinable spectroscopically by locating the 0–0 transition. If vibrational fine structure is present, the measurement is readily made for S_1 and T_1 states by recording the $S_0 \rightarrow S_1$ and $S_0 \rightarrow T_1$ absorption and/or emission spectra. In the absence of vibrational structure, the energies of the states may be estimated roughly from the position of the long wavelength tail of the absorption spectrum, recorded at low temperature to eliminate 'hot' bands.

The precise determination of the energies of states other than the lowest singlet and triplet is more difficult. Emission spectra are useless because emission almost invariably occurs from the bottom state; $S_0 \rightarrow S_n$ transitions usually lack fine structure and tend to merge into the short wavelength $S_0 \rightarrow S_1$ absorption. The same is true for upper triplets. Fortunately, for most purposes, it is the energy of the lowest states which are of greatest interest to photochemists.

2.6.1 Singlet–Triplet Splitting

The Table 2.3 records E_{S_1} and E_{T_1} for some representative systems. It is apparent that in all cases $E_{T_1} < E_{S_1}$. This result is quite general—all triplets are more stable than the corresponding singlets (i.e., singlets with the same orbital

Table 2.3. Singlet–triplet splittings of excited states

Compound	E_{S_1} in kJ (kcal) mol^{-1}	E_{T_1} in kJ (kcal) mol^{-1}	Singlet–Triplet splitting in kJ (kcal) mol^{-1}	Transition
Formaldehyde	338 (80·6)	302 (72·1)	36 (8·5)	
Acrolein	310 (74·0)	291 (69·3)	19 (4·7)	$n \rightarrow \pi^*$
Benzophenone	318 (75·8)	291 (69·4)	27 (6·4)	
Benzene	577 (137·4)	354 (84·3)	223 (53·1)	
Naphthalene	415 (99·0)	255 (60·8)	160 (38·2)	
Tricyclic, catacondensed aromatic hydrocarbons[a]			~132 (31)	
Tetracyclic, catacondensed aromatic hydrocarbons[a]			~125 (30)	$\pi \rightarrow \pi^*$
Heptacyclic, catacondensed aromatic hydrocarbons[a]			~114 (27)	
Ethylene		344 (82·1)	290 (69·2)[c]	
Butadiene		250 (59·6)	271 (64·6)[c]	
Toluene/tetra-cyanobenzene	288 (68·6)[b]	252 (60·1)[b]	36 (8·5)	Charge–transfer
Hexamethylbenzene/tetracyanobenzene	222 (52·9)[b]	206 (49·1)[b]	16 (3·8)	

[a] In catacondensed hydrocarbons, no carbon atom lies at the junction of more than 2 rings.
[b] Positions of emission maxima.
[c] From oxygen perturbation of absorption spectra.

configuration). The energy difference $(E_S - E_T)$ is known as the *singlet–triplet splitting* and has large values for (π, π^*) states and small ones for (n, π^*) and CT states. These points require explanation.

Spatial wavefunctions must be either symmetric Ψ_S or antisymmetric Ψ_A with respect to the interchange of the coordinates of two electrons. The antisymmetric wavefunction for a system of two electrons can be written

$$\Psi_A = \frac{1}{\sqrt{2}}[\psi(r_1)\psi'(r_2) - \psi(r_2)\psi'(r_1)]$$

where ψ and ψ' are the space orbitals, and r_1 and r_2 are the coordinates of the electrons. Clearly such a wavefunction is antisymmetric, because interchanging the electrons gives

$$\frac{1}{\sqrt{2}}[\psi(r_2)\psi'(r_1) - \psi(r_1)\psi'(r_2)] = -\Psi_A$$

Ψ_A becomes zero when $r_1 = r_2$, and since wavefunctions must be continuous a plot of the variation of Ψ_A against $(r_1 - r_2)$ is of the form shown in Figure 2.41.

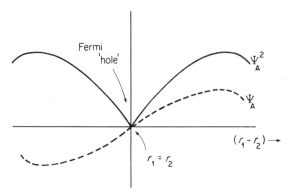

Figure 2.41. Fermi hole associated with an antisymmetric spatial wavefunction

The probability of finding one electron at r_1 and the other at r_2 is given by Ψ_A^2, and it is apparent that the probability of the electrons being in close proximity is very small (*Fermi hole*), reducing to zero when $r_1 = r_2$ and they are at the same point in space. For the symmetric wavefunction Ψ_S there is no Fermi hole. Therefore, on average, the electrons are further apart in systems described by Ψ_A than in those described by Ψ_S, with a corresponding reduction of Coulombic repulsion. Hence the energies of the spatially antisymmetric states Ψ_A will always be less than those of the corresponding symmetric states Ψ_S.

We now consider electron spin; this appears in wavefunctions as either α or β. Thus for two electrons the possible spin combinations are $\alpha(1)\alpha(2)$, $\beta(1)\beta(2)$, $\alpha(1)\beta(2)$ and $\beta(1)\alpha(2)$, of which the last two wavefunctions are unacceptable since they are neither symmetric nor antisymmetric with respect to electron interchange. They are therefore replaced by the linear combinations $(\alpha(1)\beta(2) \pm \alpha(2)\beta(1))$ which do satisfy the symmetry requirements. Three of these spin wavefunctions are symmetric and one is antisymmetric (2.25), and these are the wavefunctions for triplet and singlet states respectively. Since the Pauli Principle requires that the *total* electronic wavefunction (adequately represented as the product of space and spin functions) must be antisymmetric, it follows that triplet states will have an antisymmetric spatial wavefunction and singlets a symmetric one:

$$^1\Psi = \Psi_S \cdot \Psi_A(\text{spin}); \qquad ^3\Psi = \Psi_A \cdot \Psi_S(\text{spin})$$

Hence any triplet state will always be more stable than the corresponding singlet, simply because its space wavefunction is antisymmetric with the inevitable result that the charge centres of the two electron clouds are further apart than in the corresponding singlet.

triplet spin wavefunctions   singlet spin wavefunction

$$\alpha(1)\alpha(2)$$

$$\beta(1)\beta(2)$$

$$\frac{1}{\sqrt{2}}[\alpha(1)\beta(2) + \beta(1)\alpha(2)] \qquad\qquad \frac{1}{\sqrt{2}}[\alpha(1)\beta(2) - \beta(1)\alpha(2)] \qquad (2.25)$$

Antisymmetric   *Symmetric*

The singlet–triplet splitting, which is the difference in energy between singlet and triplet states, is thus a measure of the difference in electron distribution between singlet and triplet states. The situation just described implies that electrons with parallel spins avoid each other (*spin correlation*).

In molecular orbital theory, the singlet–triplet splitting is twice the *exchange integral* K_{ul} given by (2.26), where the subscripts u and l refer to upper and lower states concerned.[†]

$$K_{ul} = \int\!\!\int \psi_u(1)\psi_l(1)\left(\frac{e^2}{r_1 - r_2}\right)\psi_u(2)\psi_l(2)\,dr_1\,dr_2 \qquad (2.26)$$

Recalling that the products $\psi_u\psi_l$ in this expression are the integrands of overlap integrals (p. 26), it can be deduced that if the two electrons are located in orbitals which overlap but slightly, the products $\psi_u\psi_l$ will be small and the singlet–triplet splitting will be small, as with (n, π^*) states and the excited states of donor–acceptor complexes. Similarly, in extended chromophores, where the orbitals involved in the transition are large, the average values of ψ_u and ψ_l will be smaller than in more compact systems, and this leads to a reduction in K_{ul} and the singlet–triplet splitting. These expectations are all borne out by the date in Table 2.3.

[†] Strictly, $K_{ul} = \int\!\!\int \psi_u^*(1)\psi_l^*(1)\left(\dfrac{e^2}{r_1 - r_2}\right)\psi_u(2)\psi_l(2)\,dr_1\,dr_2$, but in this book the wavefunctions are assumed to be real (see footnote, p 15).

2.6.2   **Singlets, Triplets and Biradicals**[37]

The literature is replete with examples of excited states formulated as biradical structures. Thus the (π, π^*) state of ethylene is frequently written as (2.27), implying a high chemical reactivity of the carbon centres linked by the single bond. Similarly, the (n, π^*) state of formaldehyde is formulated as (2.28).

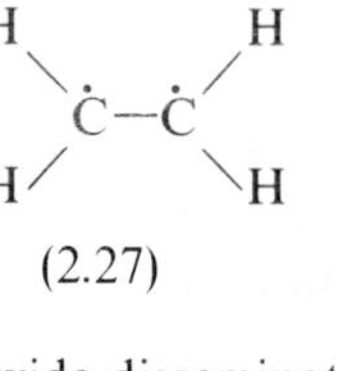

(2.27) 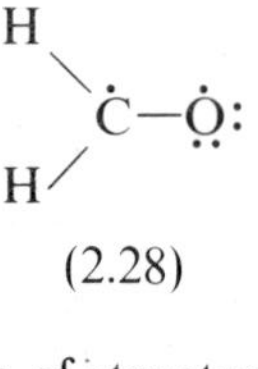   (2.28) 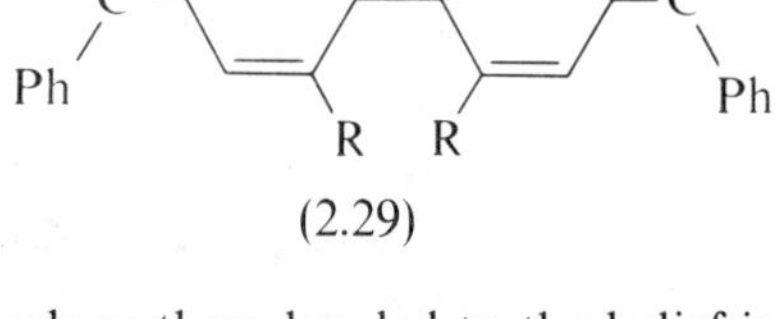   (2.29)

The wide dissemination of structures such as these has led to the belief in the biradical character of excited states. Whether excited states are biradicals or not is partly a matter of definition, but considerable added confusion has arisen because triplet states are paramagnetic (as are free radicals) but singlet states are not.

The problem may be clarified by first considering what is meant by a biradical. The ultimate biradical would be a *bifunctional* system in which the two radical centres behaved quite independently of each other. The e.s.r. spectrum of such a system would show a pair of doublets, since each electron could independently have the α or β spin. Such a situation, of which the tetrachloro derivative of Chichibabin's hydrocarbon (2.29), $R = Cl$) is a rare example, can be represented diagrammatically as in Figure 2.42.

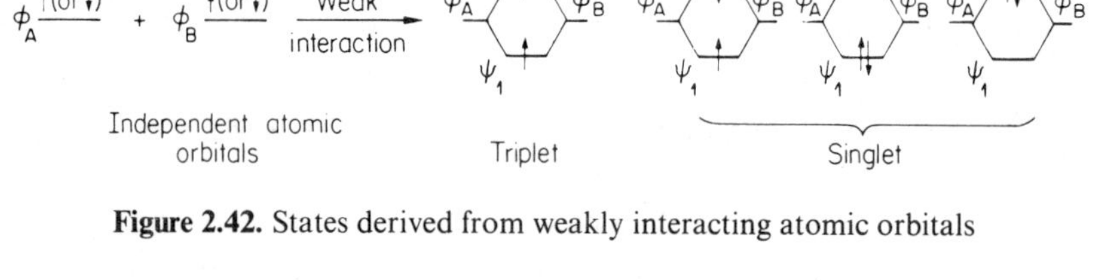

Figure 2.42. States derived from weakly interacting atomic orbitals

Case A

If, as is commonly the case, two degenerate or nearly degenerate radical centres with wavefunctions ϕ_A and ϕ_B interact weakly, then distinct molecular orbitals ψ_1 and ψ_2 embracing both centres develop. Since the interaction is weak, the energy difference $(E(\psi_1) - E(\psi_2))$ will be small, ψ_1 and ψ_2 will be almost degenerate, and the two electrons can be distributed over the two orbitals as shown in Figure 2.42. This gives a triplet biradical (detectable by e.s.r.) and a singlet biradical state whose wavefunction can only be properly described as an appropriate mixture of the three configurations in Figure 2.42.

Case B

Now suppose that the overlap between ϕ_A and ϕ_B (given by the overlap integral $S_{AB} = \langle \phi_A \phi_B \rangle$) increases. Then ψ_1 and ψ_2 separate further (Figure 2.43), the electrons therefore tend to pair off in ψ_1 for energetic reasons, and the singlet now acquires closed shell character and becomes stabler than the triplet. The singlet approximates increasingly to that state in which a covalent bond is formed between the two radical centres ϕ_A and ϕ_B.

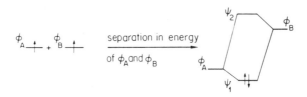

Figure 2.43. Increasing the overlap of the atomic orbitals ϕ_A and ϕ_B increases the energy splitting of the molecular orbitals ψ_1 and ψ_2

Case C

If ϕ_A and ϕ_B cease to be degenerate, then again ψ_1 and ψ_2 separate in energy (Figure 2.44), and ψ_1 approximates more and more closely to ϕ_A. Again the most stable state of the system is the singlet, but in this case it corresponds approximately to the zwitterionic state with both electrons located in ϕ_A.

Figure 2.44. Production of zwitterionic singlet from radical centres

These trends are summarized in Figures 2.45 and 2.46.

If one takes the view just advanced that biradical character occurs only when the singlet–triplet splitting is small† and when the *molecular* orbitals are close in energy, then biradicals occur on the left of these diagrams and, proceeding through an ill-defined (shaded) region, arrive at an area on the right of the diagrams where the systems in their ground state acquire closed shell (covalent or zwitterionic) character.

† If the overlap between ϕ_A and ϕ_B is small then the exchange integral and therefore the singlet–triplet splitting are also small.

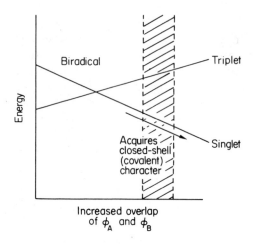

Figure 2.45. The effect of increasing orbital overlap on the closed shell covalent character of biradicals

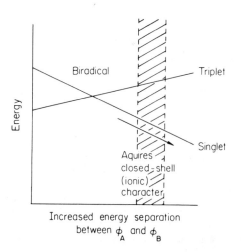

Figure 2.46. The effect of increasing the energy separation of the radical centres on the closed shell zwitterionic character of biradicals

On this basis the *planar* triplet and singlet excited states of ethylene would not qualify as biradicals because the large overlap between ϕ_A and ϕ_B means that the molecular orbitals ψ_1 and ψ_2 would have large separation. Consequently in the planar triplet state the electron in ψ_2 would be much more reactive than that in ψ_1, leading to radical but not *biradical* behaviour. The singlet state would be expected to show little radical behaviour. On the other hand the relaxed orthogonal (π, π^*) excited states of ethylene (2.30), dienes (2.31) and T_1 benzene (2.32) would be classified as biradicals.

(2.30)   (2.31)   (2.32)

It should be noted that there is, as yet, no consensus of opinion on these matters and alternative views have been advanced. For example, it has been suggested that a divalent intermediate which reacts like a radical can probably be termed a biradical, independent of whether it is of singlet, or triplet multiplicity.

If S_1 and T_1 states of a molecule had similar electron distributions one would expect their chemical reactivity† to be similar except with regard to pericyclic reactions and processes where the operation of the Wigner spin conservation rules (see Chapter 4, p. 101) might impose a difference. However, the singlet–triplet splitting, the energy difference between the singlet and corresponding triplet states, is merely a measure of this difference in electron distribution (see Chapter 2, p. 55). Therefore, in (n, π^*) states, where the singlet–triplet splitting is small, the charge distributions will be closely similar. We therefore expect the two states to have similar photochemical behaviour. Conversely, where the singlet–triplet splitting is large, as in the (π, π^*) state of benzene, singlet and triplet photochemistry is expected to be dissimilar.

2.6.3   Solvent Effects

When absorption spectra are measured in solvents of increasing polarity it is found that for some systems λ_{max} moves to longer wavelengths (*red or bathochromic shift*) and for others an inverse *blue or hypsochromic* shift is encountered. These shifts provide information on the nature of the transition and can afford estimates of the dipole moments of excited states. This is a difficult area, requiring for its proper interpretation[38] consideration of solute–solvent interactions (dipole–dipole, dipole–polarization, hydrogen bonding) and of the impact of the Franck–Condon Principle.

In a general way, one can see that for a solute having an excited state much more polar than the ground state, increase of solvent polarity will stabilize the excited state more than the ground state, thereby lowering the energy of the transition and leading to a predicted red shift. Such is observed with the merocyanine dyes which contain the chromophore (2.33) and whose first excited state is approximated by dipolar structures such as (2.34).

---

† In such systems the differing lifetimes of singlet and triplet states may be a critical factor in determining behaviour.

$$\text{\Large >N-(CH=CH)}_n\text{-C=O}$$
$$\text{(2.33)}$$

$$\text{\Large >N=(CH-CH)}_n\text{=C}\overset{\overset{\ominus}{O}}{\big<}$$
$$\text{(2.34)}$$

Conversely, the quaternary pyridinium iodide (2.35), whose absorption spectrum is remarkably solvent dependent, shows a marked *blue* shift, which has been rationalized[39] by assuming that its long wavelength absorption is due to a charge–transfer transition involving the transfer of charge from the iodide ion to the aromatic ring. Here, the excited state, with a large contribution from structure (2.36), is considerably *less* polar than the ground state. The magnitude of the solvent shifts for this compound has been proposed as an empirical scale for classifying solvent polarity.

$$\text{Et}-\overset{\oplus}{N}\!\!\left\langle\!\!\bigcirc\!\!\right\rangle\!\!-\text{CO}_2\text{Et}$$
$$[\text{I}^{\ominus}]$$
$$\text{(2.35)}$$

$$\text{Et}-\overset{\oplus}{N}\!\!\left\langle\!\!\overset{\ominus}{\bigcirc}\!\!\right\rangle\!\!-\text{CO}_2\text{Et}$$
$$[\text{I}^{\bullet}]$$
$$\text{(2.36)}$$

The blue shift associated with $(n \rightarrow \pi^*)$ transitions, first noted by Kasha,[40] has now been extensively documented. It is so general a phenomenon that it has acquired a diagnostic importance. The observation of a low-intensity transition showing a shift of λ_{max} to shorter wavelengths on passing from hydrocarbons or carbon tetrachloride to alcohols as solvents is strong presumptive evidence of an $n \rightarrow \pi^*$ transition (see Table 2.4 for data for acetone). The origin of this effect

Table 2.4. Solvent shifts: λ_{max} of Me_2CO in nanometres

Solvent	H_2O	MeOH	EtOH	$CHCl_3$	CCl_4	C_nH_{2n+2}
λ_{max}	264	270	272	276	280	280

is not completely clear. With respect to aliphatic ketones, there is a small reduction in dipole moment on excitation, and this would be expected to lead to a small blue shift. Other factors, however, and notably the Franck–Condon Principle, probably play an important role. Around a polar solute molecule in its ground state, the solvent molecules are oriented so as to minimize the total free energy of the system. Electronic excitation is so rapid a process that the solvent cage surrounding the excited species is identical with that which surrounded the ground state species, and it will not in general have the configuration appropriate to the different charge distribution in the excited state. Hence the solvation energy of the solute in its excited state may well be less than that of the solute in its ground state. Since this effect will increase with increasing polarity of the solvent molecules, a blue shift will result.[41] Differences in hydrogen bonding between solvent and ground state or excited state will also contribute to the blue shift in hydroxylic solvents.[42]

Polar solvents stabilize (π, π^*) states and destabilize (n, π^*) states with respect to the situation in hydrocarbon solvents. Thus, if molecular (π, π^*) and (n, π^*) states have comparable energies, changing the solvent may invert the energetic ordering of the levels (Figure 2.47) with dramatic results.

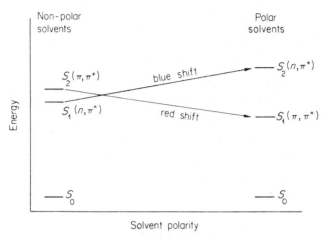

Figure 2.47. The effect of solvent polarity on the ordering of (n, π^*) and (π, π^*) states

For example, naphthalene-2-aldehyde and anthracene-9-aldehyde both fluoresce in ethanol but are non-fluorescent in heptane. The explanation probably resides in the fact that whereas $^1(\pi, \pi^*)$ states are frequently fluorescent, fluorescence from $^1(n, \pi^*)$ states is usually weak or undetectable (for reasons given in Chapter 3, p. 97). Hence, in heptane rapid internal conversion causes population of the lowest energy $^1(n, \pi^*)$ (non-fluorescent) state, whereas in ethanol it is the fluorescent $^1(\pi, \pi^*)$ state which becomes populated because it is the state of lowest energy.

2.7   IDENTIFICATION OF (n, π^*) AND (π, π^*) STATES

Since photochemical reactions are predominantly a function of the lowest excited singlet of triplet states, it is necessary to be able to identify the nature of this state. For this purpose appeal is usually made to spectroscopic data. Tables 2.5 and 2.6 set out the differences frequently encountered between the (n, π^*) and (π, π^*) states of organic molecules. It must be emphasized that many exceptions are known to the generalizations in the Tables, and for the identification of a state to be secure, it is essential that it conform to several of the criteria mentioned.

Table 2.5. Properties of singlet (n, π^*) and (π, π^*) states

	$^1(n, \pi^*)$	$^1(\pi, \pi^*)$
Intensity of absorption	*Weak.* Absorption band absent in hydrocarbon analogues	*Strong* (unless symmetry forbidden)
Solvent effect	Blue shift in polar or hydroxylic solvents. Band disappears on protonation	Red shift in polar solvents
Polarization of transition	Perpendicular to molecular plane	Parallel to molecular plane
Energy	Usually lowest energy[a] transition:	—

[a] This may not be so in conjugated compounds.

Table 2.6. Properties of triplet (n, π^*) and (π, π^*) states

	$^3(n, \pi^*)$	$^3(\pi, \pi^*)$
Phosphorescent lifetime	$< 10^{-1} - 10^{-2}$ s	> 1 s
Phosphorescence vibrational structure	Prominent	Variable
Singlet–triplet splitting	$\begin{cases} < 36 \text{ kJ (8 kcal) mol}^{-1} \\ \quad \text{for C=O} \\ < 60 \text{ kJ (14 kcal) mole}^{-1} \\ \quad \text{for azines} \end{cases}$	> 60 kJ (14 kcal) mol^{-1}
Itensity of $S_0 \rightarrow T$ transition	$f \sim 10^{-5} - 10^{-7}$	$f \sim 10^{-11}$ (in hydrocarbons) $10^{-9} - 10^{-6}$ (in haloaromatics)
External heavy atom effect on $S_0 \rightarrow T$ transition intensity	Little effect	Increases intensity
E.s.r.	No e.s.r. spectrum	Gives e.s.r. spectrum

REFERENCES

1. W. Kauzmann, *Quantum Chemistry*, Academic Press, New York (1957), ch. 16.
2. J. B. Birks and D. J. Dyson, *Proc. Roy. Soc.*, **A275**, 135 (1963); S. J. Strickler and R. A. Berg, *J. Chem. Phys.*, **37**, 814 (1962).
3. S. P. McGlynn, T. Azumi and M. Kinoshita, *Molecular Spectroscopy of the Triplet State*, Prentice-Hall, New Jersey (1969), p. 199.
4. R. S. Becker, *Theory and Interpretation of Fluorescence and Phosphorescence*, Wiley-Interscience, London, (1969), p. 82.
5. Ref. 4, p. 218.
6. D. F. Evans, *J. Chem. Soc.*, 1351 (1957).
7. J. B. Birks, *Photophysics of Aromatic Molecules*, Wiley-Interscience, London (1970), p. 495.

8. J. R. Platt, *J. Chem. Phys.*, **17**, 484 (1949).
9. W. C. Price and W. C. Tutte, *Proc. Roy. Soc.*, **A174**, 207 (1940).
10. J. E. Lennard-Jones, *Proc. Roy. Soc.*, **A158**, 280 (1937).
11. H. L. McMurry, *J. Chem. Phys.*, **9**, 231 (1941).
12. J. N. Murrell, *The Theory of the Electronic Spectra of Organic Molecules*, Methuen, London (1963), p. 161.
13. For a discussion of the optical absorption of alkyl halides, see J. G. Calvert and J. N. Pitts, *Photochemistry*, Wiley, London (1966), p. 522.
14. See ref. 7, p. 489 for relevant references.
15. G. Briegleb, *Elektronen-Donator-Acceptor-Komplexe*, Springer, Berlin (1961).
16. D. F. Evans, *J. Chem. Phys.*, **23**, 1424 (1955); *ibid.*, 1426.
17. H. A. Benesi and J. H. Hildebrand, *J. Amer. Chem. Soc.*, **71**, 2703 (1949).
18. J. N. Murrell, *J. Amer. Chem. Soc.*, **81**, 5037 (1959).
19. E. J. Bowen, *Chemical Aspects of Light*, 2nd edition, Clarendon Press, Oxford (1946).
20. For a review of chemiluminescence, see F. McCapra, *Quart. Rev.*, **20**, 485 (1966).
21. Reviewed by A. Zweig, *Adv. Photochem.*, **6**, 425 (1968).
22. A. U. Khan and M. Kasha, *J. Amer. Chem. Soc.*, **88**, 1574 (1966).
23. F. McCapra, *Chem. Commun.*, 154 (1968).
24. F. McCapra and Y. C. Chang, *Chem. Commun.*, 522 (1966).
25. (a) P. P. Sorokin, *Scientific American*, **220**, 30 (1969); (b) F. P. Schäfer, *Angew. Chem. Inter. Ed.*, **9**, 9 (1970); (c) B. B. Snavely in J. B. Birks (ed.), *Organic Molecular Photophysics*, volume 1, Wiley, London (1973), p. 239.
26. P. M. Rentzepis, *Photochem. Photobiol.*, **8**, 579 (1968).
27. D. A. Long, *Chem. in Brit.*, **7**, 108 (1971).
28. W. Kaiser and C. G. B. Garrett, *Phys. Rev. Letters*, **7**, 229 (1961).
29. R. M. Hochstrasser, H.-N. Sung and J. W. Wessel, *J. Amer. Chem. Soc.*, **95**, 8179 (1973).
30. For references and discussion, see J. N. Murrell, *The Theory of the Electronic Spectra of Organic Molecules*, Methuen, London (1963).
31. A. J. Merer and R. S. Mullikin, *Chem. Rev.*, **69**, 642 (1969).
32. J. C. D. Brand and D. G. Williamson, *Adv. Phys. Org. Chem.*, **1**, 401 (1963); ref. 3, p. 166.
33. Reviewed by E. Vander Donckt, *Progr. Reaction Kinetics*, **5**, 273 (1970); ref. 4, p. 239.
34. F. Wilkinson in G. M. Burnett and A. M. North (ed.), *Transfer and Storage of Energy by Organic Molecules*, Wiley-Interscience, London (1969), p. 114.
35. E. Lippert, *Z. Electrochem.*, **61**, 962 (1957).
36. J. Czekalla, *Z. Electrochem.*, **64**, 1221 (1960).
37. L. Salem and C. Rowland, *Angew. Chem. Inter. Ed.*, **11**, 92 (1972).
38. See ref. 7, p. 115 for relevant references.
39. E. M. Kosower, *J. Amer. Chem. Soc.*, **80**, 3253 (1958).
40. M. Kasha, *Discuss. Faraday Soc.*, **9**, 14 (1950).
41. H. McConnell, *J. Chem. Phys.*, **20**, 700 (1952); G. C. Pimentel, *J. Amer. Chem. Soc.*, **79**, 3323 (1957).
42. N. S. Bayliss and E. G. McRae, *J. Phys. Chem.*, **58**, 1002 (1954); G. J. Brealey and M. Kasha, *J. Amer. Chem. Soc.*, **77**, 4462 (1955).

Chapter 3

Excited States: Time-dependent Phenomena

3.1   INTRODUCTION

In the previous chapter the factors relating to the production of excited states and their static properties were discussed. We now wish to analyse the time-dependent evolution of excited species into other entities, which may be alternative states of the same species (photophysical processes) or of other molecules (photochemical processes). This distinction between photophysics and photochemistry may be more apparent than real, but it will be retained for pedagogical reasons.

Excited states are short-lived, for they are compelled to lose their electronic energy within a short period of time. Even if no competing process intervenes, excited molecules must collapse to their ground state by the emission of radiation. Competing physical or chemical processes can give rise to a new excited state and hence delay momentarily the total loss of electronic energy, but the ultimate and rapid fate of all excited states is collapse to a ground state system. It is with some aspects of this area that the present chapter is concerned.

3.1.1   Dissipative Pathways

The processes responsible for the dissipation of the excess energy of an excited species may be differentiated and classified as in Figure 3.1.

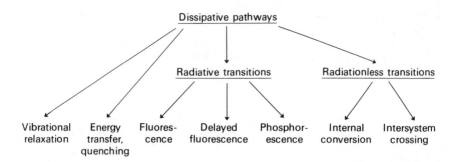

Figure 3.1. Physical pathways for the dissipation of electronic energy

It should be realized that these processes can compete with each other for the deactivation of an excited state, and the relative magnitude of the rate constants determines the contribution made by a particular pathway. To take a simple example, if an excited species can either fluoresce (rate constant k_f) or undergo some non-radiative deactivation (rate constant k_D), then if $k_D \gg k_f$ the population of excited species will be depleted predominantly by the radiationless route so that fluorescence will be weak and possibly undetectable. Conversely, if k_D and k_f are of comparable magnitude, strong fluorescence will be observed.

Vibrational Relaxation

Unless formed by a $0 \rightarrow 0$ transition,† an excited species finds itself endowed at the moment of its creation with excess vibrational (and rotational) energy in addition to the electronic energy. The rate constant for the emission of infrared photons is so small due to the operation of the 'v^3 Law' (see Chapter 2, p. 13) that loss of vibrational energy (called *vibrational relaxation* or *vibrational cascade*) is largely dependent upon collisions, as a result of which vibrational energy is converted into kinetic energy distributed between the partners in the collision. Consequently, in low pressure vapours where the time interval between collisions is greater than the lifetimes associated with radiative processes, emission is observed to come from the vibrational level populated by the absorption act.‡ As the pressure increases, collisions occur within the lifetime of the excited species and lead to a progressive loss of vibrational quanta, which is manifested by emission at longer wavelengths. At sufficiently high pressures, or in solution where the collision rate is of the order of 10^{13} s^{-1}, total vibrational relaxation is the rule and emission occurs almost exclusively from the $v' = 0$ level. There is, of course, a Boltzmann distribution over the vibrational levels, but at room temperature it is the zero-point level which is populated predominantly.

3.1.2   Radiative Transitions

In radiative transitions, represented by straight arrows on a Jablonski diagram (Figure 3.2), an excited species passes from a higher excited state to a lower one with the emission of a photon. Three processes may be distinguished:

(i) *Fluorescence* is caused by a radiative transition between states of the same multiplicity, and it is a rapid process ($k_f \sim 10^6$–10^9 s^{-1}). For the polyatomic molecules encountered in organic chemistry, the transition is usually $S_1 \rightarrow S_0$, although $S_2 \rightarrow S_0$ fluorescence is occasionally observed (e.g. with azulene), and a very weak $S_k \rightarrow S_m$ emission has recently been detected for a number of

† A transition between the zero-point vibrational levels ($v = 0$) of the ground and excited states is referred to as a $0 \rightarrow 0$ transition.

‡ This statement is not strictly correct. There is now good evidence that some systems with very sharp absorption bands, when excited under isolated molecule conditions, emit a structureless fluorescence. This phenomenon seems to mean that the initially populated vibrational level rapidly and non-radiatively redistributes its energy in a unimolecular fashion into a dense manifold of other isoenergetic levels, which then emit. Consideration of non-radiative transitions is deferred to later in this chapter.

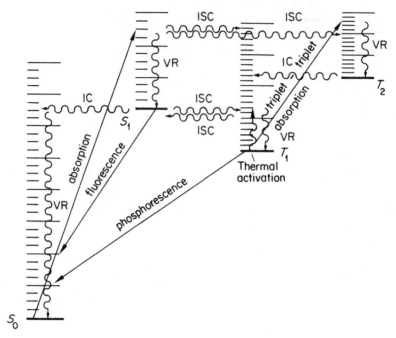

Figure 3.2. Jablonski diagram showing some of the radiative and non-radiative processes available to molecules (VR = vibrational, relaxation; IC = internal conversion; ISC = intersystem crossing)

other molecules, as has $T_n \rightarrow T_m$ fluorescence. In sharp contrast, strong $S_n \rightarrow S_m$ fluorescence is observed in diatomics.

(ii) *Phosphorescence* is the result of a transition between states of different multiplicity, typically $T_1 \rightarrow S_0$; $T_n \rightarrow S_0$ is very rare. The process, being spin forbidden, has a much smaller rate constant ($k_p \sim 10^{-2}$–10^4 s^{-1}) than that for fluorescence.

(iii) *Delayed Fluorescence* differs from ordinary fluorescence in that the measured rate of decay of emission is less than that expected from the transition giving rise to the emission.

3.1.3   Radiationless (or non-radiative) Transitions

Radiationless transitions occur between *isoenergetic* (or degenerate) vibrational–rotational levels of different electronic states. Since there is no change in the total energy of the system, no photon is emitted, and the process is represented by a horizontal line on a Jablonski diagram. Wavy arrows are used (e.g., $S_1 \rightsquigarrow T_1$) to distinguish radiationless transitions from radiative ones. If the electronic states participating in the radiationless transition are different states of the same molecule, then the transition is a *photophysical process* (e.g., internal conversion or intersystem crossing). However, a radiationless transition taking place between the excited state of one molecule and a state (usually the ground state) of another molecule gives rise to a *photochemical* transformation. Seen against this background, photochemistry is a facet of the general study of radiationless transitions (this point is discussed later).

Internal Conversion is a radiationless transition between isoenergetic states of the same multiplicity. Such transitions between upper states (e.g., $S_m \rightsquigarrow S_n$ or $T_m \rightsquigarrow T_n$) are extremely rapid, accounting for the negligible emission from upper states. Internal conversion from the first excited singlet state ($S_1 \rightsquigarrow S_0$) is so much slower that fluorescence can compete.

Intersystem Crossing is a radiationless transition between states of different multiplicity. The radiationless deactivation of the lowest triplet ($T_1 \rightsquigarrow S_0$) is a process in competition with normal phosphorescence. The intersystem crossing $S_1 \rightsquigarrow T_1$ or $S_1 \rightsquigarrow T_n$, which is competitive with (and reduces the quantum yield of) fluorescence, is the process by which the triplet manifolds are normally populated. $S_n \rightsquigarrow T_n$ has been observed but is rare because it has to compete with extremely fast internal conversion to S_1. The transition $T_1 \rightsquigarrow S_1$ requires thermal activation of T_1 to a vibrational level isoenergetic with S_1—it is the basis of one of the mechanisms leading to delayed fluorescence.

A Jablonski diagram (Figure 3.2) is frequently used to display the various exciting and dissipative pathways (compare the simplified diagram introduced in Chapter 1, Figure 1.7).

3.1.4   Kinetics, Quantum Yields and Lifetimes

Since kinetic analysis provides a powerful tool for unravelling the complexities of many photochemical phenomena, certain quantitative relations involving experimental quantities such as quantum yield and lifetime will now be established to show how the key rate constants may be extracted from such data.

First, consider a situation such as that depicted in Figure 3.3 where an excited species A* is subject to several first order or pseudo-first order deactivating processes

Figure 3.3

It follows that:

$$\frac{d[A^*]}{dt} = -[A^*](k_1 + k_2 + k_3) = -\sum k[A^*]$$

whence

$$[A^*] = [A_0^*]\, e^{-\Sigma kt}$$

The concentration of A* falls exponentially with a rate constant given by $\sum k$, and if the lifetime (τ) of A* is defined as the time taken for [A*] to fall to $1/e$ of its initial value, then

$$\tau = \frac{1}{\sum k} \qquad (3.1)$$

τ, which is the actual measured lifetime of A*, should be clearly distinguished from the radiative lifetime τ_0 (see Chapter 2, section 2.1.2), which would be the lifetime if decay occurred exclusively by emission.

Now consider the general kinetic scheme set forth in Figure 3.4, where I is the rate of absorption of photons and k_f and k_p are the rate constants for fluorescence and phosphorescence.

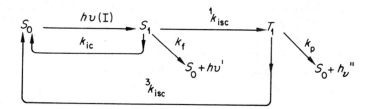

Figure 3.4. A general kinetic scheme for photophysical processes

The steady state approximation is applicable to excited states, and it follows that the rates of formation and destruction of S_1 are equal.

$$\therefore \quad I = [S_1] \sum {}^1k \quad (\text{where } \sum {}^1k = k_{ic} + k_f + {}^1k_{isc}) \qquad (3.2)$$

The quantum yield of fluorescence is given by

$$\phi_f = \frac{\text{rate of emission by } S_1}{\text{rate of absorption of photons by } S_0} = \frac{k_f[S_1]}{I} = \frac{k_f}{\sum {}^1k} \qquad (3.3)$$

and

$$\tau_f = \frac{1}{\sum {}^1k} \qquad (3.4)$$

Hence

$$k_f = \frac{\phi_f}{\tau_f} \qquad (3.5)$$

This permits calculation of k_f from experimental data. Since $k_f = A_{ul'}$ the Einstein coefficient of spontaneous emission, it may also be obtained from absorption spectra by measuring the oscillator strength f and calculating B_{ul} and hence A_{ul} (see Chapter 2, p. 13). k_f is also related to τ_0 by the equation

$$\tau_0 = \frac{1}{k_f} \qquad (3.6)$$

Therefore

$$\phi_f = \tau_f/\tau_0 \tag{3.7}$$

The steady state approximation applied to T_1 gives

$$^1k_{isc}[S_1] = \sum {}^3k[T_1] \quad (\text{where } \sum {}^3k = k_p + {}^3k_{isc}) \tag{3.8}$$

The quantum yield of phosphorescence is:

$$\phi_p = \frac{k_p[T_1]}{I} = \frac{k_p}{I} \cdot \frac{{}^1k_{isc}[S_1]}{\sum {}^3k} \quad (\text{from 3.8}) \tag{3.9}$$

$$= \frac{k_p}{\sum {}^3k} \cdot \frac{{}^1k_{isc}}{\sum {}^1k} \quad (\text{from 3.2}) \tag{3.10}$$

Hence

$$\phi_p = \theta_p \cdot \theta_{isc} \tag{3.11}$$

where θ_p and θ_{isc} are the *quantum efficiencies* of phosphorescence and intersystem crossing respectively. Quantum efficiency (to be distinguished from quantum yield) is the ratio of the rate of a process involving an excited state to the rate of production of that state. Thus

$$\theta_p = \frac{k_p[T_1]}{\text{rate of production of } T_1} = \frac{k_p[T_1]}{\sum {}^3k[T_1]} \tag{3.12}$$

because the rates of production and destruction of T_1 are equal (steady state approximation). Hence

$$\theta_p = \frac{k_p}{\sum {}^3k} \tag{3.13}$$

Similarly,

$$\theta_{isc} = \frac{{}^1k_{isc}[S_1]}{I} = \frac{{}^1k_{isc}[S_1]}{\sum {}^1k[S_1]}$$

$$= \frac{[T_1]\sum {}^3k}{[S_1]\sum {}^1k} = \frac{\sum {}^3k[T_1]}{I} \quad (\text{from 3.2 and 3.8}) \tag{3.14}$$

$$\therefore \quad \theta_{isc} = \frac{\text{rate of production of } T_1}{\text{rate of absorption of photons}} = \frac{\text{rate of destruction of } T_1}{\text{rate of absorption of photons}}$$

Hence

$$\theta_{isc} = \phi_T \quad (\text{the quantum yield for formation of triplets})$$

Since

$$\phi_T = \theta_{isc} = \frac{{}^1k_{isc}}{\sum {}^1k}$$

and

$$\tau_f = \frac{1}{\sum {}^1 k}$$

it follows that

$${}^1 k_{isc} = \frac{\phi_T}{\tau_f} \qquad (3.15)$$

Thus can ${}^1 k_{isc}$ be calculated. k_p can, in principle, be obtained knowing θ_p and τ_p, for from (3.13)

$$\theta_p = k_p \tau_p, \quad \text{where } \tau_p = \frac{1}{\sum {}^3 k}$$

Since θ_p cannot be measured directly, use is made of ϕ_p instead, because from (3.9)

$$\frac{\phi_p}{k_p} = \frac{[T_1]}{I} = \frac{\sum {}^3 k [T_1]}{\sum {}^3 k . I} = \tau_p . \phi_T$$

Hence

$$k_p = \frac{\phi_p}{\phi_T . \tau_p} \qquad (3.16)$$

ϕ_T, the quantum yield of triplet production, is obtained from triplet counting techniques (see Chapter 5, p. 156).

3.2 RADIATIVE TRANSITIONS[1,2]

3.2.1 Methods

Since upper states do not generally luminesce, the methods just described cannot be employed to obtain k_{ic} among upper states, and consequently there is little information about this rate constant. An order of magnitude estimate may be derived from an argument due to Kasha[3] that since fluorescence from upper singlets would have been detected were it 10^{-4} times as intense as the ordinary fluorescence, then k_{ic} must be $\geqslant 10^4 k_f$ (where k_f is rate constant for the unobserved fluorescence). If k_f is estimated from the ultraviolet absorption spectrum to be $> 10^8$ s^{-1}, $k_{ic} > 10^{12}$ s^{-1} and $\tau < 10^{-12}$ s. One way of obtaining the lifetime of upper states is to examine the width of absorption lines. Because of the Uncertainty Principle a short-lived state would have a relatively broad line. Currently accepted values of k_{ic} are $\sim 10^{12}$ s^{-1}.

Fluorescence Spectra

A spectrofluorimeter (Figure 3.5) is the instrument of choice for recording fluorescence spectra. A beam of monochromatic light excites the specimen in the cell, and the emission is observed and analysed at right angles to the incident beam. The output is the emission spectrum plotted by the XY recorder.

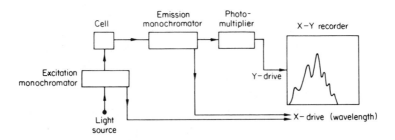

Figure 3.5. Schematic diagram of a spectrofluorimeter

Since with few exceptions (notably biacetyl) organic molecules do not phosphoresce in ordinary solvents at room temperature, the system outlined will record just the fluorescence spectrum, and since either the excitation or the emission monochromator may be coupled to the XY recorder it follows that two sorts of spectra may be obtained:

(i) Set the excitation monochromator to a particular wavelength absorbed by the sample and scan the emitted light with the emission monochromator. This affords the *fluorescence emission spectrum.*

(ii) Set the emission monochromator to a particular wavelength in the fluorescent output and scan the exciting wavelengths with the exciting monochromator. This gives rise to the *fluorescence excitation spectrum*, which normally closely resembles† the absorption spectrum in sufficiently dilute solutions (see Figure 3.8).

Fluorescence excitation spectroscopy can be used to identify and quantitatively to estimate fluorescent molecules.[4] It supplements absorption spectroscopy, but with the difference that fluorimetric analysis is often orders of magnitude more sensitive. This increase in sensitivity is primarily due to the fact that the fluorescence signal can be observed, amplified and recorded directly, whereas with absorption spectroscopy what is measured is the signal due to the *difference* between the incident and transmitted light intensities. Fluorimetry provides a powerful tool for the analysis of mixtures. Often only one component is fluorescent, and when more than one component is fluorescent it may well be possible selectively to excite just one of the components by adjusting the excitation wavelength.

Phosphorescence Spectra

These are recorded with spectrophosphorimeters, which differ from spectrofluorimeters only in the incorporation of a mechanical or optical shutter which repetitively chops the exciting and emitted light beams in such a manner that excitation occurs when the detector is cut off and emission is not observed until a definite period after excitation has ceased. This delay between excitation

† For a precise correspondence to be obtained with a single-beam instrument (Figure 3.5), the recorded spectrum should be corrected for the variation with wavelength of the emission of the light source, of the transmission of the monochromators and of the response of the photomultiplier.

and observation permits fluorescence ($\tau_f < 10^{-6}$ s) to decay to zero before the longer-lived phosphorescent emission is recorded. Since phosphorescence is most easily observed in rigid matrices which inhibit the quenching collisions between adventitious impurities and the excited triplets, the sample is usually investigated in mixed organic solvents which set to form rigid glasses when cooled in liquid nitrogen. It is important to note that the observed phosphorescence is derived from triplet species which have been formed indirectly by intersystem crossing from the singlet manifold, so that the phosphorescence *excitation* spectrum corresponds to the ordinary singlet $\rightarrow$ singlet absorption spectrum.

The extreme sensitivity associated with emission spectroscopy makes it possible to estimate triplets produced *directly* via forbidden $S \rightarrow T$ transitions. This forms the basis of *phosphorescence excitation spectroscopy*.[5] An intense light source is used to excite the $S \rightarrow T$ transition, a heavy atom solvent (e.g., ethyl iodide) being used where appropriate to increase the intensity of absorption. The production of excited triplets is then monitored by observing their phosphorescence at a particular set wavelength. The spectrum obtained by plotting the intensity of phosphorescence against the wavelength of the exciting light (at constant intensity) is the $S \rightarrow T$ absorption spectrum, because the intensity of the phosphorescence is directly proportional to the concentration of triplets which is itself proportional to the extinction coefficient for the $S \rightarrow T$ transition at the particular exciting frequency used (Figure 3.6) This technique may be used even when the compound itself is non-phosphorescent by having present a phosphorescent molecule of lower triplet energy which is excited by energy transfer (see Chapter 4) from the non-phosphorescent triplets. Some representative emission spectra are given in Figures 3.6–3.8. Figure 3.7 illustrates the general point that phosphorescence occurs at longer wavelengths

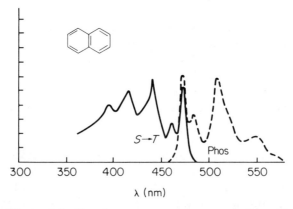

Figure 3.6. Phosphorescence spectrum and $S_0 \rightarrow T_1$ absorption spectrum from phosphorescence excitation of naphthalene (reproduced with permission from A. P. Marchetti and D. R. Kearns, *J. Amer. Chem. Soc.*, **89**, 768 (1967); copyright by the American Chemical Society)

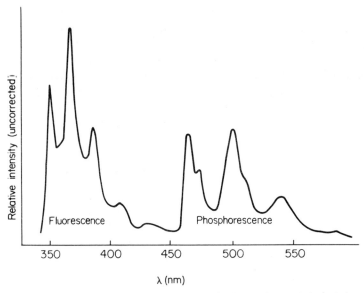

Figure 3.7. Total emission spectrum of phenanthrene (10^{-4} M) in ethanol at 77 K (from F. Wilkinson and A. R. Horrocks, in E. J. Bowen (ed.), *Luminescence in Chemistry* (1968), by permission of Van Nostrand Reinhold Co. Ltd.)

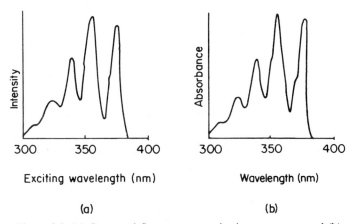

(a)   (b)

Figure 3.8. (a) Corrected fluorescence excitation spectrum, and (b) absorption spectrum, of anthracene

than fluorescence ($E_T < E_S$), and Figure 3.8 demonstrates that fluorescence excitation and absorption spectra are closely similar.

Lifetimes

Of the various methods available for measuring the actual lifetimes of excited states, the most direct is pulse fluorimetry or phosphorimetry, in which a recurrent light pulse of very short duration (~ 1 ns) is used to excite the sample. The emission is monitored (after each pulse) with a fast photomultiplier, and the output, as a function of time, is displayed on an oscilloscope screen. When this has been corrected for the decay function of the exciting flash, one is left with the decay curve of the emitting species. If this is a simple exponential function, it is merely necessary to determine the time taken for the luminescence to decay to $1/e$ of some arbitrary intensity. By this method lifetimes as short as a few nanoseconds may be readily measured. An extension of this technique is the *photon*

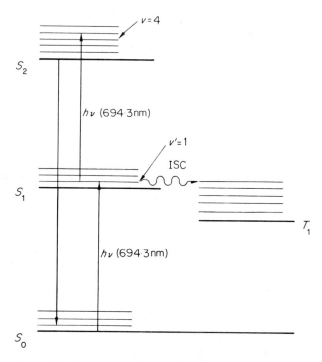

Figure 3.9. Energy level diagram for azulene excited by ruby laser

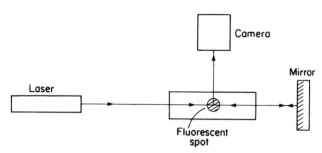

Figure 3.10. Schematic diagram of apparatus for measuring k_{isc} for azulene

sampling technique[6] which, because it depends on counting individual photons, can be used on extremely weakly luminescent substances. It permits measurement of lifetimes in the range $10^{-6}-10^{-10}$ s. Still shorter lifetimes in the picosecond range depend for time measurement on the distance travelled by light. This technique, pioneered by Rentzepis, can be illustrated by the determination of k_{isc} for azulene.[7] Azulene was selected because (exceptionally) it fluoresces from S_2 and only negligibly from S_1. The output from a Q-switched mode-locked ruby laser consisting of a train of pulses ($\bar{v} = 14\,400\,cm^{-1}$) of a few picoseconds bandwidth is passed through a solution of azulene. A given pulse, as it travels down the cell, excites azulene molecules in its path to the first vibrational level of the S_1 state (Figure 3.9). The excited azulene molecules start to decay immediately at a rate determined by the rate constant for radiationless processes ($S_1 \rightsquigarrow T_1$ and $S_1 \rightsquigarrow S_0$). Consequently the pulse leaves behind it a trail of decaying excited molecules. The pulse is reflected back along its original path by a mirror at the end of the cell (Figure 3.10). On its return journey it excites any remaining S_1 azulene molecules to the fluorescent S_2 state. A fluorescent spot is thus produced in the cell. The further the reflected pulse travels in its return journey, the less the probability of its encountering azulene molecules still remaining in the S_1 state. Hence the dimensions of the fluorescent spot, after correction for the pulse width, are related to the rate constants for radiationless processes. By these means it was estimated that the lifetime of the S_1 state of azulene was ~ 4 ps. The process primarily responsible for deactivating the S_1 state was shown to be the intersystem crossing $S_1 \rightsquigarrow T_1$. It was concluded that, in this case, $k_{isc} \sim 2 \cdot 5 \times 10^{11}\,s^{-1}$.

Quantum Yields

Although in order to obtain the quantum yield of fluorescence one has in principle merely to measure the ratio of the number of photons emitted to those absorbed, in practice grave difficulties attend the determination arising from (i) the difference in the spatial distribution of the exciting and emitted light, (ii) the polychromatic character of the emitted light, and (iii) the variation of the sensitivity of the detector with wavelength.

The last two problems can be greatly simplified by directing the incident light and fluorescent emission successively onto a *'quantum counter'* (a solution of a substance such as Rhodamine B, which, within a certain range of wavelengths, converts all absorbed light at constant quantum yield into its own fluorescent emission). The detector then receives signals of constant spectral distribution from both incident and emitted light beams.

Given that the spectral response of the light detector is known, the most rapid and accurate way of determining emission efficiency is to measure the unknown quantum yield relative to that of some substance whose absolute emission quantum yield has already been accurately measured. It is then only necessary to determine, under identical conditions of cell geometry, incident light intensity and temperature, the fluorescence spectra of dilute solutions of the unknown and of the standard. The solutions should have the same optical density at the wavelength of the exciting light so that they both capture the same number of photons. The quantum yield of the unknown relative to that of the standard is the ratio of the integrated band areas under the two fluorescence spectra (plotted on a frequency scale) after they have been corrected for the detector response function. Multiplying by the known quantum yield of the standard then gives the absolute quantum yield of the unknown. The technique has been amply described.[8]

Polarization Spectra

In Chapter 2 it was seen that both absorption and emission are polarized along particular molecular axes determined by the symmetry properties of the participating orbitals and predictable by Group Theory. The method of obtaining polarization data and their applications are now considered.

With single crystals of known and appropriate structure, the orientation of the plane of polarization with respect to the crystal axes and hence the molecular axes may be determined. This gives the absolute polarization of absorption and emission. It is easier and often more useful to determine the *relative* polarization of absorption bands, and this information is given by polarization spectra in which the polarization is plotted against the wavelength of the exciting light. Such spectra may be obtained by inserting polarizing devices (P_1 and P_2) into the incident and emitted beams of light of a spectrofluorimeter or spectro phosphorimeter (Figure 3.11). With P_1 fixed, for each exciting frequency the

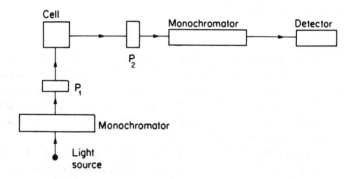

Figure 3.11. Schematic diagram of apparatus for determining polarization spectra

intensity of the emitted light is measured with the plane of polarization defined by P_2 parallel or perpendicular to that of P_1. This gives the *degree of polarization* (*P*) defined by

$$P = \frac{I_{\parallel} - I_{\perp}}{I_{\parallel} + I_{\perp}} \qquad (3.17)$$

where $I_{\parallel}$ and $I_{\perp}$ are the intensities of the parallel and perpendicular components of the emitted light.

For randomly oriented molecules in a highly viscous solvent which inhibits rotation in the time interval between absorption and emission it can be shown that

$$P = \frac{3\cos^2\alpha - 1}{\cos^2\alpha + 3} \tag{3.18}$$

where α is the angle between the directions of polarization of absorption and emission of the substrate. Since α is commonly $0°$ or $90°$, P can assume values ranging only between $+\frac{1}{2}$ and $-\frac{1}{3}$. In fact, such values are rarely achieved because of various depolarizing effects.

The absorption and fluorescence polarization spectra of phenol are shown in Figures 3.12 and 3.13. Notice that P changes sign at about 240 nm indicating that the $S_0 \rightarrow S_1$ and the $S_0 \rightarrow S_2$ transitions have different directions of polarization with respect to the molecular axes.

The apparently simple long wavelength absorption band of aniline ($\lambda_{max} \sim$ 283 nm) is revealed to be due to at least two different transitions by its polarization spectrum (Figure 3.14).

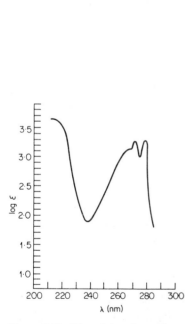

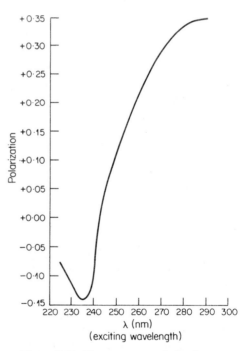

Figure 3.12. Ultraviolet absorption spectrum of phenol in cyclohexane (from R. A. Friedel and M. Orchin, *Ultraviolet Spectra of Aromatic Compounds*, (1951), John Wiley and Sons Ltd.)

Figure 3.13. Fluorescence polarization spectrum of phenol at $-70\,°C$ in propylene glycol (from G. Weber, in D. M. Hercules (ed.), *Fluorescence and Phosphorescence Analysis*, (1966), John Wiley and Sons Ltd.)

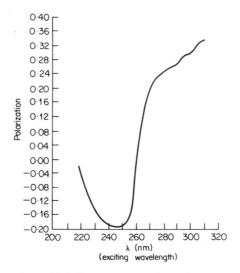

Figure 3.14. Fluorescence polarization spectrum of aniline at $-70\,°C$ in propylene glycol (from G. Weber, in D. M. Hercules (ed.), *Fluorescence and Phosphorescence Analysis*, (1966), John Wiley and Sons Ltd.)

In general, emission polarization spectroscopy gives a value for α, and, if the absorption (or emission) can be identified with a particular transition, the transition associated with the emission (or absorption) can often be assigned. The interpretation of phosphorescence polarization data depends upon a knowledge of the mixing of singlet and triplet states induced by spin–orbit coupling, and the reader is referred to reference 9 for details.

The data emerging from the application of the methods just described will now be considered.

3.2.2   **Fluorescence**

Mirror Symmetry Relation

It is commonly observed, particularly among large and rigid systems in condensed phases, that the absorption and fluorescence spectra are approximate mirror images when plotted on a frequency (energy) scale (e.g. Figure 3.15). Recalling that at room temperature absorption occurs only from the $v'' = 0$ level of the ground state and that because of rapid vibrational relaxation emission occurs only from the $v' = 0$ level of the excited state, the existence of a mirror symmetry relation must imply close similarity in the spacings of the vibrational levels in the ground and excited states. This in turn implies that only minor changes in geometry occur on excitation, which is to be expected with rigid and extended chromophores (see Chapter 2, p. 49). Furthermore, the Franck–Condon principle, which determines the shape of the absorption envelopes, must also apply to emission processes.

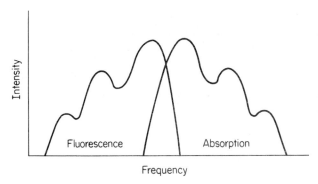

Figure 3.15. Typical mirror-symmetry relation between fluorescence and absorption spectra

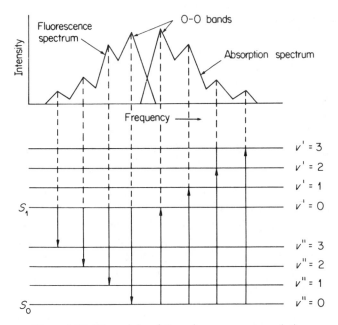

Figure 3.16. The origin of the mirror-symmetry relation

Figure 3.16 shows (i) that the vibrational spacings in the fluorescence spectrum should correspond to ground state energy levels (contrast absorption spectra), and (ii) that the two spectra should be symmetrically disposed about the 0—0 bands which indeed should be superimposed. This latter is actually observed in vapour phase spectra, but when spectra are recorded in solution there is a separation between the 0—0 bands of the fluorescence and absorption spectra, the magnitude of which is temperature and solvent dependent.

The separation arises from a Franck–Condon effect. The equilibrium solvent cages surrounding a molecule in its ground state and excited state will be different because of the changes in dipole moment and geometry occurring on excitation. Electronic excitation is so fast ($\sim 10^{-15}$ s) that after excitation the molecule is still surrounded by its ground state solvent cage, which has not had time to reorganize. The resultant solvent–solute interactions will be less stabilizing than for the ground state and may even be destabilizing. Before emission occurs, however, the solvent cage has enough time to relax to the lower energy configuration appropriate to the excited species. The energy of the $0 \rightarrow 0$ transition in emission is thus less than that in absorption, and the $0-0$ bands separate.

A similar mirror symmetry relationship holds between $S_0 \rightarrow T_1$ absorption and $T_1 \rightarrow S_0$ phosphorescence spectra (Figure 3.6).

It should be recognized that the mirror symmetry relation will only be observed if certain conditions are fulfilled:

(i) It requires that the excited molecules emit from the $v' = 0$ level. Hence it is found only with spectra from condensed phases (or gases at such pressures that vibrational relaxation is faster than emission).

(ii) There must be no large charge of geometry on excitation.

(iii) The relation exists only for the longest wavelength absorption band. If shorter wavelengths are used to excite states higher than S_1, internal conversion to S_1 is normally so rapid that emission occurs only from S_1. Hence the fluorescence emission spectrum is independent of the wavelength of the exciting radiation. The rapidity of internal conversion ($S_n \rightsquigarrow S_1$ and $T_n \rightsquigarrow T_1$) is the basis of *Kasha's Rule*, that for electronic transitions the emitting level is the lowest excited level of that multiplicity.

Temperature

The intensity of fluorescence (i.e., ϕ_f) often diminishes with increasing temperature, implying the existence of an energy barrier of some sort (commonly 4–40 kJ mol^{-1}, 1–10 kcal mol^{-1}). It is not expected that emission will be temperature-dependent, so that the energy barrier is presumably associated with a competing radiationless process. It seems[10] that this process is intersystem crossing from S_1 to a higher triplet T_n which is approximately isoenergetic (degenerate) with S_1. The effect of temperature is to increase the population of higher vibrational and rotational sub-levels of S_1 from which faster intersystem crossing may occur. Hence as the temperature rises, the rate of intersystem crossing increases, a smaller proportion of S_1 molecules have an opportunity to fluoresce, and ϕ_f drops accordingly.

Quenching

If $(\phi_f + \phi_p) < 1$, then fewer photons are emitted than are absorbed and luminescence can be thought of as having been 'quenched'. Quenching may be partial or total, and it may be ascribed to internal factors (radiationless processes leading to the ground state) or to external factors (interactions of the excited state with other molecules). These quenching processes are extremely important, but, since they are analysed in detail later in this chapter and in Chapter 4, they will not be discussed further here.

3.2.3   Phosphorescence

Because of the long radiative lifetime of triplet states (typically 10^{-4}–10^2 s) caused by the spin-forbidden nature of the emission process, they are particularly susceptible to quenching collisions with adventitious impurities. Thus phosphorescence (except that of biacetyl) is difficult to observe in the gas phase or in fluid solution. Although the use of highly purified solvents, particularly perfluorocarbons,[11] does permit useful observations of phosphorescence in fluid solution, most observations are made on solutions in rigid glasses in which the diffusion of quenchers is strongly inhibited. The commonly used phases are mixtures of organic solvents (e.g., ether, isopentane, ethanol) which, at 77 K, set to form glasses. For work at room temperature use is made of organic plastics or of melts in boric acid or other inorganic glasses.[12]

That the extremely feeble phosphorescence associated with solutions is not due to some effect on the production of the triplet species has been amply demonstrated by flash photolysis and energy transfer studies, which show that triplets can be formed in high yield in solutions whose phosphorescence is virtually non-existent.

Triplets and Phosphorescence

Although it has been implicitly assumed so far that phosphorescence is a property of the triplet state, it is only relatively recently that this identification has been established with certainty. A great deal of suggestive evidence has been collected, which has been excellently and critically reviewed by McGlynn *et al.*,[13] but the decisive observations concern the magnetic properties of the phosphorescent state.

(i) Triplet states, having two spin-parallel electrons, should be paramagnetic. Several authors have shown that fluorescein anion and other organic molecules when irradiated in a rigid glass showed both phosphorescence and paramagnetic susceptibility and that the lifetimes of both were identical (Figure 3.17).

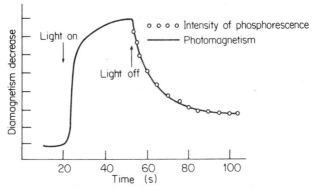

Figure 3.17. Comparison of photomagnetism and decay of phosphorescence of triphenylene in boric acid 'glass' at room temperature (from D. F. Evans, *Nature*, **176**, 777 (1955), reproduced by permission of Nature)

(ii) The triplet state has three degenerate components with the spin wave-functions given in Chapter 2 (p. 54) and magnetic quantum numbers $M_S = 1, 0, -1$. In a magnetic field the degeneracy is lifted, and transitions between the triplet sublevels become possible. These transitions can be detected by electron spin resonance (e.s.r.) The first successful application of e.s.r. was by Hutchinson and Mangum,[14] who in 1958 detected the paramagnetic resonance of the phosphorescent state of naphthalene, examined in durene as a single crystal at 77 K. Subsequently it was shown that the intensities of phosphorescent emission and of the e.s.r. signal from irradiated aromatic ketones decayed at the same rate.

Heavy Atom Effects

In Chapter 2 it was pointed out that the heavy atom effect markedly increased the rates of all singlet $\leftrightarrow$ triplet processes, both radiative and non-radiative. The consequences with respect to molecular emission are set out in the following Table 3.1.

Table 3.1. Heavy atom effect on lifetimes and quantum yields of emission

Rate process accelerated by heavy atom effect	Expected consequences
(1) $S_0 \rightarrow T_n$	Enhanced $S_0 \rightarrow T_n$ absorption (Chapter 2)
(2) ISC $(S_1 \rightsquigarrow T_n)$	τ_f and ϕ_f decreased, ϕ_T increased
(3) ISC $(T_1 \rightsquigarrow S_0)$	τ_p decreased, ϕ_p decreased
(4) $T_1 \rightarrow S_0$	τ_p decreased, ϕ_p increased

Although it is clear that τ_p, τ_f and ϕ_f should all decrease, it is difficult to predict whether the intensity of phosphorescence (ϕ_p) will increase or decrease as a result of the heavy atom effect, for processes (2) and (4) conflict with process (3).

Experimental data supporting the above predictions are given in Tables 3.2 and 3.3.

Table 3.2. Heavy atom effects in Group IV tetraphenyls at 77 K in a rigid glass[15]

Compound	$\dfrac{\phi_p}{\phi_f}$	$\tau_p(s)$
CPh_4	$\ll 0.1$	2.9
$SiPh_4$	0.1	1.1
$GePh_4$	1	0.055
$SnPh_4$	10	0.003
$PbPh_4$	$\gg 10$	< 0.001

Table 3.3. Heavy atom effect in halogenonaphthalenes in a rigid glass at 77 K[16]

Substituted Naphthalene	ϕ_p	ϕ_f	$\dfrac{\phi_p}{\phi_f}$	$\tau_p(s)$
Naphthalene	0·05	0·55	0·091	2·3
1-Fluoro-	0·056	0·84	0·067	1·5
1-Chloro-	0·30	0·058	5·2	0·29
1-Bromo-	0·27	0·0016	169	0·018
1-Iodo-	0·38	<0·0005	>760	0·002

Deuteration

Perdeuteration markedly increases the phosphorescent lifetime τ_p for many aromatic hydrocarbons. For example, for naphthalene in durene at 77 K, $\tau_p = 2·1$ s for $C_{10}H_8$ and 16·9 s for $C_{10}D_8$. A similar effect has been observed with the luminescence of rare earth ions when the water of hydration is replaced by D_2O. This phenomenon arises because deuteration markedly reduces $k_{isc}(T_1 \rightsquigarrow S_0)$, a point discussed on p. 92.

3.2.4   Delayed Fluorescence[17]

Delayed fluorescence, in which the luminescence decays more slowly than normal 'prompt' fluorescence from the same molecule, can arise by several mechanisms, of which the most closely investigated are triplet–triplet annihilation and thermally-activated delayed fluorescence.

Triplet–Triplet Annihilation

The fluorescent emission from a number of aromatic hydrocarbons (e.g., naphthalene, anthracene, phenanthrene) shows two components with identical emission spectra. One component decays at the rate for normal fluorescence, and the other has a lifetime approximately half that of phosphorescence. The implication of triplet species in this delayed fluorescence, suggested by the fact that the delayed emission can be induced by triplet sensitizers, has been confirmed by kinetic analysis. The accepted mechanism[18] is:

$$S_0 \xrightarrow{h\nu}_{I} S_1 \qquad \text{absorption} \tag{3.19}$$

$$S_1 \xrightarrow{k_f} S_0 + h\nu' \quad \text{normal fluorescence} \tag{3.20}$$

$$S_1 \xrightarrow{k_{isc}} T_1 \qquad \text{intersystem crossing} \tag{3.21}$$

$$T_1 \xrightarrow{k_p} S_0 + h\nu'' \quad \text{phosphorescence} \tag{3.22}$$

$$\left. \begin{array}{l} T_1 + T_1 \xrightarrow{k_5} X \\ X \xrightarrow{k_6} S_1 + S_0 \end{array} \right\} \text{triplet–triplet annihilation (spin–allowed)} \qquad \begin{array}{l} (3.23) \\ (3.24) \end{array}$$

$$X \xrightarrow{k_7} S_0 + S_0 \quad \text{deactivation} \tag{3.25}$$

$$S_1 \xrightarrow{k_f} S_0 + h\nu' \quad \text{delayed fluorescence} \tag{3.26}$$

The crucial steps are equations (3.23) and (3.24), in which two excited triplets, on collision, redistribute their energies via the entity X so that one is promoted to S_1 and the other collapses to the ground state. It is the S_1 state so produced that is responsible for the delayed fluorescence. Although it emits (equation 3.26) with the same rate constant as prompt fluorescence (k_f), its decay is inhibited because it continues to be regenerated via steps (3.23) and (3.24).

Application of the steady state approximation shows that under conditions of low exciting light intensity, where $[T_1]$ will be small and terms involving $[T_1]^2$ can be neglected, the intensity of delayed fluorescence (I_{DF}) is given by:

$$I_{DF} = \frac{k_5 k_6}{k_6 + k_7}\left[\frac{k_{isc}I}{k_p(k_f + k_{isc})}\right]^2 \tag{3.27}$$

Also,

$$k_{DF} = -\frac{d(\ln I_{DF})}{dt} = -\frac{d(\ln[T_1]^2)}{dt} = -2\frac{d(\ln[T_1])}{dt}$$

so that

$$k_{DF} = 2k_p \quad \text{and} \quad \tau_{DF} = \tfrac{1}{2}\tau_p \tag{3.28}$$

These predictions, (i) that the intensity of the delayed emission should be proportional to the square of the incident light intensity and (ii) that the lifetime of the delayed fluorescence should be one-half that of phosphorescence, have been abundantly confirmed. The intermediate X in the above mechanism is clearly an excited dimer or excimer (see Chapter 4), and the reaction of equation (3.24) is the reverse of that involved in the concentration quenching of fluorescence (see Chapter 4, p. 103). For anthracene, k_6 is so large that delayed fluorescence of only the monomer is observed. With pyrene, on the other hand, k_6 is much smaller, and delayed emission derives from both the monomer and (mainly) the dimer. Delayed fluorescence from the dimer, unlike that from the monomer, is not the same as normal fluorescence.

Delayed fluorescence of this type has been detected in gases, solution, rigid glasses and even crystals. In the last two phases, where the diffusion of triplets is difficult or impossible, the emission probably depends on exciton migration— a form of energy transfer in which the energy of excitation 'hops' from molecule to molecule. Ultimately two adjacent triplets are formed and delayed emission ensues.

Thermally-activated Delayed Fluorescence

This emission is characterized by the following features:

(i) The prompt and delayed fluorescence emission spectra are identical.

(ii) $\tau_{DF} = \tau_p$ (contrast triplet–triplet annihilation).

(iii) $I_{DF} \propto I$ (intensity proportional to absorbed light intensity, contrast triplet–triplet annihilation).

(iv) The ratio of the intensities of delayed fluorescence and phosphorescence decreases exponentially with the singlet–triplet splitting and with the reciprocal of the absolute temperature, i.e. $I_{Df}/I_p \propto e^{-(E_S - E_T)/RT}$. No emission is observed at low temperatures.

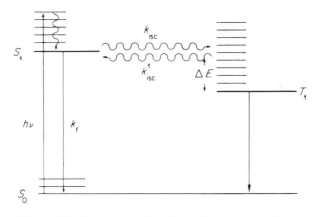

Figure 3.18. Energy levels for thermally-activated delayed fluorescence

These facts are explained by the mechanism depicted in Figure 3.18. Light absorption followed by intersystem crossing and vibrational relaxation gives triplets in their zero-point vibrational level. Thermal activation through the energy gap ΔE followed by reverse intersystem crossing $(T_1 \rightsquigarrow S_1)$ gives excited singlets, which then fluoresce. It follows that:

$$I_p = k_p[T_1]$$

and

$$I_{DF} = \theta_f . k'_{isc}[T_1] e^{-\Delta E/RT}$$

where θ_f is the quantum efficiency of fluorescence, whence

$$\frac{I_{DF}}{I_p} = \frac{\theta_f . k'_{isc}}{k_p} . e^{-\Delta E/RT} \tag{3.29}$$

On this mechanism, it would be expected that the experimentally measured activation energy ΔE should correspond to the singlet–triplet splitting obtained from spectroscopic data, and within experimental error this is found to be so. It has also been found that $k_{isc} \simeq k'_{isc}$.

It should be noted that thermally-activated delayed fluorescence seems to be largely confined to certain dyestuffs (eosin, fluorescein, acriflavine and proflavine) in which the singlet–triplet splitting is small (20–40 kJ mol^{-1}, 5–10 kcal mol^{-1}). With aromatic hydrocarbons the magnitude of the (π, π^*) splitting prohibits this mechanism for delayed fluorescence.

The importance of delayed fluorescence is that, in exploiting the extremely high sensitivity associated with emission spectroscopy, it provides a powerful tool for examining the behaviour of relatively low concentrations of triplets in solutions. The major alternative technique, flash photolysis, requires a high concentration of triplets, and this can introduce difficulties. Also, delayed emission gives information about the rates of all three intersystem crossings and permits triplet–triplet quenching to be observed directly.

3.3   RADIATIONLESS TRANSITIONS[19]

Resonance fluorescence is a term used to describe fluorescent emission of the same wavelength as that of the exciting light. It is only observed with atoms and simple molecules in the gas phase at low pressure where the time between collisions is greater than τ_f. Under these conditions the vibrational level which is populated by excitation is the one from which emission occurs. As the pressure increases, collisions degrade some of the original excitation energy into translational energy. The emission therefore occurs from lower vibrational levels, and fluorescence moves to longer wavelengths.

Benzene is probably the most complex molecule for which resonance fluorescence has been observed. With yet larger molecules (e.g., naphthalene), excitation to a higher singlet state S_n in a rarified gas leads to emission similar to, but more diffuse than, the normal fluorescence obtained by exciting S_1. It seems, then, that sufficiently large molecules in their S_n states undergo a transition to S_1 before luminescing. Since this transition ($S_n \rightsquigarrow S_1$) is not accompanied by photon emission, it is referred to as a *radiationless* or *non-radiative transition*. That is occurs in low-pressure gases indicates that it is non-collisional and therefore probably intramolecular. Since energy must be conserved, it follows that in this case the radiationless transition must be from S_n to an isoenergetic level of S_1. Phenomenologically the process can be represented by a horizontal (wavy) line on a Jablonski diagram (Figure 3.19).

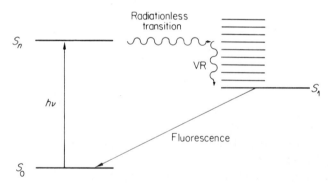

Figure 3.19. Radiationless transition to an isoenergetic level of a second state—horizontal wavy line on a Jablonski diagram

As indicated above, the emissive behaviour is a function of the size of molecules. It also depends on the environment, i.e. on whether the excited species is in solution or trapped in a dense medium or is in a rarified gas. This variation in emission reflects changes in the rates of radiationless transitions. A unified quantum-mechanical theory[19,20] is being rapidly developed at the moment which seeks to account for all the diverse phenomena associated with radiationless processes. It is impossible in a book of this sort to deal adequately with this theory. Rather we shall concern ourselves with radiationless phenomena exhibited by large molecules (i.e., benzene or larger) in condensed phases, because such systems are of greatest interest to organic photochemists and because the theoretical treatment is simpler.

3.3.1 Summary of Experimental Data on Large Molecules in Dense Media

The basic data emerging from many measurements is:

(i) Emission almost invariably occurs from S_1 or T_1, independent of the state which is initially excited (Kasha's Rule).[†]

(ii) ϕ_f does not depend on which state (S_1, S_2, S_3 etc.) is first excited (Vavilov's Law). This implies that very rapid internal conversion occurs between upper excited singlet states.

(iii) $\phi_f < 1$ and often $\phi_f \ll 1$.

(iv) The observed values of the fluorescent lifetime τ_f are less than the radiative lifetime τ_0 calculated from the oscillator strength (see Chapter 2, p. 14).

(v) The rates of radiationless transitions conform to an 'energy gap law', which will be discussed in more detail later, and according to which the rate of radiationless transitions falls very rapidly with increase in the energy difference between the $v = 0$ levels of the states concerned.

3.3.2 A Simple Model

Some understanding of radiationless processes in polyatomics in condensed phases may be obtained from the (oversimplified) model represented by Figure 3.20.

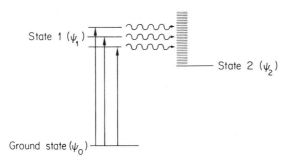

Figure 3.20. A model for radiationless transitions in
polyatomics in condensed phases

The molecule has a ground state (ψ_0), an excited state (ψ_1) into which it can be excited, and another excited state (ψ_2) of lower energy (either triplet or lower singlet, not excluding the ground state) and with a dense manifold of states provided by vibrational and rotational sublevels, some of which are isoenergetic with those of ψ_1. The essential near-degeneracy between the levels of ψ_1 and ψ_2

[†] The notable exceptions are azulene and its derivatives, which fluoresce strongly from S_2 and whose emission from S_1 is so weak that it has only recently been detected, and pyrene and 3,4-benzpyrene.

can be achieved, if necessary, by adding in small amounts of lattice energy from the medium.† In other words, solvent perturbations broaden the sublevels.

Excitation leads to population of a vibrational level or levels $v' = m$, from which a non-radiative transition to the quasi-continuum of state 2 occurs. These vibrational levels may be those initially populated, or, if k_{VR} the rate constant for vibrational relaxation is much greater than k_{nr} the rate constant for non-radiative transition, then the $v' = 0$ level of ψ_1 is that from which the radiationless transition occurs. In any case, the transition can be treated by time-dependent perturbation theory (see Chapter 2), a perturbation H' inducing a time-dependent evolution of the system from an initial state (ψ_1) into a final one (ψ_2). For internal conversion H' arises from electrostatic interactions between the electrons and nuclei, and for intersystem crossing H' is the spin–orbit interaction. Therefore internal conversion and intersystem crossing are treated similarly. Subject to certain conditions,[21] the rate constant k_{nr} for the non-radiative transition from each populated level of state 1 is then given by the Fermi golden rule:

$$k_{nr} = \frac{2\pi}{h}\langle\psi_1|H'|\psi_2\rangle^2\rho \qquad (3.30)$$

where ρ is the state-density factor which describes the number of states in the quasi-continuum isoenergetic with the levels of state 1 from which the radiationless transition occurs, and $\langle\psi_1|H'|\psi_2\rangle$ is a matrix element giving the energy of the interaction between the initial and final states induced by the perturbation H'. Invoking the Born–Oppenheimer approximation and factorizing the wavefunctions into electronic (ϕ) and vibrational (θ) components, the approximate expression (3.31) is obtained.

$$k_{nr} \propto \langle\phi_1|H'|\phi_2\rangle^2 \sum_i\sum_j \langle\rho\theta_{1i}|\theta_{2j}\rangle^2 \qquad (3.31)$$

$\langle\phi_1|H'|\phi_2\rangle$ is the electronic matrix element ($\equiv \int \phi_1 H' \phi_2 \, d\tau$), and $\langle\theta_{1i}|\theta_{2j}\rangle$ is the vibrational overlap integral (Franck–Condon factor) between the ith vibrational level of the initial state and the jth vibrational level of the second state. The double summation embraces all the populated levels of the state 1 from which radiationless transitions occur and the approximately isoenergetic levels of state 2 to which the transition occurs, and ρ is the state-density factor weighting each Franck–Condon term by the number of sublevels of each vibrational level.

Non-radiative phenomena in large molecules will now be discussed in the light of this equation.

---

† In small molecules with widely spaced vibrational levels in the gas phase at pressures where collisions are unimportant, the required degeneracy of the levels will occur only rarely, and radiationless transitions are very unlikely, so that $\phi_f \sim 1$. For sufficiently large molecules, however, with a large number of vibrational levels, ψ_2 becomes a quasi-continuum and degeneracy is readily achieved. Thus radiationless transitions would be expected to occur in large isolated molecules, as is observed to be the case.

3.3.3   **The Electronic Matrix Element**

This is the non-radiative counterpart of the electronic transition moment (see Chapter 2) and, like the latter, will have very small values unless the initial and final orbitals overlap effectively. Symmetry may, on occasion, make this integral zero, leading to a forbidden radiationless transition. It is possible to analyse such situations with the aid of group theory (see Appendix). For intersystem crossing, H' is the spin–orbit coupling operator H_{SO}, which can be resolved into three perpendicular components which transform like rotations R_x, R_y and R_z in the group character tables.

Consider the intersystem crossing $^1(n, \pi^*) \rightsquigarrow {}^3(n, \pi^*)$. Since both states are (n, π^*) they will have the same spatial wavefunction (say φ). The electronic matrix element is then $\langle \varphi | H_{SO} | \varphi \rangle$. The direct product $\Gamma\varphi \times \Gamma\varphi$ of the irreducible representations of φ will belong to the totally symmetric irreducible representation. Rotations in the point groups corresponding to the vast majority of molecules of interest to organic photochemists never belong to the totally symmetric irreducible representation. The triple product $\Gamma\varphi \times \Gamma\varphi \times \Gamma H_{SO}$ will therefore not be totally symmetric, the matrix element will be zero, $k_{nr} = 0$, and the transition will be forbidden. The same arguments apply to the intersystem crossing $^1(\pi, \pi^*) \rightsquigarrow {}^3(\pi, \pi^*)$. However, for $^1(n, \pi^*) \rightsquigarrow {}^3(\pi, \pi^*)$, the direct product $\Gamma\varphi_1 \times \Gamma\varphi_2$ is not totally symmetric, so that radiationless transitions may be allowed.

This is the theoretical basis of *El Sayed's selection rules* for intersystem crossing:

$$Allowed: \quad ^1(n, \pi^*) \leftrightarrow {}^3(\pi, \pi^*); \quad {}^3(n, \pi^*) \leftrightarrow {}^1(\pi, \pi^*)$$

$$Forbidden: \, ^1(n, \pi^*) \leftrightarrow {}^3(n, \pi^*); \quad {}^1(\pi, \pi^*) \leftrightarrow {}^3(\pi, \pi^*)$$

Processes forbidden under these rules still occur, but with rate constants 10^{-2}–10^{-3} times those for allowed intersystem crossings.

The preceding discussion, based on symmetry, has implicitly assumed the validity of the Born–Oppenheimer approximation. To be more accurate, vibronic wavefunctions ψ should be used and the value of integrals of the form $\langle \psi_i | H_{SO} | \psi_f \rangle$ computed. Since H_{SO} is an electronic operator and since the Born–Oppenheimer approximation is a close one, the values of the matrix elements $\langle \psi_i | H_{SO} | \psi_f \rangle$ and $\langle \varphi_i | H_{SO} | \varphi_f \rangle$ will be similar. However, the symmetries of ψ and φ will, in general, be different, because the symmetry of the former contains contributions from the symmetry of the vibrational components, whereas the latter (Born–Oppenheimer) wavefunction does not. Thus, when the orbital matrix element is zero, the vibronic matrix element will often be non-zero though small (cf. discussion of radiative vibronic coupling, Appendix and Chapter 2, section 2.1.8). In other words, rigorous symmetry restraints can be relaxed by including vibrational terms (vibronic coupling). There is evidence that much of the observed non-radiative decay ($S_1 \rightsquigarrow T_1$) in aromatics occurs *via* vibronic spin–orbit coupling.

The well known heavy atom effects in intersystem crossing (see Chapter 2, p. 24 and Chapter 3, p. 82) are, of course, another manifestation of the role of the electronic matrix element, which becomes magnified because $H' = H_{SO}$ is

90

greater in systems containing heavy atomic nuclei. This leads to increased values of k_{isc} with the consequences discussed earlier.

For internal conversion, the electronic matrix element is $\langle \varphi_1 | H_{ic} | \varphi_2 \rangle$ and H_{ic} is the nuclear kinetic energy operator.† Since H_{ic} belongs to the totally symmetric irreducible representation, it is immediately apparent that the integrand will be totally symmetric and the matrix element non-zero only if φ_1 and φ_2 belong to the *same* irreducible representation. What this means is that, rigorously, only states of the same symmetry should internally convert (contrast intersystem crossing), and hence, since S_1 and S_0 must have different symmetries, the internal conversion $S_1 \rightsquigarrow S_0$ is forbidden. That this phenomenon occurs is again due to vibronic coupling, for the crossing from S_1 must be to very high vibrational levels of S_0 and some of these will have the same vibronic symmetries as those of S_1.

3.3.4 Vibrational Overlap Integral (Franck–Condon Factor)

Before proceeding, the reader will find it helpful to read the material on vibrational overlap integrals in Chapter 2 (section 2.1.5).

Radiationless transitions can be visualized by thinking of the intersection of potential energy surfaces. A molecule on the potential energy surface corresponding to state 1 'crosses' at the point of intersection (X) to the potential energy surface associated with state 2. Consider the disposition of energy surfaces‡ in Figure 3.21. At the internuclear separation corresponding to the intersection, state 1 and state 2 have the same energy and the same internuclear distance. A molecule in the level AX of state 1, when it arrives at X, has merely to change the quantum numbers of one of its electrons (to rearrange the motion of one of its electrons) to be in state 2 in the level BX. Since X is a turning point

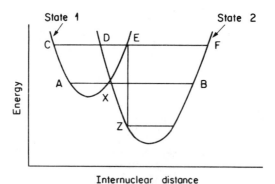

Figure 3.21. The Franck–Condon principle and radiationless transitions

† Although H_{ic} operates only on nuclei, it nevertheless affects electronic wavefunctions because of their dependence upon nuclear coordinates.

‡ The complete representation of the energy of a molecule of N atoms requires a hypersurface in $(3N-6)$ dimensions. For the purpose of visualization it is common practice to take a section through the hypersurface corresponding to changes in only one internuclear distance—in other words to treat polyatomics as though they are diatomics (this can be misleading).

in the vibrations A–X–A and B–X–B, where the kinetic energy is zero and the system is momentarily at rest, more time is available at this point than at any other for the system to cross from one state to the other. The process becomes irreversible if the molecule having arrived in the level BX undergoes rapid vibrational relaxation.

Now consider the possibility of a radiationless transition from level CE. If the transition occurs when the molecule is at the turning point E, then virtually instantaneously there must either be a change in the nuclear coordinates to arrive at points D or F in state 2, or a change in kinetic energy (equal to EZ) if the internuclear separation remains constant. Similar arguments apply to transitions occurring at (say) point D. Such rapid changes are forbidden by the Franck–Condon principle, which applies equally to radiative and non-radiative processes.

It is desirable for later purposes to recast the above classical argument into quantum-mechanical terms based on vibrational overlap integrals. In Chapter 2 vertical (radiative) transitions were so analysed. We now wish to understand the impact of such considerations on horizontal (non-radiative) transitions. If θ_1 and θ_2 are vibrational wavefunctions associated with two electronic states designated 1 and 2, then the probability of crossing between the states is proportional to $\int \theta_1 \theta_2 \, d\tau_N$. Compare the situation of Figure 3.22, where the zero-point levels of the two states have approximately the same energy, with that of Figure 3.23, where there is a considerable gap between the zero-point levels.

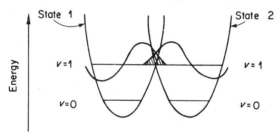

Figure 3.22. Large vibrational overlap at the crossing point of approximately degenerate electronic states

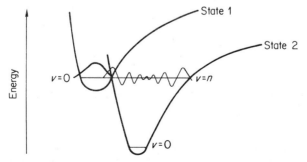

Figure 3.23. Vibrational overlap with a large energy difference between states

It is immediately apparent that the vibrational overlap integral at the crossing point in Figure 3.22 is of significant dimensions, but because of the rapidly oscillating character of the $v = n$ function in Figure 3.23, the positive contributions to the vibrational overlap integral are largely cancelled by the negative contributions, so that the integral is very small.[†]

A molecule in each of the populated levels of state 1 has therefore a definite probability of crossing to each of the many approximately isoenergetic levels of state 2, and the total probability of crossing is given by the double summation term of equation (3.31). The effect of this term is seen in the energy gap law, in the effect of deuteration on the rates of radiationless transitions, and in certain other phenomena now to be discussed.

3.3.5   Energy Gap Law

There is an inverse correlation between the rates of non-radiative transitions involving the lowest states of similar molecules and the difference in energy between the $v = 0$ levels of the states involved. In other words, the smaller the energy gap the bigger the rate. This is shown in Figures 3.24 and 3.25, where the logarithms of the rate constants for intersystem crossing ($T_1 \rightsquigarrow S_0$) and for internal conversion ($S_1 \rightsquigarrow S_0$) for a series of aromatic hydrocarbons are plotted against the energy gap.

The energy gap law is readily understood by noting that as the gap increases the radiationless transition from a given level of state 1 will be to an increasingly high vibrational level of state 2, with reduced vibrational overlap and a correspondingly reduced rate constant.

The law can be invoked to provide a simple rationalization of Kasha's Rule and Vavilov's Law. Since upper excited states are densely packed, i.e. since the energy gaps between them are small, internal conversions between them will be very rapid, so that $k_{ic} \gg k_f$ and their fluorescence will not be observed. However, the energy difference between S_0 and S_1 or T_1 is much larger, and radiationless depopulation of S_1 or T_1 will in many cases be unable to quench emission from these states. It is noteworthy that $k_{isc}(S_1 \rightsquigarrow T_1)$ is frequently very large ($\sim 10^{10}\,\mathrm{s}^{-1}$), even though intersystem crossing is a spin-forbidden process, and it may be that this is due in part to the existence of higher triplet states lying between S_1 and T_1 which thus partition the original energy gap into smaller ones.

3.3.6   Deuteration and Other Effects

Further evidence relating to the importance of Franck–Condon factors may be found in the dramatically increased phosphorescence lifetimes of the triplet states of aromatic systems which have been perdeuterated. Table 3.4 exemplifies this point.

[†] Note that since vibrational wavefunctions extend beyond the potential surfaces, it is not actually necessary for the surfaces to cross, i.e. a radiationless transition between the $v = 0$ levels in Figure 3.22 is still possible, though with strongly reduced probability (quantum mechanical 'tunnelling').

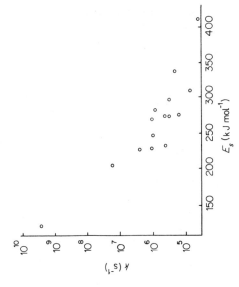

Figure 3.25. Singlet decay rate constants (k) of aromatic hydrocarbons plotted against the singlet energy (E_s). (From data in J. B. Birks, *Photophysics of Aromatic Molecules*, (1970), Wiley)

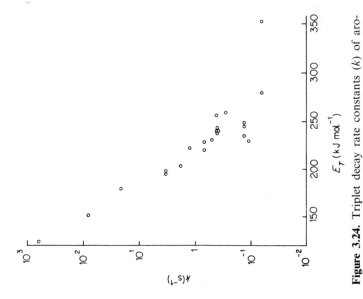

Figure 3.24. Triplet decay rate constants (k) of aromatic hydrocarbons plotted against the triplet energy (E_T). (From W. Siebrand, in A. B. Zahlan (ed.), *The Triplet State*, (1967), reproduced by permission of Cambridge University Press)

Table 3.4. Effect of deuteration on phosphorescent lifetimes of aromatic hydrocarbons[a]

Compound	Benzene C_6H_6	C_6H_5D	C_6D_6	Naphthalene $C_{10}H_8$	$C_{10}D_8$
$\tau_p(s)$	5·75	8·90	12·0	2·51	21·7

[a] Measured in 3-methylpentane glass, except $C_{10}D_8$ in E.P.A. glass.

It has also been shown that for incompletely deuterated compounds, τ_p depends upon the position of deuteration.

The deuterium effect is explained on the assumption that k_{nr} is largely determined by the C—H vibrations because, being of high frequency, they are widely spaced. The C—D vibrations are of lower frequency, so that the levels are more closely packed. Hence a molecule in a given level of T_1 crosses to a vibrational level of S_0 which has a much larger quantum number in the deuterated compound. $k_{isc}(T_1 \rightsquigarrow S_0)$ is thus reduced in the deuterated compound, and τ_p is increased accordingly. The result clearly indicates the important influence of C—H stretching modes in determining k_{nr}.

There is another effect which depends on Franck–Condon factors. With large and rigid molecules (e.g., aromatics) there is little change of geometry on excitation, so that the minimum in the S_1 surface is likely to be only slightly displaced with respect to that of S_0. In such cases, as inspection of Figure 3.26 reveals, internal conversion from the zero-point level of S_1 to S_0 will be very slow because the vibrational overlap integral will be very small. This means that fluorescence can compete with radiationless depopulation of S_1, and hence the rule that rigid systems tend to fluoresce.

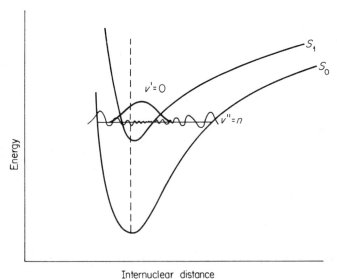

Figure 3.26. Vibrational overlap in large, rigid molecules

In condensed phases the rate of vibrational relaxation is likely to be much larger than k_{nr}, so that radiationless transitions would be expected to occur exclusively from the $v = 0$ level of state 1 (except in so far as thermal activation will induce a Boltzmannian population of higher vibrational levels). That this is indeed the case is shown by innumerable observations that the nature and life-time of the emission from large molecules in such phases is independent of the wavelength of the exciting light and therefore of which vibrational levels are initially populated. Since molecules in the various vibrational levels of state 1 have different probabilities of crossing to state 2, it follows that the emission parameters will be wavelength-dependent in a low pressure gas phase. This has been found to be so. Recent work has focussed upon exciting individual vibrational levels and determining relative values of k_{nr}.[22] A theoretical model[23] has been developed for treating the non-radiative decay of single vibrational levels of large molecules, using state-density-weighted Franck–Condon factors, which gives excellent agreement with experimental observations for benzene and its derivatives in the gas phase.

3.3.7   **Fluorescence vs. Phosphorescence**

Whether a molecule fluoresces or phosphoresces or does neither, depends on the values of k_f, k_p, $^1k_{isc}(S_1 \rightsquigarrow T_1)$ and ^1k_r and ^3k_r (the rate constants for photochemical change in excited singlet and triplet states), Figure 3.27. Neglecting

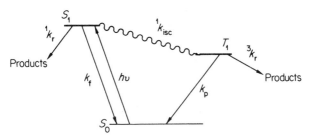

Figure 3.27. A kinetic scheme for deriving quantum yields of fluorescence and phosphorescence

internal conversion ($S_1 \rightsquigarrow S_0$) and intersystem crossing ($T_1 \rightsquigarrow S_0$) because their rate constants are too small to affect the argument, it is immediately apparent that an excited species can revert to the ground state only by photochemical reaction or luminescence. In other words, absence of luminescence means that rapid photochemical changes are occurring. However, if the photoproducts are highly labile entities which revert rapidly to starting material, then no overall change will be observed. This point is taken up in the next section.

Applying equations (3.3) and (3.10) to the situation of Figure 3.27, the quantum yields of fluorescence and phosphorescence can be expressed as:

$$\phi_f = \frac{k_f}{^1k_r + k_f + {}^1k_{isc}} \tag{3.32}$$

96

and

$$\phi_p = \frac{k_p}{^3k_r + k_p} \cdot \frac{^1k_{isc}}{^1k_r + k_f + ^1k_{isc}} \tag{3.33}$$

From these expressions it is apparent that fluorescence will be observed only if $k_f \ll (^1k_r + ^1k_{isc})$, and phosphorescence will be observed only if both $k_p \ll ^3k_r$ and $^1k_{isc} \ll (^1k_r + k_f)$. Furthermore, $^1k_{isc}$ imposes limits on the processes open to S_1. Clearly, if $^1k_{isc} \gg ^1k_r$, then any photochemical changes involving S_1 can only proceed with a very small quantum yield. Equally if $^1k_{isc} \gg k_f$, no fluorescence will be detectable.

Since $^1k_{isc}$ is normally at least comparable with k_f, triplets are always produced, and phosphorescence must occur unless $^3k_r \gg k_p$. Since k_p is small this condition is often satisfied, and phosphorescence is by no means a widespread phenomenon among organic molecules as a whole.

These considerations, illustrated in Figures 3.28 and 3.29, show why benzophenone fails to fluoresce strongly while aliphatic ketones both fluoresce and phosphoresce. The essential point is that $^1k_{isc}$ is much greater for benzophenone because of the operation of El Sayed's selection rules.

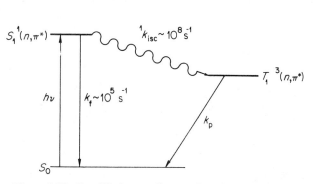

Figure 3.28. Simplified state diagram for aliphatic ketones

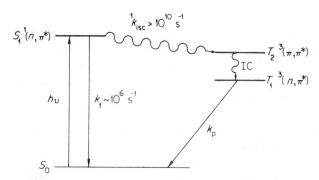

Figure 3.29. Simplified state diagram for benzophenone

Emission Rate Constants

Since $k_f = A_{ul}$, the Einstein coefficient of spontaneous emission, and A_{ul} is related to B_{ul}, the discussion in Chapter 2 on the intensity of absorption applies equally to k_f. Hence k_f will be large (of the order of 10^7–10^8 s^{-1}) for allowed or partially allowed $\pi \rightarrow \pi^*$ transitions and will be very much smaller ($\sim 10^5$ s^{-1}) for forbidden $n \rightarrow \pi^*$ transitions. Therefore, other things being equal, aromatic hydrocarbons would be expected to fluoresce more strongly than simple ketones.

With respect to phosphorescence, in Chapter 2 (p. 21) the intensities of singlet–triplet transitions were related to the extent that H_{SO}, the spin–orbit coupling operator, mixes into nominal triplet and singlet states small amounts of the state of the other multiplicity (see equation 2.12). If we can neglect the mixing of triplet states into S_0 because of the large energy gap $(E_T - E_{S0})$, then the rate of phosphorescence from a given nominal triplet level will be correlated to the amount of the singlet component of that triplet state. These concepts are now applied to aliphatic ketones and aromatic hydrocarbons.

In aliphatic ketones, T_1 is $^3(n, \pi^*)$ and in the C_{2v} point group; $\Gamma(n, \pi^*) = A_2$ (see Appendix). Bearing in mind that the components of H_{SO} transform like rotations, examination of the C_{2v} character table shows that H_{SO} mixes the triplet (n, π^*) state of the carbonyl group with the singlet states $^1(\sigma, \pi^*)(^1B_1)$, $^1(n, \sigma^*)(^1B_2)$ and $^1(\pi, \pi^*)(^1A_1)$. It follows that the nominal $^3(n, \pi^*)$ state of aliphatic ketones will have significant singlet character, leading to a relatively large rate for transition to the singlet ground state and hence to a short triplet lifetime.

Conversely, in benzene where all the lowest states are (π, π^*), appeal to the character table for the D_{6h} point group shows that the lowest triplet T_1, which belongs to the B_{1u} representation, cannot be mixed by H_{SO} with any of the singlet (π, π^*) states. Thus T_1 would be expected to be a virtually pure triplet state except in so far as vibronic spin–orbit coupling (Chapter 3, p. 89) relaxes these symmetry restraints. Radiative collapse to the ground state is thus expected to be very slow. To the extent that similar arguments apply to other aromatic hydrocarbons, it can be understood why for aromatic hydrocarbons τ_p is of the order of 10 s and for aliphatic ketones $\tau_p \sim 10^{-3}$ s.

3.3.8 Radiationless Transitions, Isomerization and Photochemistry

A photostable molecule, excited into S_1, must either fluoresce or undergo intersystem crossing or internal conversion. This implies that:

$$\phi_f + \phi_{ic} + \phi_T = 1 \quad \text{(note that } \phi_T \equiv \theta_{isc} \geqslant \phi_p) \tag{3.34}$$

Because of the large energy gap between S_0 and S_1, k_{ic} is expected to be small relative to k_f and k_{isc}, and ϕ_{ic} is expected to be negligible. Hence $(\phi_f + \phi_T) \sim 1$, and this is found to be the case for many molecules (see Table 3.5). In other systems however, $(\phi_f + \phi_T) \ll 1$. This must mean that k_{ic} and/or $^3k_{isc}(T_1 \rightsquigarrow S_0)$ is large in spite of the energy gap law, and it is tempting to ascribe the enhanced

Table 3.5. Quantum yields for fluorescence and triplet formation

Compound	ϕ_f	$\phi_T (\equiv \theta_{isc})$
Naphthalene	0·21	0·71
Anthracene	0·33	0·58
Phenanthrene	0·14	0·70
Triphenylene	0·09	0·89
1-Methoxynaphthalene	0·53	0·46
9-Phenylanthracene	0·45	0·505
Fluorescein	0·92	0·05
Eosin	0·19	0·71
Erythrosin	0·02	1·07
Pyridazine	0·01	0·00
Pyrimidine	0·0058	0·14
Pyrazine	0·0006	0·30

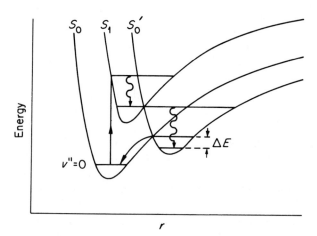

Figure 3.30. A possible role of intermediates in internal conversion

rate of radiationless decay found in such systems to the formation of unstable isomers.[24] Figure 3.30 depicts the situation.

Internal conversion from S_1 to the ground state isomer S_0' will be much faster than to S_0 (Energy gap law). Vibrational relaxation sends the molecule into the zero-point level of S_0', and if the energy barrier ΔE between S_0' and S_0 is small enough to be surmounted rapidly by thermal activation, then all that will be observed is an enhanced rate of internal conversion. If ΔE is larger than a few kT (where k is Boltzmann's constant), then the molecule will remain in S_0' as a metastable isomer. One system exhibiting this phenomenon is biacetyl, which on $n \rightarrow \pi^*$ excitation in water or methanol gives rise to the enol (3.35), which of course reverts to biacetyl. Another is benzene, which on irradiation is transformed into benzvalene (3.36), which slowly reverts to benzene at room temperature.

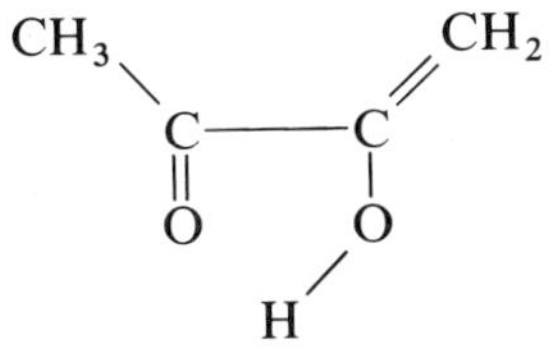

(3.35) 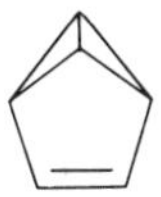   (3.36)

At this point we should pause and take stock. Radiationless transitions occurring between states of the *same* molecule have been discussed. Formally:

$$^1M^{**} \rightsquigarrow {}^1M^* \left. \begin{array}{c} \\ \\ \end{array} \right\} \text{internal} \qquad {}^1M^* \rightsquigarrow {}^3M^* \left. \begin{array}{c} \\ \end{array} \right\} \text{intersystem}$$
$$^1M^* \rightsquigarrow {}^1M \qquad \qquad {}^3M^* \rightsquigarrow {}^1M \quad \text{crossing}$$
$$^3M^{**} \rightsquigarrow {}^3M^* \left. \begin{array}{c} \\ \end{array} \right\} \text{conversion}$$

Then the concept of radiationless transitions between states of *isomeric* molecules was introduced:

$$^1M^* \rightsquigarrow {}^1M' \rightsquigarrow {}^1M$$

There seems to be no difference in principle between any of the above photophysical processes and the photochemical processes:

$$^1M^* \rightsquigarrow {}^1X \quad \text{or} \quad {}^3M^* \rightsquigarrow {}^1X$$

which involve radiationless transitions between states of *different* molecules. Radiationless transitions and photochemical reactions both involve a non-radiative electron rearrangement whereby an excited state decays into a state of lower energy. This implies a conversion of electronic energy into vibrational energy.[25] The difference between the two classes of phenomena seems to be this, that in radiationless transitions the electronic decay is into quantized vibrational levels, whereas in photochemical reactions the decay is into continuum states, for bonds are broken and/or rearranged. The electronic-vibrational energy transfer is often highly selective in that only certain vibrational modes are excited, as witnessed by the orbital symmetry rules (see Chapter 6) and the frequent observation that bonds to hydrogen, usually the strongest in the molecule, are those broken photochemically. With the current rapidly developing theoretical situation it may ultimately become possible to integrate photochemical reactions into the theory of radiationless transitions, but, for the time being, other concepts, discussed in later chapters, are used with considerable success in rationalizing the course of such reactions.

REFERENCES

1. E. J. Bowen (ed.), *Luminescence in Chemistry*, Van Nostrand, London (1968).
2. R. S. Becker, *Theory and Interpretation of Fluorescence and Phosphoresence*, Wiley, London, (1969).
3. M. Kasha, *Discuss. Faraday Soc.*, **9**, 14 (1950).
4. C. A. Parker, *Chem. in Brit.*, **2**, 160 (1966).
5. A. P. Marchetti and D. R. Kearns, *J. Amer. Chem. Soc.*, **89**, 768 (1967).
6. J. B. Birks and I. H. Munro, *Progress in Reaction Kinetics*, **4**, 277 (1967).
7. E. Drent, *Chem. Phys. Letters*, **2**, 526 (1968).
8. J. G. Calvert and J. N. Pitts, Jr., *Photochemistry*, Wiley, London (1966), p. 800.
9. S. P. McGlynn, T. Azumi and M. Kinoshita, *Molecular Spectroscopy of the Triplet State*, Prentice-Hall, New Jersey (1962), p. 204.
10. W. T. Stacey and C. E. Swenberg, *J. Chem. Phys.*, **52**, 1962 (1970); R. G. Bennett and P. J. McCartin, *ibid.*, **44**, 1969 (1966); B. Stevens and M. F. Thomasz, *Chem. Phys. Letters*, **1**, 549 (1968).
11. C. A. Parker and T. A. Joyce, *Chem. Commun.*, 749 (1968).
12. W. D. Bellamy and A. G. Tweet, *Nature*, **197**, 482 (1963); W. H. Melhuish and R. Hardwick, *Trans. Faraday Soc.*, **58**, 1908 (1962).
13. Ref. 9, Chapter 2.
14. C. A. Hutchison and B. W. Mangum, *J. Chem. Phys.*, **29**, 952 (1958).
15. S. R. La Paglia, *J. Mol. Spectr.*, **7**, 427 (1961).
16. Data extracted from ref. 2, p. 139.
17. C. A. Parker, *Adv. Photochem.*, **2**, 305 (1964).
18. C. A. Parker and C. G. Hatchard, *Proc. Roy. Soc.*, **A269**, 574 (1962).
19. J. Jortner, S. A. Rice and R. M. Hochstrasser, *Adv. Photochem.*, **7**, 149 (1969).
20. K. Freed, *Fortschr. chem. Forsch.*, **31**, 105 (1972); D. Phillips in *Chem. Soc. Spec. Reports, Photochem.*, volumes 1–4 (1970–1973); B. R. Henry and W. Siebrand in J. B. Birks (ed.), *Organic Molecular Photophysics*, volume 1, Wiley, London (1973), p. 153.
21. See D. Phillips, *Chem. Soc. Spec. Reports, Photochem.*, **4**, 59 (1973).
22. An excellent review is by D. Phillips and K. Salisbury, *Chem. Soc. Spec. Reports, Photochem.*, **4**, 228 (1973).
23. D. F. Heller, K. F. Freed and W. M. Gelbart, *J. Chem. Phys.*, **56**, 2309 (1972).
24. D. Phillips, J. Lemaire, C. S. Burton and W. A. Noyes, Jr., *Adv. Photochem.*, **5**, 329 (1968).
25. G. S. Hammond, *Adv. in Photochem.*, **7**, 373 (1969).

Chapter 4

Quenching of Excited States

4.1   INTRODUCTION

A substance which accelerates the decay of an electronically excited state to the ground state or to a lower electronically excited state is described as a *quencher* and is said to *quench* that state. Thus, if the original excited state is luminescent, *quenching* will be observed as a diminution of the intensity (quantum yield) of light emission. The process can be represented as:

$$M^* \xrightarrow{\;Q\;} M' \tag{4.1}$$

(where M' is the ground state or another excited state of M, and Q is the quencher).

It must be emphasized that a reduction in (for example) fluorescence is *prima facie* evidence of electronic quenching only if a Boltzmann distribution over vibrational levels of the emitting state is attained before emission occurs. In the gas phase, the frequently observed variation of ϕ_f with pressure is found often to be due to collisionally-induced vibrational relaxation rather than electronic quenching.

The quenching process of (4.1) is of such generality that, as might be expected, it occurs by many different mechanisms and is induced by many different substances. Of these, oxygen is the most ubiquitous and one of the most efficient in that each encounter with an excited molecule leads to quenching. For this reason it is essential in all quantitative work to reduce the concentration of dissolved oxygen to the smallest possible value, either by bubbling oxygen-free nitrogen through the solution or, better, by degassing with several 'freeze–pump–thaw' cycles.† For similar reasons rigorous standards of purity are essential in all work on luminescence. Solvents should be non-fluorescent, and substrates should be purified by chromatography, zone-refining etc. until τ_f is constant.

Quenching processes, with the exception of certain kinds of electronic energy transfer, seem to be collisional and therefore subject to the Wigner

---

† The vessel is cooled in liquid nitrogen, evacuated to pump off the supernatant gas, sealed from the pump and allowed to thaw to release dissolved gases.

spin-conservation rules.† The bimolecular entity in which the quenching occurs can be either an encounter complex or an excimer/exciplex.‡ The distinction between these two sorts of entities seems to be this, that in an encounter complex, represented in this book as (M*...Q), the components are separated by distances of the order of 7 Å or more and have random relative orientations, the only requirement being some 'significant' overlap of the orbitals of the components. On the other hand, excimers and exciplexes [represented by (MM)* and (MQ)*] are entities occupying energy minima in the excited state potential surface and therefore having definite geometries. In aromatic excimers the components are organized into parallel planes (see below).

In principle, quenching can occur in either encounter complexes or excimers/exciplexes or both. The actual entity implicated has been defined for only relatively few systems. Seen in these terms, bimolecular quenching depends on the reversible formation of an encounter complex or exciplex which subsequently relaxes to ground state entities by various modes. Figure 4.1 depicts relaxation modes open to exciplexes. Similar diagrams may be constructed for encounter complexes or excimers.

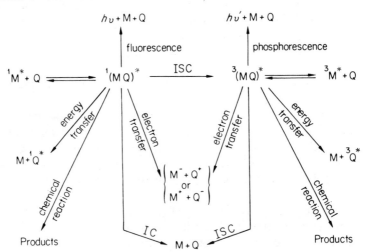

Figure 4.1. Relaxation pathways in exciplexes

† If two reactants A and B with spin S_A and S_B react on collision to give products X and Y with spins S_X and S_Y, then the spin of the transition state or collision complex can only have one of the following values:

$$(S_A + S_B), (S_A + S_B - 1), (S_A + S_B - 2)\ldots|S_A - S_B|$$

Similarly, the transition state giving rise to the products must have one of the spin values:

$$(S_X + S_Y), (S_X + S_Y - 1), (S_X + S_Y - 2)\ldots|S_X - S_Y|$$

Therefore for a collisional process to be spin-allowed, the two sequences must have a number in common.

‡ An exciplex is the excited state of a complex and is formed by the association of two different species, one excited and the other in its ground state. An excimer is an *excited dimer*, i.e. an exciplex involving the ground and excited states of the same species. These entities are discussed in the next section.

Quenching by electronic energy transfer and by the routes indicated in Figure 4.1 form the main subject matter of this chapter, but an account of the nature of excimers and exciplexes and of the kinetics of quenching is required first.

4.2   EXCIMERS[1]

It is commonly observed that an increase in the concentration of a solute is accompanied by a decrease in the intensity (quantum yield) of its fluorescence. This phenomenon is called *concentration quenching* and has been known for some time. More recently it has been found that such quenching is often accompanied by the appearance of a new emission at longer wavelengths, the intensity of which increases with concentration. For example, the violet fluorescence of pyrene (4.2) in dilute solution is gradually replaced by a blue fluorescence with increasing pyrene concentration (Figure 4.2). Förster and Kasper[2] showed that

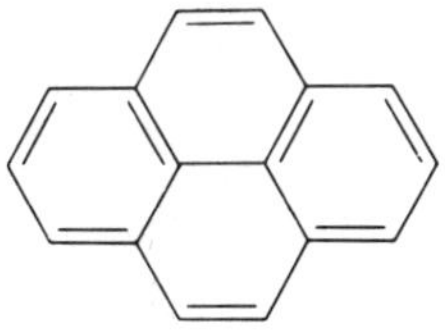

(4.2)

these phenomena could be explained by the formation and fluorescence of a pyrene excimer (4.3).

$$M \underset{\substack{\text{monomer}\\\text{fluorescence}}}{\overset{h\nu}{\rightleftarrows}} {}^{1}M^{*} \underset{}{\overset{M}{\rightleftarrows}} {}^{1}(MM)^{*} \xrightarrow[\substack{\text{fluorescence}}]{\text{excimer}} M + M + h\nu' \qquad (4.3)$$

It has since been found that excimer formation is widespread among aromatic hydrocarbons, though with simpler systems such as benzene, naphthalene and their methyl derivatives lower temperatures and higher concentrations than those necessary for pyrene are required.

Such complexes exist only in the excited state, being dissociated and therefore undetectable in the ground state. Figure 4.3 illustrates the points that emission from an excimer will be devoid of fine structure and will occur at longer wavelengths than that of its components.

Figure 4.4 shows the time-dependence of emission from pyrene monomer and excimer after flash excitation. Kinetic analysis[3] of such decay curves permits evaluation of the lifetime of the excimer and of the rate constants for its formation and dissociation. The rate constants for excimer formation (also obtainable from measurements of the quantum yield of emission as a function of concentration) are frequently found to be close to the diffusion-controlled rate constant k_{diff}, except where substituents can exercise a steric effect.

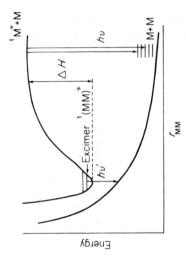

Figure 4.3. Schematic energy surfaces showing excimer formation and emission. The emission to the nonquantized ground state is structureless

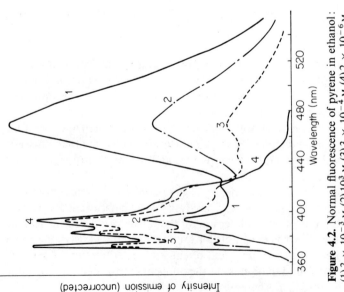

Figure 4.2. Normal fluorescence of pyrene in ethanol: (1) 3×10^{-3} M, (2) 10^3 M, (3) 3×10^{-4} M, (4) 2×10^{-6} M. The instrumental sensitivity settings for curves 1 and 4 were approximately 0·6 and 3·7 times that for curves 2 and 3. The short wavelength ends of the spectra in the more concentrated solutions are distorted by self-absorption. (From C. A. Parker and C. G. Hatchard, *Trans. Faraday Soc.*, **59**, 284 (1963), reproduced by permission of the Chemical Society)

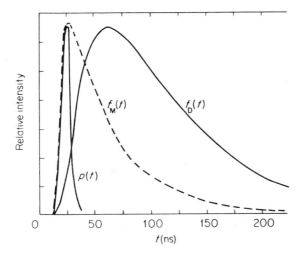

Figure 4.4. Pyrene (5×10^{-3} M) in cyclohexane. Monomer fluorescence response $f_M(t)$ and excimer fluorescence response $f_D(t)$ to excitation light pulse $p(t)$ (from J. B. Birks, D. J. Dyson and I. H. Munro, *Proc. Roy. Soc. A*, **275**, 575 (1963), reproduced by permission of the Royal Society)

From an examination of the temperature-dependence of the emission and by other methods it is possible to evaluate thermodynamic parameters. For pyrene[4] the enthalpy of dissociation (ΔH in Figure 4.3) is $\sim 40 \text{ kJ mol}^{-1}$ (10 kcal mol^{-1}), and the entropy of dissociation is $\sim 80 \text{ J K}^{-1}$ (20 cal K^{-1}), showing that this excimer has a strongly bonded and rigid structure. Excimers derived from other hydrocarbons have smaller enthalpies ($\sim 20 \text{ kJ mol}^{-1}$, 5 kcal mol^{-1}), though the dissociation entropies are usually similar.

4.2.1   Excimer Structure and Bonding

Excimers derived from aromatic hydrocarbons seem to adopt a sandwich structure with the molecular planes separated by $\sim 3 \cdot 3$ Å. Evidence supporting this point is the fact that those crystals of hydrocarbons in which the molecules are stacked in parallel planes have a fluorescence emission closely resembling that of the corresponding excimer,[5] and that in the paracyclophane series the 4,4'-compound (4.4, $n = 4$) with an interplanar distance of $3 \cdot 73$ Å is the lowest member of the series to exhibit an alkylbenzene absorption spectrum and a broad structureless intramolecular excimer emission. With the 4,5'- and 6,6'-paracyclophanes the emission returns to that of an alkyl benzene.[6]

Furthermore, photolysis of (4.5) in a low temperature glass generates (4.6) in the unstable sandwich geometry, in which form it has a fluorescent emission similar to the excimer emission naphthalenes in fluid solution. When the glass is thawed and then refrozen, permitting the sandwich system to attain the more

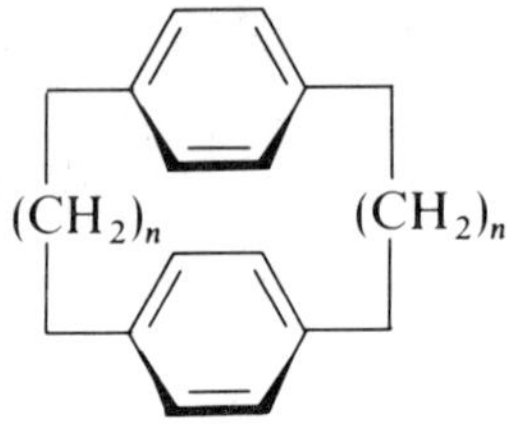

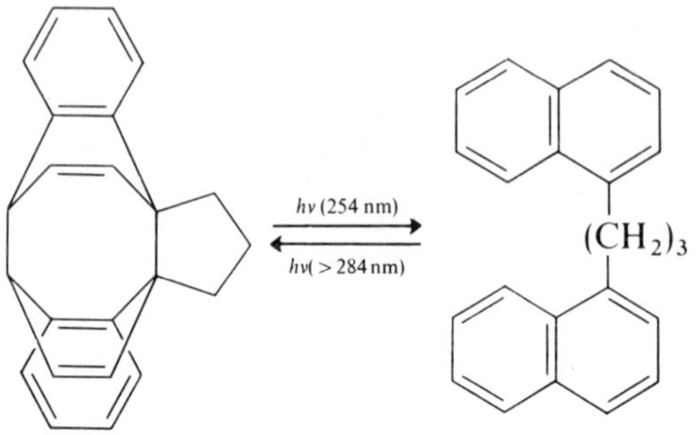

(4.4)	(4.5)	(4.6)

stable extended configuration, the emission changes to become closely similar to that of 1-methylnaphthalene.

A simple treatment of excimer bonding assumes that the wavefunction is of the form (4.7).

$$\psi_{\text{excimer}} = a\psi_{MM^*} + b\psi_{M^*M} + c\psi_{M^-M^+} + d\psi_{M^-M^+} \qquad (4.7)$$

This implies both excitin resonance [MM* ↔ M*M] and charge–transfer resonance [$M^-M^+ ↔ M^+M^-$]. Calculations[7] show that agreement between theoretical and observed values of excimer fluorescence can only be obtained if the components adopt a sandwich configuration with an interplanar distance of 3·0–3·6 Å, and if both exciton and charge–transfer resonance are invoked.

Excimer formation also occurs with non-aromatic molecules, and although the discussion has been directed to singlet excimers, the observation of excimer phosphorescence establishes the existence of triplet excimers also.

4.3 EXCIPLEXES[1]

Phenomena similar to those described for excimers are observed in solutions of mixed solutes. For example,[8] addition of diethylaniline to a solution of anthracene in toluene quenches the fluorescence of the latter and replaces it by a new structureless emission at longer wavelengths (Figure 4.5). This is ascribed to the formation of an exciplex (*exci*ted com*plex*) ^{1}(anthracene-diethylaniline)*. The term exciplex refers to an excited complex of definite stoichiometry (usually 1:1) formed between an excited species and one or more different molecules in their ground states (4.8).

$$M \xrightarrow{h\nu} M^* \xrightarrow{Q} (MQ)^* \qquad (4.8)$$
$$\text{exciplex}$$

Exciplexes are polar entities. The dipole moments of exciplexes from aromatic hydrocarbons and aromatic tertiary amines, estimated from the red shift of the emission peak with increasing dielectric constant of solvent, have values of >10 debyes.[9] Excimers, on the other hand, have zero dipole moments.

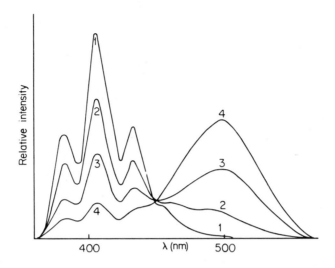

Figure 4.5. Fluorescence spectra of anthracene (3×10^{-4} M) in toluene in presence of diethylaniline at concentrations 0·000 M (1); 0·005 M (2); 0·025 M (3); 0·100 M (4). (From A. Weller, in S. Claesson (ed.), *Fast Reactions and Primary Processes in Chemical Kinetics*, (1967), John Wiley and Sons Ltd.)

A simple molecular orbital treatment due to Weller[10] provides an explanation of the large dipole moment and other properties of exciplexes. It assumes that in an exciplex an electron is transferred from one component, D (donor) to the other, A (acceptor). Figures 4.6 and 4.7 illustrate the situation when the donor and the acceptor respectively are excited.

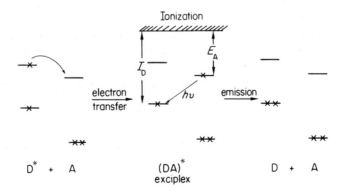

Figure 4.6. Molecular orbital treatment of exciplex formation and emission (donor excited)

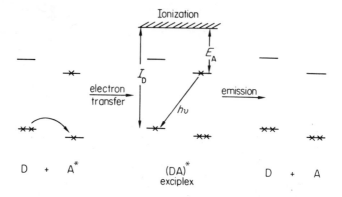

Figure 4.7. Molecular orbital treatment of exciplex formation and emission (acceptor excited)

This treatment seems to be equivalent to writing the wavefunction of the exciplex as (4.9) and neglecting all terms except the fourth.

$$\psi_{\text{exciplex}} = a\psi_{\text{D*A}} + b\psi_{\text{DA*}} + c\psi_{\text{D}^- \text{A}^+} + d\psi_{\text{D}^- \text{A}^+} \tag{4.9}$$

One can readily formulate the conditions for the formation of exciplexes on the charge–transfer model. Since the ionization potential (I) and electron affinity (E) are measures of the energies of the highest bonding and lowest antibonding molecular orbitals respectively, ΔG, the energy of formation of the separate ions A^- and D^+ from A and D in their ground states, is $I_D - E_A$. Since we actually start with (say A*, the energy of formation is diminished by the excitation energy possessed by A*. Bringing the separated ions A^- and D^+ to their equilibrium distance (r) in the exciplex reduces ΔG by an electrostatic term so that we can write

$$\Delta G = I_D - E_A - E_{00} - e^2/r \tag{4.10}$$

where E_{00} is the singlet excitation energy. The condition for exciplex formation is that ΔG be negative.

Because a molecule in its excited state is both a better electron donor and a better electron acceptor than in its ground state, exciplex formation is expected to be a widespread phenomenon.

It is also expected on this model (see Figures 4.6 and 4.7) that, neglecting solvation effects,

$$h\nu_{\text{exciplex}} = I_D - E_A + \text{constant} \tag{4.11}$$

where ν_{exciplex} is the frequency of the exciplex emission peak. Knibbe et al.[11] found that for a series of aromatic hydrocarbon acceptors with a common donor, dimethylaniline, the exciplex fluorescence maximum obeyed the relation (4.12).

$$h\nu_{\text{exciplex}} = 1{\cdot}17 - 0{\cdot}65(E_{\text{A} -/\text{A}})\,(\text{eV}) \tag{4.12}$$

Since the reduction potential $E_{A\text{ -}/A}$ is linearly related to the electron affinity E_A of the acceptor, these results support the model and clearly demonstrate the charge–transfer character of the exciplex.

It should be recognized that exciplex formation is not restricted to aromatic systems and that there is no requirement that exciplexes should necessarily luminesce. If the sum of the rate constants for radiationless processes involving the exciplex (see Figure 4.1) is sufficiently high, the lifetime of the exciplex will be so short that its emission may be indetectable. Similarly, very short-lived non-luminescent exciplexes may be generated if their binding energy is $< kT$ (where k is Boltzmann's constant). Since exciplexes/excimers do not exist in the ground state, direct evidence for their formation can only be obtained from their emission, or perhaps from their absorption, characteristics following their generation by a high-intensity light flash.

Exciplexes or excimers, though difficult to detect, seem to be implicated in many photochemical processes, e.g.

(i) Photodimerization of polyacenes[12] (4.13).

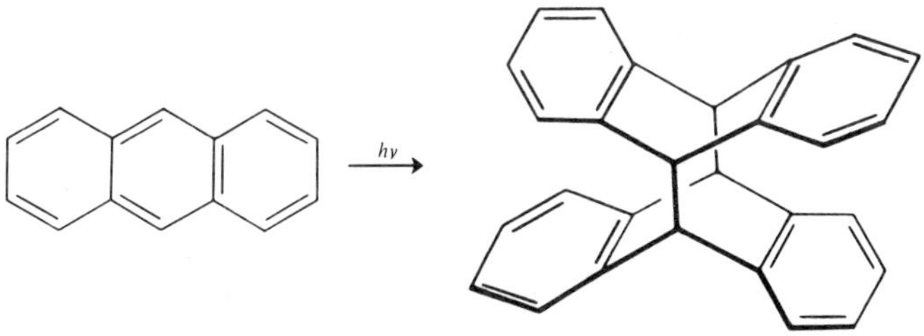

$$(4.13)$$

(ii) Photoaddition of ketones to electron-deficient olefins[13] (4.14).

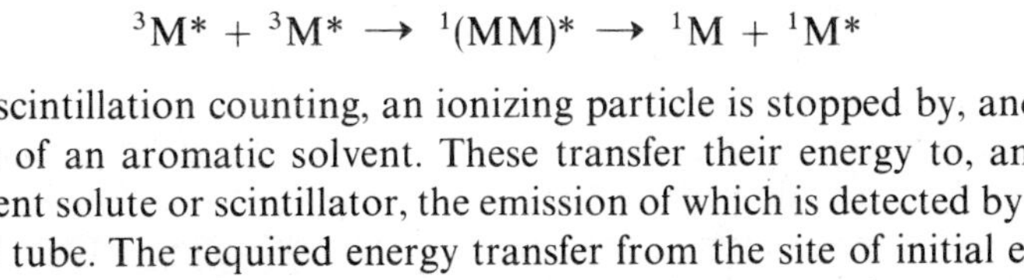

$$(4.14)$$

exciplex

(iii) Triplet–triplet annihilation (4.15, see Chapter 3, p. 83).

$$^3M^* + {}^3M^* \rightarrow {}^1(MM)^* \rightarrow {}^1M + {}^1M^* \qquad (4.15)$$

(iv) In scintillation counting, an ionizing particle is stopped by, and excites, molecules of an aromatic solvent. These transfer their energy to, and excite, a fluorescent solute or scintillator, the emission of which is detected by a photo-multiplier tube. The required energy transfer from the site of initial excitation to the scintillator could occur by non-radiative energy transfer (see Chapter 4, p. 121), or alternatively by the reversible formation of excimers derived from the solvent (4.16).

$$^1M^* + {}^1M \rightarrow {}^1(MM)^* \rightarrow {}^1M + {}^1M^* \qquad (4.16)$$

(v) Exciplexes are also probably involved in many quenching processes (see below).

THE KINETICS OF QUENCHING

Because kinetics are basic to any real understanding of quenching phenomena, a simple treatment is now given (the more complex aspects are deferred until Chapter 5, section 5.2.2). Consider the situation of (4.17), where an excited species fluoresces, decays non-radiatively or is quenched, with rate constants k_f, k_D and k_q respectively.

$$
\text{M} \xrightarrow[I]{h\nu} \text{M*}
\begin{array}{c}
\xrightarrow{k_f} \text{M} + h\nu \\
\xrightarrow{k_D} \text{M} \\
\xrightarrow[Q]{k_q[Q]} \text{M} + \text{Q}
\end{array}
\tag{4.17}
$$

Applying the steady-state approximation to the concentration of M* gives:

$$
I = [\text{M*}](k_f + k_D + k_q[\text{Q}])
$$

$$
\therefore \quad \phi_f = \frac{k_f[\text{M*}]}{I} = \frac{k_f}{k_f + k_D + k_q[\text{Q}]}
$$

The quantum yield of fluorescence in the absence of quencher is given by:

$$
\phi_f^0 = \frac{k_f}{k_f + k_D}
$$

Hence

$$
\frac{\phi_f^0}{\phi_f} = \frac{k_f + k_D + k_q[\text{Q}]}{k_f + k_D} = 1 + \frac{k_q[\text{Q}]}{k_f + k_D}
$$

and

$$
\frac{\phi_f^0}{\phi_f} = 1 + k_q\tau[\text{Q}]
\tag{4.18}
$$

where τ is the lifetime in the absence of quencher (see Chapter 3, equation 3.1). Equation (4.18) is the Stern–Volmer equation, which shows that under ideal conditions a plot of ϕ_f^0/ϕ_f against the quencher concentration gives a straight line of gradient $k_q\tau$. Hence if τ is known, then the quenching rate constant is immediately obtained.

For many systems, k_q in mobile solvents is of the order of 10^9–$10^{10}\,\text{l}\,\text{mol}^{-1}\,\text{s}^{-1}$, i.e. close to the diffusion-controlled rate constant k_{diff}. This suggests that in these cases quenching is so rapid that the rate-determining step is the diffusion of the quencher to within the 'active sphere' of M*. The position is best examined quantitatively. Suppose that bimolecular quenching cannot occur until Q and M* have formed an encounter complex/exciplex (4.19).

$$
\text{M*} + \text{Q} \underset{k_{-1}}{\overset{k_{\text{diff}}}{\rightleftharpoons}} (\text{M*} \cdots \text{Q}) \xrightarrow{k_c} \text{M} + \text{Q}
\tag{4.19}
$$

Here, k_c is the actual quenching rate constant within the complex. Then, under steady-state conditions,

$$
k_{\text{diff}}[\text{M*}][\text{Q}] = [(\text{M*} \cdots \text{Q})](k_c + k_{-1})
$$

and

$$\text{rate of quenching} = k_c[(M^* \cdots Q)]$$

$$= \frac{k_c k_{\text{diff}}}{k_c + k_{-1}}[M^*][Q]$$

$$\therefore \quad k_q \text{ (observed)} = \frac{k_c k_{\text{diff}}}{k_c + k_{-1}} \tag{4.20}$$

There are three cases of interest:

(a) If $k_c \gg k_{-1}$, then k_q (obs) $\sim k_{\text{diff}}$ and will be dependent upon solvent viscosity;

(b) If $k_c \ll k_{-1}$, then k_q (obs) $\sim k_c \cdot k_{\text{diff}}/k_{-1} = k_c K$ where K is the equilibrium constant for the formation of the complex. This means that with weak quenchers k_q (obs) will be independent of solvent viscosity, as are the rate constants of ordinary bimolecular ground state reactions;

(c) If k_c and k_{-1} are of the same order, then k_q (obs) will be less than k_{diff}.

Because diffusion dominates bimolecular quenching processes in solution, it is important to have a reliable method of estimating k_{diff}. One commonly used relation is the Debye equation (4.21).

$$k_{\text{diff}} = \frac{8RT}{3000\eta}(\text{l mol}^{-1}\text{ s}^{-1}) \tag{4.21}$$

R is the gas content, T is the temperature, and η is the viscosity (in poise). Although the derivation of this equation depends upon a number of simplifying assumptions (e.g., that the diffusing species are spherical, of the same diameter, with the same interaction radii, and that the microscopic and macroscopic viscosities of the solvent are the same), it succeeds reasonably well in predicting the maximum values of k_q in a variety of solvents. However, critical tests of its validity reveal deficiencies. For example, k_q for quenching the fluorescence of a number of polynuclear aromatic hydrocarbons by oxygen is found[14] to be several times as large as k_{diff} calculated from equation (4.21). It seems that this equation fails when the solute molecules differ considerably in size or when they are small by comparison with solvent molecules. Hence, for the quenching of naphthalene phosphorescence by 1-iodonaphthalene the ratio k_q/k_{diff} rises to 4·5 in liquid paraffin.[15] Even for molecules of comparable size in normal solvents, equation (4.21) tends to underestimate k_{diff}, and the modified relation (4.22) seems to be more satisfactory.[15]

$$k_{\text{diff}} = \frac{8RT}{2000\eta} \tag{4.22}$$

4.5 QUENCHING PROCESSES AND QUENCHING MECHANISMS

There are many quenching processes, and Figure 4.8 illustrates a possible classification scheme.

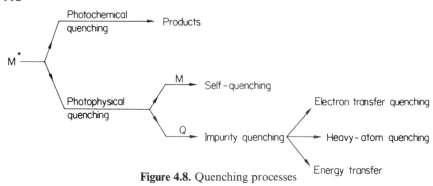

Figure 4.8. Quenching processes

Quenching by photochemical reaction forms the subject matter of the later chapters of this book and will not be discussed here. Photophysical quenching, which does not lead to new ground state products, can be divided into self-quenching (see excimers, section 4.2), in which the quenching species is M, and 'impurity' quenching, where the quencher is some other chemical species. This last category can be further split into quenching by electron transfer, by heavy-atom effects and by energy transfer, and these three processes will now be analysed.

4.5.1   Electron Transfer Quenching

The emission by an exciplex of an aromatic hydrocarbon with an amine in solvents of increasing polarity is observed not only to shift to the red but also to diminish in intensity, becoming zero in solvents with high dielectric constant such as acetonitrile.[10] Flash photolysis experiments in acetonitrile show the transient generation of the hydrocarbon radical anion and the amine radical cation, and this implies the total transfer of an electron from donor to acceptor. Similarly the products derived from the irradiation of naphthalene with tri-ethylamine in acetonitrile indicate[16] that electron transfer is a primary process (4.23).

$$\left[\text{naphthalene}\right]^* + Et_3\ddot{N} \longrightarrow \left[\text{naphthalene}\right]^{\overset{.}{-}} + Et_3N^{\overset{+}{\cdot}} \xrightarrow{H_2O}$$

$$\xrightarrow{} \xrightarrow[H^+]{-e^-}$$

$$\tag{4.23}$$

The simplest hypothesis is that in polar solvents the exciplex dissociates into solvated ions (4.24).

$$^1(AD)^* \rightarrow A_s^{\pm} + D_s^{\pm}$$ (4.24)

However, it has been shown that, for the pyrene–dimethylaniline system, increasing the dielectric constant of the solvent reduces the lifetime of the exciplex very much less than it reduces the intensity of exciplex fluorescence. These facts are explained by postulating a competing electron transfer process occurring in an encounter complex between $^1A^*$ and D, giving rise to solvated ion pairs (4.25).

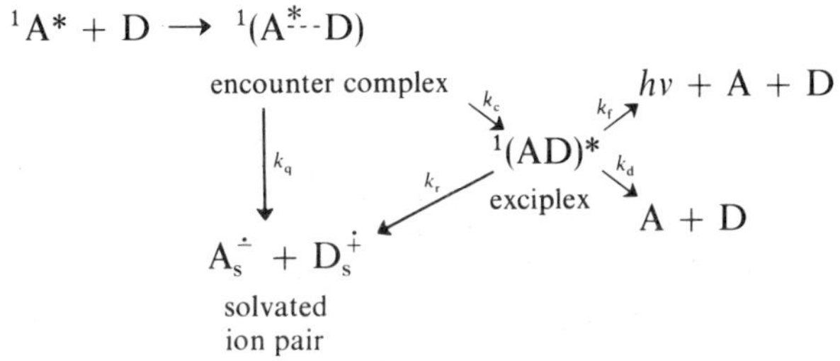

k_q and k_r will both increase with solvent polarity, but k_c will be independent of this parameter. Thus the exciplex lifetime, $\tau_e = (k_f + k_d + k_r)^{-1}$ will decrease in polar solvents. However, the quantum yield for exciplex fluorescence, which equals the probability of exciplex formation multiplied by $k_f\tau_e$, is given by (4.26).

$$\phi_f = \frac{k_c}{k_c + k_q} \cdot k_f\tau_e$$ (4.26)

Hence ϕ_f, the product of two solvent-dependent terms, will decrease with solvent polarity more rapidly than will τ_e.

The distance between the partners in an encounter complex is estimated to be ~ 7 Å or more, considerably larger than that obtaining in excimers/exciplexes, and this implies an ill-defined mutual orientation of the components and perhaps requires little more than orbital overlap somewhere in the complex. Whether such entities are to be regarded as exciplexes is a matter of definition.

Further evidence relating to the role of encounter complexes in electron transfer quenching was obtained by Rehm and Weller,[17] who proposed the scheme (4.27) to account for the kinetics of the quenching of a number of polynuclear aromatics by electron donors in acetonitrile.

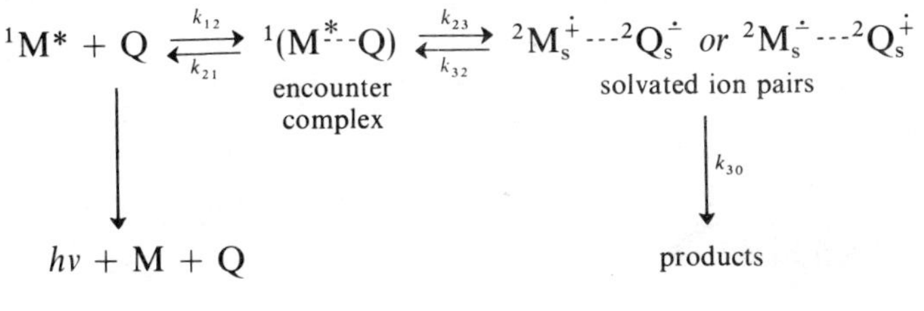

Given the linear relationship between ionization potentials, electron affinities and polarographic redox potentials, the free energy of the electron transfer process can be seen from Figure 4.6 to be:

$$\Delta G_{23} = E_{(D/D^+)} - E_{(A^-/A)} - \frac{e^2}{\varepsilon a} - \Delta E_{00} \qquad (4.28)$$

where the first two terms are redox potentials, the third term is the Coulomb energy released in bringing the ions to within the encounter distance a, ε is the dielectric constant of the medium, and ΔE_{00} is the electronic excitation energy. A kinetic analysis[17] of scheme (4.27) gives the quenching rate constant as:

$$k_q = \frac{2.0 \times 10^{10}}{1 + 0.25[\exp(\Delta G_{23})/RT + \exp(\Delta G_{23}^{\ddagger})/RT]}(l\ mol^{-1}\ s^{-1}) \qquad (4.29)$$

where $\Delta G_{23}^{\ddagger}$, which can be calculated from ΔG_{23}, is the energy of activation for the formation of the solvated ion pairs from the encounter complex. Thus k_q can be calculated from spectroscopic and electrochemical data. The experimental data shown in Figure 4.9 conform to a relationship of the type in equation (4.29), since $k_q \approx k_{diff}$ when ΔG_{23} is more exothermic than 20–40 kJ mol^{-1} (5–10 kcal mol^{-1}), then k_q falls as ΔG_{23} becomes less exothermic, and finally k_q becomes proportional to $\exp(-\Delta G_{23}/RT)$ when ΔG_{23} becomes endothermic by more than 20 kJ mol^{-1} (5 kcal mol^{-1}). The results are incompatible with electron transfer quenching within an exciplex.

The quenching of the singlet states of aromatic hydrocarbons by dienes and by quadricyclane (4.30) and of azo-compounds (e.g. 4.31) by dienes and olefins

(4.30)   (4.31)

has been shown not to be due to energy transfer nor to chemical quenching. In these systems, k_q is rather insensitive to solvent polarity, but is affected by substitution which inhibits close approach of the quencher. The quenching seems to be due to exciplex formation, even though exciplex emission is not observed. For example, the quenching of aromatic hydrocarbons by quadricyclane gives the results[18] shown in Figure 4.10. This clearly establishes electron transfer as the mechanism of quenching, because, from equation (4.28), for a constant donor and solvent and assuming a constant separation of the components of the encounter complex, the energy for electron transfer may be written as

$$\Delta G = constant - [^1\Delta E_A + E_{(A^-/A)}] \qquad (4.32)$$

where $^1\Delta E_A$ is the excitation energy of the acceptor.

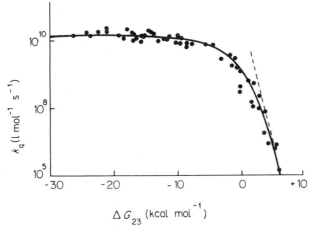

Figure 4.9. Relationship between the rate constant for quenching (k_q) of several polynuclear aromatics by electron donors, and the energy for the electron transfer process (ΔG_{23}). (From D. Rehm and A. Weller, *Ber. Bunsenges. Phys. Chem.*, **73**, 836 (1969), reproduced by permission of Verlag Chemie Gmbh.)

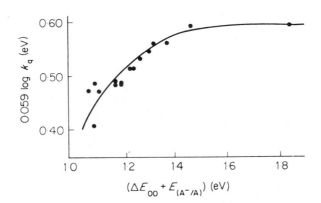

Figure 4.10. Correlation between rate constants for fluorescence quenching (k_q) by quadricyclane and electron affinities of the excited molecules (from B. S. Solomon, C. Steel and A. Weller, *Chem. Commun.*, 927 (1969), reproduced by permission of the Chemical Society)

Similarly, Evans[19] established for the quenching of the fluorescence of azo-compound (4.31) by dienes, and of naphthalene by dienes and olefins, a linear relationship between the ionization potential of the quencher and $\ln[k_q/(k_{diff} - k_q)]$, where k_q is the Stern–Volmer quenching rate constant, and further showed that such a dependence was to be expected if electron transfer were the rate-limiting step.

The current position can be summarized as follows. Electron transfer has been clearly established as an important process for the quenching of singlet excited states, but doubt exists as to the exact nature of the entities involved in the electron transfer. They may be exciplexes or encounter/collision complexes. In polar solvents electron transfer is complete and leads to radical ions. In non-polar solvents, where there is no evidence for the formation of products as a result of the quenching step, electron transfer may be incomplete, giving rise to an exciplex which may relax either by fluorescence or radiationlessly by a 'return' of the partially transferred electron to regenerate M and Q in their S_0 or T_1 states. The last cases may be thought of as exciplex-induced internal conversion or intersystem crossing.

4.5.2   Heavy Atom Quenching

Molecular fluorescence is quenched by the presence of species containing heavy atoms, and it seems that the phenomenon is due to the formation of a singlet exciplex (or encounter complex) which, because of the heavy atom effect, undergoes enhanced intersystem crossing to the triplet exciplex, followed by dissociation into its components (4.33).

$$^1M^* + Q \rightarrow {}^1(MQ)^* \xrightarrow{isc} {}^3(MQ)^* \rightarrow {}^3M^* + Q \qquad (4.33)$$

Since the exciplexes elude detection, what is observed in such systems is:

$$^1M^* + Q \rightarrow {}^3M^* + Q$$

Much evidence in support of this concept has been provided by Wilkinson and his co-workers,[20] who investigated the quenching of the fluorescence of several aromatic hydrocarbons by xenon† and by various bromine- and iodine-containing molecules and established that the kinetics of quenching were consistent with the above mechanism. Particularly important was the demonstration, by monitoring the intensity of triplet–triplet absorption of $^3M^*$ following flash excitation, that the effect of heavy atom quenchers is to increase the concentration of $^3M^*$.

It should be recognized that intersystem crossing can occur in exciplexes, as in other systems, even in the absence of heavy atoms.

4.5.3   Quenching by Oxygen[21] and Paramagnetic Species[22]

The quenching of the excited states of many organic molecules by oxygen is diffusion-controlled. With respect to singlet states, the process seems to depend upon a catalysed intersystem crossing (4.34).

$$^1M^* + {}^3O_2 \rightarrow {}^3M^* + {}^3O_2 \qquad (4.34)$$

Whereas intersystem crossing is spin-forbidden, the bimolecular process (4.34)

---

† Xenon is an ideal heavy atom quencher, since it is chemically inert, optically transparent and readily soluble in organic solvents.

is allowed† under the Wigner spin-conservation rules (see p. 101). It may be thought of as occurring *via* the sequence (4.35).

$$^1M^* + {}^3O_2 \xrightarrow{k_{diff}} {}^3(MO_2)_n^* \xrightarrow{ic} {}^3(MO_2)_1^* \xrightarrow{\text{spin-allowed}} {}^3M^* + {}^3O_2 \qquad (4.35)$$

highly excited   lowest excited
triplet exciplex   triplet exciplex

The quenching of triplet states, which could also occur by catalysed inter-system crossing to the ground state (4.36), actually seems to be dependent upon triplet energy transfer (see p. 125) generating singlet oxygen (4.37).

$$^3M^* + {}^3O_2 \rightarrow {}^3(MO_2)_n^* \rightarrow {}^3(MO_2)_1^* \xrightarrow{\text{spin-allowed}} {}^1M + {}^3O_2 \qquad (4.36)$$

$$^3M^* + {}^3O_2 \xrightarrow{\text{spin-allowed}} {}^1M + {}^1O_2^* \qquad (4.37)$$

The free radical nitric oxide (^2NO) is also a highly efficient quencher of excited singlet states. By analogy with oxygen, it probably quenches by en-hancing the rate of intersystem crossing (4.38).

$$^1M^* + {}^2NO \rightarrow {}^2(M.NO)_n^* \rightarrow {}^2(M.NO)_1^* \xrightarrow{\text{spin-allowed}} {}^3M^* + {}^2NO \qquad (4.38)$$

4.5.4   **Electronic Energy Transfer**[24]

In this phenomenon, which is of crucial importance in photochemistry, an excited donor molecule D* collapses to its ground state with the simultaneous transfer of its electronic excitation energy to an acceptor molecule A which is thereby promoted to an excited state (4.39).

$$D^* + A \rightarrow D + A^* \qquad (4.39)$$

It should be noted that the acceptor can itself be an excited state, as in triplet–triplet annihilation (4.40, see Chapter 3, p. 83).

$$^3M^* + {}^3M^* \rightarrow {}^1M + {}^1M^* \qquad (4.40)$$

What is observed in an energy transfer experiment is the quenching of the emission (or photochemistry) associated with D* and its replacement by the emission (or photochemistry) characteristic of A*. Hence, although the photons are absorbed by D, it is A which becomes excited. The processes resulting from A* generated in this manner are said to be *sensitized*. When the donor and acceptor are identical, the term energy migration is used, i.e. for:

$$M^* + M \rightarrow M + M^*$$

The principal mechanisms of electronic energy transfer are set out in Figure 4.11.

---

† The treatment of the Wigner spin rules in ref. 23, though graphic, is incorrect. It predicts that the processes (4.34) and (4.36) are spin-forbidden.

118

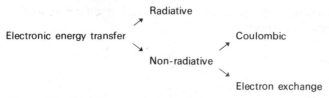

Figure 4.11. Principle mechanisms of electronic energy transfer

Although energy transfer can occur between other modes (translational, vibrational, rotational), the prime concern in this chapter will be with transfer of electronic energy. Radiative energy transfer depends on the capture, by the acceptor, of photons emitted by the donor (4.41).

$$D^* \rightarrow D + h\nu$$
$$h\nu + A \rightarrow A^*$$
(4.41)

Clearly, such energy transfer can occur over immense distances: photochemical and photobiological changes occurring on Earth under the influence of sunlight are extreme examples of long-range radiative energy transfer. Non-radiative energy transfer induced by Coulombic interactions is also long-range in that it can occur over distances (50 Å or more) much larger than molecular dimensions. The electron exchange mechanism requires close approach (~ 10–15 Å), but not necessarily the contact, of the donor and acceptor species.

Spin conservation and other factors impose restrictions upon the relative multiplicities of the initial and final excited species involved in the electronic energy transfer, and the only common situations in organic photochemistry are those of (4.42).

Singlet–singlet energy transfer: $\quad {}^1D^* + {}^1A \quad \rightarrow \quad {}^1D + {}^1A^*$

Triplet–singlet energy transfer: $\quad {}^3D^* + {}^1A \quad \rightarrow \quad {}^1D + {}^1A^*$

Triplet–triplet energy transfer: $\quad {}^3D^* + {}^1A \quad \rightarrow \quad {}^1D + {}^3A^*$
(4.42)

Oxygen quenching: $\quad {}^3D^* + {}^3O_2 \rightarrow {}^1D + {}^1O_2^*$

These phenomena will now be discussed in detail.

Radiative Energy Transfer

Since the process requires that the acceptor absorb photons emitted by the donor, it is obvious that the probability (rate) of transfer will depend upon (i) the quantum efficiency of emission by the donor θ_E, (ii) the number of acceptor molecules in the path of the emitted photon, (iii) the light-absorbing power of the acceptor, and (iv) the extent of the overlap between the emission spectrum of the donor and the absorption spectrum of the acceptor. This last requirement is expressed mathematically by the integral:

$$\int_0^\infty F_D(\bar{\nu})\varepsilon_A(\bar{\nu})\, d\bar{\nu}$$
(4.43)

where $F_D(\bar{\nu})$ is the emission spectrum of D*, and $\varepsilon_A(\bar{\nu})$ is the absorption spectrum of the acceptor A, both plotted on a wavenumber scale (see Figure 4.12).

The probability of radiative transfer, P_{rt}, in homogeneous solution can therefore be expressed as (4.44),

$$P_{rt} \propto \frac{[A]l}{\theta_E} \int_0^\infty F_D(\bar{v})\varepsilon_A(\bar{v}) \, d\bar{v} \qquad (4.44)$$

where l is the distance over which energy transfer occurs.

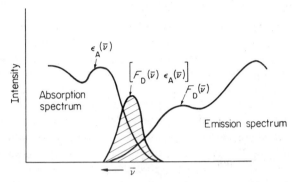

Figure 4.12. The shaded area is the integral $\int_0^\infty F_D(\bar{v})\varepsilon_A(\bar{v}) \, d\bar{v}$

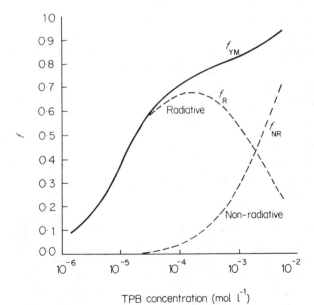

TPB concentration (mol l^{-1})

Figure 4.13. Singlet–singlet energy transfer from p-terphenyl (2.17×10^{-2} M) to tetraphenylbutadiene (TPB) in toluene solutions. Energy transfer quantum efficiency f_{YM} against TPB concentration. Radiative component, f_R; radiationless component, f_{NR}. (From J. B. Birks, *Photophysics of Aromatic Molecules*, (1970), John Wiley and Sons Ltd.)

120

The efficiency of such energy transfer depends also on the shape and size of the vessel used in the experiment, since photons emitted near the walls have less chance of capture by the acceptor than those emitted in the centre of the cell.

Radiative energy transfer, though frequently described as the 'trivial' mechanism because of its conceptual simplicity, is far from being such. It may be the dominant mechanism of energy transfer in dilute solutions (Figure 4.13), because its probability falls off only relatively slowly with dilution.

Energy migration can also occur radiatively whenever there is overlap of absorption and emission spectra. Such migration is manifested by changes in the fluorescence emission spectrum in the region where the fluorescence and absorption spectra overlap (Figure 4.14), due to the preferential reabsorption of light of these wavelengths. Emission spectra should therefore always be recorded on very dilute solutions. Radiative energy transfer to ground state species is restricted to the singlet–singlet and the triplet–singlet cases where there is no change in the multiplicity of the acceptor, because the extinction coefficient of singlet–triplet absorption is so small that the integral (4.43) is microscopic.

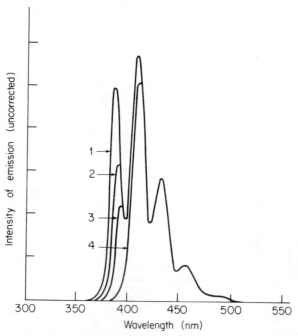

Figure 4.14. Change in the observed fluorescence spectra of anthracene in benzene solution at 20 °C due to reabsorption of the short wavelength band at higher concentrations (concentration increases 1 → 2 → 3 → 4). (From F. Wilkinson, in E. J. Bowen (ed.), *Luminescence in Chemistry*, (1968), reproduced by permission of Van Nostrand Reinhold Co. Ltd.)

Non-radiative Energy Transfer

If an acceptor molecule A has no interaction with a proximate excited donor molecule D*, then each, being 'unaware' of the other, will be in a state described by the appropriate Schrödinger equation. However, if there is an interaction between D* and A, however small, described by a perturbation Hamiltonian operator H', then the system (D* + A) is no longer a 'stationary' state of the total Hamiltonian $(H_{D*} + H_A + H')$. If there is an *isoenergetic* system (D + A*), in which the excitation now resides in the acceptor, then time-dependent perturbation theory (see Chapter 2) reveals that the effect of the interaction H' is to cause the system (D* + A) to evolve into the system (D + A*), and vice versa, with a probability given[25] by (4.45),

$$P \propto \rho \langle \psi_i | H' | \psi_f \rangle^2 \tag{4.45}$$

where ρ is the density of isoenergetic states, ψ_i describes the initial system (D* + A), and ψ_f describes the final system (D + A*).

In other words, the perturbation induces coupled radiationless transitions in both D and A so that excitation energy is shuttled back and forth between D and A and the rate of energy transfer (k_{ET}) is determined, according to equation (4.45), by the strength of the interaction H' between the components. If the coupling is sufficiently strong, as in some crystals, k_{ET} is so large that it is no longer possible to think of the excitation as being even temporarily localized on D or A, and it becomes associated with the system as a whole. This case, which requires that the absorption spectrum of the system (D + A) is not the sum of those of its components, will not be considered further.

We shall rather be concerned with the situation where H' is sufficiently small that $k_{ET} < k_{VR}$, the rate constant for vibrational relaxation. Under these circumstances, (i) D* will undergo energy transfer from its bottom vibrational level, (ii) A*, when formed, will promptly collapse to its bottom vibrational level, thereby destroying the degeneracy of (D* + A) and (D + A*) and making the energy transfer unidirectional, and (iii) k_{ET} will be given by the rate constant for forward transfer. This situation is pictorialized in Figure 4.15. The required degeneracy of the two systems (D* + A) and (D + A*) is readily achieved for organic molecules because of the availability of a multitude of vibrational and rotational sub-levels. Examination of Figure 4.15 also shows that the condition for irreversible energy transfer is that the $0 \rightarrow 0$ excitation energy of D (E_D) shall be significantly greater than that of A (E_A).

The state density factor ρ in equation (4.45) is a measure of the number of coupled isoenergetic donor and acceptor transitions. For organic molecules, which in solution exhibit broad spectra, ρ can be estimated from the overlap between the emission spectrum of the donor and the absorption spectrum of the acceptor, expressed in energy (i.e., wavenumber) units, or in other words by the integral (equation 4.43). It must be emphasized that although the rates of both radiative and non-radiative energy transfer depend upon this integral, the mechanisms are totally different. For the latter, energy transfer occurs before

122

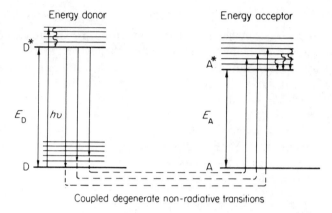

Figure 4.15. Energy level scheme showing the coupling of isoenergetic donor and acceptor transitions necessary for non-radiative energy transfer

emission takes place, and a physical interaction between D* and A is a pre-requisite.

The effect of the coupling between D* and A, represented by the matrix element $\langle \psi_i | H' | \psi_f \rangle$ in equation (4.45), will now be considered. The perturbation H' will contain several terms, among which the most important are electrostatic (Coulombic) and electron exchange interactions, each independently capable of inducing energy transfer.

Coulombic Interaction

This interaction can be expressed as a number of terms, dipole–dipole, dipole–quadripole, multipole–multipole, of decreasing significance. The contribution of the normally dominant dipole–dipole interaction to energy transfer has been treated by Förster, who derives the equation (4.46),

$$k_{ET} = 1.25 \times 10^{17} \left(\frac{\phi_E}{n^4 \tau_D r^6} \right) \int_0^\infty F_D(\bar{v}) \varepsilon_A(\bar{v}) \frac{d\bar{v}}{\bar{v}^4} \qquad (4.46)$$

where ϕ_E is the quantum yield for donor emission, τ_D is the lifetime of the emission, n is the solvent refractive index, and r is the distance in nm between D* and A. $F_D(\bar{v})$ is the emission spectrum of the donor, expressed in wavenumbers and normalized to unity (i.e., $\int_0^\infty F_D(\bar{v}) \, d\bar{v} = 1$), and $\varepsilon_A(\bar{v})$ is the decadic molar extinction coefficient of A at the wavenumber $\bar{v}$.

The distance between D* and A at which energy transfer to A and internal deactivation of D* are equally probable is known as the *critical transfer distance* R_0. Substituting $k_{ET} = \tau_D^{-1}$ into equation (4.46) gives:

$$R_0^6 = 1.25 \times 10^{17} \frac{\phi_E}{n^4} \int_0^\infty F_D(\bar{v}) \varepsilon_A(\bar{v}) \frac{d\bar{v}}{\bar{v}^4} \qquad (4.47)$$

Introduction of reasonable numerical values into these equations leads to the expectation that k_{ET} can be much larger than k_{diff} and that energy transfer by dipole–dipole interaction can be significant even at distances of the order of 100 Å (10 nm). These expectations have been amply confirmed, as Table 4.1 shows. It is also found, as expected, that k_{ET} is independent of solvent viscosity, except at very low concentrations ($c.$ 10^{-4} M) when molecules must diffuse in order to come within the critical transfer distance.

Table 4.1. Long-range Coulombic energy transfer

Donor	Acceptor	R_0 (Å) Theor.	R_0 (Å) Expt.	$10^{-10} k_{ET}$ (l mol^{-1} s^{-1}) Theor.	$10^{-10} k_{ET}$ (l mol^{-1} s^{-1}) Expt.
(a) Anthracene (S_1)	Perylene	31	54	2·3	12
(a) Perylene (S_1)	Rubrene	38	65	2·8	13
(a) 9,10-Dichloro-anthracene (S_1)	Perylene	40	67	1·7	8·0
(a) Anthracene (S_1)	Rubrene	23	39	0·77	3·7
(b) Phenanthrene-d_{10} (T_1)	Rhodamine B	45	47		
(b) Phenanthrene-d_{10} (T_1)	Phenanthrene-d_{10} (T_1)	40	35		
(c) p-Phenylbenzalde-hyde (T_1)	Chrysoidin	32	33		
(c) Triphenylamine (T_1)	Chrysoidin	34	52		
(c) Triphenylamine (T_1)	Fuchsin	29	37		

Solvents: (a) benzene at room temperature; (b) cellulose acetate at 77 K; (c) ethanol or dibutyl ether at 77 K. In benzene at room temperature, $k_{diff} \sim 10^{10}$ l mol^{-1} s^{-1}. Data taken from Wilkinson et al., ref 24(a), p. 252, and ref. 26.

The only transfer processes allowed by the Coulombic interaction are those in which there is no change in spin in either component. Thus

$$^1D^* + {}^1A \rightarrow {}^1D + {}^1A^*$$

and

$$^1D^* + {}^3A \rightarrow {}^1D + {}^3A^*(T_n)$$

are fully allowed. Triplet–singlet transfer

$$^3D^* + {}^1A \rightarrow {}^1D + {}^1A^*$$

is forbidden. Nevertheless it is observed (see Table 4.1). This is because the spin-forbidden nature of the $^3D^* \rightarrow {}^1D$ transition, whilst strongly reducing k_{ET}, also so prolongs the lifetime of $^3D^*$, particularly in a rigid medium, that the probability of energy transfer can still be high compared with the probability of deactivation of $^3D^*$. For similar reasons, triplet–triplet annihilation has been observed[27] to occur over long distances (~ 40 Å) in cellulose acetate films. Long-range energy transfer involving a change in the multiplicity of the acceptor is, of course, not expected to occur.

This discussion has so far centred on the dipole–dipole term in the Coulombic contribution to H'. Dexter[25] has analysed the effect of multipole–multipole

interactions, and in particular has shown that for dipole–quadrupole interactions

$$k_{ET} \propto r^{-8}$$

Since this rate constant falls off with distance much more rapidly than that for
dipole–dipole transfer, it is likely to be important only at distances $\ll 40\,\text{Å}$
and in those cases where the long-range dipole–dipole transfer is inhibited
because $\varepsilon_A(\bar{v})$ is small. Note that for Coulombic interactions, k_{ET} is independent
of the strength of the optical transition involving D, because the $F_D(\bar{v})$ term in
equation (4.46) is normalized to unity.

Electron Exchange Interaction

The perturbation H' responsible for energy transfer can include electron
exchange terms in addition to the Coulombic terms treated above. The effect
of such terms was also analysed by Dexter,[25] who derived the relation (4.48),

$$k_{ET} \propto e^{-2R/L} \int_0^\infty F_D(\bar{v})\varepsilon_A(\bar{v})\,d\bar{v} \tag{4.48}$$

where R is the distance between D* and A, and L is a constant. $F_D(\bar{v})$ and $\varepsilon_A(\bar{v})$
are the emission and absorption spectra of D and A respectively *both normalized
to unity*, so that k_{ET} is independent of the oscillator strength of both transitions
(contrast Coulombic interaction).

The negative exponential term in equation (4.48) shows that energy transfer
by this mechanism is a short-range phenomenon, which is to be expected since
the process involves the interchange of electrons between D* and A and therefore overlap of the orbitals of the two components.

The energy transfer can be formulated thus:

$$D^* + A \rightarrow (D\cdots A)^* \rightarrow D + A^* \tag{4.49}$$

Because of the intervention of a bimolecular intermediate, which could be an
exciplex or a collision complex, the energy transfer will be subject to conservation of electron spin. Under Wigner's spin rules (p. 101), both of the processes
(4.50) and (4.51) are allowed collisionally.

$$^3D^* + {}^1A \rightarrow {}^1D + {}^3A^* \tag{4.50}$$

$$^1D^* + {}^1A \rightarrow {}^1D + {}^1A^* \tag{4.51}$$

The former (triplet–triplet energy transfer), though allowed under the exchange interaction, is doubly forbidden under the Coulombic dipole–dipole
interaction, while the latter (singlet–singlet energy transfer) is allowed under
both interactions. These processes will now be considered in more detail.

Singlet–Singlet Collisional Energy Transfer

The enormous values of the rate constants for long-range dipole–dipole energy
transfer and the fact that the rate constant increases with the inverse sixth
power of the distance between the donor and acceptor molecules means that

collisional singlet–singlet energy transfer is likely to be rare and observable only under special conditions. Many of the examples of this phenomenon involve the use of biacetyl as quencher, because the feeble absorption of light by biacetyl over the near ultraviolet and visible regions means that the rate constant for long-range energy transfer will be small (equations 4.44 and 4.46). Exploiting this fact, Dubois and co-workers[28] have demonstrated that the addition of biacetyl to aerated solutions of several aromatic hydrocarbons (benzene, alkyl benzenes and naphthalenes) leads to a quenching of the fluorescence of the hydrocarbon and a sensitization of the fluorescence of biacetyl. The value of k_{ET} obtained from Stern–Volmer analysis was close to k_{diff} calculated from equation (4.22), as expected for exothermic energy transfer.

Triplet–Triplet Energy Transfer (Triplet Transfer)

The existence of this phenomenon was first established by Terenin and Ermolaev,[29] who showed that excitation of benzophenone in a rigid glass at 77 K caused phosphorescence of the substrate which, with increasing concentration of added naphthalene, was progressively quenched and replaced by the phosphorescence emission of naphthalene. The light used selectively excited the benzophenone. Since the energy of S_1-naphthalene is greater than that of S_1-benzophenone, energy transfer could not have occurred at the singlet level. Therefore it must have taken place between the triplet states of the two molecules. In other words, the process shown in equation (4.52) must have occurred.

$$Ph_2CO(S_1) \xrightarrow{isc} Ph_2CO(T_1) + C_{10}H_8(S_0) \rightarrow Ph_2CO(S_0) + C_{10}H_8(T_1) \quad (4.52)$$

Subsequent work[30] has revealed many other examples of the phenomenon in rigid glasses and has established that the energy transfer occurs over a distance of 10–15 Å, comparable with collisional diameters. In 'rigid' glasses it seems that diffusion is rate-determining, for a study[31] of triplet–triplet transfer from phenanthrene-d_{10} to naphthalene-d_8 in hydrocarbon glasses at 77 K shows that $(k_{ET}\eta)$ is a constant over a large range of viscosities, as expected from equations (4.21) and (4.22).

Since biacetyl, exceptionally, phosphoresces strongly in solution, it constitutes a valuable probe for examining triplet–triplet energy transfer in the *fluid* phase. Bäckström and Sandros[32] demonstrated that triplet–triplet transfer occurs from biacetyl as donor to many polynuclear hydrocarbons as acceptors, and they obtained values for both k_{ET} and k_q. Note that $k_q = k_{ET}$ only if the energy transfer is unidirectional. For example, if in the process shown in equation (4.53) the energies of M* and Q* are comparable, then Q*, once formed, can transfer its excitation back to M.

$$M^* + Q \underset{}{\overset{k_{ET}}{\rightleftarrows}} M + Q^* \quad (4.53)$$

The observed value of k_q will therefore be less than k_{ET}. These points are illustrated in Figure 4.16. Notice that when the energy transfer is more exothermic than about 15 kJ mol^{-1} (3–4 kcal mol^{-1}), $k_q \sim k_{ET} \sim k_{diff}$, and that when energy transfer is endothermic, requiring thermal activation, then k_q and

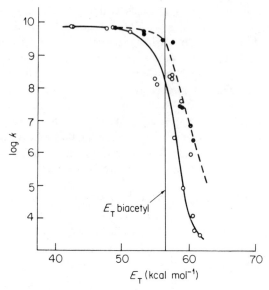

Figure 4.16. Triplet–triplet energy transfer from biacetyl to various acceptors in benzene solutions at room temperature. Rate parameter against triplet energy E_T of acceptor: ○, k_q, observed quenching rate parameter; ●, k_{ET}, energy transfer rate parameter corrected for back transfer. (From A. A. Lamola, *Photochem. Photobiol.*, **8**, 601 (1968), reproduced by permission of Pergamon Press Ltd.)

k_{ET} decrease rapidly, k_q being always less than k_{ET} because of back-transfer. When the triplet energies of M* and Q* are the same, then only the 0—0 bands of the emission spectrum of M* and the absorption spectrum of Q overlap. As predicted by equation (4.48), this leads to a decrease in k_{ET}.

Studies by Porter and Wilkinson[33] on a variety of aromatic hydrocarbons as triplet donors and acceptors, using flash photolysis to monitor donor decay,

Table 4.2. Rate constants for triplet quenching (^3k_q) in hexane solution at room temperature

Donor	Acceptor	Triplet energy gap (kcal mol^{-1}) $E_T(D) - E_T(A)$	^3k_q (l mol^{-1} s^{-1})
Phenanthrene	Iodine	27·7	$1·4 \times 10^{10}$
Triphenylene	Naphthalene	6·3	$1·3 \times 10^9$
Anthracene	Iodine	5·4	$2·4 \times 10^9$
Phenanthrene	1-Iodonaphthalene	3·1	$7·0 \times 10^9$
Phenanthrene	1-Bromonaphthalene	2·6	$1·5 \times 10^8$
Phenanthrene	Naphthalene	0·9	$2·9 \times 10^6$
Naphthalene	Phenanthrene	−0·9	$\leqslant 2 \times 10^4$

established a similar dependence of k_q upon the energy difference between the triplet energies of the donor and acceptor. Some of their data are shown in Table 4.2.

Steric Effects in Collisional Energy Transfer

If the close approach of the partners is essential to collisional energy transfer, then steric effects should be significant, and there are many indications that they are. For example, it has been shown[34] that in the collisional quenching of the fluorescence of the diazabicylooctene (4.54) in solution by the pairs of dienes (4.55), (4.56) and (4.57), the introduction of *gem*-dimethyl groups reduces k_q by a factor of 3–4.

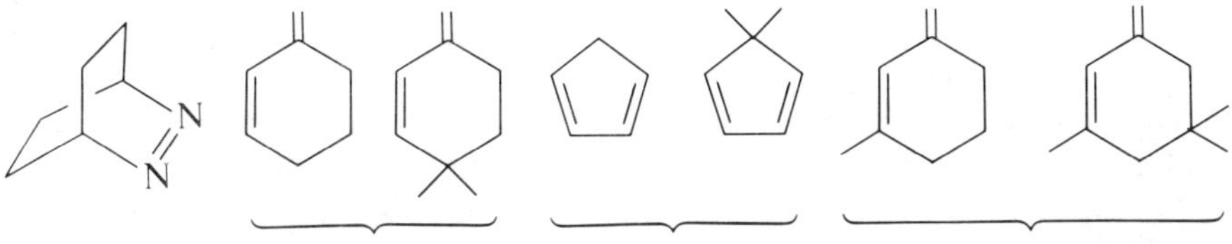

(4.54)   (4.55)   (4.56)   (4.57)

Similarly,[35] in the quenching of benzene fluorescence in the gas phase by simple ketones, increasing α-methyl substitution leads to decreasing values of k_q, as shown in (4.58).

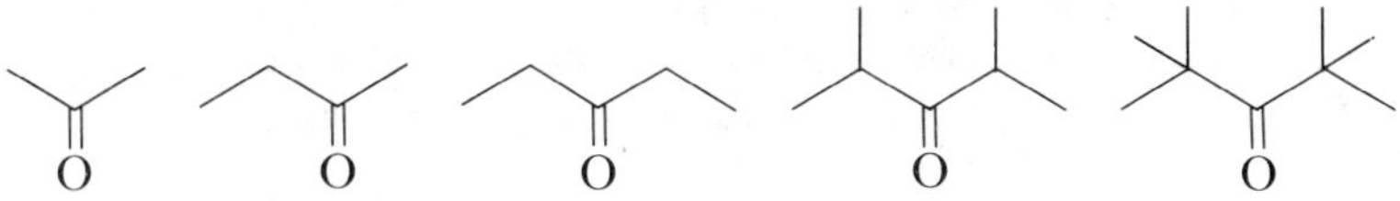

| $10^{10} k_q (l\,mol^{-1}\,s^{-1})$ | 8·3 | ~8·1 | ~9·8 | ~6·2 | ~2·0 |

(4.58)

It is also expected that a chiral sensitizer colliding with a quencher molecule would impose its chirality upon the quencher, leading, in suitable cases, to asymmetric induction and optical activity in the products. This was first demonstrated by Hammond and Cole,[36] who used the optically active amide (4.59) to photosensitize the isomerization of *trans*-1,2-diphenylcyclopropane to the *cis*-isomer (4.60). The recovered *trans*-isomer had a specific rotation of $+28°$, indicating that energy transfer was more efficient to the $(-)$-enantiomer of the *trans*-isomer than to the $(+)$-enantiomer.

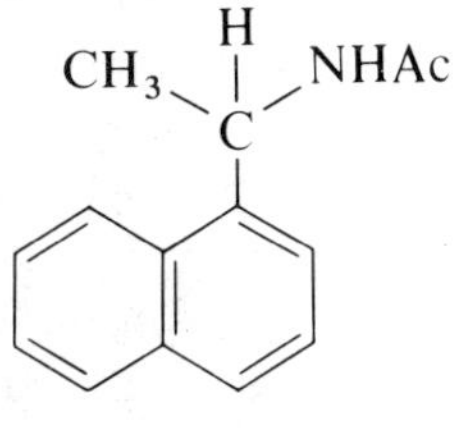

(4.59) 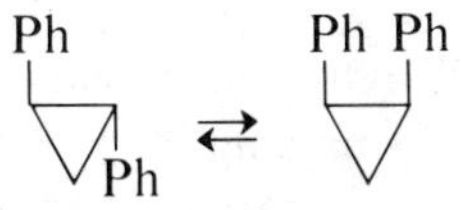   (4.60)

Similarly, it has been shown[37] that the optically active steroid (4.61) induces a partial photoresolution of penta-2,3-diene (4.62), and that the amide (4.59) when excited will induce optical activity[38] in the sulphoxide (4.63).

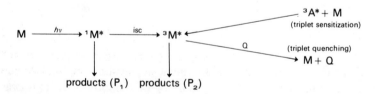

(4.61)   (4.62)   (4.63)

The observed effects are all small, in some cases[39] vanishingly so, implying that the entity in which the energy transfer occurs is a rather loose complex with the components separated by at least a few Å, though these points need clarification. Salem[40] gives a theoretical discussion of photosensitized asymmetric induction.

Photosensitized Processes

It often happens that the S_1 and T_1 states of a molecule have different photochemical reactions. For example, S_1-butadiene cyclizes to cyclobutene but T_1-butadiene dimerizes; β,γ-unsaturated ketones undergo 1,3-acyl migration from the S_1 state but 1,2-migration from the T_1 state.

$$M \xrightarrow{\ h\nu\ } {}^1M^* \xrightarrow{\ isc\ } {}^3M^*$$

$$\begin{array}{c} {}^3A^* + M \\ \text{(triplet sensitization)} \end{array}$$

$$\begin{array}{c} \text{(triplet quenching)} \\ M + Q \end{array}$$

products (P₁)   products (P₂)

Figure 4.17. Selection between the products P_1 and P_2 may be obtained by quenching or sensitizing ${}^3M^*$

Control over the nature of the products can be obtained for such systems by sensitization and/or quenching techniques. Thus for situations represented by the scheme of Figure 4.17, the formation of product P_2 may be suppressed by quenching ${}^3M^*$ with a triplet quencher Q whose triplet energy is less than that of ${}^3M^*$. Conversely, the product P_2 may be obtained free from P_1 by specifically producing ${}^3M^*$ by triplet transfer from a photosensitizer ${}^3A^*$ of higher triplet energy, thereby bypassing ${}^1M^*$. In an analogous way, sensitization and quenching studies are frequently used to define the photochemically reactive states of molecules (see Chapter 5, section 5.1.4).

For an effective triplet sensitization experiment it is essential that the sensitizer absorb all, or nearly all, of the light and that the triplet energy of the sensitizer be greater than that of the substance being sensitized. Such conditions are satisfied when the energy levels are disposed as in Figure 4.18. The filter removes the short-wavelength radiation required to excite the naphthalene $S_0 \rightarrow S_1$ transition and ensures preferential excitation of the benzophenone, which can transfer its triplet excitation to naphthalene.

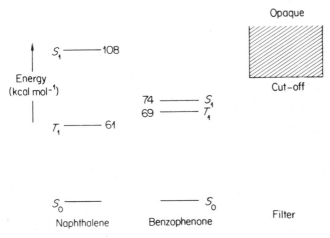

Figure 4.18. Benzophenone as a triplet sensitizer for naphthalene

The desirability of 'bracketing' the S_1 and T_1 levels of the sensitizer by those of the substrate means that ketones, with their small singlet–triplet splitting, high triplet energies and high ϕ_T, make excellent sensitizers.

If the energy levels of the singlet states cannot be disposed as in Figure 4.18, then ultraviolet absorption spectra of the substrate and proposed sensitizer should be examined in order to find a region of the spectrum where the sensitizer absorbs more strongly than the substrate; the irradiation should be conducted at this wavelength and the sensitizer should be used in a sufficiently high concentration to trap most of the available light.

A compilation of triplet energies is given in Table 4.3. Other extensive compilations may be found in references 41 and 42.

Intramolecular Energy Transfer

Intramolecular energy transfer has been observed in many molecules so constructed that the linked D and A moieties cannot collide with each other. Thus Keller[43] has demonstrated complete intramolecular transfer of singlet excitation energy from the naphthalene to the anthrone systems of (4.64) and (4.65), and also transfer of triplet excitation from the anthrone to the naphthalene moiety.

Table 4.3. Triplet energies of sensitizers and quenchers

Compound	E_T kcal(kJ) mol^{-1}	Compound	E_T kcal(kJ) mol^{-1}
Mercury	113 (472)	Isoquinoline	61 (255)
Pyridine	85 (355)	2-Naphthaldehyde	60 (251)
Benzene	84 (351)	2-Naphthyl methyl ketone	59 (247)
Phenol	82 (343)	trans-Piperylene	59 (247)
Phenyl methyl sulphone	82 (343)	p-Terphenyl	59 (247)
Toluene	82 (343)	1-Naphthoic acid	58 (242)
Anisole	81 (339)	Chrysene	57 (238)
1,4-Dichlorobenzene	80 (334)	Picene	57 (238)
Acetone	78 (326)	cis-Piperylene	57 (238)
Aniline	77 (322)	Biacetyl	56 (234)
Benzonitrile	77 (322)	1-Naphthyl methyl ketone	56 (234)
Pyrazine	76 (318)	1-Naphthaldehyde	56 (234)
Benzoin	76 (318)	Coronene	55 (230)
Xanthone	74 (309)	Fluorenone	53 (222)
Diphenylamine	72 (301)	trans-Stilbene	50 (209)
Benzaldehyde	72 (301)	Pyrene	48 (201)
Triphenylamine	70 (293)	Acridine	45 (188)
Benzophenone	69 (288)	Phenazine	44 (184)
Fluorene	68 (284)	Anthracene	43 (180)
Triphenylene	68 (284)	1,4-Diphenylbutadiene	42 (176)
Diphenyl	66 (276)	Thiobenzophenone	40 (167)
m-Terphenyl	65 (272)	9,10-Dichloroanthracene	40 (167)
Thioxanthone	65 (272)	Crystal Violet	39 (163)
Diphenylacetylene	63 (263)	Perylene	36 (150)
Anthraquinone	62 (259)	Naphthacene	29 (121)
Quinoline	62 (259)	Oxygen[a]	23 (96)
Phenanthrene	62 (259)		
Michler's Ketone	61 (255)		
Naphthalene	61 (255)		

[a] Energy of $T_1 \rightarrow S_1$ transition.

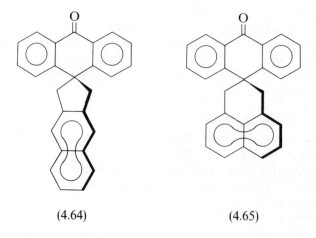

(4.64)   (4.65)

Particularly interesting in this context is the work of Filipescu,[44] who showed that whereas efficient *inter*molecular singlet–singlet energy transfer occurred from (4.66) to (4.67), in (4.68) no *intra*molecular energy transfer, either singlet–singlet or triplet–triplet, could be detected. In (4.68) the two chromophores are

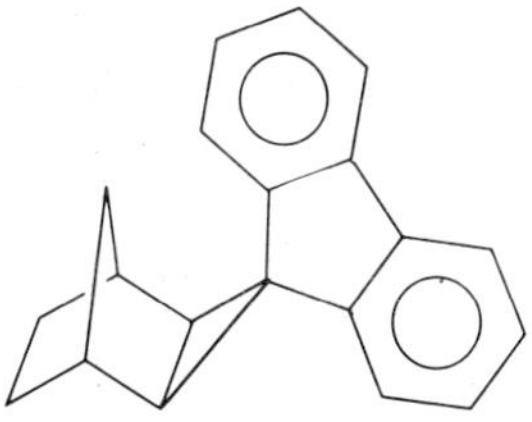

| (4.66) | (4.67) | (4.68) |

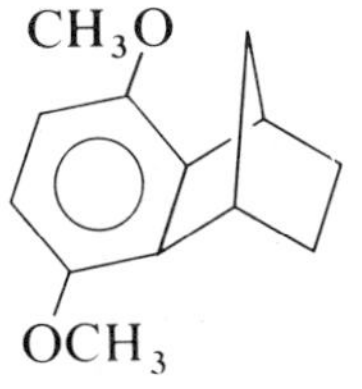

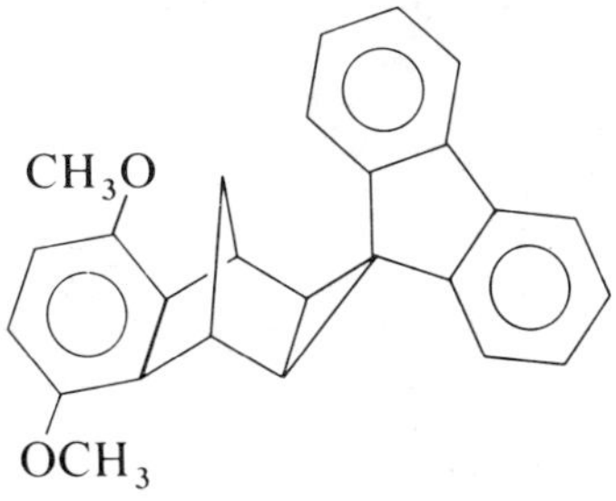

held rigidly perpendicular to each other, and since the transition moments are perpendicular to each other and to the line joining the centres of the chromophores, the Förster treatment predicts that Coulombic energy transfer will be forbidden. The absence of triplet–triplet transfer from the dimethoxybenzene system to the lower-lying triplet of the fluorene, though the two groupings are only 7 Å apart, suggests that there is an orientational factor associated with triplet–triplet transfer as well. In (4.64) and (4.65) the chromophores are not held *rigidly* perpendicular to each other.

REFERENCES

1. For an excellent review see B. Stevens, *Adv. Photochem.*, **8**, 161 (1971).
2. Th. Förster and K. Kasper, *Z. Phys. Chem. N.F.*, **1**, 275 (1954); *Z. Electrochem.*, **59**, 976 (1955).
3. J. B. Birks, D. J. Dyson and I. H. Munro, *Proc. Roy. Soc.*, **A275**, 575 (1963).
4. J. B. Birks, M. D. Lumb and I. H. Munro, *Proc. Roy. Soc.*, **A280**, 289 (1964); E. Döller and Th. Förster, *Z. Phys. Chem. N.F.*, **34**, 132 (1962).
5. B. Stevens, *Spectrochim. Acta*, **18**, 439 (1962).
6. M. T. Vala, J. Haebig and J. A. Rice, *J. Chem. Phys.*, **43**, 886 (1965).
7. E. Konijnenberg, *Diss.*, Freie Univeriteit Amsterdam, 1963; J. N. Murrell and J. Tanaka, *Mol. Phys.*, **7**, 363 (1964).
8. A. Weller, *Pure Appl. Chem.*, **16**, 115 (1968).
9. H. Beens, H. Knibbe and A. Weller, *J. Chem. Phys.*, **47**, 1183 (1967).
10. See ref. 8, p. 117.
11. H. Knibbe, D. Rehm and A. Weller, *Z. Phys. Chem. N.F.*, **56**, 95 (1967).
12. For references, see ref. 1, p. 207.
13. J. A. Barltrop and H. A. J. Carless, *J. Amer. Chem. Soc.*, **94**, 1951 (1972).
14. W. R. Ware, *J. Phys. Chem.*, **66**, 455 (1962).
15. A. D. Osborne and G. Porter, *Proc. Roy. Soc.*, **A284**, 9 (1965).
16. J. A. Barltrop and R. Owers, *Chem. Commun.*, 1462 (1970).
17. D. Rehm and A. Weller, *Ber. Bunsenges.*, **73**, 834 (1969).
18. B. S. Solomon, C. Steel and A. Weller, *Chem. Commun.*, 927 (1969).
19. T. R. Evans, *J. Amer. Chem. Soc.*, **93**, 2081 (1971).

20. T. Medinger and F. Wilkinson, *Trans. Faraday Soc.*, **61**, 620 (1965); A. R. Horrocks, A. Kearwell, K. Tickle and F. Wilkinson, *ibid.*, **62**, 3393 (1966); A. R. Horrocks and F. Wilkinson, *Proc. Roy. Soc.*, **A306**, 257 (1968).

21. For a review and references see J. B. Birks, *Photophysics of Aromatic Molecules*, Wiley–Interscience, London (1970), p. 496 *et. seq.*

22. It is worth noting that rare-earth ions also induce quenching: P. J. Wagner and H. N. Schott, *J. Phys. Chem.*, **72**, 3702 (1968); G. W. Mushrush, F. L. Minn and N. Filipescu, *J. Chem. Soc.* (*B*), 427 (1971).

23. J. G. Calvert and J. N. Pitts, *Photochemistry*, Wiley, London (1966), p. 89.

24. For reviews see F. Wilkinson, (a) *Adv. Photochem.*, **3**, 241 (1964), and (b) in E. J. Bowen (ed.), *Luminescence in Chemistry*, Van Nostrand Reinhold (1968), p. 154.

25. D. L. Dexter, *J. Chem. Phys.*, **21**, 838 (1953).

26. A. Kearwell and F. Wilkinson in G. M. Burnett and A. M. North (ed.), *Transfer and Storage of Energy by Molecules*, volume 1, Wiley–Interscience, London (1969), p. 129.

27. R. E. Kellogg, *J. Chem. Phys.*, **41**, 3046 (1964).

28. F. Wilkinson and J. T. Dubois, *J. Chem. Phys.*, **39**, 377 (1963); J. T. Dubois and R. L. Van Hemert, *ibid.*, **40**, 923 (1964); B. Stevens and J. T. Dubois in H. P. Kallmann and G. M. Spruch (ed.), *Luminescence of Organic and Inorganic Molecules*, Wiley, New York (1962), p. 115.

29. A. Terenin and V. L. Ermolaev, *Trans. Faraday Soc.*, **52**, 1042 (1956).

30. V. L. Ermolaev, *Opt. Spectrosc.*, **6**, 417 (1959); *Soviet Phys. Uspekhi*, **6**, 333 (1963).

31. B. Smaller, E. C. Avery and J. R. Remko, *J. Chem. Phys.*, **43**, 922 (1965).

32. H. L. J. Bäckström and K. Sandros, *Acta Chem. Scand.*, **12**, 823 (1958); **14**, 48 (1960); K. Sandros and H. L. J. Bäckström, *ibid.*, **16**, 958 (1962); **18**, 2355 (1964).

33. G. Porter and F. Wilkinson, *Proc. Roy. Soc.*, **A264**, 1 (1961); F. Wilkinson, *Adv. Photochem.*, **3**, 241 (1964).

34. T. R. Wright, *D.Phil. Thesis*, Oxford (1970).

35. K. Janda and F. S. Wettack, *J. Amer. Chem. Soc.*, **94**, 305 (1972).

36. G. S. Hammond and R. S. Cole, *J. Amer. Chem. Soc.*, **87**, 3256 (1965).

37. C. S. Drucker, V. G. Toscano and R. G. Weiss, *J. Amer. Chem. Soc.*, **95**, 6482 (1973).

38. G. Balavoine, S. Juge and H. B. Kagan, *Tetrahedron Letters*, 4159 (1973).

39. P. J. Wagner, J. M. McGrath and R. G. Zepp, *J. Amer. Chem. Soc.*, **94**, 6883 (1972).

40. L. Salem, *J. Amer. Chem. Soc.*, **95**, 94 (1973).

41. Ref. 21, p. 256. S. L. Murov *Handbook of Chemistry*, Dekker, New York (1973).

42. S. P. McGlynn, T. Azumi and M. Kinoshita, *Molecular Spectroscopy of the Triplet State*, Prentice-Hall, New Jersey (1969), p. 67.

43. R. A. Keller, *J. Amer. Chem. Soc.*, **90**, 1940 (1968).

44. N. Filipescu in E. C. Lim (ed.), *Molecular Luminescence*, Benjamin, New York (1969), p.697.

Chapter 5

Investigation of Reaction Mechanisms

A great deal of effort has been directed towards the elucidation of the mechanisms of organic photochemical reactions in recent years, and general methods of tackling the problems involved have been developed. Definition of a mechanism for a particular reaction requires identification of excited state intermediates and other short- or long-lived species on the reaction path, and determination of rate constants for the various chemical and physical steps in the sequence. There is obviously a considerable overlap between methods used for investigating photochemical reactions and those used for investigating thermal reactions, especially fast thermal reactions. In this chapter the methods specific to photochemistry, such as the use of excited state quenchers, are dealt with in some detail, and methods of investigation which are common to all mechanistic studies are mentioned and illustrated as appropriate. The material is divided into qualitative methods for identifying reaction products, intermediates and excited states, and quantitative methods for determination of quantum efficiencies and rate constants, but there is no such neat division in practice, and cross-reference is necessary.

5.1	**QUALITATIVE METHODS**

5.1.1	**Products**

Identification of the chemical structure of products is the first step in the definition of a reaction, and this is achieved by standard analytical, spectroscopic, chemical and degradative techniques. It is often possible to learn a considerable amount about a mechanistic pathway if the starting material can be 'labelled' in a particular position, and the position of the label in the product can be determined. Isotopic labelling is a valuable method for such investigations since it can reasonably be assumed that the mechanism is substantially the same for reactions of compounds which differ only in the nuclear mass of a small number of atoms. For example, 17-ketosteroids on irradiation in aqueous solution give a reasonable yield of ring-opened carboxylic acid (5.1). The mechanism may involve a direct 4-centre addition of water or may involve a keten as intermediate (5.2), and the use of a deuterium labelled reactant[1] distinguishes between the two pathways (see Chapter 7, p. 185).

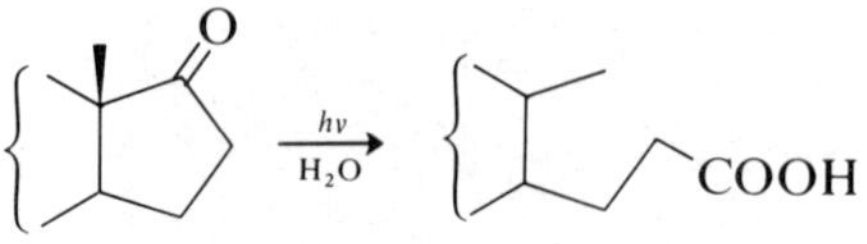

$$(5.1)$$

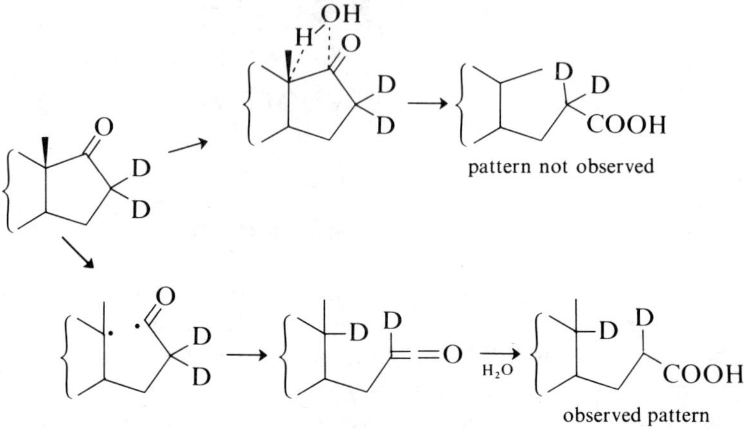

pattern not observed

$$(5.2)$$

observed pattern

Alternatively, a position in the reactant can be 'labelled' by substitution with an alkyl or other chemically inert group. If no alkyl shift occurs during the course of the reaction, the position of the substituent in the product can be used to differentiate between alternative mechanistic routes, as for the isomerization of the bicylic ketone (5.3).[2]

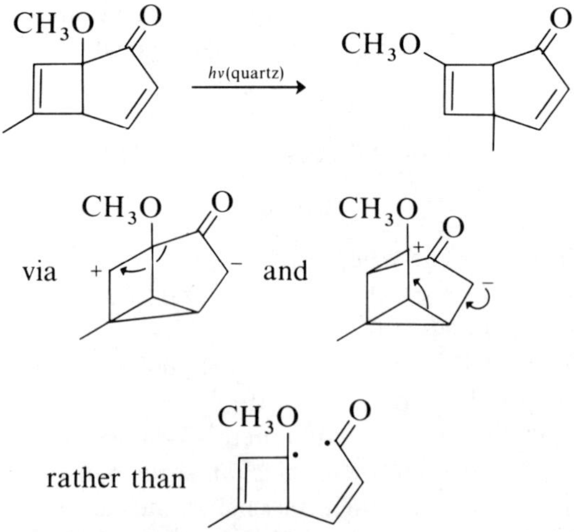

$$(5.3)$$

Stereochemical features in the product can often help in a complete definition of a reaction mechanism. If geometrical or optical stereochemistry in the reactant is fully preserved in the product, the mechanism will probably have some concerted character, whereas loss of such stereochemistry in a reaction sequence points to the existence of intermediates which are planar or which exhibit free rotation about single bonds or rapid inversion at asymmetric centres. In principle it is possible that there are two concerted pathways leading at similar rates to isomeric products, and some theoretical support for such a situation in certain cycloaddition reactions has been put forward, but it is likely to be an infrequent occurrence.

5.1.2   **Intermediates**

One of the advantages of the use of photochemical reactions in synthesis is that the large amount of energy supplied in a specific (electronic) manner can effect conversions to high energy systems which are not readily accessible by thermal processes. These high energy products may be stable under the conditions of reaction, but some are thermally unstable or chemically reactive, and the detection and identification of such intermediates is important in the elucidation of the overall reaction pathway leading to the isolated products. The methods employed depend to a large extent on the lifetime of the species under investigation. For relatively long-lived species ($\tau > 10^1$–10^2 s) it is possible, for instance, to irradiate a sample directly in the cell of a spectrometer. Bands characteristic of the unstable intermediate may then be observed, as for the enol form of acetone (5.4) which is detected when pentan-2-one is irradiated in the cell of an infrared spectrometer.[3]

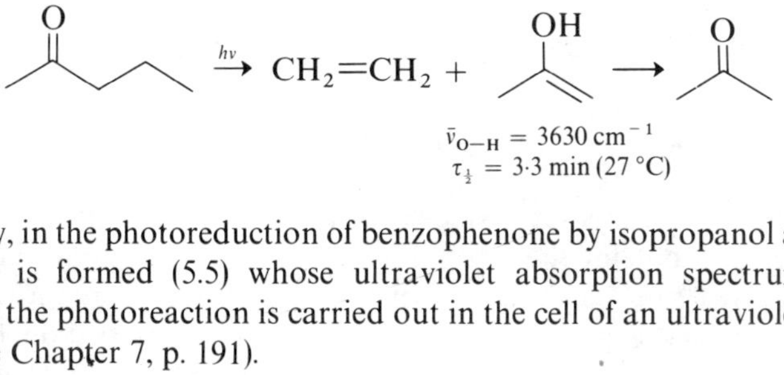

$$\bar{v}_{O-H} = 3630 \text{ cm}^{-1}$$
$$\tau_{\frac{1}{2}} = 3\cdot3 \text{ min} (27\,°C)$$

$$(5.4)$$

Similarly, in the photoreduction of benzophenone by isopropanol a coloured compound is formed (5.5) whose ultraviolet absorption spectrum can be recorded if the photoreaction is carried out in the cell of an ultraviolet spectrometer[4] (see Chapter 7, p. 191).

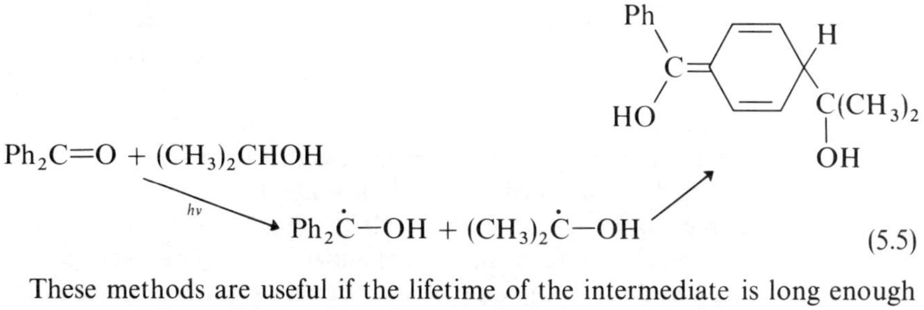

$$(5.5)$$

These methods are useful if the lifetime of the intermediate is long enough under normal conditions either for a sufficiently high stationary state concentration to be built up and detected under continuous irradiation conditions, or for a sufficiently high concentration to be achieved for detection immediately after interruption of the irradiation. Extension of the useful range of such methods can be achieved by the use of multiple reflections to give a longer effective pathlength in a spectrometer cell, or by the use of modulated photolysis ('chopping') to increase the signal-to-noise ratio,[5] but the detection of shorter-lived intermediates demands the employment of different techniques.

Low Temperature Photolysis

For intermediates which are short-lived at normal temperatures, it may be possible to conduct a photolysis at low temperature and to determine 'at leisure' the spectral or chemical characteristics of the intermediate formed. This method may lead to characterization of a primary photoproduct and establishment of the subsequent occurrence of a secondary, thermal reaction to give products which are those normally observed on irradiation at room temperature. An example of this is seen in the photochemical isomerization of *trans*- to *cis*-9,10-dihydronaphthalene (5.6). Low-temperature irradiation gives rise to a (10)-annulene which cyclizes to *cis*-9,10-dihydronaphthalene on warming, and it is therefore likely that the room temperature isomerization proceeds by way of the intermediate annulene.[6]

$$(5.6)$$

(or the all-*cis* isomer)

The usefulness of this technique is limited by the temperature dependence of the rate constants for the secondary reactions of the intermediate (i.e., can the thermal reaction be slowed down sufficiently?), and by the temperature limit below which experimental manipulation and observation become too difficult. A few experiments have been carried out at temperatures as low as 4 K, and at these temperatures and somewhat higher temperatures when very reactive intermediates such as free radicals are involved the method most generally employed is that of matrix isolation. The primary photoproduct is produced and trapped in a rigid matrix and its spectral characteristics observed. The main purpose of the matrix is to inhibit diffusion, and therefore the matrix material must be rigid and well below its melting point. In addition the material must be transparent to the radiation used, adequate as a solvent for the compounds involved, and inert as far as the system under investigation is concerned. Nitrogen or argon at 20 K is a good matrix material[7]—xenon often enhances intersystem crossing processes by a heavy atom effect (see Chapter 3, p. 82), and this interference with the photochemical reaction pathway is undesirable in mechanistic studies. Techniques of this type have allowed the identification of cyclobutadiene as a product in the photochemical reaction of the bicyclic lactone from α-pyrone (5.7),[8] and of benzyne in the reaction of the bicyclic dione (5.8).[9]

$$(5.7)$$

$$(5.8)$$

The results of experiments in solid matrices need to be treated with some caution, in the first instance because a matrix environment can affect the spectra of the species produced, and more generally because it is not always possible to extrapolate results obtained for low temperature, solid–state photochemistry to provide information about the same systems in the liquid or vapour phase at ambient temperature. These considerations apart, much valuable information has been obtained from low–temperature studies, and at extremely low temperatures it has been possible to reduce the rate of chemical decay of electronically excited states so that their lifetime is long enough to allow a second photon to be absorbed by a significant proportion of the excited states. The reactions of very high energy excited states thus produced by the absorption of two photons can be studied.

Flash Photolysis

A different approach to the characterization of short-lived or very short-lived intermediates (with lifetimes down to 10^{-11} or 10^{-12} s) is that involved in the applications of flash photolysis.[10] Flash techniques have developed rapidly over the past few years with the introduction of laser sources, and they now represent one of the most powerful tools for studying qualitatively and quantitatively the excited states and intermediates in photochemical processes. In principle flash photolysis involves the generation of a high concentration of a short-lived intermediate using a very high intensity pulse of radiation of very short duration. At a short time interval after the generating pulse the system is analysed, usually by observing its emission or absorption characteristics. A summary of flash techniques is given here, and for greater detail the reader is referred to reviews.[11,12]

A basic flash system in diagrammatic form is shown in Figure 5.1. The conventional flash sources are based on gas discharge lamps (flash durations down to 1 ns), spark discharge sources (flashes down to a few μs) or exploding wire sources (flashes down to a few hundred μs). Detection techniques vary according to the nature of the system and the information required, but in all cases the time resolution of the method is limited by the duration of the initial flash. The emission spectrum of an intermediate can be photographed using a spectrograph, or the visible absorption spectrum can be similarly recorded if an

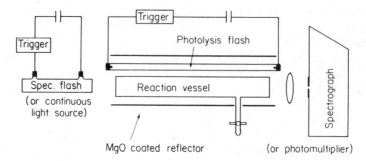

Figure 5.1. Basic flash photolysis arrangement (from D. N. Hague, *Fast Reactions*, (1971), John Wiley and Sons Ltd.)

analytical beam passing through the reaction cell is triggered to flash at a predetermined time interval after the initial flash. Alternatively, the processes can be followed kinetically by monitoring the emission or absorption at a particular wavelength and coupling to an oscilloscope with a time-based sweep. These methods give some indication from the spectrum obtained as to the nature of the species generated, and they allow direct estimation of the lifetime of an intermediate. The difficulties of interpretation lie largely in the possibility of having more than one emitting or absorbing species. In gas phase systems further uncertainties arise because the generating flash can cause a considerable local temperature rise. Short-lived free radicals such as those produced in the photoreduction of aromatic ketones (5.9, see Chapter 7, section 7.5.1) are among the many intermediates which can be detected by conventional flash methods.

$$
\underset{\substack{Ar \quad Ar \\ (T_1)}}{\overset{O}{\underset{\|}{C}}} + \underset{CH_3 \quad CH_3}{\overset{OH}{\underset{|}{CH}}} \rightarrow \underset{Ar \quad Ar}{\overset{OH}{\underset{|}{\overset{C}{\cdot}}}} + \underset{CH_3 \quad CH_3}{\overset{OH}{\underset{|}{\overset{C}{\cdot}}}} \tag{5.9}
$$

The polychromatic nature of the radiation from conventional gas discharge lamps is a disadvantage which has been largely eliminated by the introduction of lasers for flash photolysis. The radiation from laser sources is unique in its monochromatic nature, and Q-switching allows for the pulse to be of very short duration and highly reproducible (see Chapter 2, section 2.4). A wide range of lasers is available,[13,14] though the most commonly used are solid-state lasers employing rods of such material as ruby or neodymium glass. Often it is necessary to subject the pulse to the process of 'frequency doubling' so that it emerges with a wavelength in the ultraviolet rather than the red or infrared region.

The use of such short duration generating pulses requires a modification of the monitoring procedure for lifetime measurement if any advantage is to be gained over the use of discharge lamps except for monochromaticity. Various methods are employed. An oscilloscope trace of the decay of an intermediate can be obtained if a high intensity (flash discharge) monitoring beam is used to overcome the problems caused by background noise with lower intensity sources. More tediously, background noise can be reduced by repeating the measurement many times and averaging the results, and this can be advantageous if it is important to avoid supplying excess energy to the system under study. It is also possible to build up a picture of the decay of an intermediate by recording a series of spectra at different, predetermined time intervals after the excitation flash or to build up a complete spectrum of the intermediate point by point by changing the setting of the monitoring monochromator in a series of readings at a fixed time interval after the initial flash. Laser flash photolysis has been particularly widely used to study the excited singlet states of aromatic hydrocarbons in solution.[12]

A further stage in the improvement of time resolution in laser studies is achieved by the use of mode-locked laser sources, which provide a chain of laser pulses each of a few picoseconds (10^{-12} s) duration. Techniques using these sources enable processes with rate constants around 10^{11}–10^{12} s^{-1} to be studied, and this limit is sufficiently high for vibrational relaxation processes to be measured.

CIDNP

A recently developed technique for studying products formed in rapid radical combination steps is based on the phenomenon of chemically induced dynamic nuclear spin polarization (CIDNP).[15] A spin-paired product from combination of two radicals is formed initially in a non-equilibrium distribution of nuclear spin states, and the intensity of microwave absorption in an applied magnetic field is different from that normally observed for the compound. The effect is investigated by irradiating a sample in the cavity of an n.m.r. spectrometer, and the intensity of the signals for the protons attached to the atoms forming the new bond in the product is observed. The signals may be of higher or lower intensity than normal, and may be negative (i.e., microwave emission occurs). The 'normal' spectrum is obtained within a few seconds of interrupting the illumination. As an example, on irradiation of benzophenone in ethylbenzene, part of the n.m.r. signal for the —CH— proton in the alcohol produced (5.10) is negative under conditions of steady illumination (see Figure 5.2).[16]

$$Ph_2C{=}O + PhCH_2CH_3 \xrightarrow{h\nu} Ph_2\dot{C}{-}OH + Ph\dot{C}H{-}CH_3 \rightarrow$$

$$
\underset{\underset{CH_3}{|}}{\overset{\overset{OH}{|}}{Ph_2C{-}CH{-}Ph}} \qquad (5.10)
$$

The best description of the origin of the effect is as a consequence of electron spin polarization accompanying interaction of radical pairs within a solvent cage.[17] Singlet–triplet mixing occurs via hyperfine interaction in radical pairs, and this is coupled to the nuclear spin states. Singlet–triplet transitions, which

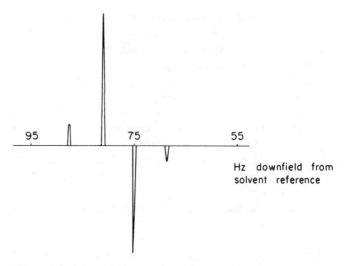

95   75   55

Hz downfield from
solvent reference

Figure 5.2. Adapted from G. L. Closs and L. E. Closs, *J. Amer. Chem. Soc.*, **91**, 4550 (1969), and reproduced by permission of the American Chemical Society. CIDNP spectrum from the irradiation of benzophenone in ethylbenzene. Copyright by the American Chemical Society

are involved because of the need for a singlet radical pair for combination, therefore occur with nuclear-spin dependent probability. Subsequent radical reaction rates depend on the extent of singlet–triplet mixing, and this results in selective population of nuclear spin configurations in the product.

It is possible to distinguish between products formed within the solvent cage and those formed subsequent to radical separation, and also between the spin multiplicities of different precursors to the radical pairs if the signs of the nuclear spin coupling constants are known. The spectrum observed differs according to whether the radical pair is formed from a singlet state, from a triplet state, or from a coming together of radicals produced separately. Such a difference is observed in the spectra of 1,1,2-triphenylethane produced by the three methods shown (5.11) for generating phenylmethyl and diphenylmethyl radicals.[18] Rules for the qualitative interpretation of spectra in terms of the state of the radical pair reactants (singlet, triplet or free, and cage or non-cage), electron–electron exchange coupling constant, and electron-nuclear hyperfine splitting have been developed.[19]

$$Ph_2CH-N{=}N-CH_2Ph \xrightarrow{h\nu} Ph_2CH^{\cdot} + PhCH_2^{-} \quad \text{(singlet)}$$

$$Ph_2C{=}\overset{+}{N}{=}\overset{-}{N} + PhCH_3 \xrightarrow{\text{heat}} Ph_2CH^{\cdot} + PhCH_2^{-} \quad \text{(triplet)} \qquad (5.11)$$

$$Ph_2CH_2 + PhCH_3 \xrightarrow{\text{peroxide}} Ph_2CH^{\cdot} + PhCH_2^{-} \quad \text{(separate)}$$

The technique has been used to study a number of photochemical reactions which occur by way of free radicals, including radical ions produced by electron transfer in such reactions as those between amines and the excited states of ketones (5.12).[20] Biradicals can also be investigated, although for effective singlet–triplet mixing the radical centres must be at least five atoms apart. The major drawback of CIDNP is its great sensitivity, so that there is the danger that an observed effect may represent only a minor part of the product-forming reaction pathway, and that some signals may be caused by a minor by-product.

$$Ar_2C{=}O + R_3N \xrightarrow{h\nu} Ar_2C{=}O^{\cdot\,-} + R_3N^{\cdot\,+} \qquad (5.12)$$

More recently, electron spin resonance emission has been observed[21] from photochemically produced radicals, and this CIDEP (chemically induced dynamic electron spin polarization) can provide information about the triplet state and its relaxation. The effect is thought to arise from initial unequal intersystem crossing rates to the sub-levels of the triplet excited state.

Trapping

The existence of short-lived intermediates can often be inferred from the results of experiments using chemical 'trapping' agents. These are selected to be reactive towards the suspected intermediate but unreactive towards the starting material, products, or any other intermediate or excited state in the reaction sequence. The basic principle is the same as for excited state quenching (see Chapter 5, p. 144) in that inhibition of the formation of normal products and the appearance of other products derived from the trapping agent can be taken as evidence for a particular intermediate, about which information can be gathered by characterization of the 'new' products.

The intermediacy of short-lived free radicals in a photochemical process can be demonstrated in favourable cases by converting the radical to a long-lived nitroxide radical using an alkyl nitroso-compound (5.13)[22] or nitrone as radical scavenger, or by conversion to a fully electron-paired species using iodine, an alkanethiol (5.14) or a stable free radical such as nitric oxide (5.15) or diphenyl-picrylhydrazyl. The long-lived products can be identified at leisure, and this identification in conjunction with the observed inhibition of the photochemical reaction provides strong evidence for a reaction pathway in a particular system.

$$R^{\cdot} + Bu^t{-}N{=}O \rightarrow \underset{Bu^t}{\overset{R}{\diagdown}} N{-}O^{\cdot} \qquad (5.13)$$

$$R^{\cdot} + Bu{-}SH \rightarrow R{-}H + Bu{-}S^{\cdot} \qquad (5.14)$$

$$R^{\cdot} + NO^{\cdot} \rightarrow R{-}NO \qquad (5.15)$$

An attempt to trap radical intermediates can complicate a situation if the trapping agent or the product of trapping is photochemically active, and there is always the possibility that the scavenger reacts with an electronically excited state of the reagent rather than with a subsequent intermediate to produce the observed quenching of reaction. This seems to be a more commonly encountered situation than has sometimes been imagined. For instance, nitroxide radicals such as $Bu_2^t NO$ can trap radical intermediates and also quench excited states. A cautious approach to the interpretation of results from radical trapping experiments is required.

Different types of chemical trapping agent can be employed according to the chemical nature of the suspected intermediate. For instance, o-methylbenzophenone is not photoreduced efficiently by isopropanol, whereas the m- and p-isomers and o-tert-butylbenzophenone undergo efficient reduction under similar conditions. The unreactivity of the o-isomer is attributed to the rapid and reversible formation of a photo-enol in an intramolecular hydrogen abstraction step (5.16, see Chapter 7, section 7.5.3). The enol can be trapped with added dienophile such as maleic anhydride, and the structure of the adduct is consistent with the proposed reaction scheme.[23]

$$(5.16)$$

5.1.3 Lowest Energy Excited States

The excited state responsible for photochemical reaction is often, though not always, either the lowest excited singlet state or the lowest triplet state of the reactant, and characterization of these states is based on spectroscopic data.

The longest wavelength band normally seen in the ultraviolet absorption spectrum arises from singlet → singlet absorption and corresponds to the lowest energy electronic transition. The nature of this transition is usually apparent from the magnitude of the extinction coefficient, the effect of solvent polarity on the position of the maximum (more strictly, on the position of the 0—0 vibrational band, since the band shape may vary with solvent), and the direction of polarization of the absorption band (see Chapter 2, p. 24). For $n \rightarrow \pi^*$ transitions ε_{max} is usually small (10^1–10^2 l mol^{-1} cm^{-1}), the band shows a large 'blue shift' (i.e., a shift to shorter wavelength) in a more polar solvent, and the transition is polarized in a direction perpendicular to the molecular plane. For fully allowed $\pi \rightarrow \pi^*$ transitions in alkenes, dienes or carbonyl compounds ε_{max} is large (10^3–10^4 l mol^{-1} cm^{-1}), the band shows a small 'red shift' (i.e., a shift to longer wavelength) in a more polar solvent, and the direction of polarization is parallel to the molecular plane. For intramolecular charge-transfer transitions, which can be regarded as extreme cases of $\pi \rightarrow \pi^*$ transitions, ε_{max} is very large (10^4–10^5 l mol^{-1} cm^{-1}), and the band shows a pronounced red shift as the solvent polarity increases.

It is not normally possible to observe singlet–triplet absorption spectra by conventional techniques because of the very low extinction coefficient associated with these transitions. The intensity of $S_0 \rightarrow T_1(\pi, \pi^*)$ absorption can be increased by the use of high pressures of added oxygen, and triplet (π, π^*) energies have been determined for conjugated dienes using this method.[24] Triplets other than (π, π^*) which may be of lower energy are not detected. Another technique which enhances the relative intensity of $S \rightarrow T$ transitions is that of phosphorescence excitation (see Chapter 3, p. 71), in which the variation of the intensity of phosphorescence emission at a given wavelength is monitored as a function of the wavelength of the exciting radiation. The method has greater sensitivity than normal absorption techniques, and it is usually possible to observe the singlet $\rightarrow$ triplet absorption bands. Aromatic ketones such as acetophenone show two singlet $\rightarrow$ triplet bands, one corresponding to a $n \rightarrow \pi^*$ and one to a $\pi \rightarrow \pi^*$ transition, which can be distinguished by the fact that the latter is enhanced in intensity by a heavy-atom solvent (see Chapter 2, p. 24).

Decay processes of the excited states of a molecule which involve emission of visible or ultraviolet radiation have already been described, as well as the methods of identifying the emitting state (see Chapter 2, p. 61). Emission spectra sometimes show more detailed fine structure than absorption spectra, and it is possible to assign the nature and energy of the emitting state. This is normally the lowest energy excited state of the molecule, either singlet or triplet, although exceptions are known in which emission originates from a higher excited state—azulene, for instance, fluoresces from the second excited singlet state.

5.1.4   Reactive Excited States

The methods described in the previous section are for the identification of the lowest energy or emitting excited states of a molecule, but these may not be the states directly responsible for photochemical reaction. Higher energy or non-emitting states may be involved. Sensitization and quenching methods are the most generally useful for characterization of reacting states. The theoretical aspects of energy transfer are described in Chapter 4, applications in qualitative mechanistic studies are dealt with in this section, and applications in quantitative investigations of reaction mechanism in the next section.

Most qualitative studies have involved triplet sensitizers or triplet quenchers, partly because the results of such studies are more easily interpreted than those involving singlet sensitizers and quenchers (singlet state sensitizers and quenchers are usually also triplet state sensitizers and quenchers, but the reverse is not true). Triplet sensitizers must be chosen carefully for use with a particular system.[25] The sensitizer must absorb radiation strongly at the wavelength employed so that absorption by the substrate is minimized. The sensitizer must have a high quantum efficiency for intersystem crossing, and its triplet state so formed must be long-lived so that transfer of its energy to another molecule can be efficient. This means that there must be no rapid loss of electronic energy by intersystem crossing to ground state, nor must there be rapid photochemical reaction in the triplet state of the sensitizer. On this latter count dibenzyl ketone $(PhCH_2)_2C{=}O$ is ruled out as a high energy $(330\,kJ\,mol^{-1}$, $79\,kcal$ $mol^{-1})$ triplet sensitizer because it undergoes a very rapid bond-cleavage reaction in the triplet state. The most widely used compounds which are good triplet sensitizers without at the same time being singlet sensitizers are aromatic ketones such as acetophenone or benzophenone.

The qualitative use of triplet sensitizers is of value in two respects. First, if triplet sensitization leads to products which are different from those obtained on direct irradiation (i.e., in the absence of sensitizer), then the reactive state in the direct irradiation is not the same as the triplet state obtained by sensitization. If triplet sensitization produces the same products as direct irradiation, then it is proved that reaction *can* occur through the triplet state. If in addition the ratio of different products is the same whether the reaction is sensitized or not, then it is very likely that both sensitized and unsensitized reactions proceed through the same excited state.

Secondly, if a reaction can be triplet sensitized it may be possible to determine the approximate triplet energy of the reactive state by using a range of sensitizers of different triplet energies. Normally those sensitizers with an energy greater than that of the reactive state lead to relatively efficient reactions, and those with an energy lower than that of the reactive state lead only inefficiently to products. The changeover in efficiency occurs at about a triplet energy equal to that of the reactive state of the substrate. The value so obtained is approximate, and in the region where sensitizer and substrate have similar energies reverse energy transfer from substrate triplet state to sensitizer ground state can occur efficiently, and this is manifest in a dependence of the efficiency of sensitization on the concentration of sensitizer. Self-quenching by the sensitizer shows itself in the same way, and this process is particularly important for ketones with (π, π^*) triplet states, such as thioxanthone, when used as sensitizers. In assigning triplet energies by this method it should be remembered that sensitizers may be able to act through a higher excited state than the lowest or emitting state. Anthracene is one such compound, which acts as a triplet sensitizer through an upper excited triplet state.[26]

Quenching techniques can in a similar manner yield information about the multiplicity and energy of the reactive excited state in a photochemical reaction. Sensitization and quenching are different aspects of the same phenomenon, namely energy transfer. The process is termed sensitization when it is described from the point of view of the energy acceptor and its reactions, and is termed quenching when described from the point of view of the energy donor and its reactions.

The most generally useful triplet state quenchers are conjugated dienes, which have relatively low triplet state energies $(220–250\,kJ\,mol^{-1}$, $53–59\,kcal$ $mol^{-1})$ and lowest singlet states of much higher energy $(>400\,kJ\,mol^{-1}$, $100\,kcal\,mol^{-1})$ so that singlet quenching is of less importance. If the photochemical reaction of a compound can be quenched efficiently by a known triplet quencher, then there is strong evidence for a reactive triplet excited state. Again, it may be possible to obtain an approximate value for the energy of the reactive triplet state by using a range of quenchers of different triplet energy and noting the region of triplet energy in which the efficiency of quenching changes markedly.

Singlet quenching and mixed singlet/triplet quenching techniques can be employed in mechanistic studies, and a use of the latter is seen[27] in the study of the photochemistry of hexan-2-one (5.17). Only part of the photochemical formation of acetone can be quenched by 1,3-diene triplet quenchers, but in the presence of excess 1,3-diene the remainder of the reaction can be quenched by added biacetyl which is able to quench both triplet and singlet states of the ketone. It is apparent, therefore, that both singlet and triplet states of hexan-2-one play a part in the production of acetone.

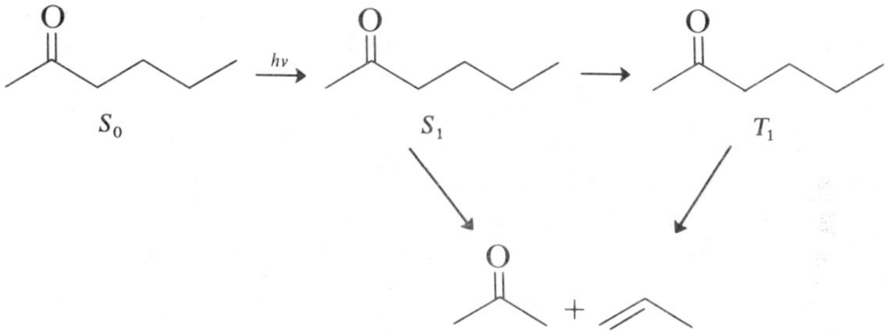

$$(5.17)$$

A comparison of the effect of quencher on product yield with the effect on the intensity of luminescence (fluorescence or phosphorescence) can be particularly useful in identifying a reactive excited state. If a quencher reduces equally the quantum yield of product formation and the quantum yield of luminescence under the same conditions, then it is likely that the emitting state is a precursor of the photochemical product. If the quencher has no effect on light emission, the emitting state is not the immediate precursor for the chemical reaction. Similarly, if the quenching of a photochemical reaction is accompanied by fluorescence or phosphorescence emission from the quencher, the quenched state can be identified as a precursor (though not necessarily the immediate precursor) of the product.

Flash photolysis techniques, described in an earlier section, provide a powerful method for detecting excited state intermediates in a photochemical reaction and for determining their lifetimes. The methods can be adapted to measure a lifetime in the presence of excited state quencher, and this gives a direct measure of the rate constant for the quenching process. Such measurements carried out for a range of quenchers enable a more precise assignment of excited state energies to be made, and the understanding of some aspects of the mechanism of the *cis–trans* isomerization of stilbene has been aided considerably in this way.[28]

5.2   **QUANTITATIVE METHODS**

A more complete description of a reaction mechanism than an identification of the final products and of the intermediates on the reaction pathway requires quantitative data on the efficiency of reaction and on the rate constants for the steps involved, particularly for the primary processes undergone by the electronic excited states of the reactant. The processes are often very fast under normal conditions (first-order rate constants greater than $10^9 \, s^{-1}$ are not unusual), and to some extent the general methods of investigating fast thermal reactions can be applied to the study of photochemical processes. In particular, competition methods have been extensively used in which the rate constants for excited state reactions are estimated by comparison with known rate constants for alternative bimolecular processes, usually energy transfer processes. Direct measurement of excited state lifetimes, especially by flash photolytic methods, is becoming increasingly important.

5.2.1   **Quantum Yields**

The efficiency with which the supplied radiation is employed in converting starting material to product is measured by the quantum yield of product formation. This quantum yield is in itself a useful function in that it allows optimum reaction times to be calculated, and for commercial processes the contribution of electrical energy costs to the unit cost requires a knowledge of product quantum yield. In the U.K. this cost of electrical power is often the prohibitive factor in the commercial utilization of photochemical processes, and at present only processes with very high quantum yield (such as the photo-initiated chlorination of alkanes) or processes for which the energy cost is small by comparison with reagent costs (such as the conversion of corticosterone to aldosterone) are economically viable.

The quantum yield (ϕ) for formation of a particular product is given by:

$$\phi = \frac{\text{amount of product formed (in moles or molecules)}}{\text{amount of radiation absorbed (in einsteins or photons)}}$$

$$= \frac{\text{rate of formation of product}}{\text{rate (i.e., intensity) of absorption of radiation}}$$

The quantum yield can be related to the rate constants in a particular system, but normally it is not a useful direct measure of the reactivity of an excited state, that is, of the rate constant for the primary chemical step from the excited state. This is because the quantum yield is a ratio of rate constants and sums of rate constants. For the simple system (5.18) a steady-state analysis (5.19) enables an expression for the quantum yield for formation of product B to be derived (5.20).

$$A \xrightarrow[\text{rate } I]{hv} A^* \begin{array}{c} \overset{k_r}{\nearrow} B \\ \underset{k_{-1}}{\searrow} A \end{array} \tag{5.18}$$

$$\frac{d[A^*]}{dt} = I - (k_r + k_{-1})[A^*] = 0 \tag{5.19}$$

$$\phi_{\text{B}} = \frac{k_r[\text{A}^*]}{I} = \frac{k_r}{k_r + k_{-1}} \qquad (5.20)$$

The quantum yield measurement provides a value for k_r (a true measure of reactivity) only if the excited state lifetime $[\tau_0 = (k_r + k_{-1})^{-1}]$ is known.

The measurement of quantum yields is, in principle, straightforward, and requires the measurement of the product formed in a given time and the radiation absorbed by the system in that time. The estimation of stable products presents no difficulties beyond those normally encountered in chemical analysis, although for accurate work it is best to restrict the extent of reaction (to less than 5% or even less than 2%), and this may necessitate particular care in analysis. This is one reason for the widespread use of gas chromatography in quantitative photochemical studies. The direct measurement of radiation absorbed presents greater problems. In the first instance the radiation is required to be monochromatic, or to contain only a narrow range of wavelengths, since quantum yields are sometimes wavelength-dependent, and, more importantly, it is unusual for a reagent and an actinometer system to be equally sensitive to the same range of radiation wavelengths. The use of chemical filter solutions, glass filters or interference filters allows the selection of individual bands from a low or medium pressure mercury arc source.[29] Conventional grating or prism monochromators can be used, but they have the disadvantage of having a very low power output at any particular wavelength. Whatever method is chosen for isolating a narrow band of radiation wavelengths, it is possible that the source output will vary with time, even when a stabilized power supply is used, and for high accuracy a beam-splitter (e.g., a partially-silvered mirror) should be used so that photolysis and actinometry can be carried out simultaneously (Figure 5.3).

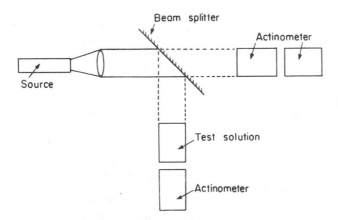

Figure 5.3. A beam splitter used to allow simultaneous photolysis and actinometry

'Direct' physical measurement of radiation intensity[30] can be carried out by radiometric methods (thermopiles, thermistors or bolometers), or by photo-electric methods (phototubes, or photovoltaic cells), but these techniques are not widely used in routine measurement, largely because of constructional difficulties in that the measuring device should ideally be sited in the exact position normally occupied by the photolysis cell (unless a beam-splitter is employed) and it should be such a shape and of such material that the radiation incident on it and recorded by the instrument is the same as that absorbed by the solution under photolysis. These conditions are difficult to achieve satis-factorily, and it is more usual to measure light intensity by chemical actinometry. This employs as a secondary standard a chemical reaction whose quantum yield is known and can be reproduced accurately and which is fairly constant over a useful range of wavelengths (or whose variation with wavelength is known). This type of actinometer is readily adapted to the same photolysis cell as the sample under investigation. Quantum yield measurement then requires only two chemical analyses, one for reaction product and one for actinometer product. The most commonly used liquid phase chemical actinometers employ transition metal oxalates which undergo redox reactions on irradiation, such as potassium trisoxalatoferrate(III), uranyl oxalate, or vanadium(V) iron(III) oxalate.[31] For vapour phase reactions the production of carbon monoxide from acetone is a common standard system.

It is often advantageous to use as chemical actinometer a reaction which has itself been calibrated against a 'primary' chemical actinometer standard. For instance, in a study of the photochemistry of an alicyclic ketone the photo-reaction of hexan-2-one to give acetone (estimated by g.l.c.) may be used as actinometer. The value of this is (i) that the actinometer solution has absorption characteristics as near as possible to those of the solution under study, (ii) that the solvent is the same in the two solutions so that errors caused by reflectance at the glass/solvent interface are minimized (these errors can be particularly high when cylindrical vessels are used[32]), and (iii) that the quantum yields of reaction in the two solutions may be similar in magnitude.

An apparatus which allows actinometry to be carried out simultaneously with several photochemical reactions (or photochemical reaction with a range of concentrations of excited state quencher) is a 'merry-go-round' (Figure 5.4). The tubes are equidistant from the radiation source (a central lamp or a sur-rounding bank of lamps), and the tube holder rotates about the central axis. Errors caused by an uneven radial distribution of radiation are minimized. Each tube receives the same amount of radiation, and the quantum yields are proportional to the extent of chemical reaction in each.[33]

5.2.2   **Kinetics of Quenching**

The lifetime of an excited state which undergoes measurable physical or chemical change (such as luminescence or product formation) can in principle be deter-mined by the employment of a suitable 'quencher' which interacts with the excited state in an energy transfer process whose rate constant is known. This

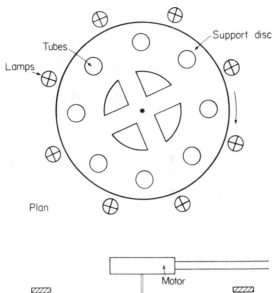

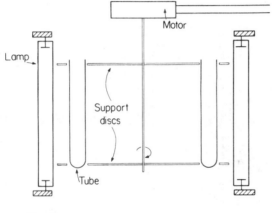

Figure 5.4. 'Merry-go-round' apparatus

is illustrated for the simple system (5.21) in which a single chemical product (**B**) is formed in a unimolecular reaction from the first-formed excited state (A*) of the starting material, and in which A* can be quenched by an added species (Q).

$$
\begin{array}{ll}
A \xrightarrow{h\nu} A^* & \text{rate } I \, (= \text{intensity}) \\
A^* \rightarrow A & k_{-1} \, [A^*] \\
A^* \rightarrow B & k_r \, [A^*] \\
A^* + Q \rightarrow A + Q^* & k_q \, [A^*][Q]
\end{array}
\qquad (5.21)
$$

k_{-1} is a composite first-order rate constant which takes into account radiative and radiationless deactivation of A*. Using the steady-state approximation (cf. 5.19), the quantum yield for product formation in the absence of quencher (ϕ_B^0) can be expressed as:

$$\phi_B^0 = \frac{k_r}{k_r + k_{-1}}$$

and the quantum yield with added quencher (ϕ_B) as:

$$\phi_B = \frac{k_r}{k_r + k_{-1} + k_q[Q]}$$

Hence

$$\frac{\phi_B^0}{\phi_B} = 1 + \frac{k_q[Q]}{k_r + k_{-1}} = 1 + k_q\tau_0[Q] \qquad (5.22)$$

where τ_0 is the lifetime of the excited state in the absence of quencher. This expression is linear in [Q] and is the form of the normal 'Stern–Volmer' quenching plot[34] often obtained in photochemical studies (Figure 5.5).

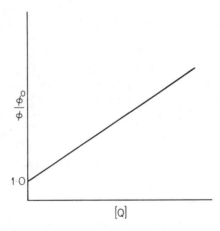

Figure 5.5

The assumptions made in the derivation of (5.22), apart from those always implicit in the application of the steady-state treatment, are (i) that the conditions are such that the intensity (I) of radiation absorbed by the reagent is effectively unchanged over the period of reaction, (ii) that the quencher interacts only with the reactive excited state involved in the product-forming process, it interacts only by energy transfer, and its concentration is not significantly altered by chemical reaction, and (iii) that the product is formed from only one excited state of A, which is the one quenched by Q. These assumptions will be considered in turn.

For the intensity of radiation absorbed by A to be constant over the period of reaction it is necessary either that the reduction in [A] occasioned by reaction is insignificant by comparison with the original concentration of A, or that A has a much higher extinction coefficient at the wavelength employed than any products formed and is present in sufficient concentration to absorb all (99–100%) of the radiation throughout the reaction. The use of low percentage conversion is usually preferred, and this has the advantage of discouraging interference by the reaction products with the processes in the photochemical reaction pathway. Products and quencher do sometimes absorb a significant amount of the incident radiation, and it is possible to apply a correction for situations where quencher absorbs a part of the radiation or where the percentage conversion is such that absorption by products builds up significantly during the course of the reaction. The second method of ensuring a constant high percentage absorption by the reagent is to use a sufficiently high initial concentration of A for absorption due to A to be greater than 99% throughout the reaction. This must not be taken too far, or another source of error becomes apparent: if the absorbance of the solution is very high the bulk of the incident radiation is absorbed in the first fraction of the pathlength of the reaction cell, and product builds up in this volume if diffusion or mixing within the cell is not rapid. This inhomogeneous generation of product can affect quantitative results. The problems associated with absorption of the incident radiation include also the differences which arise because of reflectance at the surfaces of cells of different shape or containing different solvents.[32]

If the quencher interacts with more than one excited state of A, the effect on product formation may be different from that expressed in (5.22). Interaction of quencher with an excited state formed by non-radiative transition *from* the reactive (product-forming) excited state does not affect product formation unless a serious depletion of Q in a chemical reaction ensues. Interaction with an excited state earlier on the reaction pathway than the reactive excited state does have an effect. Such a situation (5.23) leads to the quenching expression given in (5.24).

$$
\begin{array}{lll}
A \xrightarrow{h\nu} A^* & \text{rate } I & \\
A^* \rightarrow A & k'_{-1}[A^*] & \\
A^* \rightarrow A^{\ddagger} & k_{ic}[A^*] & \\
A^* + Q \rightarrow A + Q^* & k'_q[A^*][Q] & \quad (5.23)\\
A^{\ddagger} \rightarrow A & k_{-1}[A^{\ddagger}] & \\
A^{\ddagger} \rightarrow B & k_r[A^{\ddagger}] & \\
A^{\ddagger} + Q \rightarrow A + Q^* & k_q[A^{\ddagger}][Q] &
\end{array}
$$

In the absence of quencher,

$$
\phi_B^0 = \frac{k_r}{k_r + k_{-1}} \cdot \frac{k_{ic}}{k_{ic} + k'_{-1}}
$$

and with added quencher,

$$
\phi_B = \frac{k_r}{k_r + k_{-1} + k_q[Q]} \cdot \frac{k_{ic}}{k_{ic} + k'_{-1} + k'_q[Q]}
$$

Hence,

$$\frac{\phi_B^0}{\phi_B} = \left(1 + \frac{k_q[Q]}{k_r + k_{-1}}\right)\left(1 + \frac{k'_q[Q]}{k_{ic} + k'_{-1}}\right)$$

$$= (1 + k_q\tau_0[Q])(1 + k'_q\tau'_0[Q]) \tag{5.24}$$

This expression is quadratic in [Q], and the form of such a quenching curve is shown in Figure 5.6. This situation is frequently encountered when a reaction occurs from a triplet state and a quencher is employed which quenches both the triplet state and the singlet state from which the triplet state arises by inter-system crossing.

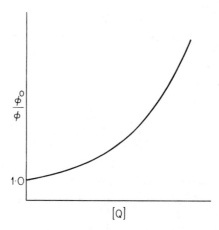

Figure 5.6

To examine the third assumption made in the simple scheme (5.21) it is necessary to consider the situation in which product is formed from two excited states of A, one or both of which may be affected by the quencher. The complete scheme (5.25) gives rise to the general expression (5.26).

$$
\begin{array}{lll}
A \xrightarrow{h\nu} A^* & \text{rate } I \\
A^* \rightarrow A & k'_{-1}[A^*] \\
A^* \rightarrow A^{\ddagger} & k_{ic}[A^*] \\
A^* \rightarrow B & k'_r[A^*] \\
A^* + Q \rightarrow A + Q^* & k'_q[A^*][Q] \\
A^{\ddagger} \rightarrow A & k_{-1}[A^{\ddagger}] \\
A^{\ddagger} \rightarrow B & k_r[A^{\ddagger}] \\
A^{\ddagger} + Q \rightarrow A + Q^* & k_q[A^{\ddagger}][Q]
\end{array} \tag{5.25}
$$

In the absence of quencher,

$$\phi_B^0 = \frac{1}{(k_{ic} + k'_r + k'_{-1})}\left(k'_r + \frac{k_r k_{ic}}{k_r + k_{-1}}\right)$$

With added quencher,

$$\phi_B = \frac{1}{(k_{ic} + k_r' + k_{-1}' + k_q'[Q])}\left(k_r' + \frac{k_r k_{ic}}{k_r + k_{-1} + k_q[Q]}\right)$$

Hence,

$$\frac{\phi_B^0}{\phi_B} = (1 + k_q'\tau_0'[Q])(1 + k_q\tau_0[Q])\left(\frac{k_r' + k_r k_{ic}\tau_0}{k_r'(1 + k_q\tau_0[Q] + k_r k_{ic}\tau_0)}\right) \tag{5.26}$$

where $\tau_0' = (k_{ic} + k_r' + k_{-1}')^{-1}$ and $\tau_0 = (k_r + k_{-1})^{-1}$.

If $k_q' = 0$, that is if the quencher affects only the second of the reactive states, the quenching curve tends to a horizontal asymptote given by

$$\frac{\phi_B^0}{\phi_B} = \left(1 + \frac{k_r k_{ic}\tau_0}{k_r'}\right)$$

This can also be expressed as

$$\frac{\phi_B^0}{\phi_B} = \left(1 + \frac{(\varphi_B^0)}{(\varphi_B^0)'}\right)$$

where (φ_B^0) and $(\varphi_B^0)'$ are the contributions to ϕ_B^0 from states $A^\ddagger$ and A^* respectively.

The form of such a quenching curve is shown in Figure 5.7. This is a situation commonly encountered when a photochemical reaction occurs through both singlet and triplet excited states of a compound and the system is studied using a triplet state quencher.

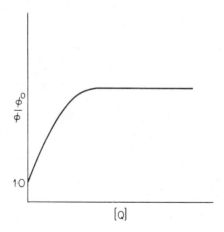

Figure 5.7

If both excited states are involved and both are quenched, the full expression (5.26) applies, and the quenching curve may take different forms. The gradient may decrease from its initial value and approach a constant positive non-zero value (Figure 5.8), or it may increase from its initial value and approach a

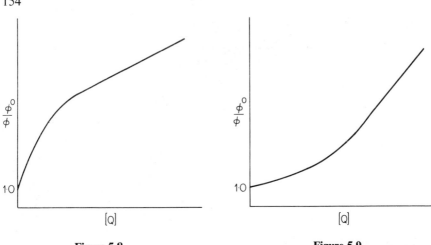

Figure 5.8   Figure 5.9

constant positive non-zero value (Figure 5.9) according to whether $k'_q\tau'_0(1 + (\varphi^0_B)/(\varphi^0_B)')$ is less than or greater than $k_q\tau_0$. It is possible for a linear plot to be obtained if $k'_q\tau'_0(1 + (\varphi^0_B)/(\varphi^0_B)') = k_q\tau_0$, and such a situation has already been reported. Its diagnosis is not difficult, because a different quencher can be found which will interact with only one excited state, and this will differentiate between this complex situation and the very simple one where only excited state is involved and quenched.

It should be apparent from the above treatment that the interpretation of the results of quenching data should be approached with care, but that if the data are sufficiently precise, $k_q\tau_0$ values for the reactive excited state(s) can be extracted. A statistical treatment of data may be more accurate than a graphical method.[35] The value of the quenching rate constant k_q depends on the mechanism by which the quencher operates, and sometimes on the difference between the energy of the excited state quenched and of the excited state to which the quencher is raised. Many quenchers employed in such studies operate by a collisional energy transfer mechanism, and if the quencher is such that its excited state energy is more than a few kcal mol^{-1} below that of the donor excited state, then it can be assumed that k_q is the diffusion-controlled quenching rate constant.[36] The value of this varies with solvent viscosity and with temperature, but on the whole it is independent of the nature of either the donor or the quencher. Representative values are given in Table 5.1.

Diffusion-controlled quenching rate constants are not always known with great precision, and they can vary to some extent with the structures of the species involved, and therefore absolute lifetime values calculated on this basis are subject to some error. The relative lifetimes for a series of similar compounds can be obtained with greater confidence because the value of k_q will not vary greatly from one compound to another in a series. Certain quenchers, particularly cisoid conjugated dienes such as cyclopentadiene, seem to quench at rates in excess of those expected on the basis of the 'normal' k_q values. This may reflect

Table 5.1. Values for the diffusion-controlled quenching rate constant (k_q)

Solvent	$k_q/l\ mol^{-1}\ s^{-1}$	a
Benzene	5×10^9	b
Pentane	13×10^9	b
Hexane	11×10^9	b
Hexadecane	4.5×10^9	b
t-Butanol	2×10^9	c
Acetonitrile	11×10^9	c

[a] At 25 °C.
[b] P. J. Wagner and I. Kochevar, *J. Amer. Chem. Soc.*, **90**, 2232 (1968).
[c] P. J. Wagner, *J. Amer. Chem. Soc.*, **89**, 5898 (1967).

a molecular geometry which is particularly favourable for collisional energy transfer. Another possible source of error arises at high concentrations of quencher, when k_q is higher than the low concentration value because of a 'nearest neighbour' quenching effect.[37] This arises because the excited state is often formed with a quencher molecule in its immediate environment if there is a high concentration of quencher in the solution. Diffusion is therefore not required for energy transfer to take place, and the overall rate constant for quenching is higher than the diffusion-controlled value. This effect shows itself as a deviation from linearity at high concentration in a Stern–Volmer quenching plot (see 5.22 and Figure 5.5).

Up to this point it has been assumed that only one quencher is involved, but as described in an earlier section (see Chapter 5, p. 145) 'double' quenching experiments have been usefully employed in which one reactive excited state of A is totally quenched by excess quencher, and the quenching of the second reactive excited state is then studied using a second quencher which is more effective than the first for this particular excited state.

5.2.3 Excited State Lifetimes

The quenching techniques outlined in the previous section provide values of excited state lifetimes τ. Such lifetimes can also be measured directly either by following the luminescence decay in conventional emission spectroscopy (see Chapter 3, section 3.1.4), or by following absorption or emission decay using flash photolysis techniques (see Chapter 5, p. 137). For conventional methods to be useful the intensity of fluorescence or phosphorescence must be high enough to be monitored with reasonable accuracy, and the measurements should be made under the same reaction conditions as for other quantitative data with which the τ value is to be combined. This is particularly relevant to phosphorescence studies, since the most usual conditions for measurement of phosphorescence lifetime involve a rigid glass at liquid nitrogen temperature. A lifetime under these conditions is often much longer than the lifetime measured in liquid solution at room temperature. This difference is attributed to the presence of very small but significant amounts of quenching impurities or to the variation with temperature of rate constants for chemical reactions of the excited state, particularly reaction with the solvent.

The measured lifetime of an excited state can provide an estimate of the rate constant for reaction in the excited state. The lifetime is defined by the expression:

$$\tau = (k_r + k_{-1})^{-1} \qquad (5.27)$$

where k_r is the rate constant for chemical reaction and k_{-1} is the sum of rate constants for physical decay processes (phosphorescence and intersystem crossing to ground state for a lowest triplet state, or fluorescence, radiationless decay to ground state and intersystem crossing to triplet state for a lowest excited singlet state). In those instances where an estimate of k_{-1} can be made, a value can be assigned for k_r from equation (5.27). The most general way of estimating rate constants (k_x) for emission and for $S \rightarrow T$ intersystem crossing is to measure the quantum efficiency (ϕ_x) for a process and to combine this value with the measured lifetime (τ) according to equation (5.28).

$$\phi_x = \tau k_x \qquad (5.28)$$

Quantum efficiencies for formation of triplet states can be estimated by 'triplet counting' methods in which excess triplet quencher is employed and the number of triplet states formed is calculated from the number of product molecules arising from chemical reaction of the triplet state of the quencher. For example, through its triplet state an acyclic conjugated diene undergoes cis–trans isomerization and gives dimers (see Chapter 8, pp. 234–260) and for a given diene the quantum efficiencies for these processes can be measured.[38] When excess diene is employed to quench a triplet state in an unknown system, the extent of chemical reaction of the diene provides a direct measure of the extent of formation of the triplet state of the donor, and hence a measure of the quantum efficiency for triplet formation. The isomerization of cis-penta-1,3-diene and the dimerization of cyclohexa-1,3-diene are the two most commonly used triplet counters.

The rate constant (k_r) for chemical reaction of an excited state is a direct measure of the reactivity of that state. It is normally obtained indirectly from values of excited state lifetime and other rate constants. Because a chemical reaction is often not a simple one-step process it is therefore not always possible to estimate the rate constant for reaction from the quantum yield of product formation according to the expression (5.29).

$$\phi_B = \tau k_r \qquad (5.29)$$

This relationship breaks down if there is more than one pathway open to any intermediate on the reaction pathway from the electronic excited state to the product, and it holds only if the product is formed in a single-step process or in a series of consecutive steps each (after the first) occurring with unit efficiency.

Photochemical reactions very often involve intermediates which can undergo reaction to regenerate starting material. As an example, the intramolecular elimination reaction of the triplet state of aromatic ketones is a two-stage process (see Chapter 7, p. 198), and after the initial hydrogen abstraction has taken place, a reverse hydrogen transfer can occur to regenerate the ground state of the ketone (5.30). In this system the value of the rate constant for the primary reaction step is greater than that predicted by equation (5.29) because some of the biradical intermediate has not given rise to the measured product.

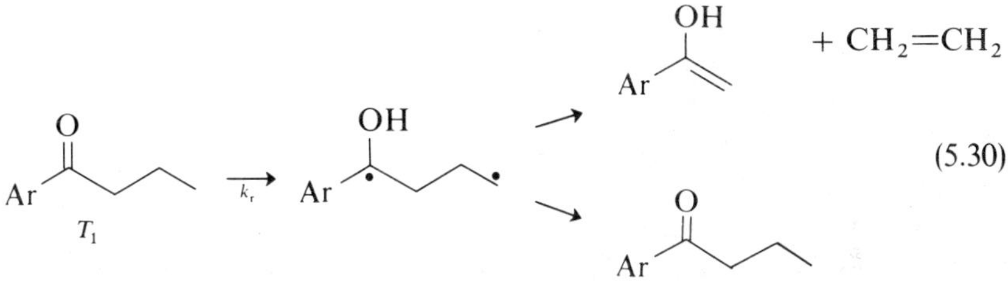

$$(5.30)$$

5.2.4   Controlled Variables

The establishment of a photochemical reaction mechanism can be aided by the study of the effects on the reaction of controlled changes in the intensity or nature of the radiation employed or in the pressure or temperature of the system.

The effect of intensity can be used to determine the number of quanta of radiation required to effect a particular chemical change. Under 'normal' circumstances, where one molecule of product arises from only one singly excited state of the reactant, the rate of formation of the product is directly proportional to the intensity of the absorbed radiation. A different dependence on intensity can arise in various ways. First, if the reactive excited state is a 'doubly' excited state which requires two photons for its formation, then the rate of formation of product is proportional to the square of the intensity. Reactions from such excited states have not been widely reported, but they do occur when a high concentration of first excited state can be generated, either by the use of a laser source or by the use of low temperature matrices in which the excited state lifetime is relatively long.

A second situation in which the rate of product formation is proportional to the square of the intensity arises when two singly excited states are involved in the formation of one molecule of product. This is not usually the direct interaction of two excited states, but rather of two intermediates formed from separate excited states. On this basis a distinction can be made between two of the proposed schemes for the formation of benzpinacol by photoreduction of benzophenone in propan-2-ol (5.31). The observed dependence of the rate of product formation on the first power of the intensity rules out the first of the mechanisms.

$$Ph_2C=O \xrightarrow{I} Ph_2C=O* \xrightarrow{(CH_3)_2CHOH}$$

$$Ph_2\overset{\cdot}{C}-OH + (CH_3)_2\overset{\cdot}{C}-OH \quad \left. \right\} \quad \text{rate} \propto I^2.$$

$$2Ph_2\overset{\cdot}{C}-OH \rightarrow \text{pinacol}$$

$$Ph_2C=O \xrightarrow{I} Ph_2C=O* \xrightarrow{(CH_3)_2CHOH}$$

$$Ph_2\overset{\cdot}{C}-OH + (CH_3)_2\overset{\cdot}{C}-OH$$

$$(CH_3)_2\overset{\cdot}{C}-OH + Ph_2C=O \xrightarrow{\hspace{2cm}}$$

$$Ph_2\overset{\cdot}{C}-OH + (CH_3)_2C=O$$

$$2Ph_2\overset{\cdot}{C}-OH \rightarrow \text{pinacol}$$

$$\left. \right\} \quad \text{rate} \propto I$$

(5.31)

Changes in the wavelength of the radiation employed can affect a photo-chemical reaction in three ways. First, for a compound which has two or more accessible (singlet) excited states the initial population of states depends on the wavelength of the exciting radiation, and in many instances it is possible to populate one excited state selectively. This may have no effect on the photo-chemistry, because an initially formed higher energy excited state can be rapidly deactivated to give the lowest energy state from which reaction then occurs. However, if a chemical reaction of the higher energy state can compete effectively with deactivation, a wavelength effect will be observed. Several systems are known which exhibit this type of wavelength-dependence of chemical products, such as the cycloaddition reaction of thioketones with certain alkenes (5.32). One type of product is obtained with radiation of wave-length > 500 nm, but a different type of product with 366 nm radiation (see Chapter 7, p. 220).

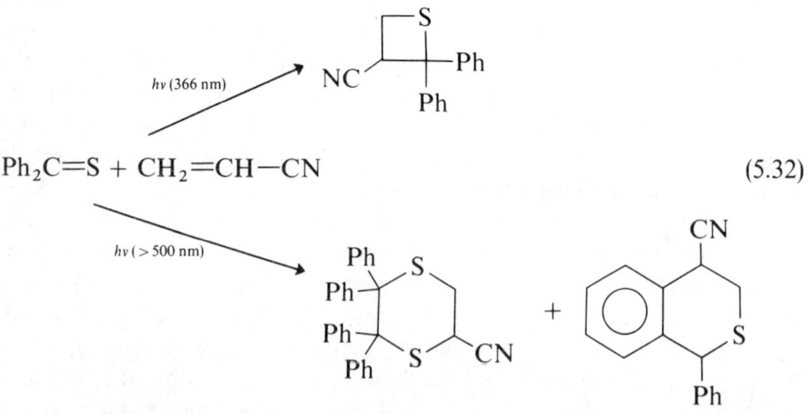

(5.32)

Secondly, a change in wavelength may lead to a change in the observed major products because the first-formed product is itself photochemically

$A \times A = S$, and $S \times A = A$. By these means the state correlation diagram shown in Figure 6.7 is obtained for the disrotatory cyclization of butadiene.

Since, according to the *orbital* diagram (Figure 6.6), the ψ_1 and ψ_2 orbitals of butadiene transform into σ and π^* orbitals of cyclobutene, it follows that S_0-butadiene is correlated with a highly excited state S_m of cyclobutene. This and other correlations are exhibited on the state diagram (Figure 6.7) by dotted lines. Because of the non-crossing rule (that lines connecting states of like symmetry may not cross), the actual correlations are those given in Figure 6.7 by the full lines. Hence, since in a state correlation diagram one is plotting some ill-defined reaction co-ordinate as horizontal axis against energy, it follows that the disrotatory interconversion of butadiene and cyclobutene is therefore forbidden. On the other hand, no such barrier is apparent in the S_1-state. The state correlation diagram therefore provides a crude representation of a particular cross-section of the relevant multi-dimensional energy surfaces.

A minor technical difficulty arises in some reactions when the orbitals involved are apparently neither symmetric nor antisymmetric with respect to a particular symmetry operation. This is the case for the σ-orbitals of cyclohexene formed in the Diels–Alder reaction (Figure 6.8).

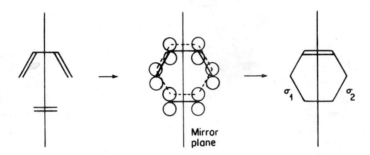

Figure 6.8. The Diels–Alder reaction

Reflection of σ_1 in the symmetry plane converts it into σ_2 and not into σ_1 or $-\sigma_1$. This sort of problem is overcome by mixing σ_1 and σ_2 to form multicentre orbitals $(\sigma_1 + \sigma_2)$ and $(\sigma_1 - \sigma_2)$, respectively symmetric and antisymmetric to the mirror plane (Figure 6.9).

Figure 6.9. Symmetric and antisymmetric molecular orbitals of the σ-bonds formed in the Diels–Alder reaction between butadiene and ethylene

This device then permits the construction of the orbital correlation diagram of Figure 6.10, which shows that the Diels–Alder addition is allowed in the ground state.

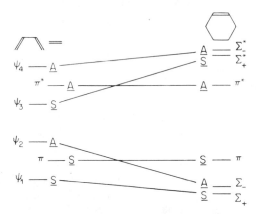

Figure 6.10. Orbital correlation diagram for the Diels–Alder cycloaddition

6.3.2   The Woodward–Hoffmann Rule

This rules states that 'a ground state pericyclic change is symmetry-allowed when the total number of $(4q + 2)_s$ and $(4r)_a$ components is odd'. In using this rule it is to be understood (i) that pericyclic reactions are treated as generalized cycloadditions; (ii) that the terms $(4q + 2)$ and $(4r)$ refer to the number of electrons from each component involved in forming the transition state; (iii) that suprafacial and antarafacial, signified by the subscripts s and a, have the extended meanings implied by Figure 6.4; and (iv) that terms of the type $(4q + 2)_a$ and $(4r)_s$ are ignored.

A few examples will clarify the procedure. Consider the dimerization of olefins (6.9). The *cis–cis* cyclodimerization takes the form $(\pi^2 s + \pi^2 s)$, corresponding to two $(4q + 2)_s$ components $(q = 0)$. The number of relevant components is even, and the reaction will be forbidden in the ground state. Since, for reasons to be discussed later, reactions thermally forbidden are allowed photochemically, we can conclude that this mode of cycloaddition will occur on irradiation. Conversely, the *cis–trans* cycloaddition, being of the type $(\pi^2 s + \pi^2 a)$, will be thermally allowed (the second term is ignored).

The disrotatory cyclization of butadiene (6.11) can be formulated as $\pi^4 s$ or as $(\pi^2 s + \pi^2 s)$. This leads to an even value (0 or 2) for the number of relevant components and to the prediction of a forbidden ground state reaction and therefore an allowed photochemical process.

Sigmatropic reactions are also easily treated. For example, the (1, 3) migration of an alkyl group suprafacially and with inversion of configuration (see equation 6.8) is a $(\pi^2 s + \sigma^2 a)$ process (Figure 6.11). The number of relevant components is odd, and the reaction is allowed in the ground state.

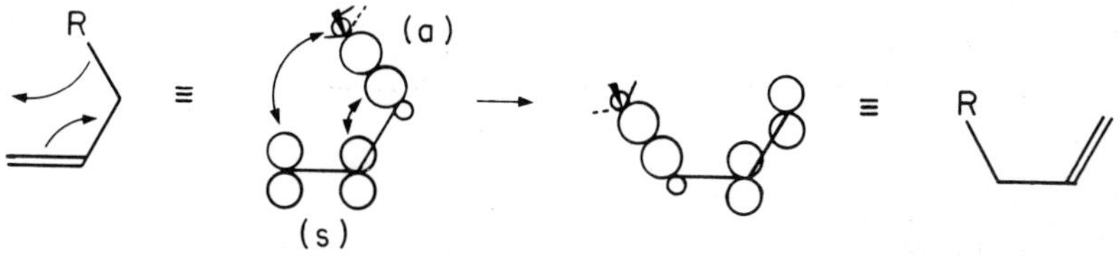

Figure 6.11. A suprafacial (1,3)-sigmatropic migration with inversion of the migrating group

6.4 ORBITAL SYMMETRY AND PHOTOCHEMICAL REACTIONS

Extensive investigations have shown that pericyclic reactions proceed in opposite senses in the ground and excited states (6.12).

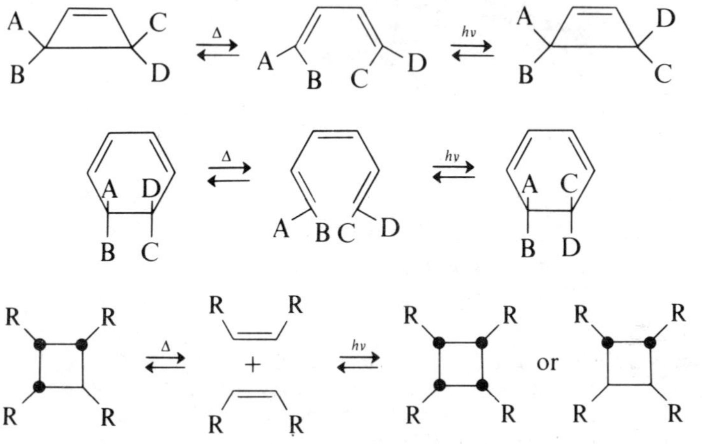

$$(6.12)$$

In order to understand the origin of this dichotomy between thermal and photochemical reactions, and in order to illustrate the nature of the problem, the electrocyclization butadiene $\rightleftharpoons$ cyclobutene will be reconsidered. From an examination of the state correlation diagram (Figure 6.7) it is seen that in the disrotatory mode the ground states do not correlate and hence are separated by an energy barrier, while the S_1 states do correlate. It might therefore be thought that herein lies the explanation of why the disrotatory cyclization of butadiene occurs photochemically but not thermally. This, however, would be a delusion, for although the disrotatory conversion of S_1-butadiene into S_1-cyclobutene is allowed by orbital symmetry, it cannot possibly occur because S_1-cyclobutene is about 200–250 kJ mol^{-1} (50–60 kcal mol^{-1}) more endothermic than S_1-butadiene. It follows that the reaction sequence cannot be

$$S_0\text{-butadiene} \rightarrow S_1\text{-butadiene} \rightarrow S_1\text{-cyclobutene} \rightarrow S_0\text{-cyclobutene}$$

172

It must involve the radiationless transition†

$$S_1\text{-butadiene} \rightsquigarrow S_0\text{-cyclobutene}$$

The essential point seems to be that state correlation diagrams are only two-dimensional diagrams and thus give extremely crude representations of energy profiles. A much better picture may be obtained by detailed calculation. According to valence-bond calculations,[7] the energy profiles for the disrotatory cyclization of butadiene actually have the form of Figure 6.12, and all now becomes clear.

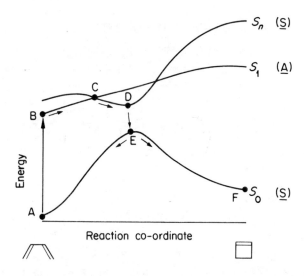

Figure 6.12. Energy profile for the disrotatory cyclization of butadiene (see text for explanation of lettering) (adapted from W. Th. A. M. van der Lugt and L. J. Oosterhoff, *Chem. Commun.*, 1235 (1968), and reproduced by permission of the Chemical Society)

S_0-Butadiene (A) is excited to the S_1-state (B), and molecular vibrations enable it to progress up the slight energy hill to the point (C) where it can cross to the symmetric S_n-state and coast downhill to the energy minimum (D). Here the molecule can undergo rapid radiationless transition (small energy gap) to S_0 (E), where it has the choice of returning to butadiene or proceeding to (F) and forming cyclobutene. In the conrotatory mode (Figure 6.13) such a sequence is impossible.

The above considerations have very general implications. Referring to Figure 6.7, the energy barrier which makes the ground state reaction forbidden

† It should be noted that apart from proton transfer reactions $MH^* \rightleftharpoons M^* + H^+$, which give excited state products and therefore occur adiabatically on the S_1-energy surface, the preponderant majority of photochemical reactions involve radiationless transitions so that the excited starting material goes directly to ground state products (and *not* via the excited state of the products).

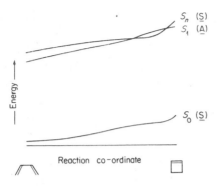

Figure 6.13. Energy profile for the conrotatory cyclization of butadiene (adapted
from W. Th. A. M. van der Lugt and L. J.
Oosterhoff, *Chem. Commun.*, 1235 (1968),
and reproduced by permission of the
Chemical Society)

arises because the two ground states are correlated with upper excited states.
Inevitably there must be a corresponding minimum in the energy surface of an
upper state of the same symmetry. Since upper states are densely packed, it will
normally be the case that the upper state with the minimum will cross lower
excited states and probably also S_1. Therefore correlation diagrams for reactions
forbidden in the ground state will usually be of the form of Figure 6.14 (for
example, see Figure 6.12) and not that of Figure 6.15.

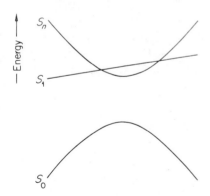

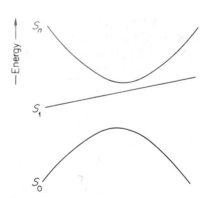

Figure 6.14. Schematic correlation
diagram showing a crossing of upper
singlet states

Figure 6.15. Schematic correlation
diagram where upper singlets do not
cross

From this it follows that, in general, when orbital symmetry forbids a reaction
in the ground state, that same reaction will be allowed photochemically. It
should always be borne in mind that reactions allowed by orbital symmetry
do not necessarily occur—there may, for example, be insuperable geometric

restraints, as in the symmetry-allowed antarafacial sigmatropic rearrangements in cyclic systems or in the thermal disrotatory cyclization of butadiene to *trans*-cyclobutene.

It has recently been shown[8] that there are photochemical reactions in which the excited reagent correlates *directly* with the ground state product. Consider the hydrogen abstraction step in the Norrish type 2 reaction (see Chapter 7, p. 193). It is assumed that the abstraction occurs in the plane of the keto group, which is taken to be the symmetry plane of the system. The electrons of the n and π^* orbitals of the carbonyl group and the electrons of the C—H σ-bond are considered and classified as σ or π with respect to the symmetry plane (Figure 6.16).

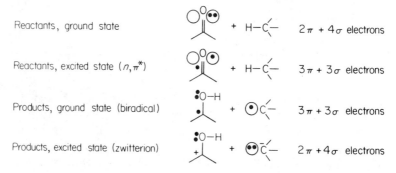

Figure 6.16. Classification of states in the reactants and products for hydrogen abstraction

This leads to the correlation diagram of Figure 6.17, from which it can be seen that the excited state of the reagents correlates *directly* with the ground state of the product biradical. The crossing is allowed because the states have different symmetries. Similar correlations have been discovered in other systems.

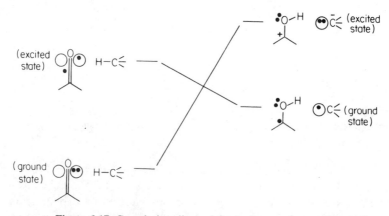

Figure 6.17. Correlation diagram for hydrogen abstraction

REFERENCES

1. An excellent introduction is T. L. Gilchrist and R. C. Storr, *Organic Reactions and Orbital Symmetry*, Cambridge University Press, London (1972).
2. W. A. Goddard, *J. Amer. Chem. Soc.*, **94**, 793 (1972).
3. Reviewed by K. Fukui and H. Fujimoto in B. S. Thyagarajan (ed.), *Mechanisms of Molecular Migration*, volume 2, Interscience, London (1969), p. 117.
4. H. C. Longuet-Higgins and E. W. Abrahamson, *J. Amer. Chem. Soc.*, **87**, 2045 (1965).
5. M. J. S. Dewar, *Molecular Orbital Theory of Organic Chemistry*, McGraw-Hill, New York (1969); H. E. Zimmermann, *J. Amer. Chem. Soc.*, **88**, 1564 and 1566 (1966); H. E. Zimmermann, *Angew. Chem. Intern. Ed.*, **8**, 1 (1969).
6. R. B. Woodward and R. Hoffmann, *The Conservation of Orbital Symmetry*, Verlag Chemie, Weinheim (1970).
7. W. Th. A. M. van der Lugt and L. J. Oosterhoff, *Chem. Commun.*, 1235 (1968).
8. L. Salem, W. G. Dauben and N. J. Turro, *J. Chim. Phys.*, **70**, 694 (1973); L. Salem, *J. Amer. Chem. Soc.*, **96**, 3486 (1974).

Chapter 7

C=O Chromophores

Compounds containing the carbonyl (C=O) chromophore are the most widely and intensively investigated in photochemistry. More information is available for ketones than for other groups of compounds with this functional group, and this balance is reflected in the choice of examples throughout the chapter. Thiocarbonyl (C=S) compounds are included here rather than with other sulphur chromophores, because of their similarity to carbonyl compounds.

7.1 SPECTRA AND EXCITED STATES

Aliphatic ketones and aldehydes show a weak ($\varepsilon \sim 10\text{--}30\,l\,mol^{-1}\,cm^{-1}$) ultraviolet absorption band around 280–300 nm, corresponding to the symmetry and overlap forbidden $n \rightarrow \pi^*$ transition (see Appendix, p. 358). Strong absorption occurs in the vacuum ultraviolet region, and there are bands centred around 195, 170 and 155 nm in the spectrum of saturated aliphatic ketones. The last of these corresponds to a $\pi \rightarrow \pi^*$ transition. The lowest energy excited states for these compounds are undoubtedly the (n, π^*) singlet and triplet states, and these are responsible for the observed photochemistry.

Acetone exhibits only weak fluorescence ($\phi_f \sim 0.001$, $\tau_f = 2.1 \times 10^{-9}$ s, $E_S = 350\,kJ\,mol^{-1}$, 84 kcal mol^{-1}) in solution at room temperature, because the rate constant for intersystem crossing to the triplet state is very high. Phosphorescence ($\phi_p = 0.043$, $\tau_p = 0.33$ ms, $E_T = 310\,kJ\,mol^{-1}$, 74 kcal mol^{-1}) is observed at 77 K.[1] Alkyl substitution alpha to the carbonyl group increases the radiative lifetime and efficiency, and for di-t-butyl ketone[2] $\phi_f = 0.0044$, $\phi_p^{77\,K} = 0.89$, and $\tau_p = 8.6$ ms. The fact that ($\phi_f + \phi_p$) is much less than unity for acetone shows that radiationless decay processes play an important part in the behaviour of (n, π^*) excited states. The triplet energy derived from phosphorescence measurements represents the difference in energy between the lowest energy (non-planar) triplet state and the ground state with the same (non-planar) geometry. The energy difference between planar ground state and non-planar triplet is higher (320–5 kJ mol^{-1}, 77–8 kcal mol^{-1}), and that between planar ground state and planar triplet, which can be obtained from $S_0 \rightarrow T_1$ absorption spectra, is higher still (335 kJ mol^{-1}, 80 kcal mol^{-1}).†

† Note that three different values for the triplet energy can be obtained, which correspond to different geometries of ground and excited state.

In unsaturated ketones the π and π^* orbitals of the individual chromophores interact to produce two π and two π^* orbitals of different energy. The n orbitals are of different symmetry and are not perturbed significantly by the linking of the chromophores. The orbitals of unsaturated ketones are therefore ordered as in Figure 7.1, from which we expect that there will be a large bathochromic shift of the $\pi \rightarrow \pi^*$ transition and a smaller bathochromic shift of the $n \rightarrow \pi^*$ transition in these compounds as compared with the corresponding transitions in the isolated chromophores.

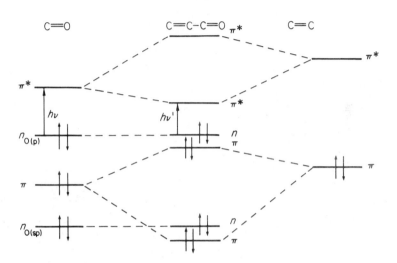

Figure 7.1. Molecular orbital diagram for a conjugated unsaturated ketone

The ultraviolet absorption spectrum of enones shows both weak $n \rightarrow \pi^*$ and strong $\pi \rightarrow \pi^*$ bands at wavelengths longer than 200 nm, as seen in the spectrum of 4-methylpent-3-en-2-one (Figure 7.2). In some instances the weak $n \rightarrow \pi^*$ band is hidden under the much more intense $\pi \rightarrow \pi^*$ absorption.

Both fluorescence and phosphorescence of conjugated unsaturated carbonyl compounds are very weak (for acrolein, $\phi_f = 0.007$ and $\phi_p = 0.00004$).[3] Aromatic ketones similarly exhibit no fluorescence because intersystem crossing is very rapid, but phosphorescence is strong at 77 K (for benzophenone, $\phi_p = 0.90$, $\tau_p = 5$ ms, and $E_T = 285$ kJ mol^{-1}, 68 kcal mol^{-1}). In completely inert solvents such as perfluorohydrocarbons, benzophenone phosphorescence can be observed even at room temperature.[4]

The major difference from the photochemical point of view between saturated and unsaturated carbonyl compounds is that for the unsaturated compounds the (n, π^*) states are not always the lowest in energy. For some the lowest triplet state is definitely (π, π^*) in nature (e.g., acylnaphthalenes, $E_T = 245$ kJ mol^{-1}, 58 kcal mol^{-1}), and for others the (n, π^*) and (π, π^*) triplet states are

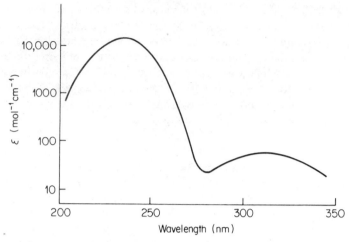

Figure 7.2. Ultraviolet spectrum of 4-methylpent-3-en-2-one in ethanol

Table 7.1. Ultraviolet absorption data for C=O compounds

	λ_{max} (nm)	ε_{max} (l mol^{-1} cm^{-1})
$CH_3-CO-CH_3$	280	15
$CH_3-CO-OH$	204	50
$CH_3-CO-OC_2H_5$	204	50
$CH_3-CO-Cl$	235	53
$CH_3-CO-NH_2$	214	

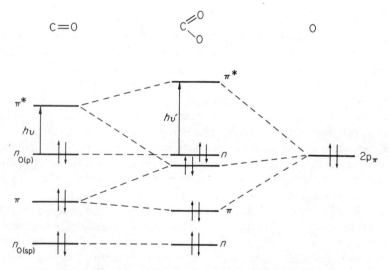

Figure 7.3. Molecular orbital diagram for a carboxylic acid derivative

very close in energy (e.g., simple acyclic enones or p-methoxyphenyl ketones). Because of the considerable difference in the reactions of (n, π^*) and (π, π^*) states, this has a profound effect on the observed photochemistry.

Saturated carboxylic acids and their derivatives such as esters or amides exhibit weak $n \rightarrow \pi^*$ absorption in the normal ultraviolet range, but the position of the band is at much shorter wavelength than that of the corresponding ketone (Table 7.1). This can be rationalized in terms of the effect of the adjacent heteroatom on the orbital energies of the carbonyl group (Figure 7.3).

Emission from saturated acid derivatives is not well documented, but unsaturated and aromatic compounds have received more attention. Benzoate esters show only very weak fluorescence, but naphthoate esters fluoresce more strongly. Benzoate and naphthoate esters and other derivatives exhibit intense, long-lived phosphorescence at 77 K. The emitting triplet states are (π, π^*), and for simple benzoates[5] $\tau_p = 2 \cdot 5$ s, and $E_T = 325$ kJ mol^{-1} (78 kcal mol^{-1}).

7.2   **REACTIONS OF C=O (n, π^*) STATES**

The (n, π^*) excited state of a carbonyl compound does not have such an electron-deficiency near the carbon atom as does the ground state, because an electron is transferred in the excitation process from a non-bonding orbital localized mainly on oxygen to an anti-bonding π^* orbital which covers both carbon and oxygen nuclei. This makes nucleophilic attack at the carbon atom a relatively unimportant reaction for the excited state. Instead, the outstanding feature of the electron distribution in the (n, π^*) state of a carbonyl compound is the unpaired electron in a p-type orbital on oxygen (7.1).

$$\begin{array}{c} \diagdown \\ \diagup \end{array} C{=}O \qquad\qquad (7.1)$$

This electronic feature gives the species a resemblance to an alkoxy radical, which also has an odd electron on oxygen in an orbital with considerable p-character (7.2).

$$\begin{array}{c} \diagdown \\ \diagup \end{array} C{-}O \qquad\qquad (7.2)$$

It is therefore not surprising that the primary steps in many of the major reactions of (n, π^*) states of carbonyl compounds resemble reactions of alkoxy radicals. Such reactions are α-cleavage (7.3), hydrogen abstraction (7.4), and

$$\begin{array}{ccc} \overset{R}{\underset{R}{\diagdown}} C{=}O \rightarrow R{-}\dot{C}{=}O & \text{cf.} & \overset{R}{\underset{R}{\diagup}} R{-}C{-}O^{\cdot} \rightarrow \overset{R}{\underset{R}{\diagdown}} C{=}O \\ {\scriptstyle (n,\, \pi^*)} & +R^{\cdot} & +R^{\cdot} \end{array} \qquad (7.3)$$

$$
\begin{array}{ccc}
\underset{R}{\overset{R}{\diagdown}}C=O \xrightarrow{\text{R'H}} & \underset{R}{\overset{R}{\diagdown}}\dot{C}-OH & \text{cf.} \quad R-\underset{R}{\overset{R}{|}}C-O\cdot \xrightarrow{\text{R'H}} R-\underset{R}{\overset{R}{|}}C-OH \\
{\scriptstyle (n,\,\pi^*)} & +\,{}^{\cdot}R' & +\,{}^{\cdot}R'
\end{array} \qquad (7.4)
$$

addition to alkenes (7.5). Differences arise in the secondary reaction steps because a radical reaction generates another radical, whereas an excited state generates two radicals.

$$
\underset{R}{\overset{R}{\diagdown}}C=O \longrightarrow R \quad \text{cf.} \quad R-\underset{R}{\overset{R}{|}}C-O\cdot \longrightarrow \underset{R}{\overset{R}{|}}R-C-O \qquad (7.5)
$$

7.3   α-CLEAVAGE

7.3.1   Acyclic Compounds

In order to be effective in promoting bond cleavage, the electronic excitation energy of the carbonyl group in an excited state must be converted into vibrational energy localized mainly in the bond which is broken in the primary photochemical process. In the gas-phase photochemistry of saturated carbonyl compounds this most frequently involves localization of vibrational energy in one or other of the bonds adjacent to the carbonyl group, and cleavage of this bond to give two radical fragments occurs (the 'Norrish type 1' process). Thus a ketone gives rise to an alkyl and an acyl radical (7.6), and an acyl chloride gives an acyl radical and a halogen atom (7.7) in the primary process.

$$
CH_3-\overset{\overset{\displaystyle O}{\|}}{C}-CH_3 \xrightarrow{h\nu} {}^{\cdot}CH_3 + CH_3-\dot{C}=O \qquad (7.6)
$$

$$
CH_3-\overset{\overset{\displaystyle O}{\|}}{C}-Cl \xrightarrow{h\nu} CH_3-\dot{C}=O + Cl^{\cdot} \qquad (7.7)
$$

When the two α-bonds are not identical, it is the weaker bond which breaks preferentially, as seen for the acyl halide above (7.7). For butan-2-one $(CH_3-CO-C_2H_5)$ the ratio of ethyl to methyl radicals produced in the primary steps is about 40:1 when 313 nm radiation is used, in agreement with this generalization. The ratio varies with wavelength in the gas phase, and when 254 nm radiation is employed the ratio of ethyl to methyl radicals is only 2·4:1. This wavelength effect is accounted for in the following way. The C—C bond which is broken in the primary cleavage process has a bond strength (~ 325 kJ mol^{-1}, 78 kcal mol^{-1}) which is very close to the (n, π^*) excited state energies of simple aliphatic ketones. The cleavage is therefore relatively slow from the $v = 0$ vibrational level of the excited state, but the rate constant for cleavage increases as higher energy (shorter wavelength) radiation or a higher temperature of reaction provides for significant population of higher vibrational levels. This explains the effect of either wavelength or temperature on the rates of both singlet and triplet state reaction in the gas phase, and also the increasing percentage of singlet contribution to the overall (singlet + triplet) reaction of acetone as the temperature increases. On the same basis, the selectivity of cleavage for a compound with two different α-cleavage modes will be highest for reaction from the $v = 0$ level, and lower for reaction from higher vibrational levels, as observed for butan-2-one.

The secondary reactions of the radicals produced in α-cleavage follow the pattern expected for free radical processes. Fragmentation (e.g., loss of carbon monoxide from an acyl radical), combination or disproportionation of two radicals, and radical substitution (e.g., abstraction of a hydrogen atom from the solvent) are all observed. The photoreactions of di-t-butyl ketone (7.8) illustrate all but the last of these reactions.[6]

$$(7.8)$$

In addition to the simple α-cleavage process, aliphatic aldehydes exhibit a different primary process in which a molecular elimination of carbon monoxide occurs. Acetaldehyde irradiated in the gas phase with 334 nm radiation gives largely hydrogen atoms and acetyl radicals (7.9) through the triplet (n, π^*) state, but when irradiated with 254 nm radiation the primary products are mainly methane and carbon monoxide (7.10) by reaction through the (n, π^*) singlet state. It is thought that the molecular elimination pathway is concerted or nearly so, with formation of the C_α—H bond occurring at the same time as, or very shortly after, cleavage of the $C(=O)$—H bond.[7]

$$CH_3CH=O \xrightarrow{\ h\nu\ (334\ nm)\ } H\cdot + CH_3-\overset{\cdot}{C}=O \qquad (7.9)$$

$$CH_3CH=O \xrightarrow{\ h\nu\ (254\ nm)\ } CH_4 + CO \qquad (7.10)$$

In solution the α-cleavage processes are of much less importance for acyclic ketones, especially if there are other processes, such as the Norrish type 2 reaction (see Chapter 7, p. 193), which the excited state can undergo. This has two causes, the first of which is related to the wavelength effect discussed above for gas phase reactions. In the gas phase, vibrational deactivation is relatively slow, and α-cleavage from a higher vibrational level of the electronically excited state competes effectively with deactivation, whereas in solution vibrational deactivation is much faster because of the involvement of solvent molecules.

In solution α-cleavage must therefore occur very largely from the $v = 0$ vibrational level, and this is relatively slow. The second factor contributing to the reduced importance of α-cleavage processes in solution is that the initially formed radicals are produced in a 'cage' of solvent molecules, and a major secondary reaction is recombination to form ground state starting material. This provides in effect a pathway for radiationless deactivation of the excited state and results in a lowering of the quantum yield for product formation as compared with gas phase photochemistry.

The importance of α-cleavage in solution phase photochemistry increases with increasing temperature,[8] as population of upper vibrational levels becomes significant. For instance,[9] the quantum yield for the Norrish type 1 process for heptan-4-one ($CH_3CH_2CH_2CO-CH_2CH_2CH_3$) in solution increases from 0·01 at 20 °C to 0·30 at 96 °C. Thus in both gas and liquid phase photochemistry the Norrish type 1 process is considerably affected by changes in temperature, but other intramolecular processes, and in particular the Norrish type 2 reaction, are not as markedly affected except insofar as the type 1 cleavage is in competition with the other processes.

This is the general state of affairs, but the situation is always governed by the relative magnitudes of the various rate constants, and where there is a particularly weak α-bond, α-cleavage plays an important part in the photochemistry in solution at room temperature. This is true for t-butyl ketones (7.11)[10] and for benzyl ketones (7.12).[11]

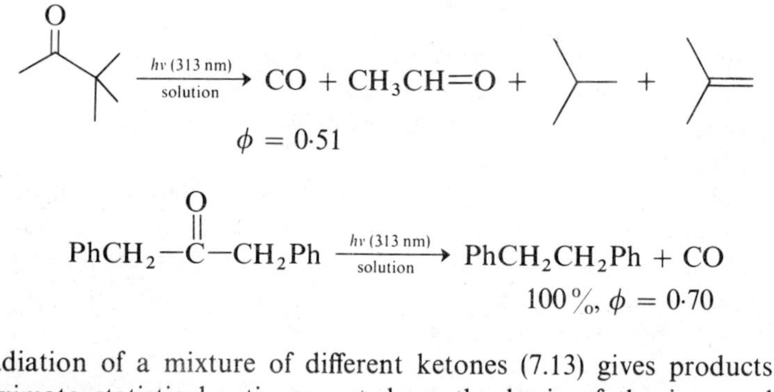

$$\text{(structure)} \xrightarrow[\text{solution}]{h\nu\ (313\ nm)} CO + CH_3CH{=}O + \text{(structure)} + \text{(structure)} \qquad (7.11)$$

$$\phi = 0·51$$

$$PhCH_2-\overset{O}{\overset{\|}{C}}-CH_2Ph \xrightarrow[\text{solution}]{h\nu\ (313\ nm)} PhCH_2CH_2Ph + CO$$

$$100\%,\ \phi = 0·70 \qquad (7.12)$$

Irradiation of a mixture of different ketones (7.13) gives products in the approximate statistical ratio expected on the basis of the intermediacy of completely free radicals,[12] and this is strong evidence in favour of a step-wise free radical mechanism. In other cases the initially produced radicals have been observed directly by e.s.r. spectroscopy.[13]

$$
\begin{array}{c}
PhCH_2-CO-CH_2Ph \\
+ \\
Ph_2CH-CO-CHPh_2
\end{array}
\xrightarrow{h\nu}
\begin{array}{c}
PhCH_2CH_2Ph \\
+ \\
PhCH_2CHPh_2 \\
+ \\
'Ph_2CH-CHPh_2
\end{array}
\quad 93\%,\ 1{:}2{:}1 \qquad (7.13)
$$

Similar results with benzyl arylacetates (7.14) indicate that decarboxylation of esters is also a step-wise free radical process.[14]

$$ArCH_2CH_2Ar'$$
$$+$$

$$ArCH_2CO_2CH_2Ar' \xrightarrow{hv\ (254\ nm)} ArCH_2CH_2Ar \quad 2.5:1:1, \phi \sim 0.2 \quad (7.14)$$

(Ar = p-CH$_3$C$_6$H$_4$—

$$+$$

Ar' = p-CH$_3$OC$_6$H$_4$—) $$Ar'CH_2CH_2Ar'$$

Aromatic ketones such as benzophenone (Ph—CO—Ph) or acetophenone (Ph—CO—CH$_3$) do not give rise to significant amounts of product by α-cleavage on irradiation, possibly because there is less electronic energy available than with alkyl ketones for the cleavage process (E_T for benzophenone is 285 (68), for acetophenone is 305 (72), and for acetone is 325 (77) kJ (kcal) mol^{-1}), and as noted earlier the cleavage process is finely balanced in terms of enthalpy. Aryl ketones with weaker than normal C(=O)—C$_\alpha$ bonds do give α-cleavage products on irradiation. Such systems are aryl t-butyl ketones[15] and benzoin ethers (7.15).[16]

$$\underset{\underset{OCH_3}{|}}{Ph-\overset{\overset{O}{\|}}{C}-CH-Ph} \xrightarrow{hv\ (Pyrex)} PhCH=O + PhCO-COPh + \underset{\underset{OCH_3}{|}}{Ph\overset{\overset{OCH_3}{|}}{C}H-CHPh}$$

$$60–70\% \qquad (7.15)$$

Aliphatic aldehydes do lose carbon monoxide more readily in solution than do ketones, but the efficiency of the reaction is reduced as compared with that of the gas phase process. Acyl chlorides and bromides are much less affected than are ketones by the change from gas to solution phase, and decarbonylation can be efficient in solution (7.16).[17] Carboxylic acids with a β-heteroatom undergo a sensitized decarboxylation reaction on irradiation in solution with benzophenone (7.17). This process involves a charge–transfer exciplex as an intermediate, and the effect of the heteroatom may be to promote α-cleavage by stabilizing the radical formed.[18]

$$C_3F_7-\overset{\overset{O}{\|}}{C}-Br \xrightarrow[solution]{hv} C_3F_7Br \qquad (7.16)$$

$$94\%$$

$$PhOCH_2COOH \xrightarrow[Ph_2CO]{hv} PhOCH_3 + CO_2 \qquad (7.17)$$

7.3.2   **Cyclic Compounds**

Cyclic ketones, esters (lactones), amides (lactams) and anhydrides undergo α-cleavage to give a biradical species on photolysis in gas or solution phase. The biradical can then recombine to give starting material (or its C$_\alpha$-epimer if appropriate), or it can fragment to lose carbon monoxide or carbon dioxide. From cyclopentanone[19] this leads to hydrocarbon products (7.18). Cyclobutanones are different in that direct irradiation leads to an alkene and a ketene (7.19), whilst triplet sensitization gives a cyclopropane and carbon monoxide in a non-concerted process (7.20). The singlet state reaction is stereospecific, and it may be a concerted rather than a two-step reaction.[20]

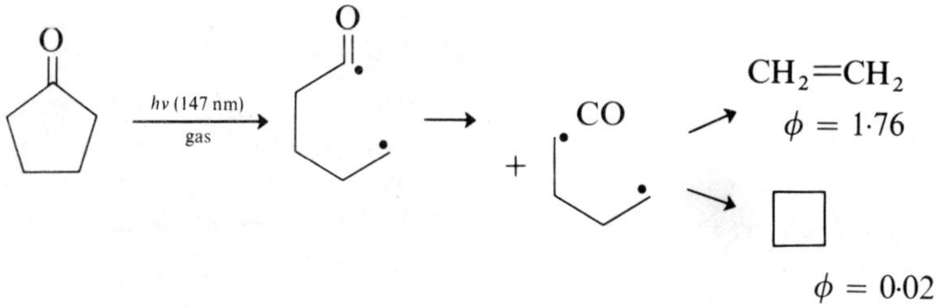

$$\phi = 1\cdot76 \qquad (7.18)$$

$$\phi = 0\cdot02$$

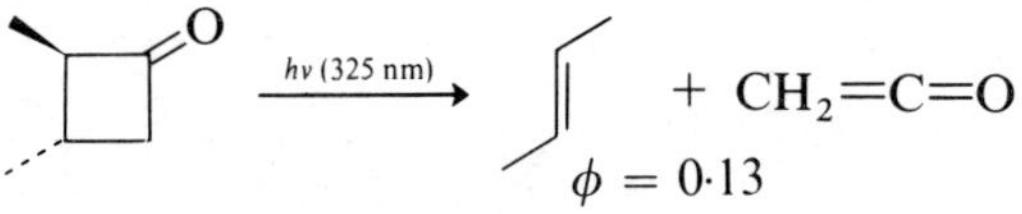

$$+ \; CH_2{=}C{=}O \qquad (7.19)$$

$$\phi = 0\cdot13$$

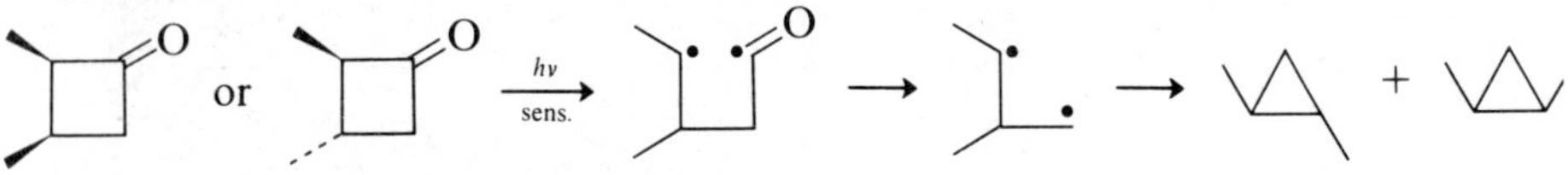

$$2\cdot4:1\cdot0$$

$$(7.20)$$

The recombination process is important even when hydrogen transfer reactions occur (see below), but loss of carbon monoxide in solution phase photochemistry is a major reaction pathway only when the alkyl radical centres are stabilized by alkyl substitution, by β,γ-unsaturation, or by cyclopropyl conjugation. This reflects an increase in the rate of loss of carbon monoxide from the acyl–alkyl biradical in these systems, and whereas 2,6-dimethylcyclohexanone (7.21) gives no carbon monoxide on photolysis in solution at room temperature, 2,2,6,6-tetramethylcyclohexanone (7.22) gives carbon monoxide in a yield greater than 70%. 7,7,9,9-Tetramethylbicyclo-[4.3.0]non-1(6)-en-8-one gives a 100% yield of carbon monoxide and hydrocarbon products (7.23).[21]

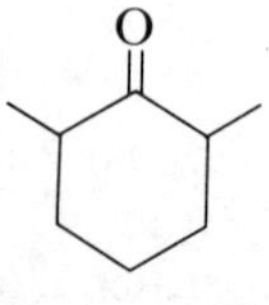

$$(7.21) \qquad\qquad (7.22)$$

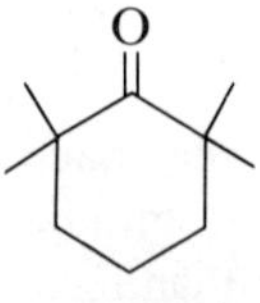

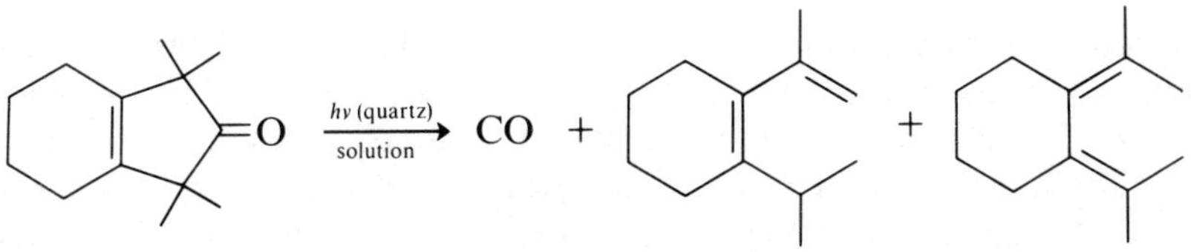

100 %   55 %   45 %   (7.23)

That the decarbonylation process is non-concerted is indicated[22] by the production of the same hydrocarbon product mixture from (−)-thujone or (+)-isothujone on photolysis (7.24).

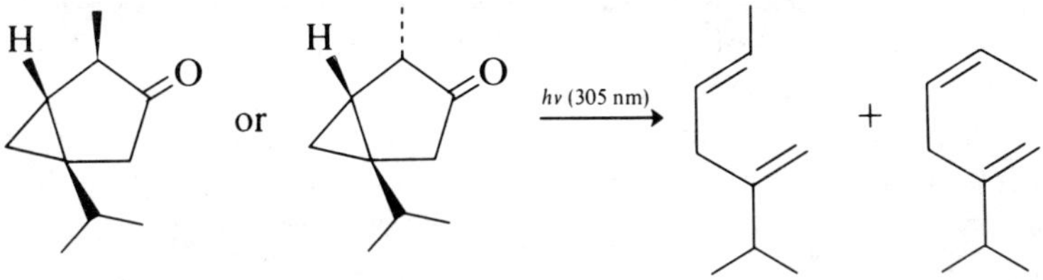

87 : 13   (7.24)

Of further interest in the cyclic systems are those intramolecular secondary processes which have no counterpart in the acyclic systems. First, the biradical can undergo one of two hydrogen-transfer processes via a cyclic transition-state, in which a hydrogen atom is transferred to one radical centre from the atom adjacent to the other radical centre. The products are an unsaturated aldehyde (7.25) or a ketene (7.26) from a cyclic ketone,[23] and an unsaturated formate (7.27) or a dialdehyde (7.28) from a lactone.[24] The ketene is normally isolated as an adduct with an added nucleophile.

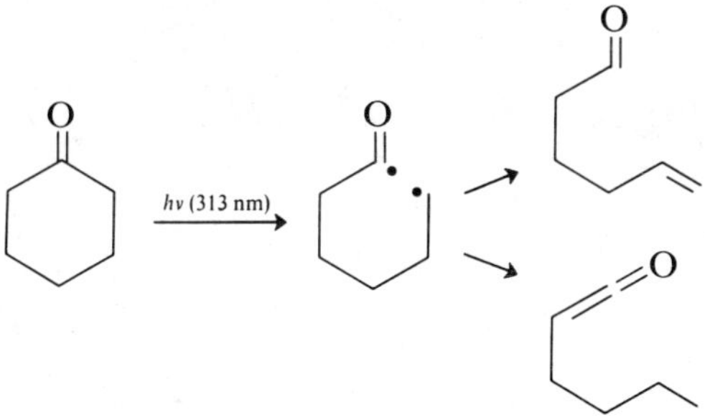

$\phi = 0.09$ (7.25)

(7.26)

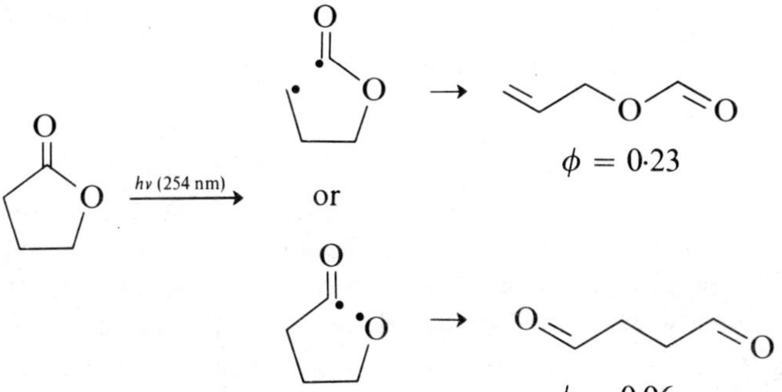

$\phi = 0.23$ (7.27)

$\phi = 0.06$ (7.28)

These processes are most important for 5- and 6-membered cyclic carbonyl compounds, that is those for which the hydrogen transfer involves a 5- or 6-membered cyclic transition-state, although they do occur for compounds with other ring sizes. The ratio of products (aldehyde:ketene) for mono- and polycyclic cyclopentanones and cyclohexanones can be rationalized qualitatively in terms of the effect of angle strain and steric interaction in the two alternative transition-states through which the biradical passes to give products.[25]

When the hydrogen-transfer processes are inhibited by ring size or other steric factors, as with cyclobutanones or certain tricyclic ketones such as fenchone (7.29), another secondary route is revealed which involves ring-closure of the biradical with formation of a C—O bond. The oxacarbene produced is isolated as a cyclic ketal in methanol solvent, or it can be trapped with added cyclohexene.[26] For cyclobutanones this process is in competition with cleavage to alkene and ketene in solution, and both reactions give products[27] in which the stereochemistry of the cyclobutanone has been retained (7.30).

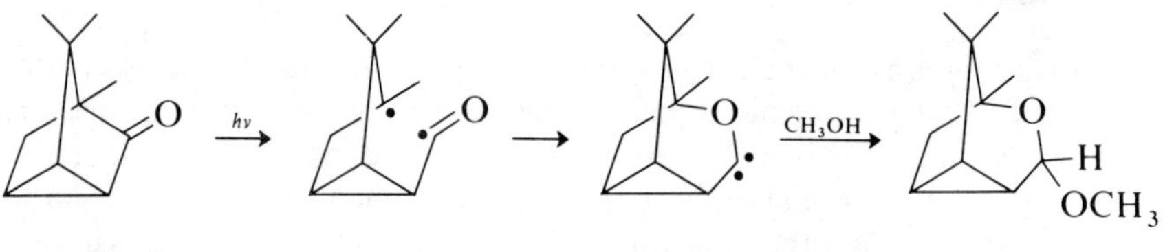

$$(7.29)$$

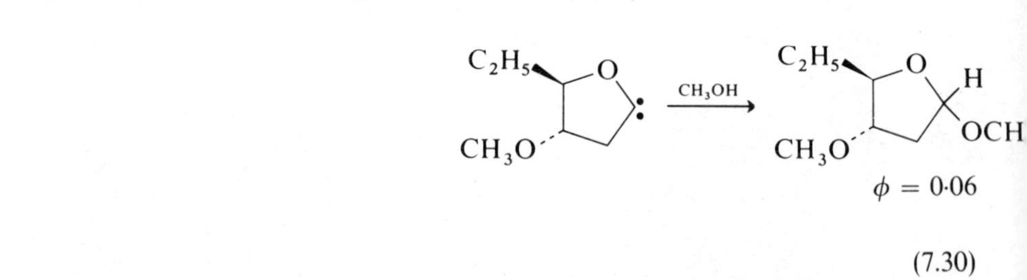

$$\phi = 0.03 \qquad \phi = 0.10$$

$$\phi = 0.06$$

$$(7.30)$$

Mechanistic studies indicate that the primary α-cleavage step in the photoreactions of cyclic ketones other than cyclobutanones occurs mainly from the triplet state (compare acyclic ketones, for which both singlet and triplet states seem to be involved). There is a good deal of evidence that the postulated biradical species is a true intermediate for the triplet state reactions, and the nature of this evidence can be illustrated using the 2,3-dimethylcyclohexanones (7.31) as an example. Each isomer undergoes photoepimerization to the other isomer, and each gives rise to the same ratio of products even before any significant amount of photoepimerization has occurred. In addition, the kinetics of product formation and of ketone disappearance are consistent with a reaction scheme involving the formation of a common biradical intermediate from either ketone.[28]

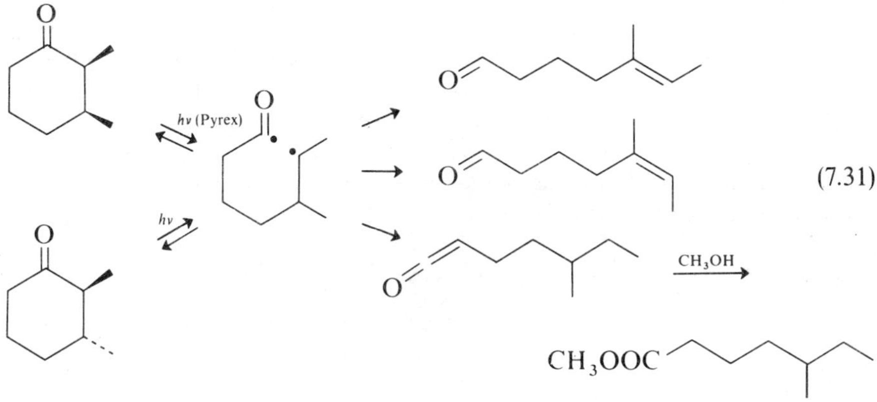

(7.31)

7.3.3   β,γ-Unsaturated Ketones

β,γ-Unsaturation in a carbonyl compound promotes α-cleavage, formally at least because of allylic stabilization of the radical produced. Two modes of overall reaction are observed. The first involves a 1,3-acyl shift resulting in the formation of an isomeric β,γ-unsaturated compound, as with cyclo-oct-3-enone (7.32)[29] or 6,7-dimethylbicyclo[3.2.0]hept-6-en-2-one (7.33).[30]

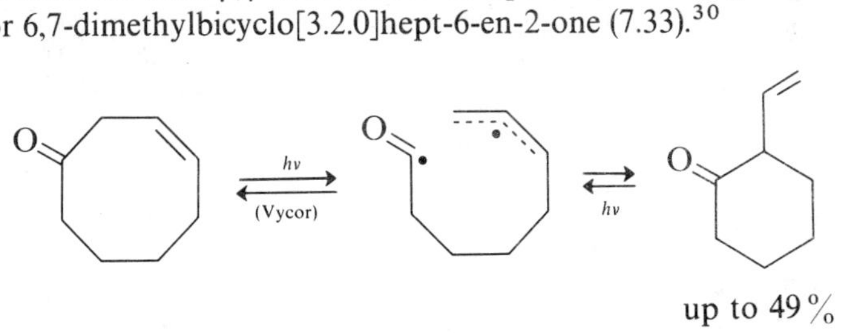

(7.32)

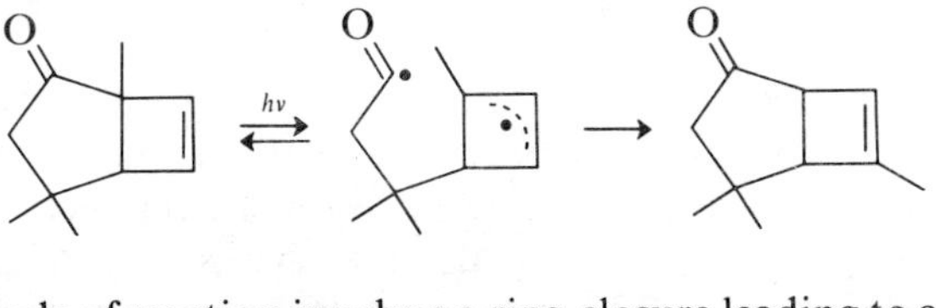

up to 49%

(7.33)

The second mode of reaction involves a ring-closure leading to a cyclopropyl ketone, as with bicyclo[4.4.0]dec-1(6)-en-3-one (7.34). It seems[31] that this enone reacts via the lowest triplet excited state, and the rearrangement parallels the triplet di-π-methane rearrangement of 1,4-dienes (see Chapter 8, section 8.4.4), which is similarly non-concerted.

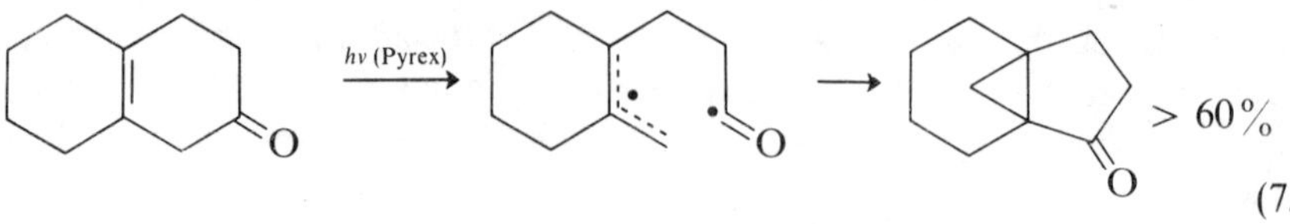

$$> 60\% \qquad (7.34)$$

The closely related bicycloundecenone (7.35) reacts via the lowest singlet excited state to give an isomeric β,γ-unsaturated ketone, possibly by way of a concerted 1,3-acyl shift. The ultraviolet absorption spectrum of this enone shows that interaction between the two chromophores occurs, since $\lambda_{max} = 292$ nm and $\varepsilon_{max} = 250\,1\,mol^{-1}\,cm^{-1}$, as compared with $\lambda_{max} = 281$ nm and $\varepsilon_{max} = 38\,1\,mol^{-1}\,cm^{-1}$ for the bicyclodecenone in (7.34), and this interaction may be related to the existence of a concerted pathway for rearrangement.

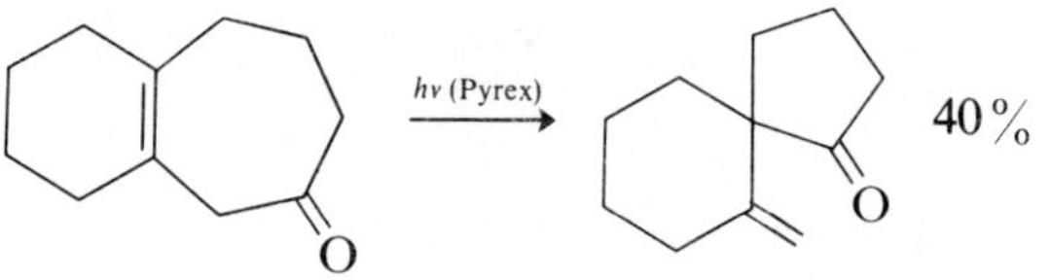

$$40\% \qquad (7.35)$$

Many β,γ-unsaturated ketones undergo the 1,3-acyl shift on direct irradiation and the 1,2-acyl shift with ring-closure on sensitization. When the 1,3-shift occurs through the singlet (n, π^*) state it is concerted and stereospecific (7.36).[32] An attempt has been made[33] to rationalize the difference in behaviour between singlet and triplet states of β,γ-unsaturated ketones on the basis of a lowest singlet (n, π^*) state and a lowest triplet (π, π^*) state.

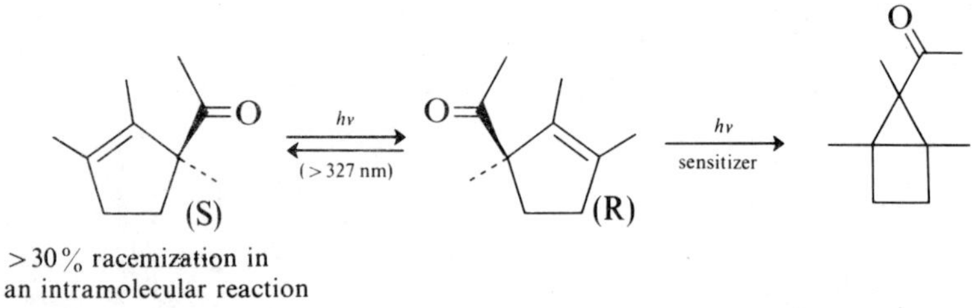

$$(7.36)$$

> 30% racemization in
an intramolecular reaction

7.4   β-CLEAVAGE REACTIONS

Some classes of compound have relatively weak C_α—C_β bonds which can undergo cleavage as a result of electronic excitation of the carbonyl group. Cyclopropyl ketones are one such class, and evidence for interaction between the carbonyl and cyclopropyl groups, which provides a mechanism by which energy may be transferred from the carbonyl group to the bond which is broken, is found in the ultraviolet absorption spectrum. The $n \rightarrow \pi^*$ absorption band is at longer wavelength and is increased in intensity as compared with the band for saturated ketones. Photolysis of acetylcyclopropane leads to cleavage of the cyclopropane ring,[34] and this is followed by a hydrogen shift to give pent-4-en-2-one as the major product (7.37).

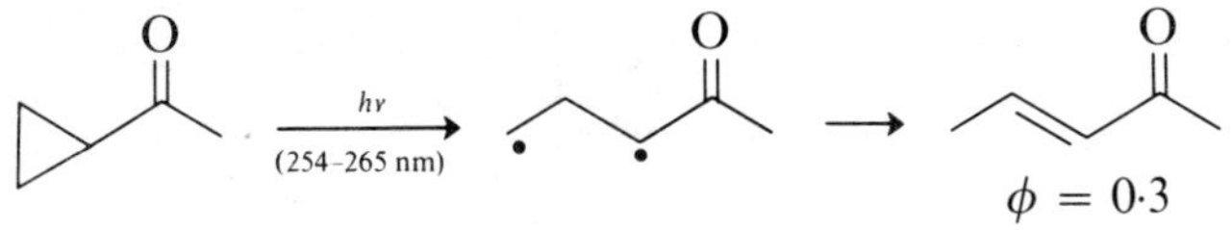

$$\phi = 0.3 \qquad (7.37)$$

In a similar way, bicyclo[4.1.0]heptan-2-ones undergo cleavage of one of the cyclopropyl C—C bonds (7.38) as well as cleavage of the (C-2)—(C-3) bond (this is an α-cleavage step).[35] Cleavage of the (C-1)—(C-2) bond, the alternative α-cleavage process, does not occur to any appreciable extent because of its higher bond strength. Substitution of methyl groups in the bicyclic system affects the ratio of products formed by the three different modes of cleavage in a manner which can be rationalized in terms of the weakening of C—C bonds by increased alkyl substitution.

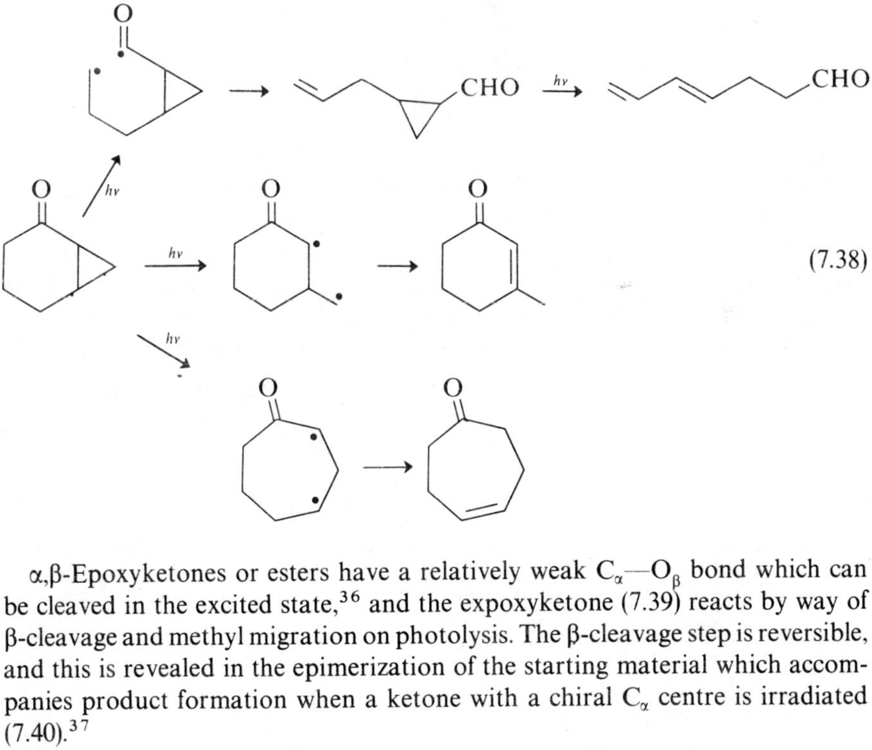

$$(7.38)$$

α,β-Epoxyketones or esters have a relatively weak C_α—O_β bond which can be cleaved in the excited state,[36] and the expoxyketone (7.39) reacts by way of β-cleavage and methyl migration on photolysis. The β-cleavage step is reversible, and this is revealed in the epimerization of the starting material which accompanies product formation when a ketone with a chiral C_α centre is irradiated (7.40).[37]

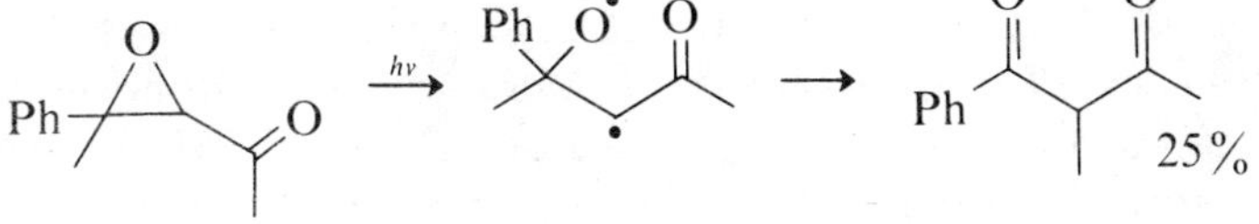

$$(7.39)$$

25%

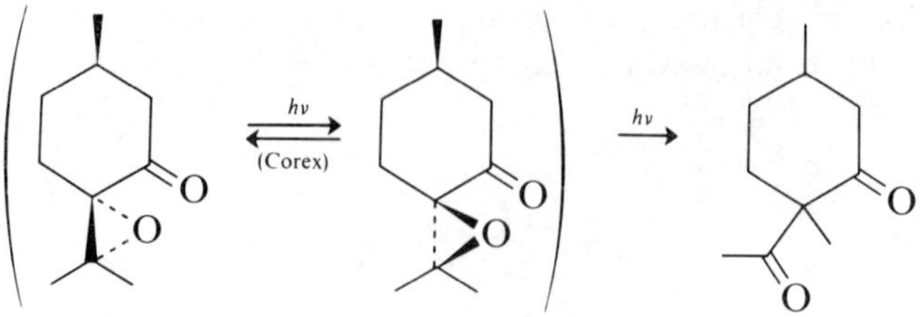

$$(7.40)$$

The cyclic γ,δ-unsaturated ketone cyclo-oct-4-enone provides a further example[38] of a system in which cleavage of a relatively weak C_α—C_β bond is an efficient process in the excited state (7.41). Acyclic γ,δ-unsaturated ketones undergo intramolecular cycloaddition (see Chapter 7, p. 208).

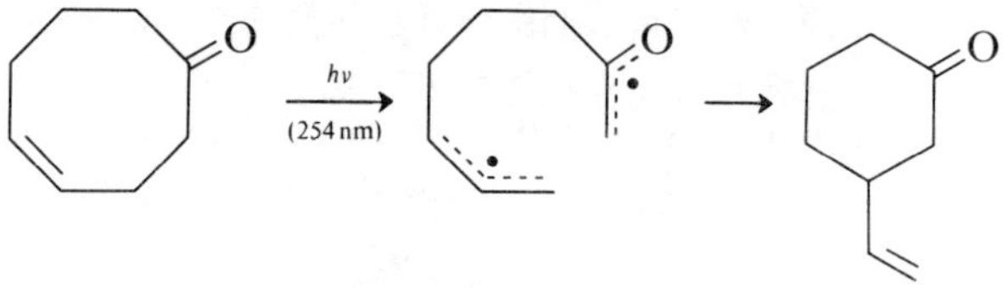

$$(7.41)$$

7.5   HYDROGEN ABSTRACTION

7.5.1   Intermolecular Photoreduction

Photoreduction of carbonyl compounds is one of the best known of photoreactions, and in its initial steps the mechanism involves transfer of a hydrogen atom to the oxygen atom of the carbonyl excited state from a donor molecule which may be solvent, an added reagent, or a ground state molecule of reactant. In this the behaviour of the (n, π^*) excited state is again similar to that of an alkoxy radical. The expected secondary radical reaction steps occur, and they are exemplified by the reaction of acetone and cyclohexene (7.42).[39]

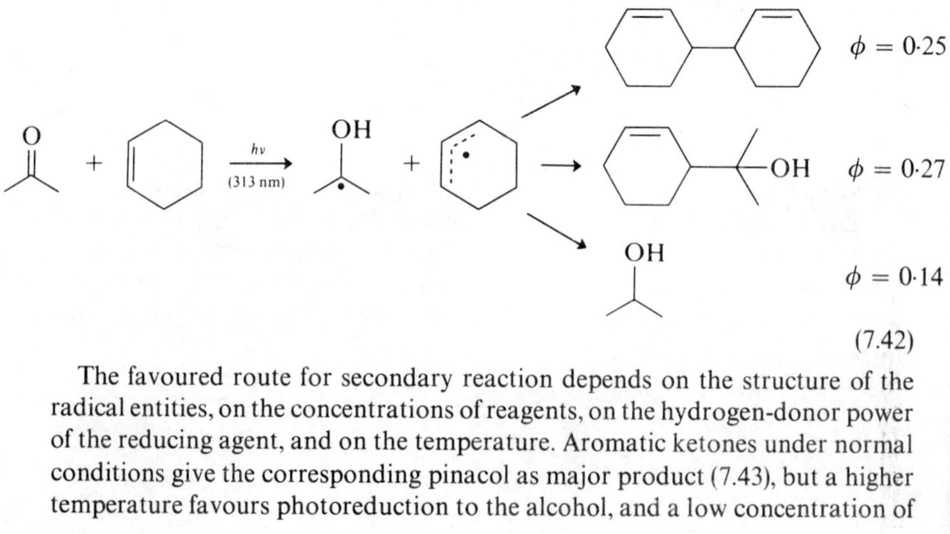

$$(7.42)$$

The favoured route for secondary reaction depends on the structure of the radical entities, on the concentrations of reagents, on the hydrogen-donor power of the reducing agent, and on the temperature. Aromatic ketones under normal conditions give the corresponding pinacol as major product (7.43), but a higher temperature favours photoreduction to the alcohol, and a low concentration of ketone promotes formation of the mixed pinacol as major product (7.44). The reaction steps involved may not be simple radical recombination processes,[40] and at least some of the reaction between benzophenone and propan-2-ol goes by way of an enol intermediate (7.45).

$$Ph_2C{=}O + CH_3OH \text{ or } (CH_3)_2CH{-}OH \xrightarrow[20\,°C]{h\nu\,(313\,nm)}$$

$(0.2\ \text{M})$

$$\underset{\displaystyle \overset{\displaystyle OH}{|}}{Ph_2C}\overset{\displaystyle OH}{\underset{\displaystyle |}{-}}CPh_2 \qquad \sim 100\% \qquad (7.43)$$

$$Ph_2C{=}O + CH_3OH \xrightarrow[20\,°C]{h\nu\,(313\,nm)} Ph_2\overset{\displaystyle OH}{\underset{\displaystyle |}{C}}{-}CH_2OH \qquad >90\%$$

$(0.0001\ \text{M})$

$$Ph_2C{=}O + (CH_3)_2CH{-}OH \xrightarrow[100\,°C]{h\nu\,(313\,nm)} Ph_2CH{-}OH \qquad (7.44)$$

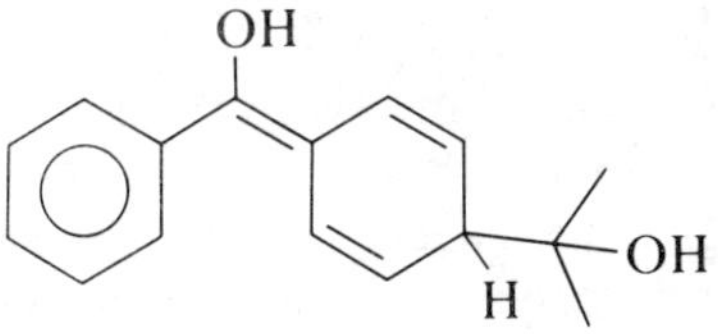

$$(7.45)$$

The excited state responsible for the hydrogen abstraction reactions of ketones is the (n, π^*) triplet state. The singlet state is too short-lived to undergo efficient intermolecular reaction, and there is some evidence that singlet state reactivity is lower than that of the triplet state. This is attributed to a requirement for closer approach of the oxygen and hydrogen atoms in the transition-state for singlet abstraction than in the transition-state for triplet abstraction.[41] For aryl ketones (Ar—CO—Ar' or Ar—CO—R) and for p-benzoquinones, those compounds which have a lowest (n, π^*) triplet state undergo efficient inter-molecular photoreduction, but those which have a lowest (π, π^*) triplet state undergo photoreduction with a much lower rate constant (for a given hydrogen donor) and usually with a much lower overall efficiency (Table 7.2).[42]

Table 7.2. Rate constants (k_r) for hydrogen abstraction by ketone triplet states

Ketone	Lowest triplet state	Hydrogen donor	$k_r/\text{l mol}^{-1}\,\text{s}^{-1}$
Cyclohexanone	(n, π^*)	Propan-2-ol	2×10^6
Cyclohexanone	(n, π^*)	$Bu_3Sn{-}H$	2×10^9
Benzophenone	(n, π^*)	Propan-2-ol	1×10^6
Benzophenone	(n, π^*)	$Bu_3Sn{-}H$	5×10^7
p-Phenylbenzophenone	(π, π^*)	Propan-2-ol	1×10^3
2-Acetylnaphthalene	(π, π^*)	$Bu_3Sn{-}H$	2×10^6

The effect of the difference in electron distribution between (n, π^*) and (π, π^*) states on rate constants for hydrogen abstraction can be described pictorially. The (n, π^*) state has a half-filled non-bonding orbital on oxygen because of electron promotion to the delocalized π^* orbital, and the species therefore resembles an alkoxy radical to some extent and is a good hydrogen abstractor. A (π, π^*) state, on the other hand, has no half-filled non-bonding orbital on oxygen, and as in the ground state the oxygen atom is electron-rich. Others factors which may affect reactivity are that the lowest excited triplet state may not be the reacting triplet (higher energy states can be populated by thermal activation[43] or during the intersystem crossing process from the singlet excited state initially produced by absorption[44]), and that there may be a significant amount of 'mixing' of states, so that a lowest nominally (π, π^*) triplet state may have some (n, π^*) character.[45] Reaction through a higher energy state does not account for the large difference in reactivity between (π, π^*) and charge–transfer (extreme π, π^*) states of substituted benzophenones of similar energy, so that electron distribution does play an important part in determining reactivity.

In systems where inter- or intra-molecular hydrogen abstraction occurs by an excited carbonyl function from a position adjacent to a nitrogen atom, the mechanism of hydrogen transfer involves two steps. First, electron transfer occurs from the nitrogen atom to the excited carbonyl group, and then proton transfer takes place to give the radical species as expected for direct hydrogen atom transfer.[46] For instance, on irradiation of benzophenone in the presence of a tertiary amine, radical cation and radical anion species (7.46) can be detected by e.s.r. spectroscopy.[47]

$$Ph_2C{=}O + Ph_3N \xrightarrow[CH_3CN]{h\nu} Ph_2CO^{\cdot\,-} + Ph_3N^{\cdot+} \qquad (7.46)$$

Tertiary amines are most effective in bringing about chemical change, and the ketone can be reduced to the pinacol or, particularly in aqueous and therefore alkaline solution, to the alcohol by way of the ketyl radical anion $(Ar_2C{-}O^{\cdot\,-})$. The amines normally reduce 2 mol of ketone, and the nitrogen-containing products of reaction vary according to the nature of the amine used (see, for example, Chapter 10, p. 309). One feature of this reaction which distinguishes it from photoreduction by other types of hydrogen donor is that it is equally effective for ketones with (n, π^*) or with (π, π^*) lowest triplet states. This reflects the charge–transfer nature of the first step of the reaction, which does not depend on electron distribution in the region of the oxygen atom in the ketone excited state.

In principle the electron transfer mechanism could apply to hydrogen abstraction from a position adjacent to a heteroatom other than nitrogen. There is evidence that it does play a major part when the heteroatom is sulphur,[48] but not when the heteroatom is oxygen (i.e., not with such reducing agents as propan-2-ol). Exceptionally, the photoreduction of benzoquinone by ethanol gives rise to the semiquinone anion as well as the semiquinone radical (7.47).[49]

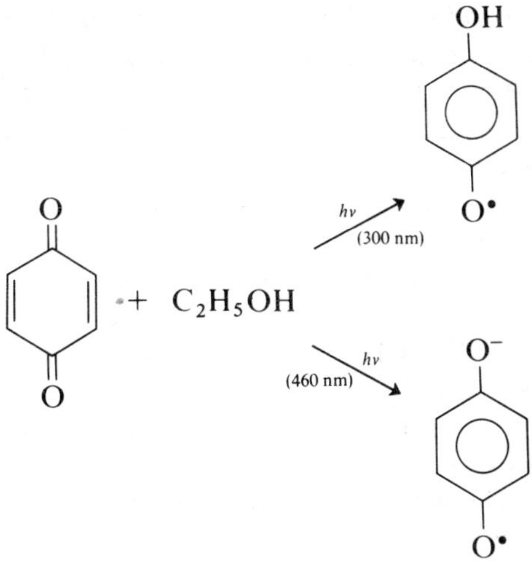

$$+ \; C_2H_5OH \tag{7.47}$$

7.5.2   Intramolecular Hydrogen Abstraction

Photolysis of a carbonyl compound with an accessible hydrogen atom in the γ-position gives products which are not easily accounted for by the α-cleavage free radical mechanism, but which can be explained in terms of intramolecular hydrogen abstraction by the excited carbonyl group from the γ-position. These products are a shorter-chain carbonyl compound and an alkene, formed by an elimination mechanism generally referred to as the 'Norrish type 2' process, and a cyclobutanol formed by a ring-closure mechanism. The products can be envisaged as arising from an initially formed biradical which can undergo cyclization or bond cleavage (7.48). The carbonyl product is formed as the enol tautomer, and the enol has been observed by infrared spectroscopy.[50] In other systems its intermediacy is indicated by the results of deuterium exchange experiments, such as those shown for hexan-2-one (7.49).[51]

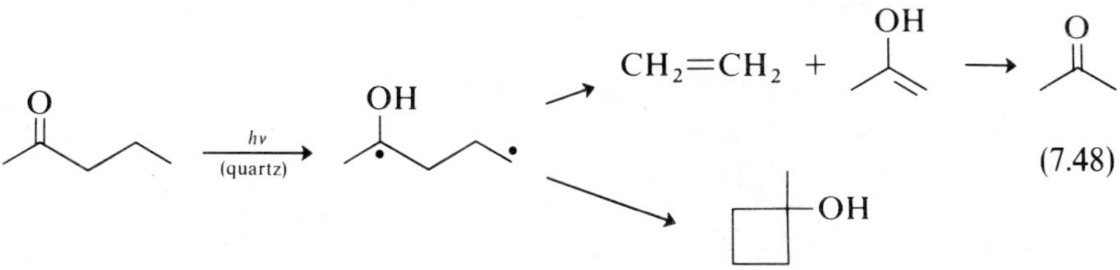

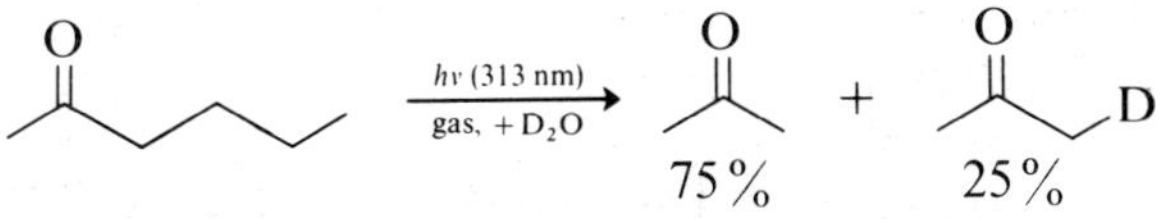

Both singlet and triplet (n, π^*) states of aliphatic ketones or aldehydes undergo the Norrish type 2 photoreaction. There is no great difference in *reactivity*

between (n, π^*) singlet and triplet excited states for a given compound, and since the rate constant for singlet reaction is often of the same order of magnitude as the rate constant for intersystem crossing, the *efficiency* of singlet state reaction is reasonably high. The relative efficiencies of singlet and triplet state reactions depend on the particular compound involved, since the values of the rate constants for hydrogen abstraction are sensitive to structure, and particularly to the strength of the C—H bond broken in the abstraction step, whereas the rate constants for intersystem crossing are not so variable. In the series[52] of ketones pentan-2-one $(CH_3-CO-CH_2CH_2CH_3)$, hexan-2-one $(CH_3-CO-CH_2CH_2CH_2CH_3)$, and 5-methylhexan-2-one $[CH_3-CO-CH_2CH_2CH(CH_3)_2]$ the C—H bond in the γ-position becomes progressively weaker and the rate constant for hydrogen abstraction consequently higher (Table 7.3). Mechanistic intepretation is complicated by the possibility of reverse hydrogen transfer in the biradical to give ground state of the starting carbonyl compound, and it is essential to compare reactivities in terms of rate constants rather than overall quantum yields.

Table 7.3. Intramolecular hydrogen abstraction for aliphatic ketones

Ketone	Rate constant for hydrogen abstraction (s^{-1})		Quantum yield for type 2 reaction	
	singlet	triplet	singlet	triplet
$CH_3-CO-CH_2CH_2CH_3$	$1{\cdot}8 \times 10^8$	$0{\cdot}13 \times 10^8$	$0{\cdot}03$	$0{\cdot}25$
$CH_3-CO-CH_2CH_2CH_2CH_3$	$9{\cdot}9 \times 10^8$	$1{\cdot}0 \times 10^8$	$0{\cdot}10$	$0{\cdot}23$
$CH_3-CO-CH_2CH_2CH(CH_3)_2$	21×10^8	$3{\cdot}8 \times 10^8$	$0{\cdot}10$	$0{\cdot}09$

The cyclization reaction occurs largely from the triplet excited state of saturated carbonyl compounds. Generally the yields of cyclization product are only 10–25% of those of the elimination product, but for certain compounds cyclization is very efficient. Some of these are systems where there are steric factors promoting cyclization, as with cyclododecanone (7.50).[53] In others there

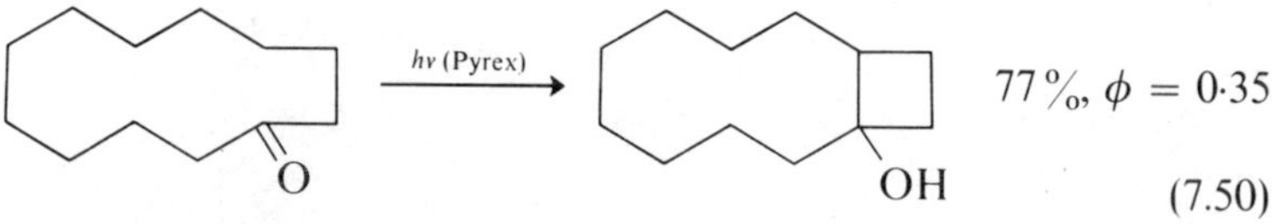

$$hv \text{ (Pyrex)}$$

$$77\%, \phi = 0{\cdot}35$$

$$(7.50)$$

are energetic factors working against the cleavage of the C_α—C_β bond, as with α-diketones (7.51)[54] or β,γ-unsaturated ketones (7.52),[55] in both of which the

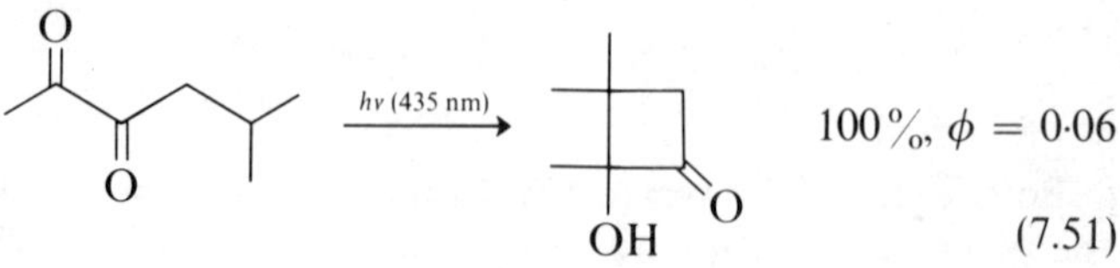

$$hv \text{ (435 nm)}$$

$$100\%, \phi = 0{\cdot}06$$

$$(7.51)$$

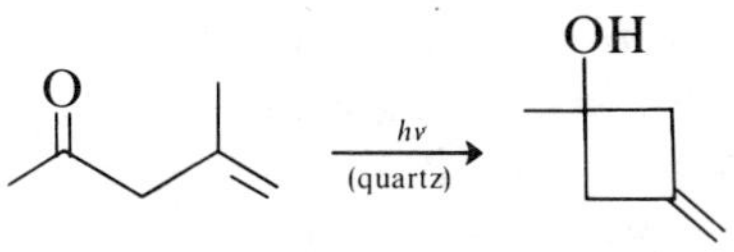

$$\text{(7.52)} \quad 57\%$$

C_α—C_β bond is $C(sp^2)$—$C(sp^3)$ and therefore slightly stronger than a $C(sp^3)$—$C(sp^3)$ bond, or the norbornane derivative (7.53)[56] for which the alkene product of elimination is highly strained.

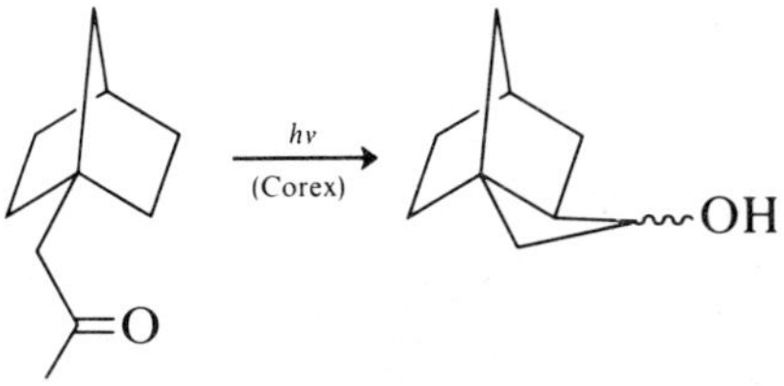

$$\text{(7.53)} \quad 45\%$$

The use of photochemical cyclization followed by oxidative ring-opening of the cyclobutanol is an efficient method of functionalizing C-18 or C-19 methyl groups in 20- or 11-ketosteroids respectively (7.54).[57]

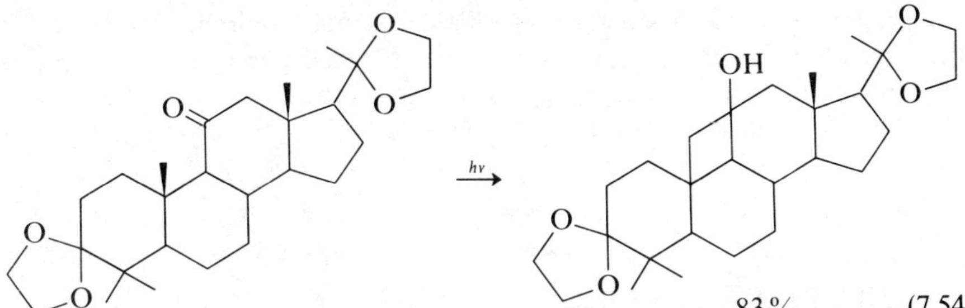

$$\text{(7.54)} \quad 83\%$$

The singlet state photoelimination reaction occurs with a high degree of stereospecificity in *threo*- and *erythro*-2-(3-methylpentyl) phenylacetates (7.55)[58] and in related ketone systems,[59] but the triplet state reaction is much less

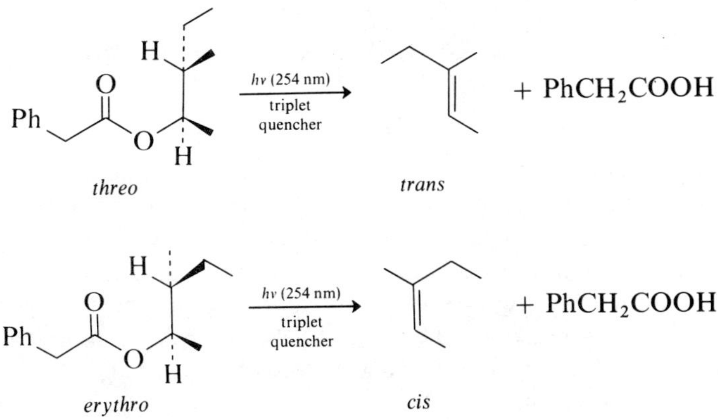

$$\text{(7.55)}$$

194

stereospecific. The cyclization reaction has not been studied in as much detail, but there is some retention of optical activity in the cyclobutanol formed from 5(R),9-dimethyldecan-2-one (7.56).[60] The singlet state reactions may be con-certed processes, but it is probable that both singlet and triplet state reactions proceed by way of a biradical intermediate, and the stereospecificity of the process is decided by the relative values of the rate constants for further re-action (cyclization or bond cleavage) and those for rotation about single bonds in the biradical. Calculations suggest that there is a much higher barrier to rotation in singlet 1,4-biradicals than in triplet.[61]

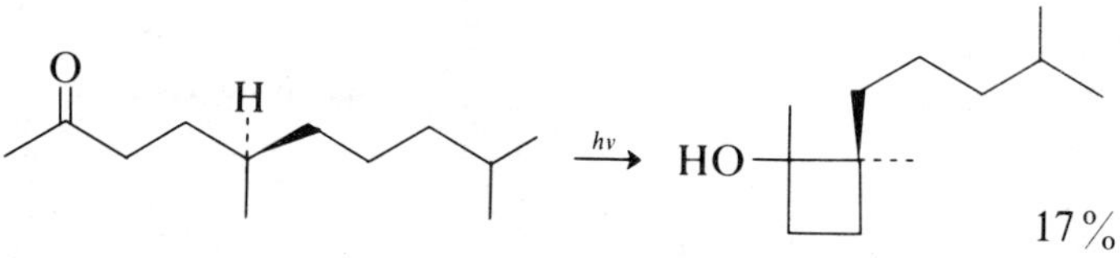

$$(7.56)$$

That a biradical is involved in the cyclization and elimination reactions of hexan-2-one is supported by the observation[62] that pyrolysis of one isomer of 1,2-dimethylcyclobutanol gives the isomeric cyclobutanol, hexan-2-one, acetone and propene as major products (7.57). This shows that 1,4-biradicals do undergo the reactions expected on the basis of the proposed photochemical reaction mechanism, though it is not proof that the photochemical reaction does pro-ceed by way of such a biradical.

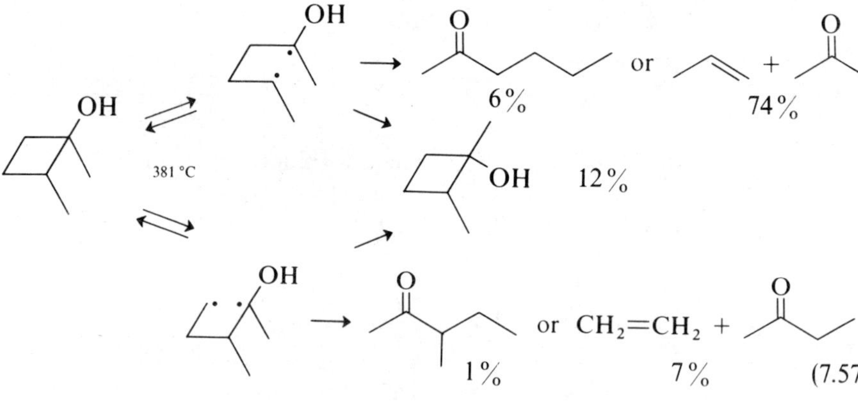

$$(7.57)$$

The competition between ring-opening and photoelimination processes for 2-alkylcyclohexanones has provided valuable information about both reactions. The *cis* isomer of 4-*t*-butyl-2-*n*-propylcyclohexanone gives detectable amounts of type 2 elimination product (7.58), but the *trans* isomer does not,[63] and this suggests that there is a 'stereoelectronic' requirement for the hydrogen abstrac-tion step. In the *trans* isomer the γ-hydrogen atom cannot approach the half-filled non-bonding orbital on the oxygen atom of the excited carbonyl group as readily as in the *cis* isomer.

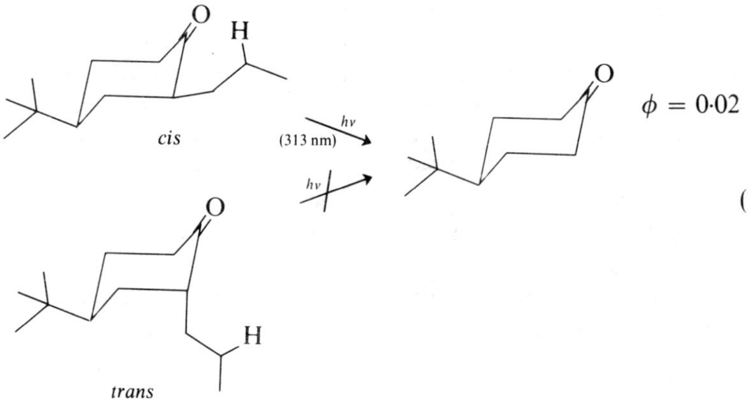

$$\phi = 0.02 \tag{7.58}$$

The type 2 elimination reaction in this and other 2-alkylcyclohexanones is inefficient, and the cause of this seems to be the unfavourable alignment of the C(2p) orbitals and the orbitals of the C_α—C_β bond in the biradical (7.59). For maximum development of double bond character in the transition-state these orbitals should all be parallel, but this cannot be achieved in the relatively inflexible cyclohexanone system, and the biradical undergoes preferential reverse hydrogen transfer to give starting ketone.

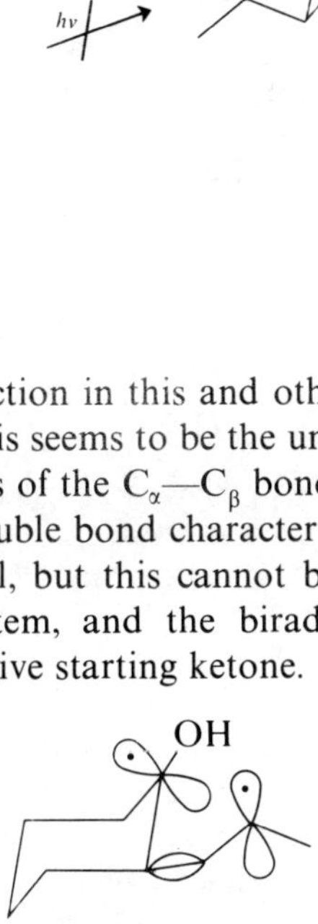

$$\tag{7.59}$$

Aromatic ketones such as valerophenone (Ph—CO—$CH_2CH_2CH_2CH_3$) differ from their aliphatic counterparts in that the rate constants for inter-system crossing are about two orders of magnitude higher, and all observable photochemistry occurs from the triplet state. The rate constants for intra-molecular hydrogen abstraction in the triplet state are very similar[64] for aromatic ketones as for aliphatic (Table 7.4).

Table 7.4. Rate constants for intramolecular hydrogen abstraction

Ketone	Rate constant for triplet hydrogen abstraction (s^{-1})	
	R = CH$_3$	R = Ph
R—CO—CH$_2$CH$_2$CH$_3$	0.13×10^8	0.08×10^8
R—CO—CH$_2$CH$_2$CH$_2$CH$_3$	1.0×10^8	1.3×10^8
R—CO—CH$_2$CH$_2$CH(CH$_3$)$_2$	3.8×10^8	4.8×10^8

For many aromatic ketones the (n, π^*) and (π, π^*) triplet states are similar in energy, and when the (n, π^*) triplet is lower the type 2 reaction proceeds with moderate to high efficiency ($\phi \sim 0.1$–0.9), but when the (π, π^*) triplet is lower the efficiency of the reaction is normally much smaller ($\phi < 0.01$). This state of affairs is very similar to that for intermolecular photoreduction, and it underlines the difference in the electronic nature of the two types of excited state.

The nature of the solvent has a considerable effect in the photochemistry of these aromatic ketones.[65] In non-polar solvents such as benzene or cyclohexane the quantum yield of product formation from valerophenone is about 0·4, but in polar solvents such as t-butanol the quantum yield is near to 1·0. This effect is understood in terms of solvation of the hydroxyl group of the biradical intermediate (7.60). In hydrogen-bonding solvents the hydroxyl group is more strongly solvated, and reverse hydrogen transfer from oxygen to carbon to give ground state valerophenone is hindered. Product formation therefore increases in efficiency.

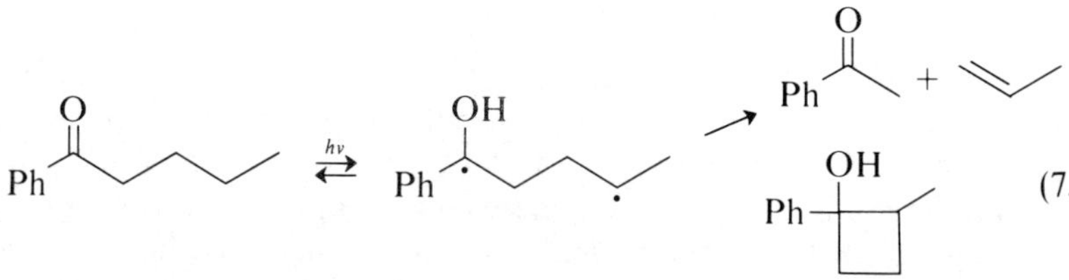

$$(7.60)$$

This reverse hydrogen transfer is the process which in non-polar solvents accounts for the inefficiency of reaction, and in conjunction with the initial hydrogen abstraction it constitutes a pathway for radiationless deactivation of the excited state. Ketones which have an asymmetric carbon atom at the γ-position undergo photoracemization in non-polar solvents.[66] A similar but smaller effect is found for the triplet, but not the singlet, type 2 reactions for aliphatic ketones. The singlet biradical has a shorter lifetime than the triplet and is not affected by hydrogen-bonding solvents, so that reverse hydrogen transfer is always a feature of the chemistry in the singlet excited state, whatever the solvent.

Nitrogen-substituted ketones such as N,N-dimethylaminoacetophenone (7.61) undergo a type 2 cleavage reaction,[67] though in this case the reaction occurs through the singlet excited state and by way of initial electron transfer. Likewise, aromatic esters such as n-propyl benzoate give type 2 elimination products (7.62) with low efficiency.[68] The reaction occurs from both singlet and triplet excited states of the esters, and the lowest states are (π, π^*) in nature.

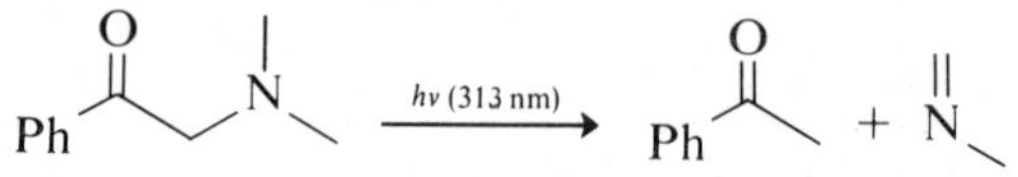

$$(7.61)$$

$$\phi = 0.05$$

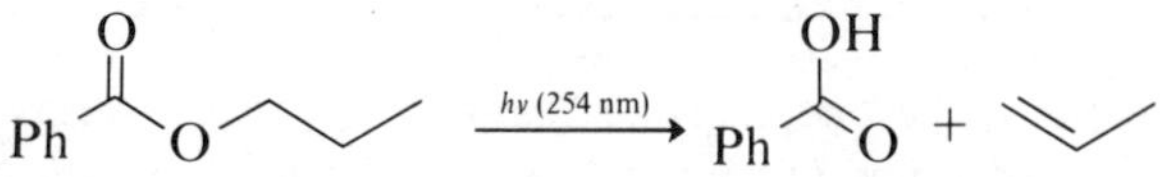

$$(7.62)$$

$$\phi = 0.006$$

In general the most favourable intramolecular hydrogen abstraction involves a six-membered cyclic transition-state in the abstraction step, but in some systems there are factors which make abstraction from other positions an important process. The methoxy-substituted ketone (7.63) has a particularly weak C—H bond at the δ-position, and a cyclopentanol is formed together with the cyclobutanol and cleavage products.[69]

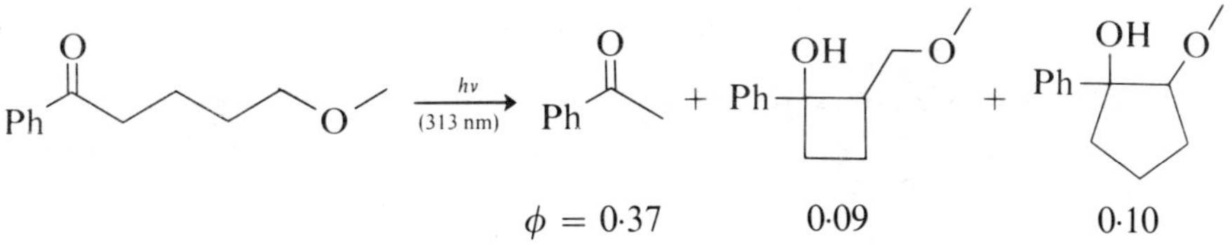

$$\phi = 0.37 \qquad\qquad 0.09 \qquad\qquad 0.10$$

(7.63)

An example of a system in which structural/conformational preference plays an important part is cyclodecanone,[70] where abstraction from C-6 is predominant, and bicyclo[4.4.0]decan-1-ol is the major product (7.64). Highly specific intramolecular hydrogen-abstraction reactions have been employed in steroidal systems in biomimetric chemistry for compounds where structural effects lead to reaction at a remote site.[71]

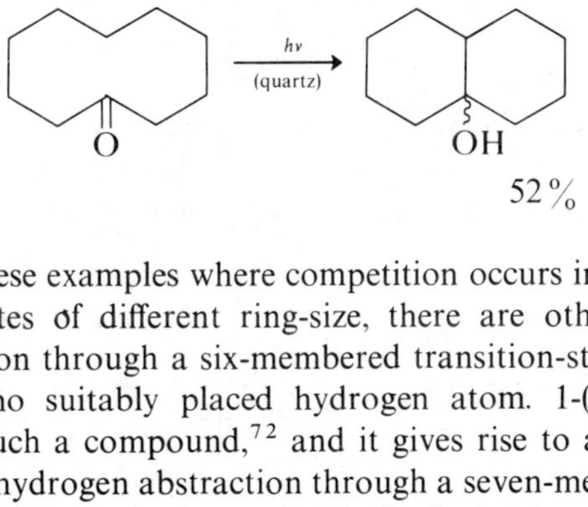

(7.64)

52 %

In addition to these examples where competition occurs involving pathways with transition-states of different ring-size, there are other systems where hydrogen abstraction through a six-membered transition-state is not possible because there is no suitably placed hydrogen atom. 1-(8-Benzylnaphthyl) phenyl ketone is such a compound,[72] and it gives rise to a photocyclization product by way of hydrogen abstraction through a seven-membered transition state in the (π, π^*) triplet excited state (7.65). Similarly, the morpholino-ketone (7.66) gives a cyclopropanol by way of hydrogen abstraction through a five-membered transition-state.[73]

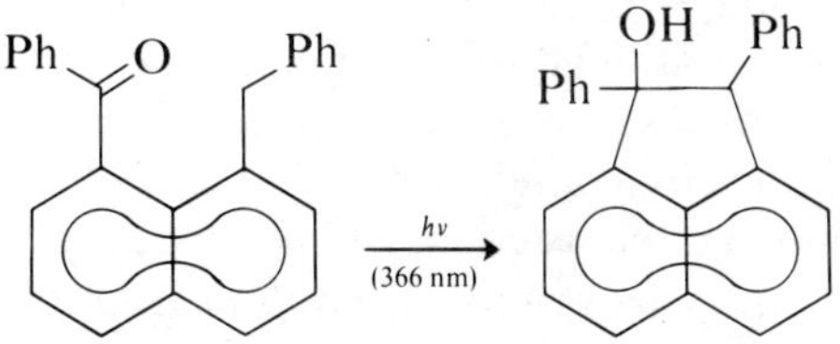

91 %, $\phi = 0.30$

(7.65)

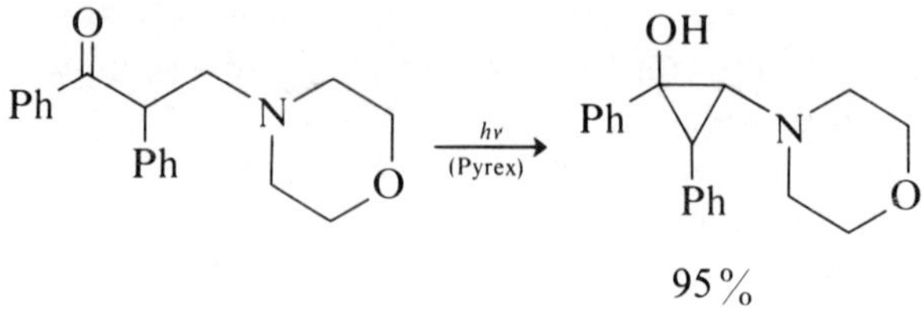

$$(7.66)$$

95%

7.5.3   Photo-enolization

Aromatic carbonyl compounds with an alkyl group in the ortho position under-go rapid and reversible photo-enolization by way of hydrogen abstraction through a six-membered cyclic transition-state. The photo-enol can be trapped with added dienophile (7.67).[74]

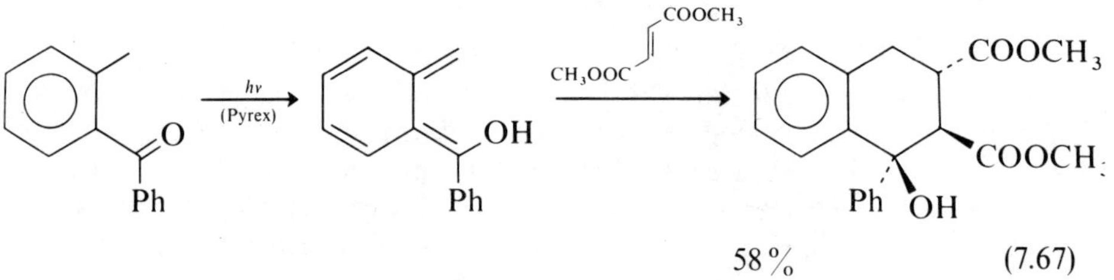

58%   (7.67)

In some systems the enol, or a product derived from it in a subsequent reaction, is coloured, and this constitutes one of the major classes of photo-chromism. A photochromic compound is one which undergoes a colour change with visible or ultraviolet radiation which is reversible in a thermal or photo-chemical reaction. In principle the cycle of colour changes is capable of indefinite repetition, though in practice there are always side reactions which lead to 'fatigue'. An example of this class of compound is 2-benzyl-3-benzoylchromone (7.68).[75]

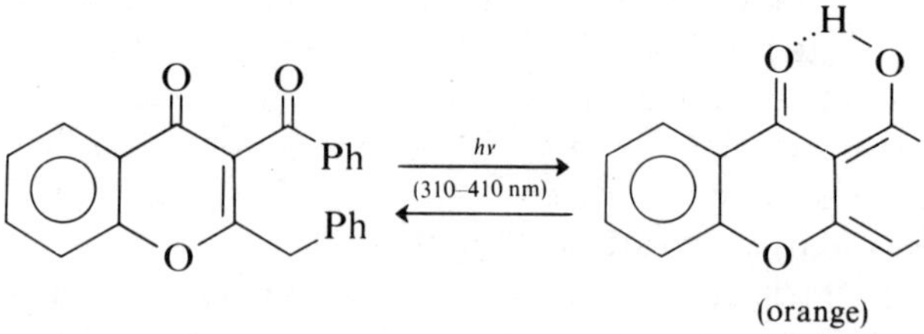

(orange)

$$(7.68)$$

7.5.4   α,β-Unsaturated Compounds

α,β-Unsaturated carbonyl compounds on irradiation undergo *cis–trans* iso-merization and a photostationary state is reached (7.69).[76] The process can be sensitized by triplet sensitizers such as acetophenone, and the reactive excited state is a triplet. Acyclic compounds and cyclic compounds with eight or more atoms in the ring undergo the reaction, and it is thought that it also occurs for 6- or 7-membered ring systems, though the *trans* isomers cannot be isolated (see Chapter 8, p. 237).

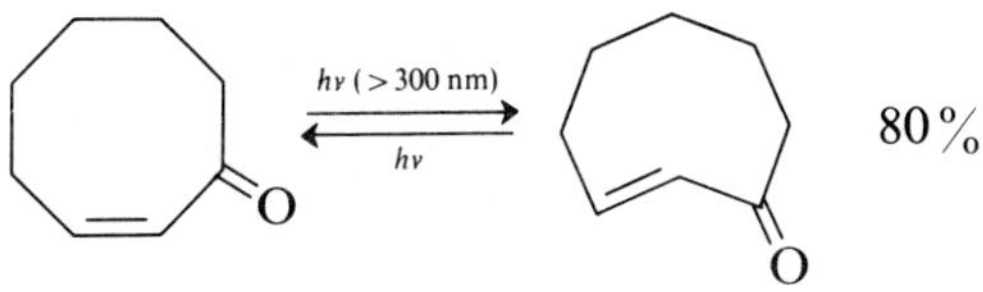

$$ \text{(cyclooctenone)} \xrightarrow[hv]{hv\,(>300\,\text{nm})} \text{(cycloheptenone)} \qquad 80\% \qquad (7.69) $$

α,β-Unsaturated compounds with a hydrogen atom in the γ-position give rise to the β,γ-isomer in an inefficient process on irradiation.[77] The reaction has a stereochemical requirement, that the carbonyl group and the hydrogen-bearing carbon atom must be *cis* with respect to each other, and *trans–cis* isomerization may precede deconjugation (7.70). Dienones and related compounds such as sorbic acid (7.71) undergo an analogous reaction to give an allene.[78]

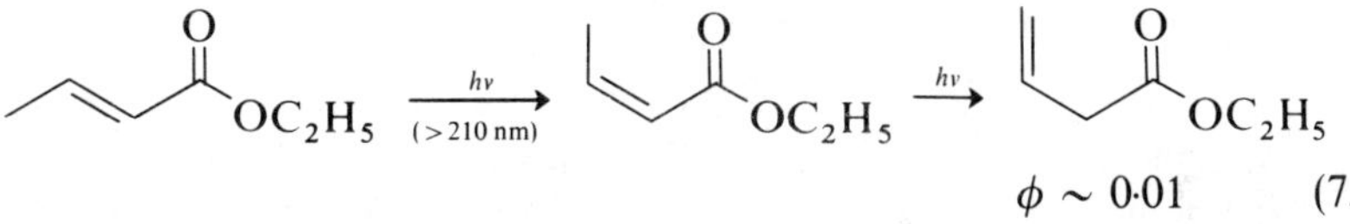

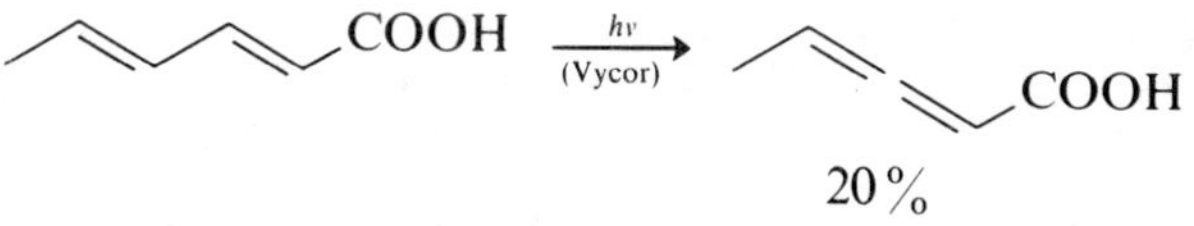

$$ \xrightarrow[(>210\,\text{nm})]{hv} \qquad \xrightarrow{hv} \qquad \phi \sim 0.01 \qquad (7.70) $$

$$ \text{COOH} \xrightarrow[(\text{Vycor})]{hv} \text{COOH} \qquad (7.71) $$

$$ 20\% $$

The reactive excited state in deconjugation is not the same as that responsible for *cis–trans* isomerization, and the reaction occurs from the (n, π^*) singlet state of the carbonyl compound. The first-formed product is a dienol, and in *O*-deuteriomethanol as solvent deuterium is incorporated in very high yield in the α-position of the product (7.72). Deuterium is also incorporated into the starting material, since the dienol can ketonize to give either α,β- or β,γ-unsaturated isomer.[79] For 1-acetylcyclo-octene (7.73) the enol products formed are stable enough to be isolated at room temperature.[80]

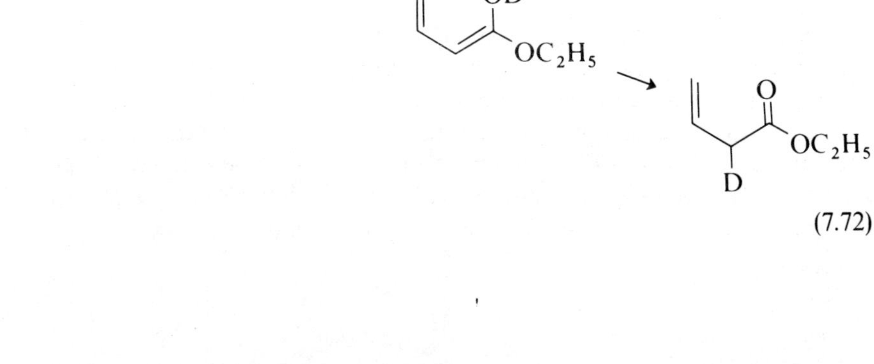

$$ (7.72) $$

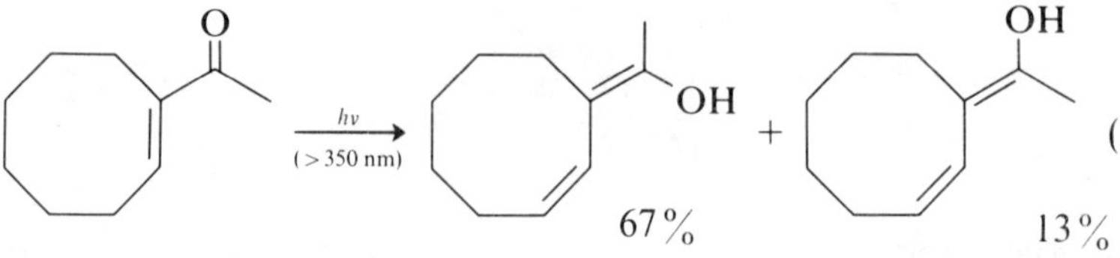

$$\text{(7.73)}$$

67%   13%

Deconjugation is also observed in cyclic systems in which hydrogen abstraction by way of a cyclic transition-state is made impossible by the rigidity of the molecule (7.74). In these compounds an alternative mechanism must operate.

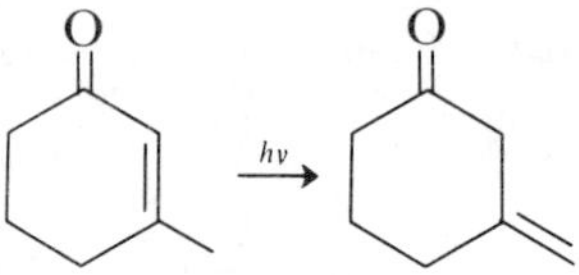

$$\text{(7.74)}$$

Although the double bond shift is observed for a wide range of α,β-unsaturated carbonyl compounds, a few simple methyl-substituted enones such as 3,4-dimethylpent-3-en-2-one are apparently photostable. This may be a result of reversible intramolecular cycloaddition to give an oxetene (see Chapter 7, p. 211). t-Butyl-substituted compounds can give rise to a cyclopropylmethyl product[81] in a competing reaction pathway (7.75), and an alternative mode of reaction predominates[82] for cyclopropyl conjugated α-β-unsaturated carbonyl compounds, leading to a cyclopentenyl product (7.76).

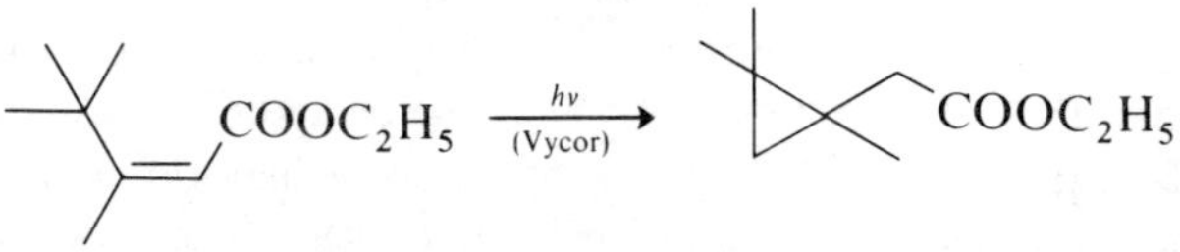

$$\text{(7.75)}$$

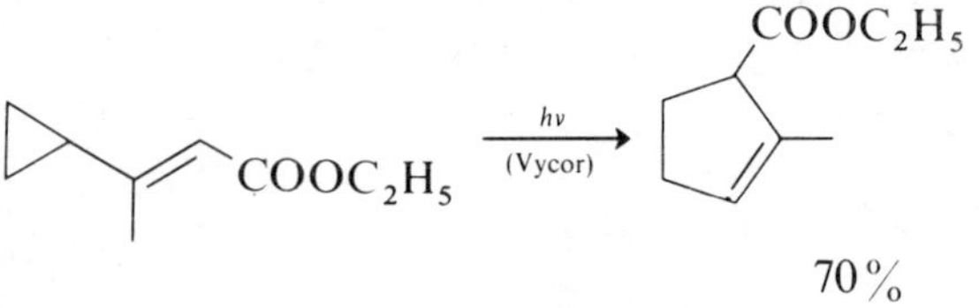

$$\text{(7.76)}$$

70%

7.6 PHOTOREARRANGEMENTS OF CYCLOHEXENONES AND CYCLOHEXADIENONES

Cyclic conjugated enones and dienones undergo a number of photochemical rearrangements which have no parallel in the photochemistry of acyclic compounds, but which are major reaction pathways for the cyclic compounds. The reactions involve the formation of bicyclo[3.1.0]hexane compounds, with migration of either a C-4 substituent or of the C-5 ring carbon atom.[83] In the reactions of 4,4-dialkylcyclohexenones the ring carbon atom migrates (7.77), but with 4,4-diarylcyclohexenones an aryl group migrates (7.78).

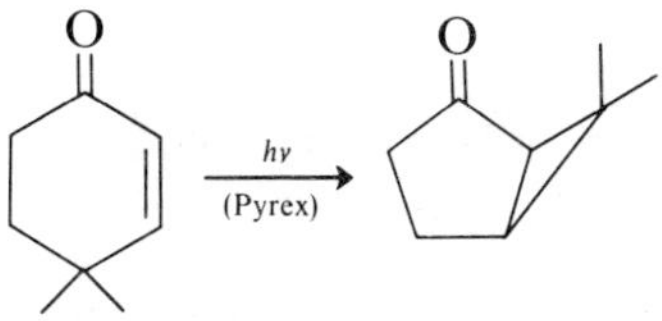

$$60\%, \phi = 0.0065 \qquad (7.77)$$

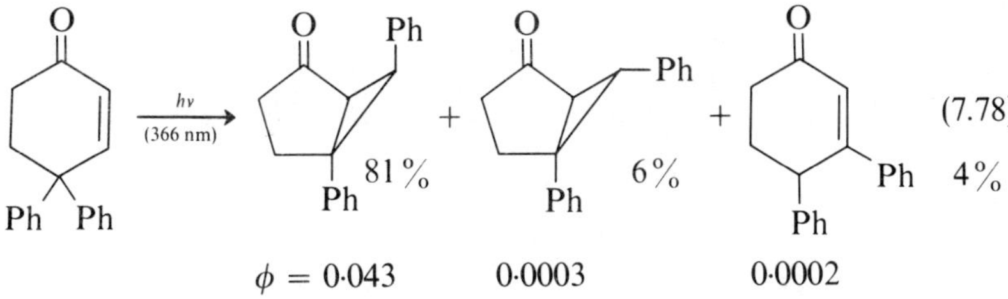

$$81\% \qquad 6\% \qquad Ph \quad 4\% \qquad (7.78)$$

$$\phi = 0.043 \qquad 0.0003 \qquad 0.0002$$

Both classes of reaction can be sensitized using propiophenone or quenched using naphthalene,[84] but the rearrangement with ring migration occurs from a (π, π^*) triplet excited state, whereas rearrangement with aryl migration occurs from an (n, π^*) triplet excited state. This is supported by the observation[85] that a 4-alkyl-4-arylcyclohexenone undergoes rearrangement with aryl migration in benzene as solvent, but with ring migration in aqueous methanol (7.79). The triplet states are very close in energy for this compound, and whereas the (n, π^*) triplet is lower in a non-polar solvent, the (π, π^*) triplet is lower in a polar solvent (see Chapter 2, p. 61).

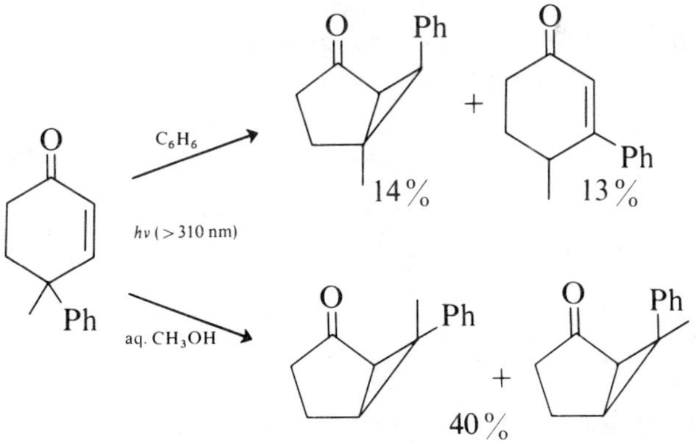

$$14\% \qquad 13\% \qquad (7.79)$$

$$40\%$$

It is thought that the reaction from the (π, π^*) excited state proceeds through a zwitterionic intermediate, and this intermediate can lead to the bicyclic product by ring-closure, or it can be intercepted by acetic acid[86] to give mono-cyclic products (7.80). Optical activity at C-4 is at least partially retained in the product in some systems, and this suggests that the bond-forming steps in the reaction sequence occur at the same time as, or very shortly after, the bond-breaking steps.

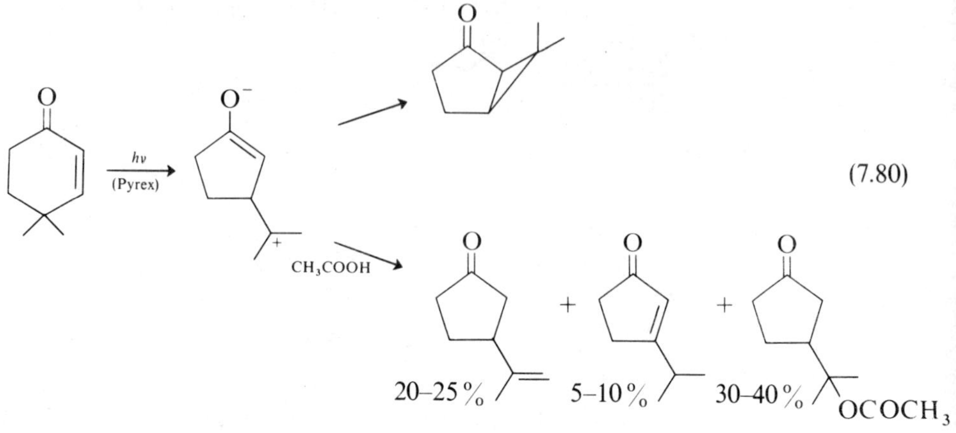

$$(7.80)$$

20–25%   5–10%   30–40%   OCOCH$_3$

The rearrangement of the diaryl enones resembles the di-π-methane rearrangement of the corresponding diarylcyclohexenes (see Chapter 8, equation 8.73), but the latter reaction occurs through a (π, π^*) singlet excited state, whereas the reaction of the cyclohexenones goes through an (n, π^*) triplet state, and it is not stereospecific but gives a mixture of stereoisomers.

4,4-Diarylcyclohexa-2,5-dienones undergo a rearrangement (7.81) which is formally analogous to that of the dialkylcyclohexenones, but the rate constant for the rearrangement step is four orders of magnitude higher, and the overall efficiency of the reaction is much greater.[87] The reactive excited state is the (n, π^*) triplet state of the dienone, and the reaction is thought to involve (C-3)—(C-5) bond formation in the excited state, followed by intersystem crossing and electron demotion to give a zwitterionic intermediate which rearranges to the observed bicyclic product.[88]

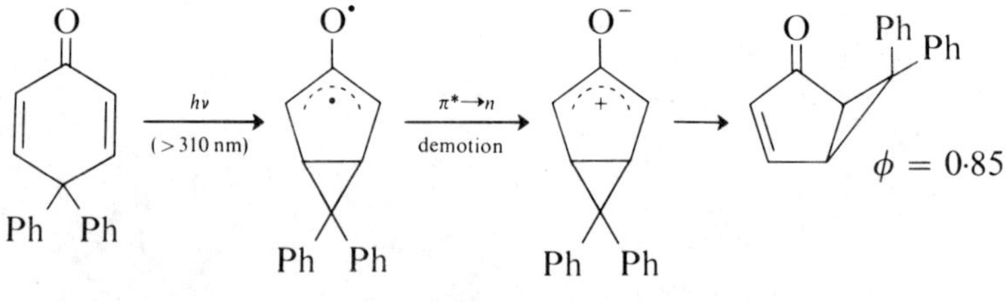

$$\phi = 0.85$$

$$(7.81)$$

The necessity for a $\pi^* \rightarrow n$ electron demotion step arises because the π-system in the (n, π^*) states is electron-rich, but the observed rearrangements are characteristic of migrations to electron-deficient centres. The order of events in the sequence (C-3)—(C-5) bond formation, intersystem crossing, electron demotion and rearrangement is not established with certainty, and there are several proposed schemes for representing the rearrangements, none of which is completely satisfactory. The scheme involving a dipolar (zwitterionic) intermediate as outlined above does account for the observed rearrangement product, and it also accounts for the subsequent photoreactions[89] of the bicyclo[3.1.0]hexenone (7.82) and for the products formed from 4-aryl-4-hydroxycyclohexadienones (7.83).[90]

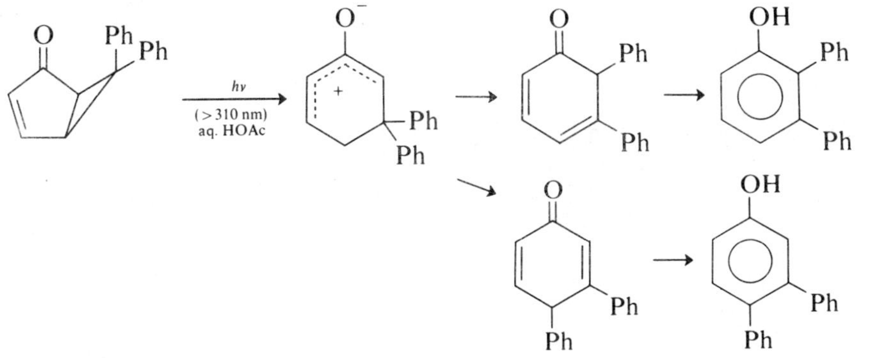

(7.82)

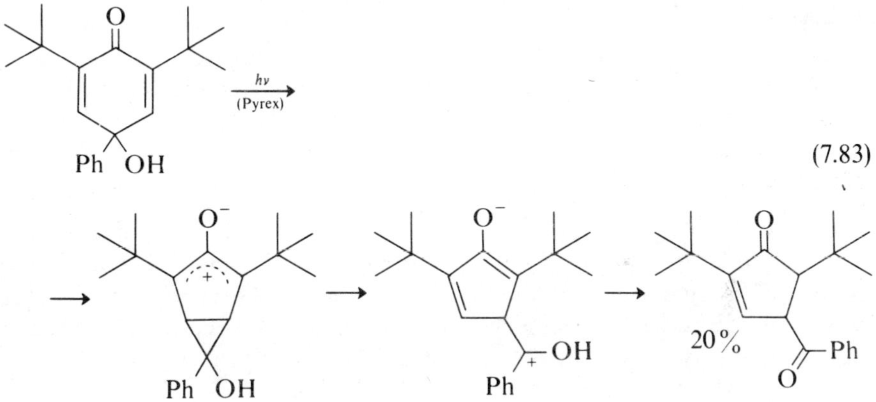

(7.83)

Similar rearrangements of simple 4,4-dialkylcyclohexa-2,5-dienones are known, and a large number in more complex systems such as the sesquiterpenoid compound α-santonin (7.84), which was the subject of early investigations.[91]

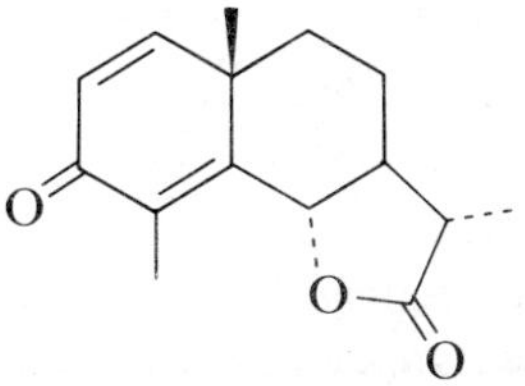

(7.84)

The different types of product obtained according to the solvent used or the substitution pattern of the dienone can be rationalized[92] in terms of the intermediate zwitterion (7.85).

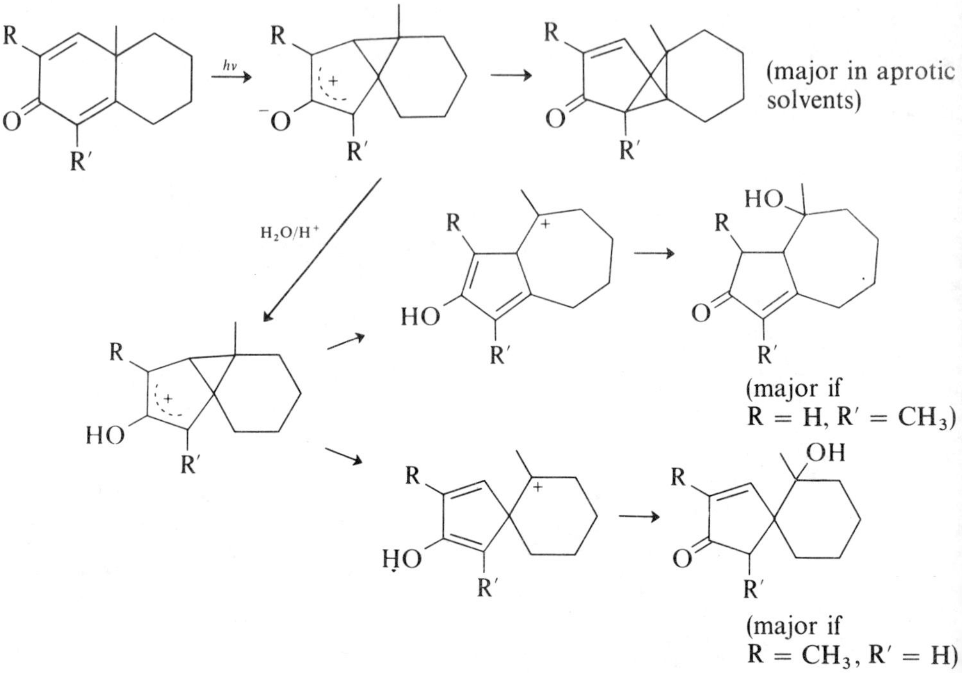

$$(7.85)$$

In view of the high energy of ionic intermediates in the absence of solvation, it is likely that the rearrangement of 4,4-dimethylcyclohexa-2,5-dienone in the gas phase[93] occurs by way of biradical intermediates (7.86). The rearrangement closely resembles the di-π-methane rearrangement of 1,4-dienes, and specifically that of 3-methylene-6,6-dimethylcyclohexa-1,4-diene (see equation 8.71 in Chapter 8), which occurs on direct (π, π^*) irradiation.

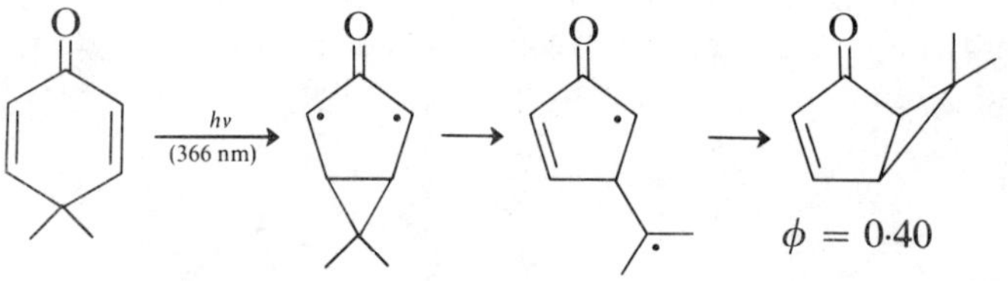

$$\phi = 0.40 \qquad (7.86)$$

A cyclohexa-2,5-dienone with a good radical leaving group on C-4 undergoes a different reaction in solvents with a readily abstracted hydrogen atom (7.87). Intermolecular hydrogen abstraction followed by elimination of the group from C-4 leads to a phenolic product.[94] Radical intermediates are also involved in the photoreaction of the spirocyclopropyldienone (7.88), which similarly gives a phenolic product in a solvent which is a good source of hydrogen atoms.[95]

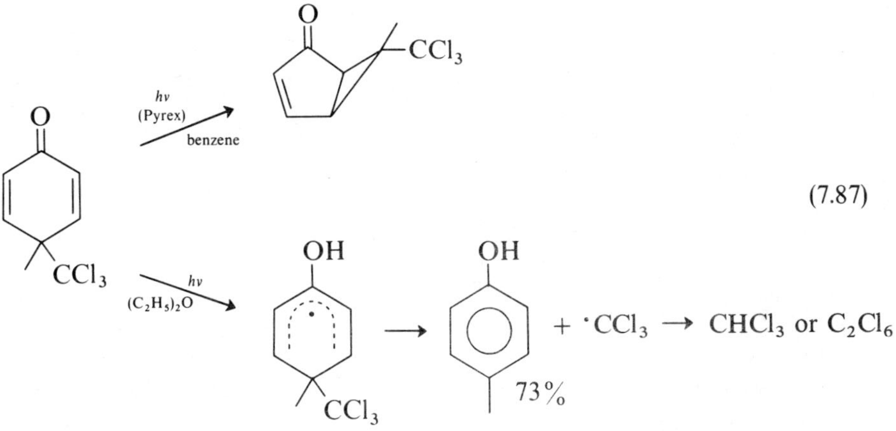

$$(7.87)$$

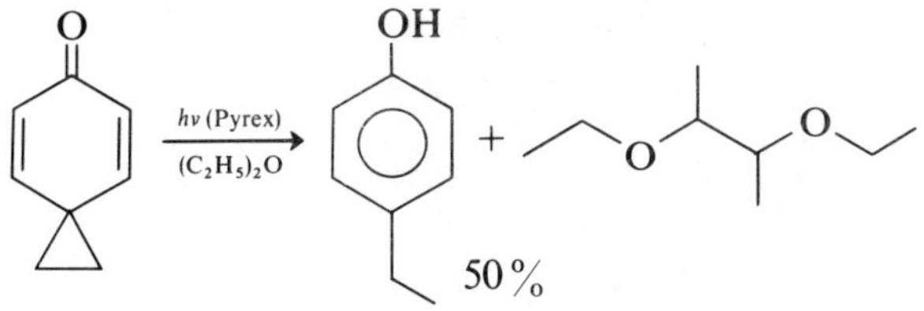

$$(7.88)$$

Cyclohexa-2,4-dienones undergo a ring-opening reaction[96] leading to a dienyl ketene through the (n, π^*) singlet excited state of the dienone (7.89). The ketene has been characterized by spectroscopic methods at low temperature, and normally it is trapped as an adduct with added nucleophile. The process could be a concerted electrocyclic ring-opening analogous to that observed for cyclohexa-1,3-dienes (see Chapter 8, p. 245). If the ketene is not trapped, it undergoes thermal reversion to the original dienone or thermal reaction to give the bicyclic enone characteristic of (π, π^*) chemistry (see equation 7.90).

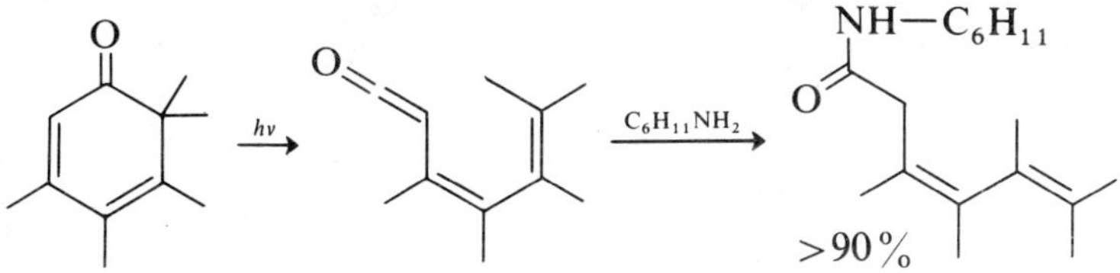

$$(7.89)$$

In the presence of silica gel the relative energy of the (n, π^*) and (π, π^*) singlet states of many cyclohexa-2,4-dienones is reversed,[97] and reaction occurs through the lowest (π, π^*) singlet to give a bicyclic rearrangement product (7.90). In this reaction the behaviour of the dienone parallels that of β,γ-unsaturated ketones (see Chapter 7, p. 187).

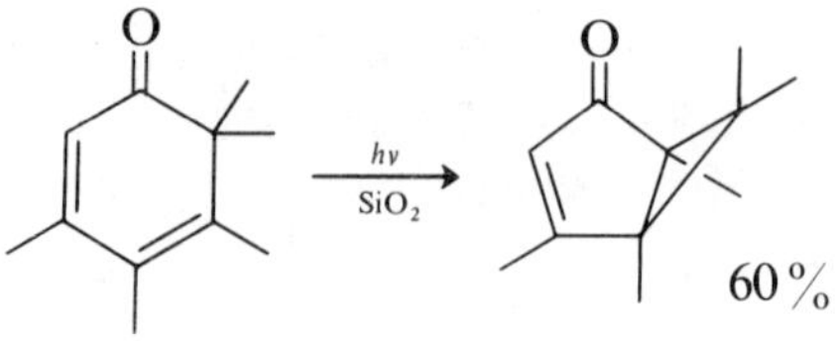

$$\text{(7.90)}$$

60%

7.7   PHOTOCYCLOADDITION

7.7.1   Oxetane Formation

On photolysis in the presence of alkenes, dienes or alkynes, carbonyl compounds undergo cycloaddition reactions to give 4-membered ring oxygen heterocycles.[98] For example, benzophenone with isobutene gives a high yield of an oxetane (7.91). Most reported studies have employed ketones or aldehydes, but carboxylic acid derivatives can also undergo this reaction if the C(=O)—X bond is reasonably strong, as with some esters, acyl fluorides or acyl nitriles. The reaction can also occur in an intramolecular manner if the functional groups are suitably oriented, as with acyclic γ,δ-unsaturated ketones (7.92).[99]

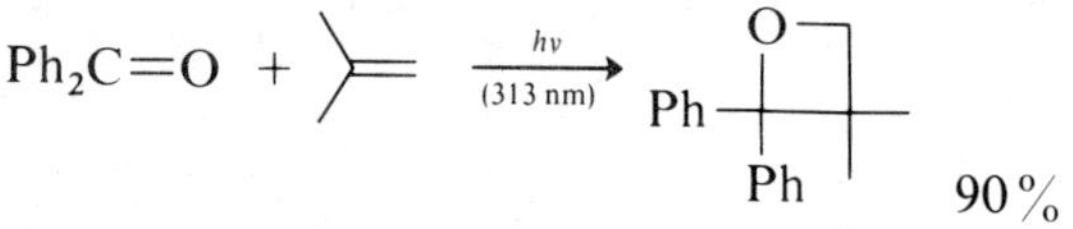

$$\text{Ph}_2\text{C}{=}\text{O} + \quad \xrightarrow[\text{(313 nm)}]{h\nu} \qquad \text{(7.91)}$$

90%

$$\xrightarrow[\text{(> 260 nm)}]{h\nu} \qquad \text{(7.92)}$$

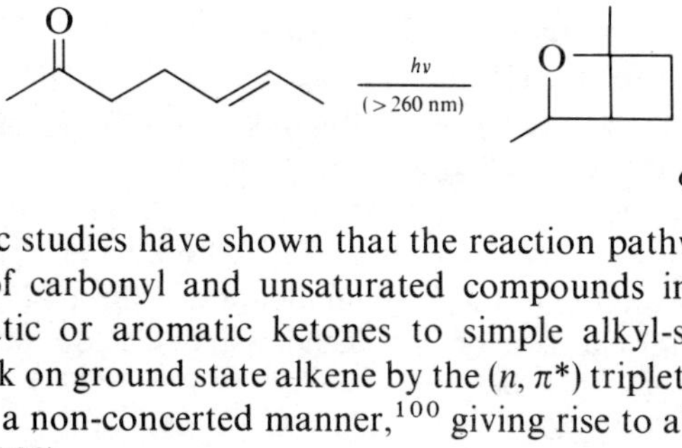

$$\phi = 0{\cdot}015$$

Mechanistic studies have shown that the reaction pathway varies according to the type of carbonyl and unsaturated compounds involved. Addition of simple aliphatic or aromatic ketones to simple alkyl-substituted ethylenes involves attack on ground state alkene by the (n, π^*) triplet state of the carbonyl compound in a non-concerted manner,[100] giving rise to all possible isomers of the oxetane (7.93).

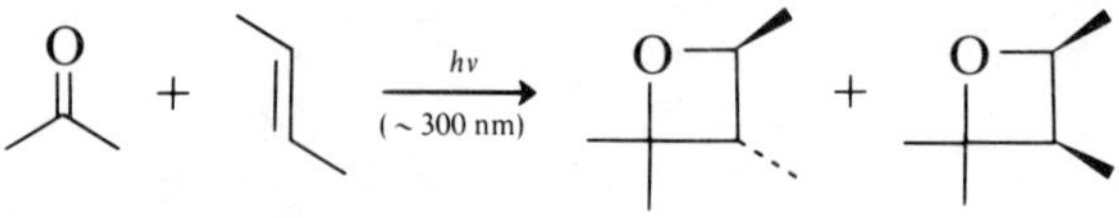

$$\quad + \quad \xrightarrow[\text{(~ 300 nm)}]{h\nu} \qquad + \qquad \text{(7.93)}$$

2:1

Although the reaction is not stereospecific, there is a preference for one orientation of addition, which can be rationalized in terms of initial attack on the alkene by the oxygen atom of the excited carbonyl group to give a biradical intermediate. In this the (n, π^*) state once again exhibits similarities to an alkoxy radical in its chemical behaviour, adding to an alkene through an electron-deficient oxygen atom. The more energetically stable of the two possible biradicals is formed more readily, and if the proportion of biradicals which undergoes ring-closure to give oxetane rather than cleavage to give starting materials is similar for the two biradicals, the preferential formation of one oxetane is accounted for (7.94).

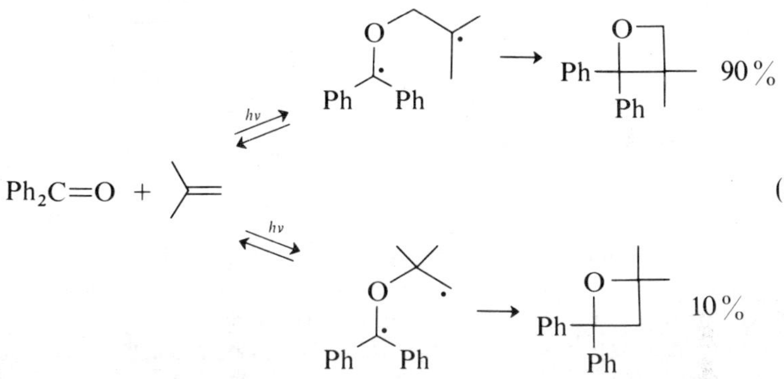

$$Ph_2C=O \;+\; \Big\rangle\!=\!\!\Big\langle \tag{7.94}$$

The consideration of biradical stability is certainly applicable to the prediction of the major product of the cycloaddition, but the precise details of the mechanism are not settled. A complex between the excited state of the ketone and the ground state of the alkene (i.e., an exciplex) is thought to be involved,[101] although an oxetane in a high vibrationally excited level has also been suggested as an intermediate.

Photocycloaddition of aliphatic ketones to alkenes substituted with electron-withdrawing groups, and in particular with cyano groups, involves addition of singlet (n, π^*) excited ketone to ground state alkene.[102] The stereochemistry of the alkene is retained in the product oxetane (7.95).

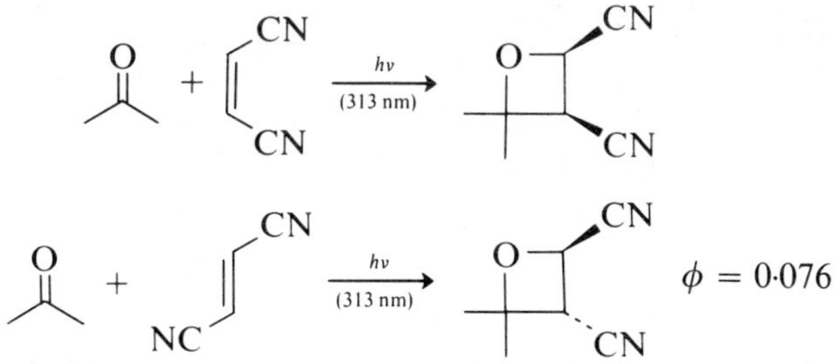

$$\tag{7.95}$$

It is suggested that formation of an exciplex between ketone excited state and alkene ground state plays an important part in this type of cycloaddition. The exciplex has considerable charge–transfer character (from the half-filled π^* orbital of the ketone excited state to the empty π^* orbital of the electron-deficient alkene), and the stereospecific formation of products is accounted for if both new bonds are formed simultaneously in the complex, or if the second is formed after the first at a rate faster than the rate of bond rotation. There is again a preference for one orientation of addition (7.96), but this is the opposite of that expected on the basis of the most stable biradical intermediate. The preference reflects the preferred orientation in the exciplex, which is governed by charge densities as illustrated.

$$\underset{^1(n,\,\pi^*)}{\overset{\delta+}{\underset{\delta-}{\bigwedge}}} + \overset{\delta-}{\underset{\delta+}{\bigwedge}}\overset{CN}{} \longrightarrow \left[\overset{CN}{\bigwedge} \right] \longrightarrow \overset{CN}{\bigwedge} \tag{7.96}$$

Aromatic ketones do not give oxetanes with α,β-unsaturated nitriles. The singlet state lifetime of an aromatic ketone is very short because of the much higher rate constant for intersystem crossing than in an aliphatic ketone, and the triplet state is unreactive because unsaturated nitriles are electron-deficient at the double bond and this discourages attack by the electron-deficient oxygen atom of the ketone triplet $(n,\,\pi^*)$ state. The rate constant for intersystem crossing can play an important part in determining the mechanism of cycloaddition. With aldehydes, if intersystem crossing is fast (as for benzaldehyde) the aldehyde undergoes non-stereospecific oxetane formation through the triplet state on irradiation with a simple alkene. If intersystem crossing is slower (as for naphthaldehyde or an aliphatic aldehyde) the compound undergoes stereo-specific reaction through the singlet state.[103]

The cycloaddition of aliphatic ketones to electron-rich alkenes, such as those substituted with alkoxy groups, is more complicated.[104] The reaction involves attack of either singlet or triplet $(n,\,\pi^*)$ state of the ketone on ground state alkene, and there is a high degree of retention of stereochemistry in the singlet state process, but not in the triplet state reaction (7.97).

$$\underset{^3(n,\,\pi^*)}{\bigwedge} + \underset{CH_3O}{\overset{C_2H_5}{\bigwedge}} \; or \; \underset{OCH_3}{\overset{C_2H_5}{\bigwedge}} \longrightarrow \overset{C_2H_5}{\underset{OCH_3}{\bigwedge}} + \overset{C_2H_5}{\underset{OCH_3}{\bigwedge}}$$

$$1:1$$

$$\underset{^1(n,\,\pi^*)}{\bigwedge} + \underset{OCH_3}{\overset{C_2H_5}{\bigwedge}} \longrightarrow \overset{C_2H_5}{\underset{OCH_3}{\bigwedge}} \qquad \phi \sim 0\cdot 1 \tag{7.97}$$

An attempt to rationalize theoretically the mechanisms of photochemical oxetane formation has been made using simple perturbational molecular orbital methods.[105] The correlation between predicted and observed orientation and stereospecificity is good, which is perhaps surprising in view of the assumptions and simplifications made.

The photoreactions of aromatic ketones with conjugated dienes to give vinyloxetanes (7.98) is a non-concerted, non-stereospecific process which occurs through the $(n,\,\pi^*)$ triplet state of the ketone.[106] The reaction is in competition with efficient triplet energy transfer from ketone triplet to the diene, and quantum yields of product formation are very low. The very high rate constant for inter-system crossing in aromatic ketones is again in evidence in this reaction with dienes, since aliphatic ketones react with conjugated dienes (7.99) through the singlet excited state.[107] Similar reactions involving unsymmetrical dienes are stereospecific at high concentration of diene.[108] The fact that conjugated dienes react with singlet states of ketones must be remembered when dienes are used as triplet quenchers in mechanistic studies of the photoreactions of ketones.

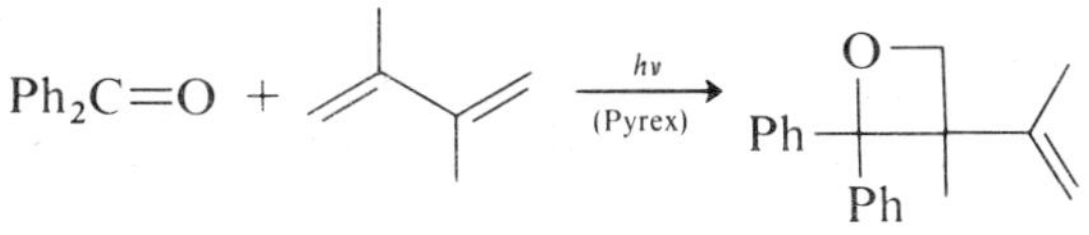

$$Ph_2C=O + \qquad \xrightarrow[\text{(Pyrex)}]{hv} \qquad \qquad 40\% \qquad (7.98)$$

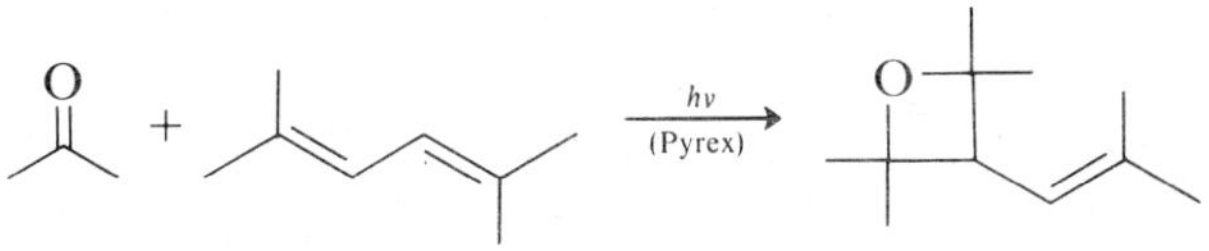

$$\xrightarrow[\text{(Pyrex)}]{hv} \qquad \qquad 28\% \qquad (7.99)$$

The different mechanistic patterns proposed for the various combinations of carbonyl and alkene types underline the fact that the observed reaction pathway depends on the relative values of the rate constants for competing processes—singlet reaction, intersystem crossing and decay to ground state for the excited singlet state, triplet reaction, triplet decay and energy transfer for the triplet state, and formation of product or regeneration of starting materials for an intermediate. In some systems the relative magnitudes of the rate con-stants are such that products are formed very largely by one pathway only, but in others there are two or more major reaction routes.

Carbonyl compounds undergo photochemical cycloaddition to alkynes to give oxetenes,[109] which are usually not isolated but isomerize to α,β-unsaturated carbonyl compounds in a subsequent thermal reaction (7.100).

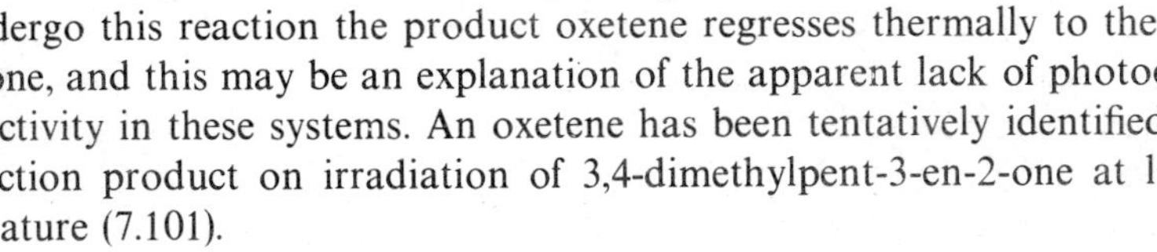

$$PhCH=O + \quad -\equiv- \quad \xrightarrow[-78\,°C]{hv\,(\text{Pyrex})} \qquad \xrightarrow{\text{warm}} \qquad 43\% \qquad (7.100)$$

Similar oxetenes may also be formed by an intramolecular cycloaddition reaction of α,β-unsaturated carbonyl compounds. For those compounds which undergo this reaction the product oxetene regresses thermally to the starting enone, and this may be an explanation of the apparent lack of photochemical reactivity in these systems. An oxetene has been tentatively identified[110] as a reaction product on irradiation of 3,4-dimethylpent-3-en-2-one at low tem-perature (7.101).

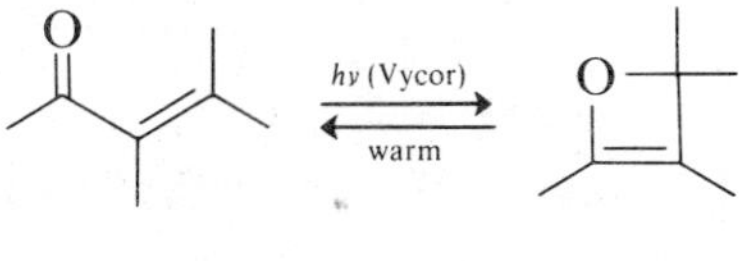

$$\xrightarrow[\text{warm}]{hv\,(\text{Vycor})} \qquad \qquad 70\% \qquad (7.101)$$

For α-phenyl-α,β-unsaturated carboxylic acids and amides the oxetene is formed and tautomerises to a β-lactone or a β-lactam (7.102). It is claimed[111] that this reaction is an electrocyclic process involving the (π, π^*) excited state of the carbonyl compound, by analogy with the reactions of conjugated dienes (see Chapter 8, p. 244).

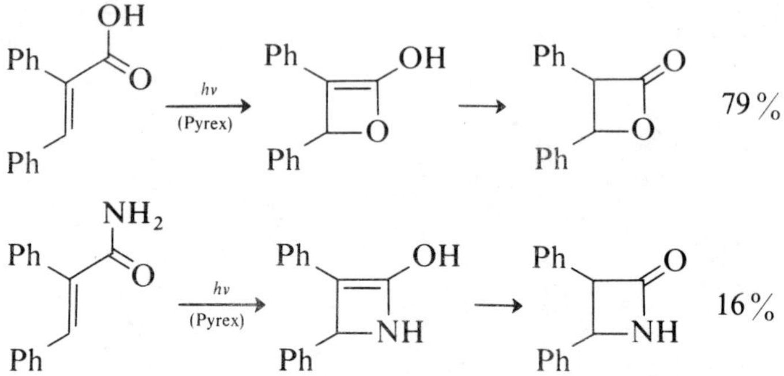

(7.102)

7.7.2   Cyclobutane Formation

Intermolecular photocycloaddition reactions of conjugated unsaturated carbonyl compounds can lead to products in which the addition has occurred either to the carbon–carbon double bond or to the carbon–oxygen double bond. Photodimerization of unsaturated compounds to give cyclobutanes is exemplified by the reactions of cyclopent-2-enone (7.103)[112] and of thymine (7.104).[113] Photo-inactivation of DNA involves such dimerization of adjacent thymine units. Cross-addition involving an unsaturated carbonyl compound and an alkene can also lead to cyclobutane adducts (7.105),[114] and these reactions have found many synthetic applications.[115]

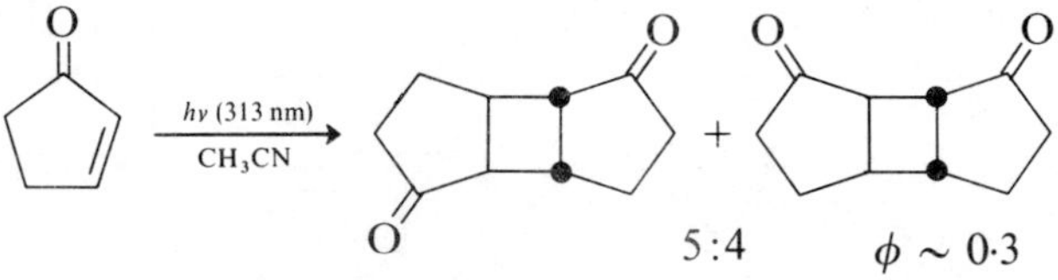

(7.103)

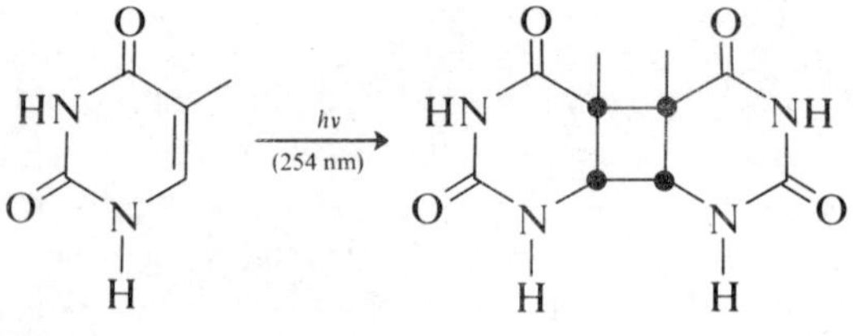

$5:4 \qquad \phi \sim 0.3$

(major isomer)

(7.104)

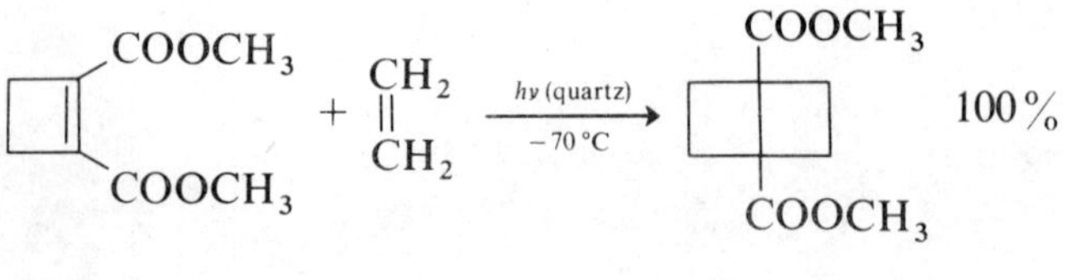

100% (7.105)

Many of the examples in the following discussion are of cyclic rather than acyclic α,β-unsaturated carbonyl compounds, and this reflects the balance of reported reactions. Acyclic enones seem to undergo photocycloaddition reactions much less readily than their cyclic counterparts, and this is probably because the triplet state of an acyclic compound has a twisted geometry and undergoes rapid intersystem crossing to the ground state. Cyclic compounds are less flexible, and intersystem crossing to ground state is slower, so that cyclo-addition reactions can compete more effectively with deactivation.

The reactive excited state in most cases is a triplet, as seen from the results of sensitization and quenching studies, although there is evidence for the involve-ment of a singlet state as well as a triplet state in the dimerization of thymine. The photodimerization is not selective as far as orientation in the product is concerned. Cyclopent-2-enone gives similar amounts of 'head-to-head' and 'head-to-tail' dimers (7.103), and cyclohex-2-enone gives a number of adducts with each orientation. The ratio is affected by substituents on the double bond and by the dielectric constant of the solvent. There is a greater degree of selec-tivity in the orientation of some cross-additions, and cyclohex-2-enone reacts with isobutene (7.106)[116] or with methyl vinyl ether (7.107)[117] to give major products with one particular orientation.

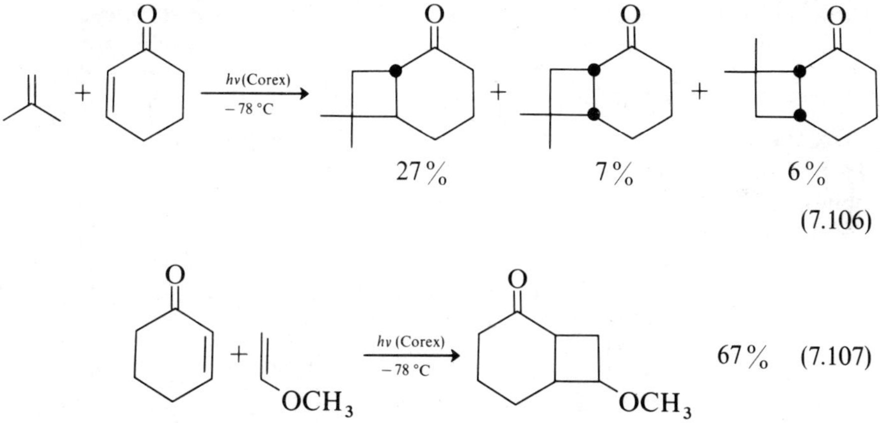

$$27\% \qquad 7\% \qquad 6\%$$

$$(7.106)$$

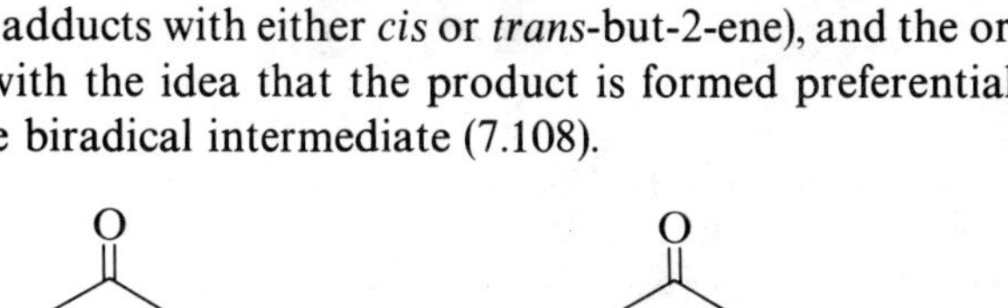

$$67\% \quad (7.107)$$

The reaction is not stereospecific (cyclohex-2-enone gives the same mixture of isomeric adducts with either cis or trans-but-2-ene), and the orientation is not in accord with the idea that the product is formed preferentially through the more stable biradical intermediate (7.108).

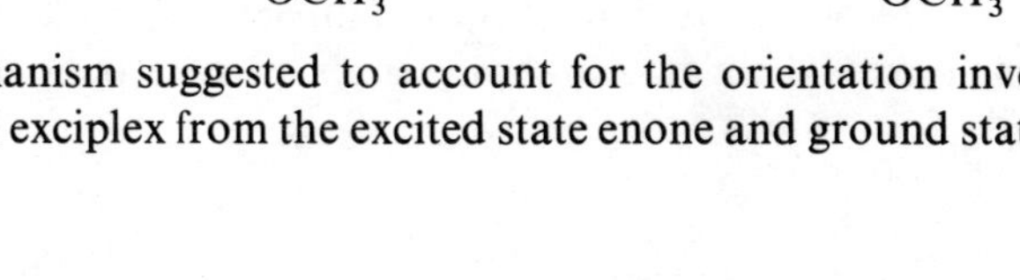

$$(7.108)$$

One mechanism suggested to account for the orientation involves the for-mation of an exciplex from the excited state enone and ground state alkene, and the geometry of the exciplex determines the preferred orientation of addition (7.109). This gives an adequate rationalization in many cases, but in others it seems necessary to consider a collision complex between two ground state molecules, with the orientation of photoaddition determined by the permanent dipoles, and there seems as yet to be no complete answer.[118]

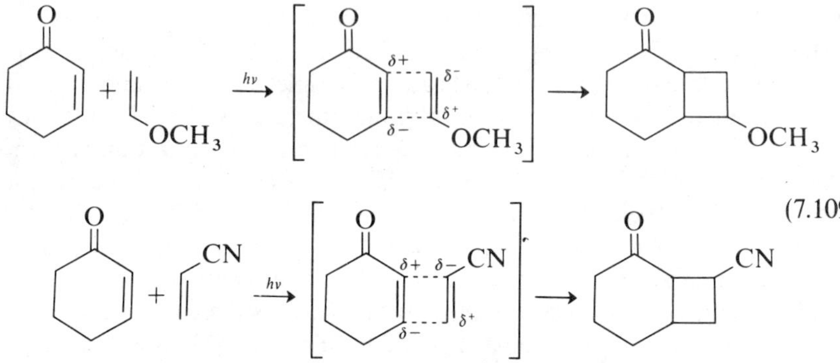

$$(7.109)$$

A mixture of stereoisomers of adduct is normally formed, containing all those expected on the basis of biradical intermediates which have sufficiently long lifetime for bond-rotation to occur before ring-closure. In certain cases other factors are involved which lead to a stereochemical preference. In the solid-state photodimerization of cinnamic acid, 'topochemical' control operates, and the stereochemistry of the product is governed by the crystal structure of the solid.[119] Two of the three crystal modifications of cinnamic acid give single dimeric photoproducts (7.110).

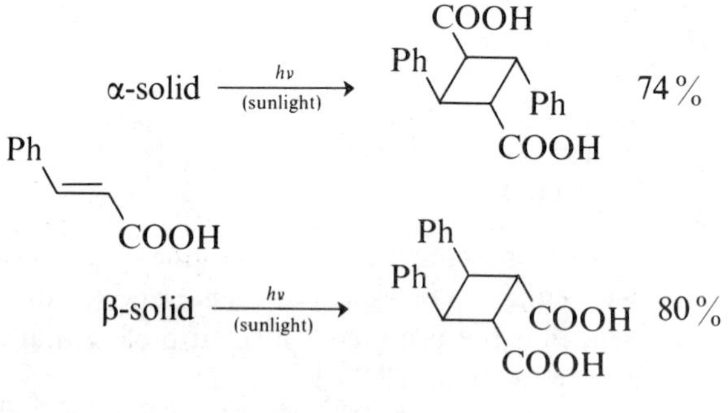

$$(7.110)$$

In another example of stereoselectivity, the photoaddition of maleic anhydrides or maleimides to cyclohexene is thought to occur by way of a ground state charge–transfer complex, which absorbs light at longer wavelength than either the alkene or the maleic acid derivative.[120] This leads to a certain degree of stereoselectivity (7.111) as compared with the adducts from cyclohexene and dimethyl maleate. In the latter system no ground state complex is formed.

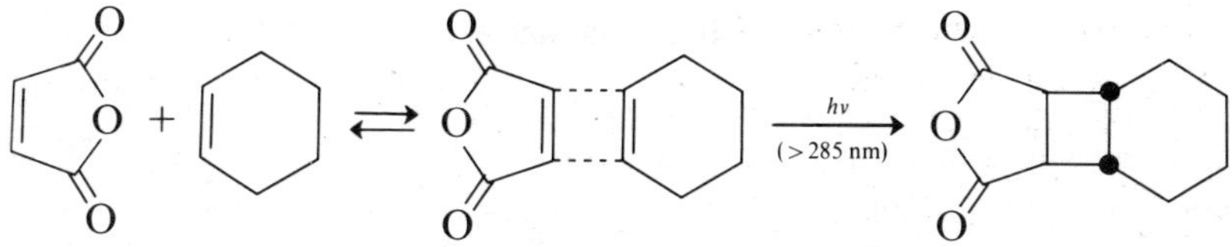

$$(7.111)$$

Cyclic conjugated carbonyl compounds add in good yield to 1,3-dienes (7.112)[121] and to alkynes (7.113)[122] on irradiation, and both these reactions have synthetic utility. The enol form of a β-diketone can also act as an unsaturated carbonyl compound in photochemical addition to an alkene,[123] and the reaction gives rise to a δ-diketone by way of an unstable cyclobutane intermediate (7.114).

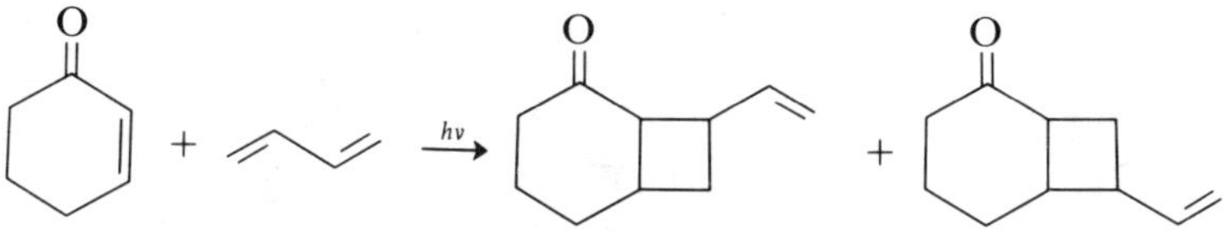

$$72\%, \phi = 0\cdot31 \qquad (7.112)$$

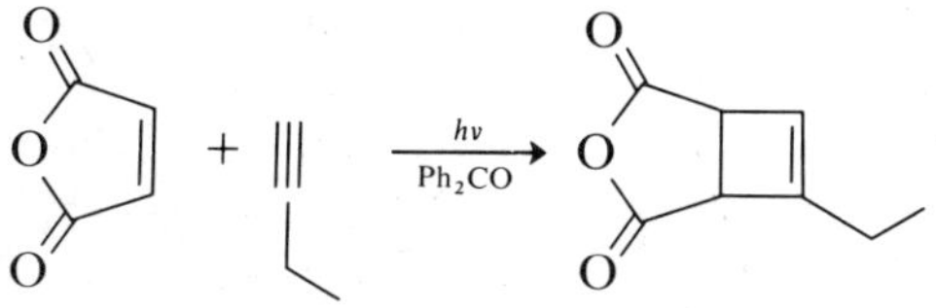

$$75\% \qquad (7.113)$$

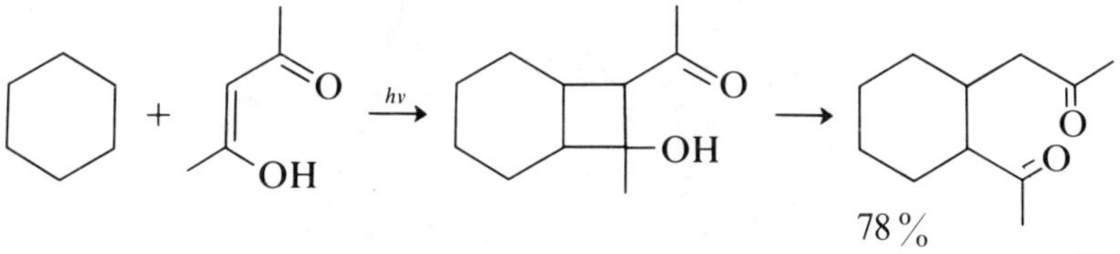

$$78\% $$

$$(7.114)$$

In a few cases, photocycloaddition of an alkene to an α,β-unsaturated carbonyl compound gives substantial amounts of both an oxetane and a cyclobutane (7.115).[124] The factors governing the competition are thought to be electronic, and they are discussed for p-benzoquinones in the following section.

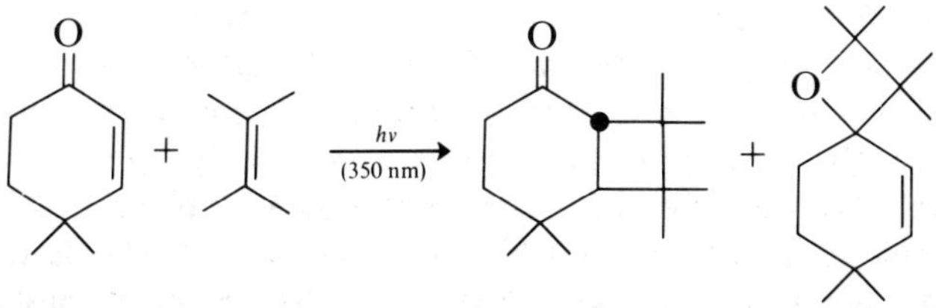

$$42\% \qquad 47\% \qquad (7.115)$$

216

_p_-Benzoquinones

_p_-Benzoquinones undergo photocycloaddition reactions with alkenes, dienes or alkynes. The primary product of reaction with an alkene is a bicyclo[4.2.0]-octene (7.116) formed by addition across the C=C of the quinone, or a spiro-oxetane (7.117) formed by addition across the C=O double bond.[125] Further irradiation of the cyclobutane adduct with excess alkene can lead to a 2:1 alkene:quinone product. The reactions can be sensitized by triplet sensitizers and quenched by triplet quenchers, and the preferred mode of addition is determined by the electronic nature of the lowest excited triplet state of the quinone. Those quinones with a lowest (n, π^*) triplet state (such as _p_-benzoquinone itself) give largely oxetane product, whereas those with a lowest (π, π^*) triplet state (such as duroquinone) give largely cyclobutane adduct.

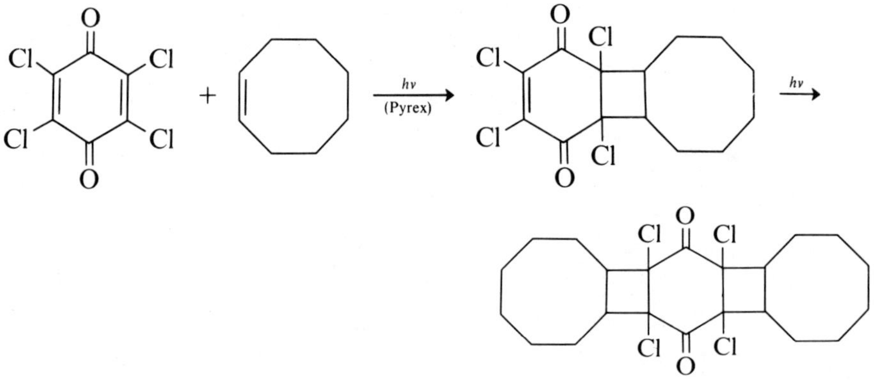

43%   (7.116)

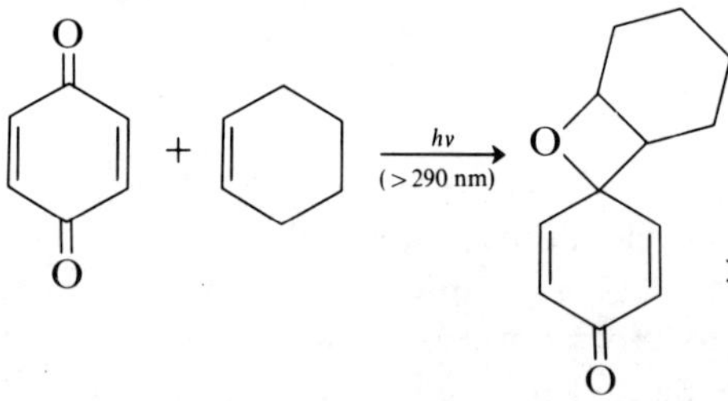

>90%   (7.117)

The reason for the different behaviour lies in the distribution of electrons in the excited states. In the (n, π^*) state the electron-deficient oxygen atom is the site of attack, whereas in the (π, π^*) state it is the carbon atoms of the ring which are relatively electron-deficient.

The adducts of quinones with conjugated dienes are again of two types, vinylcyclobutanes and spiro-dihydropyrans (7.118).[126] The latter products result from 1,4-addition to the diene rather than from 1,2-addition which normally predominates in photochemical reactions. The vinylcyclobutane adducts undergo an intramolecular photocycloaddition on prolonged irradiation,[127] and this is a valuable route to certain strained systems (7.119).

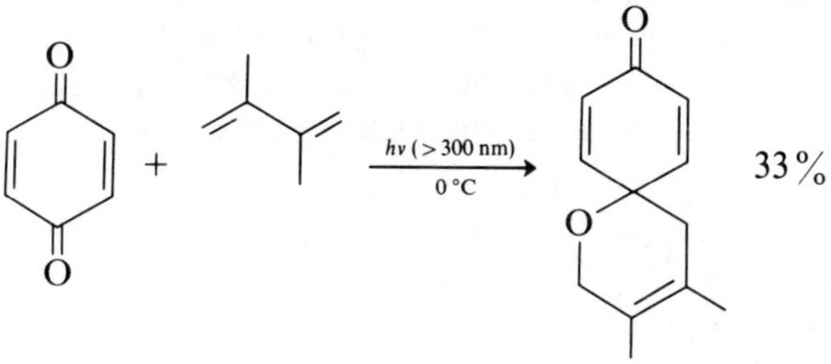

(7.118)

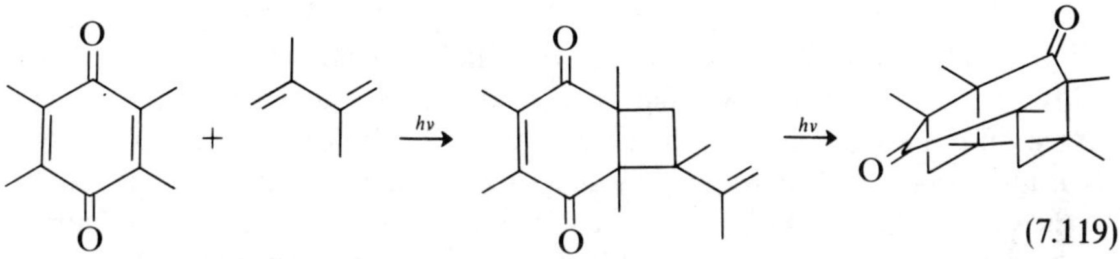

(7.119)

With alkynes,[128] quinones undergo photoaddition to give a cyclobutene or an adduct which arises by decomposition of the first-formed spiro-oxetene (7.120).

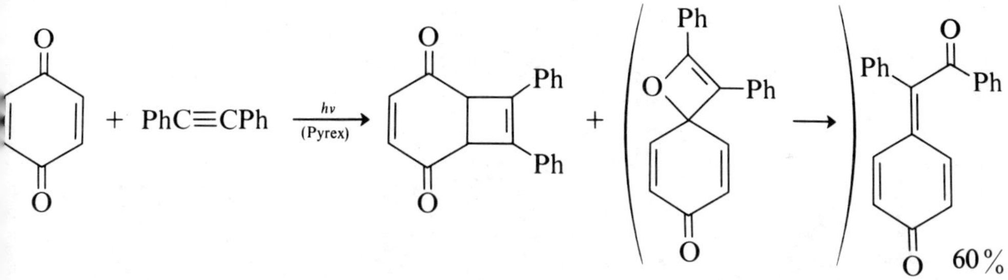

(7.120)

7.7.4   Dioxene Formation

α-Diketones or o-quinones add photochemically to alkenes to give either the expected spiro-oxetane or a dioxene (7.121). Both products are formed in a two-step non-stereospecific reaction initiated by attack of the diketone triplet state on ground state alkene.[129]

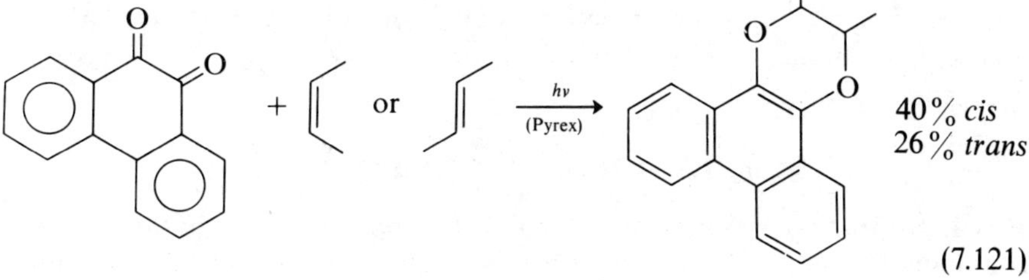

40% cis
26% trans

(7.121)

The excited states of aliphatic ketones are susceptible to attack by molecular oxygen. Acetone undergoes isotope exchange in high quantum yield when irradiated in the gas phase with 'heavy' oxygen, $^{18}O_2$ (7.122). The proposed mechanism involves 1,4-biradical intermediates.[130]

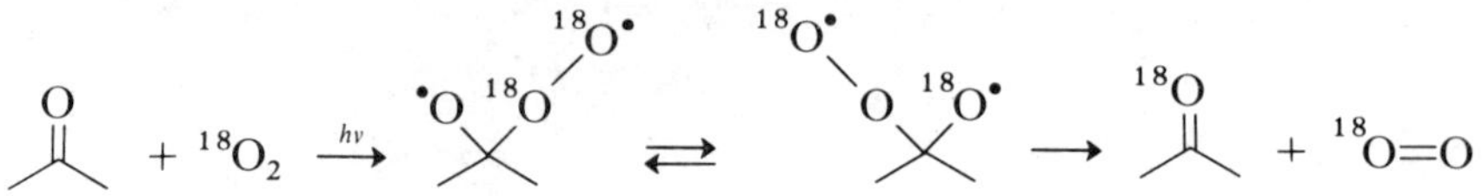

$$(7.122)$$

Cyclic ketones undergo an oxidation reaction which leads to unsaturated acids and keto-acids on irradiation in oxygen-saturated solution.[131] Initial attack on the excited state of the ketone by molecular oxygen occurs, followed by ring-cleavage and hydrogen-transfer to give a peroxyacid, which is reduced to the carboxylic acid (7.123). This acid is not the acid which would be formed by oxidation of the unsaturated aldehyde obtained on photolysis of the cyclic ketone in the absence of oxygen (7.124).

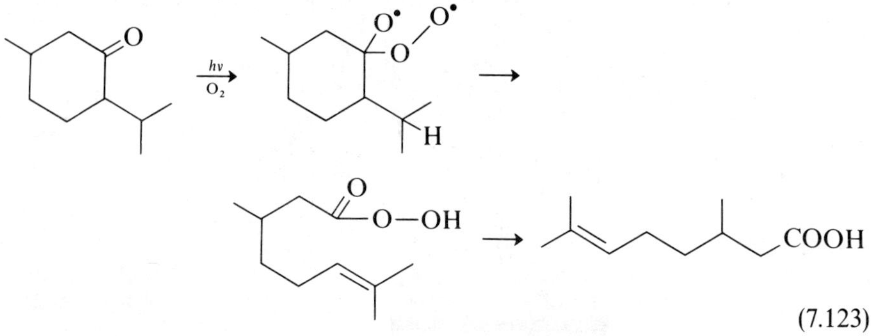

$$(7.123)$$

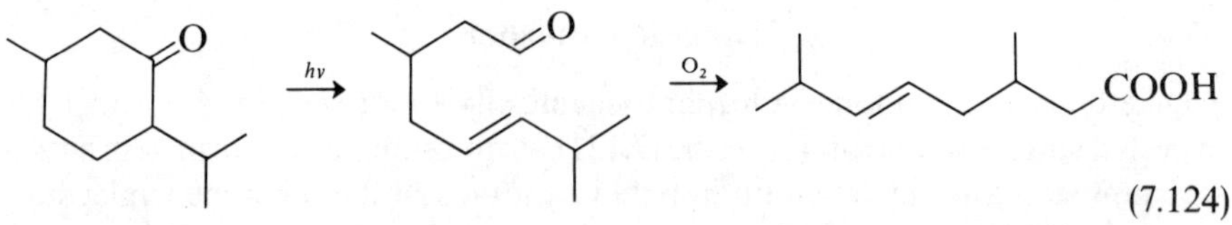

$$(7.124)$$

These oxidation reactions involve attack by the excited state of the ketone on ground state oxygen, and therefore they differ from the photo-oxidation of alkenes (see Chapter 8, section 8.6), which are reactions of singlet excited oxygen and ground state alkene.

7.9 **THIOCARBONYL COMPOUNDS**

It is appropriate to describe at this point the photochemistry of thiocarbonyl compounds because of the similarities between the C=O and C=S groups.

Simple aliphatic thioketones have not been widely studied because they can be isolated only at low temperature, and the most extensively studied thioketone is thiobenzophenone, The ultraviolet absorption spectrum of this compound shows three well separated bands at 599 (181), 316·5 (15 800) and 235 nm (900 l mol^{-1} cm^{-1}). The lowest energy band corresponds to an $n \rightarrow \pi^*$ transition, and the second band to a $\pi \rightarrow \pi^*$ transition.

Thiobenzophenone is photoreduced by ethanol or by propan-2-ol to give diphenylmethanethiol and other products.[132] The primary step involves abstraction of a hydrogen atom from the alcohol by the sulphur atom of the excited state to give a thioketyl radical (7.125) in a process analogous to that involved in the photoreduction of ketones (see Chapter 7, section 7.5.1). The first-formed product is diphenylmethanethiol, and this undergoes further reaction with an excited state of thiobenzophenone to give sulphide and disulphide products (7.126). In some cases the reaction is more complicated because the thioketone acts as a trapping agent for radicals derived from the hydrogen donor.[133]

$$Ph_2C{=}S + (CH_3)_2CH{-}OH \xrightarrow{\ h\nu\ (589\ nm)\ }$$
$$Ph_2\dot{C}{-}SH + (CH_3)_2\dot{C}{-}OH \qquad (7.125)$$

$$Ph_2C{=}S + (CH_3)_2CH{-}OH \xrightarrow{\ h\nu\ (589\ nm)\ }$$
$$Ph_2CH{-}SH + (CH_3)_2C{=}O$$
$$63\% \qquad\qquad 88\%$$

$h\nu, Ph_2CS$

$$Ph_2CH{-}S{-}CHPh_2 + Ph_2CH{-}S{-}S{-}CHPh_2 \qquad (7.126)$$
$$30\% \qquad\qquad\qquad 3\%$$

The reactive excited state of the thioketone is an (n, π^*) state, since the photo-reduction can be effected with radiation of wavelength 589 nm. Originally it was thought that the carbon atom rather than the sulphur atom in the excited state abstracted the hydrogen atom from the alcohol, but this is ruled out by the results of deuterium-labelling studies.

Intramolecular photochemical hydrogen abstraction by thiocarbonyl compounds has been observed. 2-Benzylthiobenzophenone is reported[134] to incorporate deuterium from O-deuteriomethanol into the benzylic position (7.127), and this probably occurs by way of a photoenolization reaction which

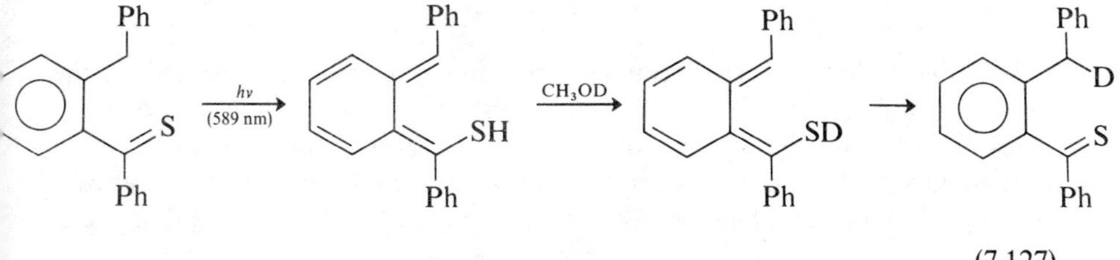

(7.127)

has a direct parallel in ketone photochemistry. *O*-Alkyl thiobenzoates undergo an intramolecular hydrogen abstraction and elimination reaction to give alkene and thiobenzoic acid (7.128),[135] and this process is analogous to the Norrish type 2 reaction of carbonyl compounds.

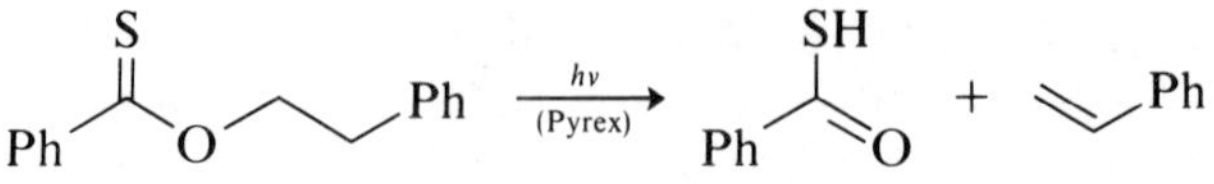

$$> 90\% \qquad (7.128)$$

The sulphur analogues of fenchone and camphor undergo what is apparently an intramolecular hydrogen abstraction from the β-position to give a cyclo-propanethiol (7.129),[136] and in this their chemistry is in contrast to that of the ketones themselves.

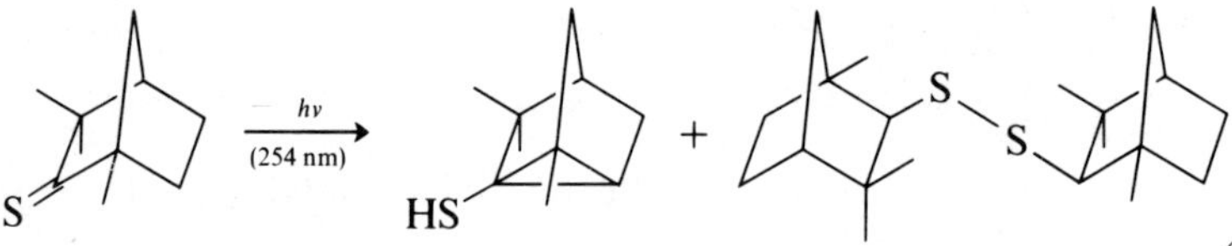

$$(7.129)$$

The second major class of photochemical reactions of thiocarbonyl compounds is cycloaddition.[137] An excited state of thiobenzophenone adds to electron-rich alkenes to give thietanes (7.130) or 1,4-dithianes (7.131) in reactions which resemble those leading to oxetanes from benzophenone. The reactive excited state is again the lowest (n, π^*) state of the thioketone, and the reaction can be brought about using either 313/366 or 589 nm radiation.

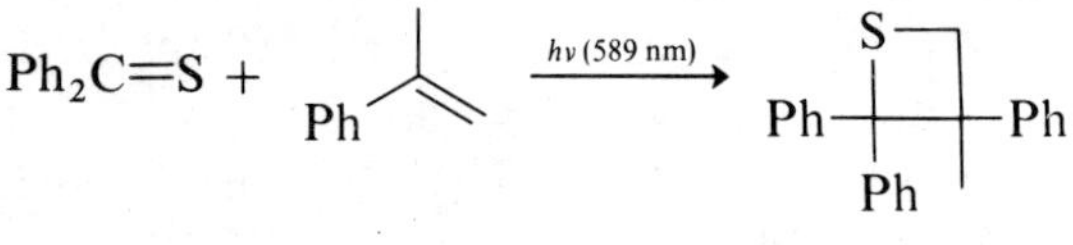

$$90\% \qquad (7.130)$$

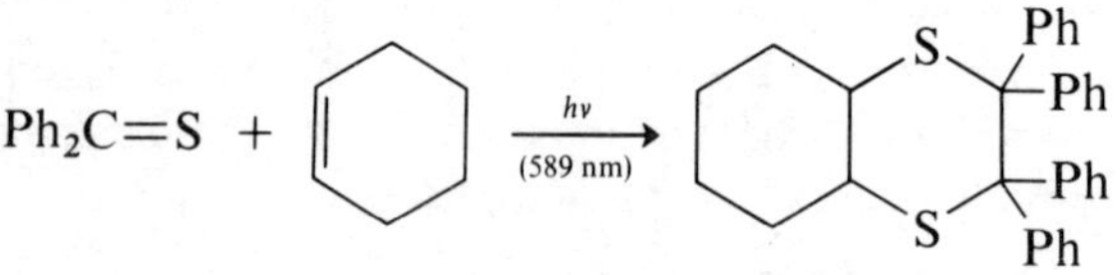

$$96\% \qquad (7.131)$$

Whether a thietane or a dithiane is the major product is governed by the relative rates of ring-closure and of attack on a second molecule of thio-benzophenone in the intermediate biradical. Steric crowding can play a part in hindering attack on another thioketone molecule, and the concen-tration of thiobenzophenone also affects the product ratio, as has been demonstrated[138] for addition to styrene (7.132).

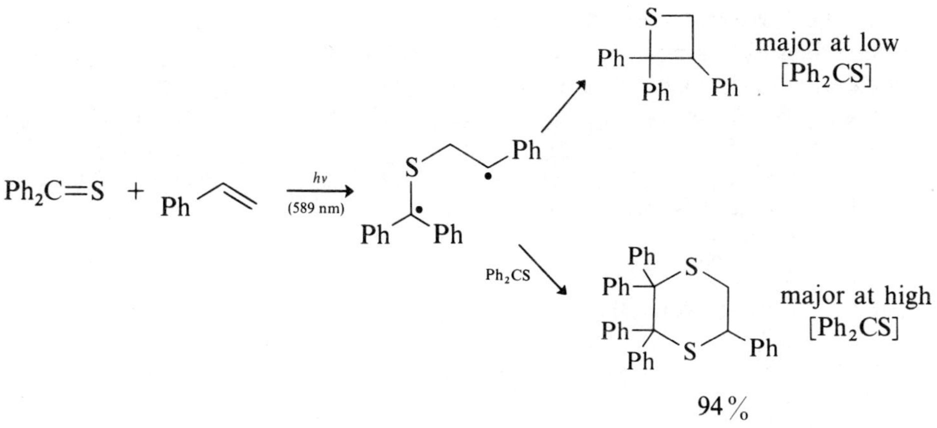

$Ph_2C=S$ + [Ph alkene] $\xrightarrow[\text{(589 nm)}]{h\nu}$ [structures]

major at low [Ph_2CS]

major at high [Ph_2CS]

94 %

Electron-deficient alkenes also give rise to thietane products, but only with the shorter wavelength radiation.[139] This process therefore occurs through a (π, π^*) state of the thioketone, and like the corresponding reaction of carbonyl compounds it is stereospecific (7.133). The situation is not completely clear, however, because acrylonitrile is reported[140] to add to thiobenzophenone to give a thietane if short wavelength radiation is used, but to give a 1,4-dithiane with long wavelength radiation (see Chapter 5, equation 5.33).

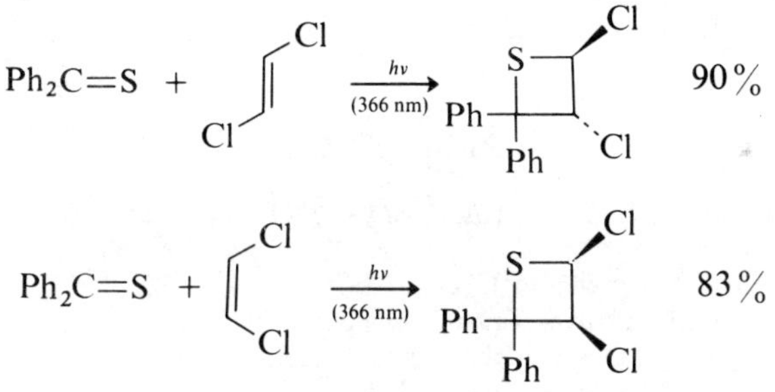

$Ph_2C=S$ + [alkene with Cl] $\xrightarrow[\text{(366 nm)}]{h\nu}$ [thietane structure] 90 %

$$(7.133)$$

$Ph_2C=S$ + [alkene with Cl] $\xrightarrow[\text{(366 nm)}]{h\nu}$ [thietane structure] 83 %

The (n, π^*) excited state of thiobenzophenone adds to alkynes[141] to give isothiochromene derivatives (7.134), and this must involve attack by an intermediate radical centre on one of the phenyl rings.

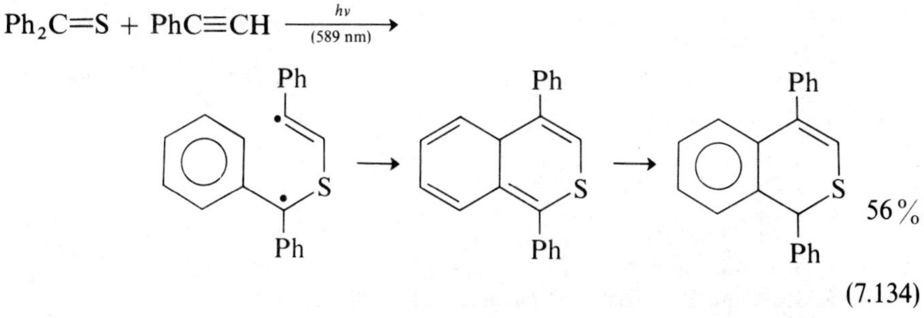

$Ph_2C=S + PhC\equiv CH \xrightarrow[\text{(589 nm)}]{h\nu}$ [structures] 56 %

$$(7.134)$$

Cycloaddition reactions have been reported for thiocarbonyl compounds other than aromatic thioketones. For instance, O-ethyl thiobenzoate gives a thietane on irradiation with cyclo-octa-1,3-diene (7.135), and thioadamantanone gives thietanes with alkenes such as 2-phenylpropene.[142] The latter thioketone in the absence of alkene undergoes photodimerization to give a 1,3-dithietane (7.136). A similar reaction had been reported previously[143] for a thioacyl fluoride (7.137).

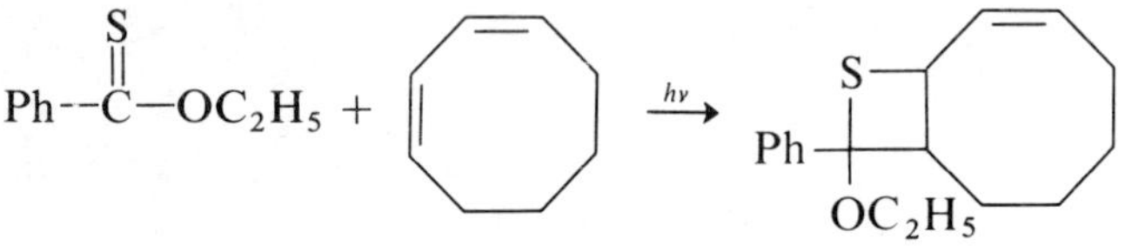

$$Ph\text{--}\overset{\displaystyle S}{\overset{\|}{C}}\text{--}OC_2H_5 + \quad \xrightarrow{h\nu} \quad \tag{7.135}$$

$$\xrightarrow[(254\ nm)]{h\nu} \tag{7.136}$$

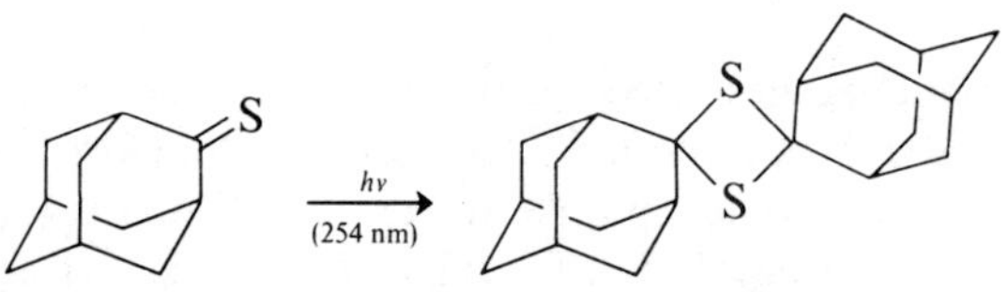

$$CF_2Cl\text{--}\overset{\displaystyle S}{\overset{\|}{C}}\text{--}F \xrightarrow{h\nu\ (quartz)} \qquad 83\% \tag{7.137}$$

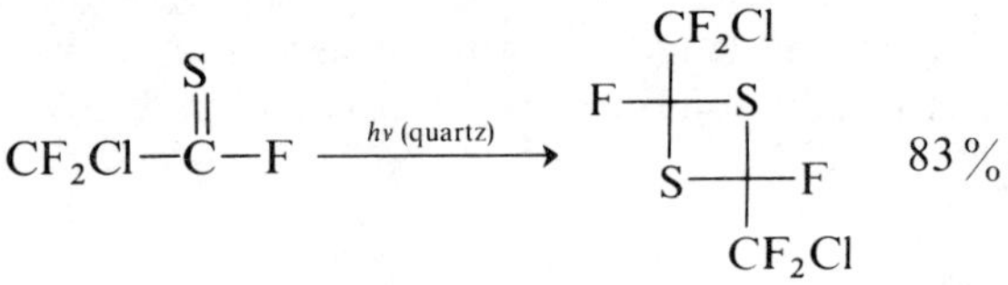

REFERENCES

1. R. F. Borkman and D. R. Kearns, *J. Chem. Phys.*, **44**, 945 (1966).
2. M. O'Sullivan and A. C. Testa, *J. Amer. Chem. Soc.*, **92**, 258 (1970); *ibid.*, 5842.
3. R. S. Becker, K. Inuzuka and J. King, *J. Chem. Phys.*, **52**, 5164 (1970).
4. C. A. Parker and T. A. Joyce, *Chem. Commun.*, 749 (1968).
5. J. A. Barltrop and J. D. Coyle, *J. Chem. Soc.* (*B*), 251 (1971).
6. N. C. Yang, E. D. Feit, M. H. Hoi, N. J. Turro and J. C. Dalton, *J. Amer. Chem. Soc.*, **92**, 6974 (1970).
7. K. Schaffner, *Chimia*, **19**, 575 (1965).
8. F. J. Golemba and J. E. Guillet, *Macromol.*, **5**, 63 (1972).
9. C. H. Bamford and R. G. W. Norrish, *J. Chem. Soc.*, 1544 (1938).
10. N. C. Yang and E. D. Feit, *J. Amer. Chem. Soc.*, **90**, 504 (1968).
11. P. S. Engel, *J. Amer. Chem. Soc.*, **92**, 6074 (1970); W. K. Robbins and R. H. Eastman, *ibid.*, **92**, 6076 (1970).
12. G. Quinkert, K. Opitz, W. W. Wiersdorff and J. Weinlich, *Tetrahedron Letters*, 1863 (1963).
13. H. Paul and H. Fischer, *Helv. Chim. Acta*, **56**, 1575 (1973).
14. R. S. Givens and W. F. Oettle, *J. Org. Chem.*, **37**, 4325 (1972).
15. H.-G. Heine, *Ann.*, **732**, 165 (1970).
16. H.-G. Heine, *Tetrahedron Letters*, 4755 (1972).
17. J. F. Harris, *J. Org. Chem.*, **30**, 2182 (1965).
18. R. S. Davidson and P. R. Steiner, *J. Chem. Soc. Perkin II*, 1357 (1972).
19. A. A. Scala and D. G. Ballan, *Canad. J. Chem.*, **50**, 3938 (1972).

20. H. A. J. Carless, J. Metcalfe and E. K. C. Lee, *J. Amer. Chem. Soc.*, **94**, 7221 (1972).
21. J. E. Starr and R. H. Eastman, *J. Org. Chem.*, **31**, 1393 (1966).
22. R. S. Cooke and G. D. Lyon, *J. Amer. Chem. Soc.*, **93**, 3840 (1971).
23. J. C. Dalton, K. Dawes, N. J. Turro, D. S. Weiss, J. A. Barltrop and J. D. Coyle, *J. Amer. Chem. Soc.*, **93**, 7213 (1971).
24. R. Simonaitis and J. N. Pitts, *J. Amer. Chem. Soc.*, **91**, 108 (1969).
25. J. D. Coyle, *J. Chem. Soc.* (*B*), 1736 (1971).
26. P. Yates, *Pure Appl. Chem.*, **16**, 93 (1968).
27. N. J. Turro and D. M. McDaniel, *J. Amer. Chem. Soc.*, **92**, 5727 (1970); see also G. Quinkert, P. Jacobs and W.-D. Stohrer, *Angew. Chem. Intern. Ed.*, **13**, 197 (1974).
28. J. A. Barltrop and J. D. Coyle, *Chem. Commun.*, 1081 (1969).
29. L. A. Paquette and R. F. Eizember, *J. Amer. Chem. Soc.*, **89**, 6205 (1967); J. K. Crandall, J. P. Arrington and J. Hen, *ibid.*, **89**, 6208 (1967).
30. G. Büchi and E. M. Burgess, *J. Amer. Chem. Soc.*, **82**, 4333 (1960).
31. J. R. Williams and G. M. Sarkisian, *Chem. Commun.*, 1564 (1971); P. S. Engel, M. A. Schexnayder, H. Ziffer and J. I. Seeman, *J. Amer. Chem. Soc.*, **96**, 924 (1974).
32. K. Schaffner, *Pure Appl. Chem.*, **33**, 329 (1973).
33. K. N. Houk, D. J. Northington and R. E. Duke, *J. Amer. Chem. Soc.*, **94**, 6233 (1972).
34. L. D. Hess, J. L. Jacobson, K. Schaffner and J. N. Pitts, *J. Amer. Chem. Soc.*, **89**, 3684 (1967).
35. W. G. Dauben, G. W. Shaffer and E. J. Deviny, *J. Amer. Chem. Soc.*, **92**, 6273 (1970).
36. A. Padwa in O. L. Chapman (ed.), *Organic Photochemistry*, Volume 1, Dekker, New York (1967), p. 91.
37. C. K. Johnson, B. Doming and W. Reusch, *J. Amer. Chem. Soc.*, **85**, 3894 (1963).
38. J. K. Crandall, C. F. Mayer, J. P. Arrington and R. J. Watkins, *J. Org. Chem.*, **39**, 248 (1974).
39. P. Borrell and J. Sedlar, *Trans. Faraday Soc.*, **66**, 1670 (1970).
40. N. Filipescu and F. L. Minn, *J. Amer. Chem. Soc.*, **90**, 1544 (1968); but see S. A. Weiner, *ibid.*, **93**, 425 (1971).
41. J. Michl, *Mol. Photochem.*, **4**, 257 (1972).
42. Most data are from J. C. Scaiano, *J. Photochem.*, **2**, 81 (1973/4).
43. P. J. Wagner, *J. Amer. Chem. Soc.*, **95**, 5604 (1973).
44. S. Hirayama, *Rev. Phys. Chem. Japan*, **42**, 49 (1972).
45. N. C. Yang and R. L. Dusenbery, *J. Amer. Chem. Soc.*, **90**, 5899 (1968).
46. S. G. Cohen, A. Parola and G. H. Parsons, *Chem. Rev.*, **73**, 141 (1973).
47. R. F. Bartholomew, R. S. Davidson, P. F. Lambeth, J. F. McKellar and P. H. Turner, *J. Chem. Soc. Perkin II*, 577 (1972).
48. J. Guttenplan and S. G. Cohen, *Chem. Commun.*, 247 (1969).
49. H. Yoshida, K. Hayashi and T. Warashina, *Bull. Chem. Soc. Japan*, **45**, 3515 (1972).
50. G. R. McMillan, J. G. Calvert and J. N. Pitts, *J. Amer. Chem. Soc.*, **86**, 3602 (1964).
51. R. Srinivasan, *J. Amer. Chem. Soc.*, **81**, 5061 (1959).
52. N. C. Yang, S. P. Elliott and B. Kim, *J. Amer. Chem. Soc.*, **91**, 7551 (1969); J. C. Dalton and N. J. Turro, *Ann. Rev. Phys. Chem.*, **21**, 499 (1970).
53. K. Schulte-Elte, B. Willhalm, A. F. Thomas, M. Stoll and G. Ohloff, *Helv. Chim. Acta*, **54**, 1759 (1971).
54. N. J. Turro and T.-J. Lee, *J. Amer. Chem. Soc.*, **91**, 5651 (1969).
55. N. C. Yang and D.-M. Thap, *Tetrahedron Letters*, 3671 (1966).
56. R. R. Sauers, M. Gorodetsky, J. A. Whittle and C. K. Hu, *J. Amer. Chem. Soc.*, **93**, 5520 (1971).
57. J. Iriarte, K. Schaffner and O. Jeger, *Helv. Chim. Acta*, **47**, 1255 (1964).
58. J. E. Gano, *Mol. Photochem.*, **3**, 79 (1971).
59. C. P. Casey and R. A. Boggs, *J. Amer. Chem. Soc.*, **94**, 6457 (1972).
60. I. Orban, K. Schaffner and O. Jeger, *J. Amer. Chem. Soc.*, **85**, 3033 (1963).

224

61. L. M. Stephenson and T. A. Gibson, *J. Amer. Chem. Soc.*, **94**, 4599 (1972).
62. E. D. Feit, *Tetrahedron Letters*, 1475 (1970); see also L. M. Stephenson and T. A. Gibson, *J. Amer. Chem. Soc.*, **96**, 5624 (1974).
63. K. Dawes, J. C. Dalton and N. J. Turro, *Mol. Photochem.*, **3**, 71 (1971).
64. P. J. Wagner and A. E. Kemppainen, *J. Amer. Chem. Soc.*, **90**, 5896 (1968).
65. P. J. Wagner, *J. Amer. Chem. Soc.*, **89**, 5898 (1967); J. A. Barltrop and J. D. Coyle, *ibid.*, **90**, 6584 (1968).
66. P. J. Wagner, *Accts. Chem. Res.*, **4**, 168 (1971).
67. A. Padwa, W. Eisenhardt, R. Gruber and D. Pashayan, *J. Amer. Chem. Soc.*, **93**, 6998 (1971).
68. Ref. 5; see also J. G. Pacifici and J. A. Hyatt, *Mol. Photochem.*, **3**, 267 (1971).
69. P. J. Wagner, P. A. Kelso, A. E. Kemppainen and R. G. Zepp, *J. Amer. Chem. Soc.*, **94**, 7500 (1972).
70. M. Barnard and N. C. Yang, *Proc. Chem. Soc.*, 302 (1958).
71. R. Breslow, *Chem. Soc. Rev.*, **1**, 553 (1972).
72. A. G. Schultz, C. D. De Boer, W. G. Herkstroeter and R. H. Schlessinger, *J. Amer. Chem. Soc.*, **92**, 6086 (1970).
73. H. J. Roth and M. H. El Raie, *Tetrahedron Letters*, 2445 (1970); *Arch. Pharm.*, **307**, 584 (1974).
74. E. Block and R. Stevenson, *J. Chem. Soc. Perkin I*, 308 (1973).
75. W. A. Henderson and E. F. Ullman, *J. Amer. Chem. Soc.*, **87**, 5424 (1965).
76. P. E. Eaton and K. Lin, *J. Amer. Chem. Soc.*, **86**, 2087 (1964).
77. R. R. Rando and W. von E. Doering, *J. Org. Chem.*, **33**, 1671 (1968).
78. K. J. Crowley, *J. Amer. Chem. Soc.*, **85**, 1210 (1963).
79. J. A. Barltrop and J. Wills, *Tetrahedron Letters*, 4987 (1968).
80. R. Noyori, H. Inoue and M. Kato, *J. Amer. Chem. Soc.*, **92**, 6699 (1970).
81. M. J. Jorgenson, *J. Amer. Chem. Soc.*, **91**, 198 (1969).
82. M. J. Jorgenson, *J. Amer. Chem. Soc.*, **91**, 6432 (1969).
83. P. J. Kropp in O. L. Chapman (ed.), *Organic Photochemistry*, Volume 1, Marcel Dekker, New York (1967), p. 1.
84. H. E. Zimmerman and K. G. Hancock, *J. Amer. Chem. Soc.*, **90**, 3749 (1968).
85. W. G. Dauben, W. A. Spitzer and M. S. Kellogg, *J. Amer. Chem. Soc.*, **93**, 3674 (1971).
86. O. L. Chapman, T. A. Rettig, A. A. Griswold, A. I. Dutton and P. Fitton, *Tetrahedron Letters*, 2049 (1963).
87. H. E. Zimmerman and D. I. Schuster, *J. Amer. Chem. Soc.*, **84**, 4527 (1962).
88. H. E. Zimmerman, *Angew. Chem. Intern. Ed.*, **8**, 1 (1969).
89. H. E. Zimmerman, *Adv. Photochem.*, **1**, 183 (1963).
90. E. R. Altwicker and C. D. Cook, *J. Org. Chem.*, **29**, 3087 (1964).
91. D. H. R. Barton, P. de Mayo and M. Shafiq, *J. Chem. Soc.*, 3314 (1958).
92. A. J. Waring, *Adv. Alicyclic Chem.*, **1**, 241 (1966).
93. J. S. Swenton, E. Saurborn, R. Srinivasan and F. I. Sontag, *J. Amer. Chem. Soc.*, **90**, 2990 (1968).
94. D. J. Patel and D. I. Schuster, *J. Amer. Chem. Soc.*, **90**, 5137 (1968).
95. D. I. Schuster and C. J. Polowczyk, *J. Amer. Chem. Soc.*, **86**, 4502 (1964).
96. G. Quinkert, *Pure Appl. Chem.*, **33**, 285 (1973).
97. J. Griffiths and H. Hart, *J. Amer. Chem. Soc.*, **90**, 3297 (1968); *ibid.*, **90**, 5296 (1968).
98. D. R. Arnold, *Adv. Photochem.*, **6**, 301 (1968).
99. S. R. Kurowsky and H. Morrison, *J. Amer. Chem. Soc.*, **94**, 507 (1972).
100. H. A. J. Carless, *Tetrahedron Letters*, 3173 (1973).
101. R. A. Caldwell, G. W. Sovocool and R. P. Gajewski, *J. Amer. Chem. Soc.*, **95**, 2549 (1973).
102. J. C. Dalton, P. A. Wriede and N. J. Turro, *J. Amer. Chem. Soc.*, **92**, 1318 (1970).
103. N. C. Yang, M. Kimura and W. Eisenhardt, *J. Amer. Chem. Soc.*, **95**, 5058 (1973).

225

104. N. J. Turro and P. A. Wriede, *J. Amer. Chem. Soc.*, **92**, 320 (1970).
105. N. D. Epiotis, *J. Amer. Chem. Soc.*, **94**, 1946 (1972).
106. J. A. Barltrop and H. A. J. Carless, *J. Amer. Chem. Soc.*, **93**, 4794 (1971).
107. J. A. Barltrop and H. A. J. Carless, *J. Amer. Chem. Soc.*, **94**, 8761 (1972).
108. R. R. Hautala and N. J. Turro, *Tetrahedron Letters*, 1229 (1972).
109. L. E. Friedrich and J. D. Bower, *J. Amer. Chem. Soc.*, **95**, 6869 (1973).
110. L. E. Friedrich and G. B. Schuster, *J. Amer. Chem. Soc.*, **94**, 1193 (1972).
111. O. L. Chapman and W. R. Adams, *J. Amer. Chem. Soc.*, **90**, 2333 (1968).
112. P. J. Wagner and D. J. Buchech, *J. Amer. Chem. Soc.*, **91**, 5090 (1969).
113. G. M. Blackburn and R. J. H. Davies, *Chem. Commun.*, 215 (1965).
114. D. C. Owsley and J. J. Bloomfield, *J. Org. Chem.*, **36**, 3768 (1971).
115. P. de Mayo, *Accts. Chem. Res.*, **4**, 41 (1971).
116. E. J. Corey, R. B. Mitra and H. Uda, *J. Amer. Chem. Soc.*, **86**, 485 (1964).
117. E. J. Corey, J. D. Ban, R. LeMahieu and R. B. Mitra, *J. Amer. Chem. Soc.*, **86**, 5570 (1964).
118. Ref. 115; W. C. Herndon, *Mol. Photochem.*, **5**, 253 (1973).
119. G. M. J. Schmidt, *Pure Appl. Chem.*, **27**, 647 (1971); M. D. Cohen, G. M. J. Schmidt and F. I. Sonntag, *J. Chem. Soc.*, 2000 (1964).
120. R. Robson, P. W. Grubb and J. A. Barltrop, *J. Chem. Soc.*, 2153 (1964).
121. T. S. Cantrell, *J. Org. Chem.*, **39**, 3063 (1974).
122. W. Hartmann, *Chem. Ber.*, **102**, 3974 (1969).
123. P. de Mayo and H. Takeshita, *Canad. J. Chem.*, **41**, 440 (1963).
124. P. J. Nelson, D. Ostrem, J. D. Lassila and O. L. Chapman, *J. Org. Chem.*, **34**, 811 (1969).
125. J. M. Bruce, *Quart. Rev.*, **21**, 405 (1967).
126. J. A. Barltrop and B. Hesp, *J. Chem. Soc.*, 5182 (1965).
127. G. Koltzenburg, K. Kraft and G. O. Schenck, *Tetrahedron Letters*, 353 (1965).
128. D. Bryce-Smith, G. I. Fray and A. Gilbert, *Tetrahedron Letters*, 2137 (1964).
129. Y. L. Chow, T. C. Joseph, H. H. Quon and J. N. S. Tam, *Canad. J. Chem.*, **48**, 3045 (1970).
130. R. Srinivasan, *Adv. Photochem.*, **1**, 89 (1963).
131. G. Quinkert, *Angew. Chem. Intern. Ed.*, **4**, 211 (1965).
132. A. Ohno and N. Kito, *Intern. J. Sulfur Chem. A*, **1**, 26 (1971).
133. N. Kito and A. Ohno, *Bull. Chem. Soc. Japan*, **46**, 2487 (1973).
134. N. Kito and A. Ohno, *Chem. Commun.*, 1338 (1971).
135. D. H. R. Barton, M. Bolton, P. D. Magnus, K. G. Marathe, G. A. Poulton and P. J. West, *J. Chem. Soc. Perkin I*, 1574 (1973); see also P. de Mayo and R. Suau, *J. Amer. Chem. Soc.*, **96**, 6807 (1974).
136. D. S. L. Blackwell and P. de Mayo, *Chem. Commun.*, 130 (1973).
137. A. Ohno, *Intern. J. Sulfur Chem. B*, **6**, 183 (1971).
138. A. Ohno, Y. Ohnishi, G. Tsuchihashi and M. Fukuyama, *J. Amer. Chem. Soc.*, **90**, 7038 (1968).
139. A. Ohno, Y. Ohnishi and G. Tsuchihashi, *Tetrahedron Letters*, 161 (1969).
140. P. de Mayo and H. Shizuka, *Mol. Photochem.*, **5**, 339 (1973); *J. Amer. Chem. Soc.*, **95**, 3942 (1973).
141. A. Ohno, T. Koizumi and Y. Ohnishi, *Bull. Chem. Soc. Japan*, **44**, 2511 (1971).
142. C. C. Liao and P. de Mayo, *Chem. Commun.*, 1525 (1971).
143. W. J. Middleton, E. G. Howard and W. H. Sharkey, *J. Org. Chem.*, **30**, 1375 (1965).

Chapter 8

C=C Chromophores

SPECTRA AND EXCITED STATES

The lowest energy electronic transition in many compounds containing only C=C chromophores is from the highest occupied π molecular orbital to the lowest unoccupied π^* orbital. This is an allowed transition (for ethylene, $^1B_{1u} \leftarrow \,^1A_{1g}, f \sim 0.3$) and absorption is intense. For simple, non-conjugated alkenes the absorption maximum is around 170–180 nm in the vacuum ultra-violet region, and the lowest excited singlet state is of relatively high energy. The absorption spectrum of ethylene itself (see Chapter 2, Figure 2.15) shows an intense band in the region 145–190 nm with a long weak tail to ~ 207 nm. Super-imposed on this band are a number of sharp lines which are the first members of a Rydberg series terminating at ionization. With increasing alkyl substitution on ethylene the absorption tail extends to longer wavelength, and the first Rydberg transition ($\pi \rightarrow 3\sigma$) becomes of progressively lower energy. *trans*-Alkenes normally have a larger extinction coefficient for absorption than the *cis*-isomers because of the greater transition dipole moment for the former, but the wave-length of maximum absorption is sometimes longer for the *trans*-isomer (e.g., for the stilbenes) and sometimes shorter.

Ethylene $\pi \rightarrow \pi^*$ triplet absorption is very weak ($\varepsilon \sim 10^{-4}\,\mathrm{l\,mol^{-1}\,cm^{-1}}$) and extends[1] to about 350 nm. The difference in energy[2] between planar singlet ($\sim 650\,\mathrm{kJ\,mol^{-1}}$, 150 kcal mol^{-1}) and planar triplet ($\sim 350\,\mathrm{kJ\,mol^{-1}}$, 80 kcal mol^{-1}) (π, π^*) states is large because of the large spatial overlap integral associated with the π and π^* orbitals of an alkene. This large energy difference results in a very low rate constant for intersystem crossing, and explains why direct irradiation of alkenes gives rise to products derived from the singlet (π, π^*) state but not from the triplet state. Triplet state reactions of alkenes can normally be studied only by using triplet sensitizers as the source of electronic energy.

Non-conjugated dienes have similar spectral properties to those of mono-alkenes except when the double bonds are in a position to interact as in bicyclo[2.2.1]hepta-2,5-diene (norbornadiene, 8.1). The spectrum of this diene shows absorption at longer wavelength and of higher intensity than would be expected for two isolated C=C double bonds, and this is attributed[3] to orbital overlap caused by the geometrical constraints of the molecule (8.2).

(8.1)   (8.2)

For conjugated systems the ultraviolet absorption is at longer wavelength, with the energy of the singlet state becoming progressively lower as the extent of conjugation is increased. The reason for this can be seen by considering the interaction of the orbitals of two $C{=}C$ double bonds as in buta-1,3-diene (Figure 8.1).

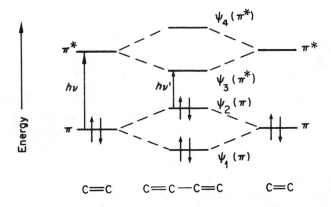

Figure 8.1. Molecular orbital diagram for a conjugated diene

The energy gap between the highest filled and lowest unfilled molecular orbitals is reduced when the double bonds are in conjugation, and the absorption is therefore at longer wavelength. In the ground state of butadiene ($\psi_1^2\psi_2^2\psi_3^0\psi_4^0$) the C(2)—C(3) bond has little double bond character, but in the first excited state ($\psi_1^2\psi_2^1\psi_3^1\psi_4^0$) this bond has considerable double bond character. The calculated (**HMO**) bond orders are

	S_0	S_1
C(1)—C(2)	0·89	0·44
C(2)—C(3)	0·45	0·72

This means that there are two isomeric first excited singlet states of butadiene corresponding to vertical excitation of the *s-cis* and *s-trans* conformations of ground state butadiene. These states can be crudely represented by the valence bond structures (8.3), though such representations should be used with caution.

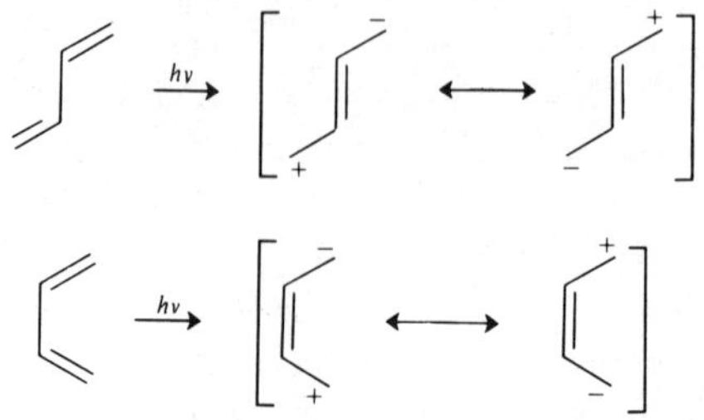

$$(8.3)$$

The major ultraviolet absorption characteristics of conjugated dienes are the result of vertical (Franck–Condon) excitation, which for acyclic dienes is normally excitation of s-*trans* conformers of the ground state, and for cyclopentadienes and cyclohexa-1,3-dienes is excitation of the rigidly held s-*cis* conformer.

	λ_{max} (nm)	ε_{---} (l mol^{-1} cm^{-1})
s-*trans*	210–250	10–30 $\times$ 10^3
s-*cis*	250–290	5–15 $\times$ 10^3

Because the geometry of the lowest energy conformation of the (π, π^*) excited states of dienes is very different from the ground state geometry (see next section), the maximum in the absorption spectrum corresponds to a transition to a high vibrational level of the electronic excited state. The position of the $(0, 0)$ vibrational band is in doubt since tail absorption is very weak, and very few dienes fluoresce. In all this the dienes resemble simple alkenes. Singlet state energies of dienes have been estimated at 380–420 kJ mol^{-1} (90–100 kcal mol^{-1}) for the s-*cis* geometry, and around 450 kJ mol^{-1} (110 kcal mol^{-1}) for s-*trans*.

Conjugated dienes, like mono-alkenes, have a large singlet–triplet splitting, and intersystem crossing is inefficient. The (π, π^*) triplet states can be obtained by sensitization, and again there are two of them, with s-*trans* and s-*cis* geometry. They are normally represented as biradical species (8.4), because the electrons are, on average, further apart than in the corresponding singlet state, but care should be taken to avoid reading unwarranted meaning into the representation. Estimates[4] for the triplet state energies of buta-1,3-diene are around 250 and 225 kJ mol^{-1} (60 and 53 kcal mol^{-1}) respectively for s-*trans* and s-*cis* geometry.

$$(8.4)$$

s-*trans*   s-*cis*

8.1.1   Geometry of Excited States

The excited states initially derived from the planar ground state of alkenes by absorption of a photon have the same geometry as the ground state (Franck–

Condon principle), but in both singlet and triplet (π, π^*) manifolds the lowest energy conformation is one in which the central bond is rotated through 90° from the ground state geometry (8.5). In these orthogonal or non-vertical (once called 'phantom') states electronic interaction is at a minimum, and alleviation of strain is possible because there is, in effect, no π bond in the (π, π^*) excited state. The very great difference in geometry between the ground state and the lowest energy singlet state may account for the long, weak tail in the ultraviolet absorption spectrum:

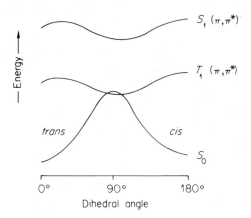

$$\text{Ground state} \qquad\qquad \text{First excited state} \tag{8.5}$$

The electronic states of alkenes and the variation in energy with dihedral angle (θ) are often represented on a simple energy diagram (Figure 8.2). For the singlet states the *cis-* and *trans-*excited states obtained by excitation from geometrically isomeric ground states are distinct. There is some doubt as to whether or not the lower energy orthogonal state is readily reached, and it appears[5] that there is an energy barrier to its formation from the vertical excited states, though the barrier may sometimes be very small from the *cis-*excited state (see next section). The orthogonal triplet state plays an important role in the sensitized isomerization of alkenes, and also in the direct photoisomerization of those alkenes, such as 1,2-dichloroethylene, for which intersystem crossing from the excited singlet state is efficient.

Figure 8.2. Potential energy diagram for electronic states of an alkene RCH=CHR

An important feature of the state diagram is the crossing (or near-crossing) of the T_1 and S_0 surfaces. This feature explains the very rapid intersystem crossing from T_1 to S_0, and the consequent very short triplet lifetime[6] of alkenes (10^{-4} to 10^{-5} s). Because of the short lifetime, phosphorescence is usually not observable.

The energy of the orthogonal triplet state for simple alkenes is not known with certainty, but values as low as $210 \, \text{kJ mol}^{-1}$ ($50 \, \text{kcal mol}^{-1}$) for dideuterio-ethylene[7] and $260 \, \text{kJ mol}^{-1}$ ($62 \, \text{kcal mol}^{-1}$) for disubstituted ethylenes[8] have been estimated.

The twisted geometry of the non-planar states is readily achieved for most acyclic alkenes and for medium and large ring cycloalkenes, but small ring (C_3 to C_5) cycloalkenes are unable to accommodate the required geometry. The excited states of these alkenes differ from those of other cycloalkenes in certain properties as a result of this, because they have no rapid relaxation pathway from T_1 to S_0.

In the excited states of dienes and polyenes a similar situation obtains, and the lowest energy conformation has the 90° twisted geometry about one of the double bonds. Theoretical studies[9] of the *cis–trans* isomerization in polyenes suggest that the lowest energy (π, π^*) triplet state has one twisted 'double' bond, preferably not a terminal bond. Overall isomerization about one *or two* bonds can occur through this state,[10] however, in contrast to the isomerization through the singlet state which takes place at only one bond (8.6). These results also confirm the low efficiency of intersystem crossing in the systems studied.

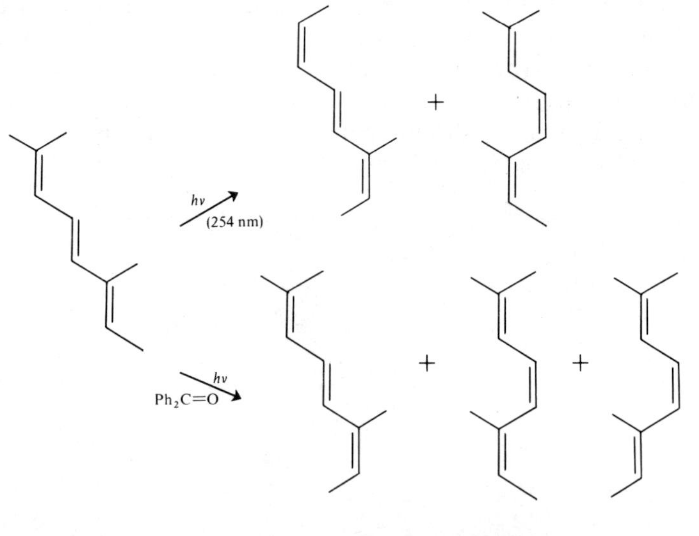

$$\text{(8.6)}$$

8.2   *cis–trans* ISOMERIZATION

The photochemical interconversion of *cis-* and *trans*-isomers of alkenes and polyenes can be achieved by direct irradiation or by the use of triplet photo-sensitizers. These methods have considerable synthetic value,[11] and the *cis–trans* isomerization of visual pigments is an important reaction in the processes of vision.[12] The mechanisms of the interconversion have been extensively investigated, but there are still details which are not fully elucidated.

Direct irradiation gives initially the vertical excited singlet states, *cis-S*$_1$ from *cis* ground state, and *trans-S*$_1$ from *trans* ground state. Decay of these spectro-scopic states to ground state isomers may occur directly (Figure 8.3) or through a common lower energy state (Figure 8.4). The evidence favours the latter, with the non-planar singlet state as the common intermediate.

The composition of the mixture when a photostationary state is reached depends on the relative extinction coefficients of *cis*- and *trans*-alkene at the wavelength employed, and on the partitioning ratios for decay to ground states

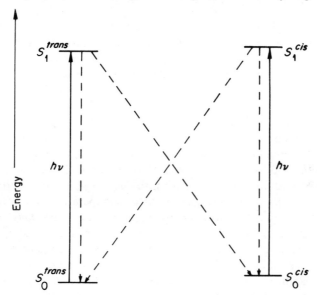

Figure 8.3. Pathway for stilbene *cis–trans* isomerization on direct irradiation, with no orthogonal state involved

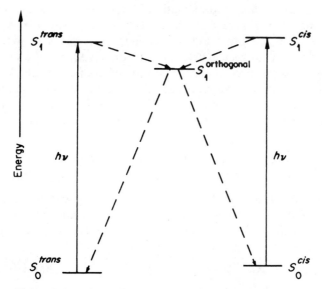

Figure 8.4. Pathway for stilbene *cis–trans* isomerization on direct irradiation, involving an orthogonal state

of each participating excited state. Simple alkenes absorb only at short wavelength, but *cis–trans* isomerization does occur together with fragmentation reactions (8.7), and the reactive excited state is the (π, π^*) singlet state rather than the Rydberg state.[13]

$$\text{(structure)} \underset{(203 \text{ nm})}{\overset{hv}{\rightleftharpoons}} \text{(structure)} \tag{8.7}$$

The stilbenes, however, provide one of the best illustrations. At long wavelengths *trans*-stilbene has a higher extinction coefficient than *cis*-stilbene, and since the partitioning ratio in the excited state(s) is close to unity, the photostationary state contains a high proportion (93%) of *cis*-stilbene as a result of selective isomerization of the *trans*-isomer.

The *cis* $\rightarrow$ *trans* isomerization of stilbene by direct irradiation is more or less independent of temperature, but the rate constant for the *trans* $\rightarrow$ *cis* isomerization is strongly temperature dependent. The quantum yield for *trans* $\rightarrow$ *cis* isomerization is 0·50 at 20 °C, but only 0·001 at -150 °C, at which temperature the quantum yield for fluorescence from the *trans*-isomer is near to unity. The decreasing rate of *trans* $\rightarrow$ *cis* reaction as the temperature is lowered is interpreted[14] in terms of a significant energy barrier between the *trans* excited state and the common intermediate (non-planar singlet) state. There is no such barrier between the *cis* excited state and the intermediate, and fluorescence is not observed from *cis*-stilbene. The same ideas explain the temperature dependence of the photoaddition of stilbene to alkenes,[15] and as well as explaining these temperature variations, the small energy barrier between S_1^{tr} and a lower energy state is in accord with the observed vibrational fine structure in the ultraviolet absorption spectrum of *trans*-stilbene which is absent in that of *cis*-stilbene. Calculation (SCF, Pople approximation)[16a] of energy surfaces for stilbene gives a picture (like Figure 8.2) which fits very well with the experimental observations. There is a low energy barrier on the S_1 surface near to the *trans*-S_1 vertical singlet, but no such barrier near to the *cis*-S_1 vertical singlet. These results have, however, been challenged on the basis of other calculations.[16b]

Although the common intermediate state in the direct isomerization of stilbene is probably not the non-planar triplet state, isomerization of 1-phenylpropene on direct irradiation (8.8) does seem to proceed through such a triplet state.[17]

$$\text{Ph} \underset{(285 \text{ nm})}{\overset{hv}{\rightleftharpoons}} \text{Ph} \qquad \begin{aligned} \phi_{cis \rightarrow trans} &= 0\cdot33 \\ \phi_{trans \rightarrow cis} &= 0\cdot22 \end{aligned} \tag{8.8}$$

1,2-Dichloroethylene has been irradiated directly in the singlet–triplet absorption band using a high pressure of oxygen to raise the extinction coefficient to a useful value.[18] Isomerization (8.9) takes place through triplet

$$\text{Cl}\overset{}{=}\text{Cl} \underset{}{\overset{hv}{\rightleftharpoons}} \text{Cl}\cdots\text{Cl}\,\text{H} \underset{hv}{\overset{}{\rightleftharpoons}} \text{Cl}\text{—}\text{Cl} \qquad (300\text{–}380 \text{ nm})$$

$$\tag{8.9}$$

states, and the sum of the quantum yields for *cis* → *trans* and *trans* → *cis* isomerization is close to unity ($\phi_{t\to c} = 0.61 \pm 0.07$, $\phi_{c\to t} = 0.45 \pm 0.06$). This points to the existence of a common intermediate triplet state for the isomerization processes.

8.2.1   Sensitized *cis–trans* Isomerization

The triplet sensitized *cis–trans* isomerization of alkenes or polyenes leads to a photostationary state whose composition is dependent on the triplet energy of the sensitizer. Some sensitizers with insufficient energy to populate the vertical *cis or trans* triplet excited states of the alkene are nonetheless efficient sensitizers of the isomerization, and in these cases energy transfer occurs to give a non-planar triplet state directly. The energy levels are illustrated in Figure 8.5 for stilbene, and the effect of sensitizer triplet energy on the *cis*:*trans* ratio in the photostationary state is seen in Figure 8.6.

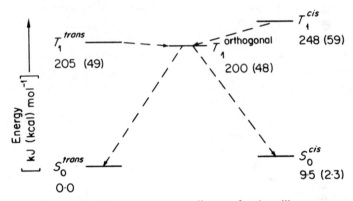

Figure 8.5. Triplet state energy diagram for the stilbenes

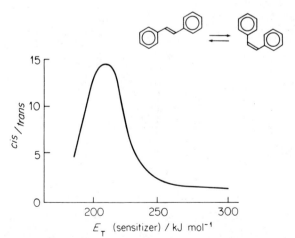

Figure 8.6. Photosensitized *cis–trans* isomerization of stilbene

234

High energy sensitizers ($E_T^{sens} > E_T^{tr}$ or E_T^c) populate both *cis* and *trans* excited triplet states, and the stationary state composition equals the branching ratio for decay of the orthogonal state to isomeric ground states. Sensitizers with a triplet energy intermediate between those of *cis*- and *trans*-stilbene ($E_T^c > E_T^{sens} > E_T^{tr}$) are able to populate the *trans*-triplet efficiently, but not the *cis*-triplet. This leads to selective isomerization of *trans*-stilbene, since the *cis*-isomer reacts more slowly, and a stationary state with a higher proportion of *cis*-stilbene is achieved. Sensitizers with a lower triplet energy than either *cis*- or *trans*-stilbene but with a higher energy than the orthogonal state ($E_T^c > E_T^{tr} > E_T^{sens} > E_T^{orth}$) populate a non-planar triplet state directly, and no new mechanism is required for this to occur.[19] The stationary state now becomes less rich in *cis*-stilbene as population of the vertical triplet states diminishes in importance. Around this point the efficiency of sensitized isomerization is markedly reduced because energy transfer to give any triplet state is slow ($E_T^{orth} > E_T^{sens}$). Flash spectroscopic measurements of the rate constants for energy transfer to *cis*- and *trans*-stilbene confirm this pattern.[20]

The situation is reversed for conjugated dienes such as penta-1,3-diene, and in the intermediate range of sensitizer triplet energies the ratio of *cis*:*trans* diene at photoequilibrium is lower than in the high energy range (Figure 8.7). The reason for this is in part that it seems that the vertical *cis*-triplet has a lower energy than the vertical *trans*-triplet (Figure 8.8), whereas for stilbenes the *trans*-triplet has the lower energy. Another factor is that there are isomeric *s-cis* and *s-trans* triplets for dienes, and selective formation of the *s-cis* triplet by lower energy sensitizers can also account in part for the observed isomer ratios at photoequilibrium,[21] since the branching ratio for decay of the *s-cis* orthogonal

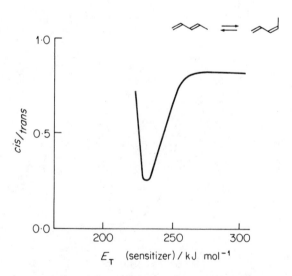

Figure 8.7. Photosensitized *cis–trans* isomerization of penta-1,3-diene

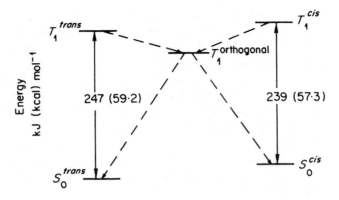

Figure 8.8. Triplet state energy diagram for the penta-1,3-dienes

triplet to ground state *trans*- or *cis*-isomer (0·72) is higher than that for the *s-trans* orthogonal triplet (0·55).

The quantum yield for sensitized isomerization of *cis*-penta-1,3-diene is concentration dependent,[22] and increases in value from 0·55 at concentrations less than about 0·08 M to a value of about 0·90 at 10 M. An intermolecular process leading to production of *trans*-diene as major product is competing with the intramolecular isomerization, and it is suggested that biradical intermediates are formed by attack of diene triplet on diene ground state, and these biradicals break down to give two molecules of ground state diene with the thermodynamically favoured *trans*-diene as major product (8.10). The process is closely allied to the formation of dimers which accompanies the isomerization (see Chapter 8, p. 260), since ring-closure of the biradicals competes (inefficiently) with cleavage to give two diene molecules.

$$(8.10)$$

The efficient quenching by conjugated dienes of higher energy triplet states is utilized in mechanistic studies for identifying the excited state responsible for photochemical reaction (see Chapter 5, p. 144), and a knowledge of the partitioning ratio for decay of the orthogonal triplet state to *cis* and *trans* ground states can be employed to 'count' the number of sensitizer triplets present under particular conditions and so estimate the quantum yield for intersystem crossing in the sensitizer molecule. Caution must be taken in interpreting the results in view of the ability of conjugated dienes to quench *singlet* excited states of aromatic hydrocarbons and carbonyl compounds.

Cis–trans isomerization of conjugated dienes and trienes follows a different pattern according to the multiplicity of the excited state involved.[23] Direct irradiation proceeds through singlet states to give isomerization in the primary processes about one bond only. It seems that there is *no* common intermediate species for the singlet state isomerization, since the sum of the quantum yields for *cis* → *trans* and *trans* → *cis* isomerization is much less than unity, and the deficit is not accounted for by excited state decay or chemical reaction. Sensitized isomerization occurs by way of triplet states to give isomerization about one *or two* bonds in the primary process (see equation 8.6). This 'two-bond' isomerization does not occur in the triplet sensitized reaction of non-conjugated dienes.[24]

The vertical triplet states of mono-alkenes have very high energies, but *cis–trans* isomerization can be sensitized[25] by triplet sensitizers such as acetone, benzene, $Hg(^3P_1)$ or acetophenone (8.11).

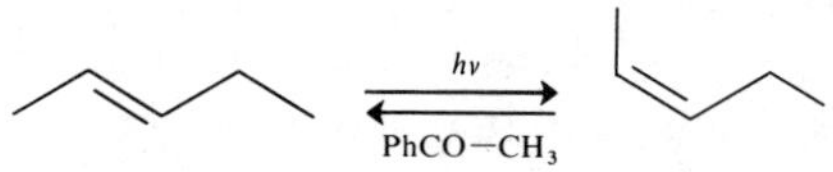

$$\text{(8.11)}$$

The quantum yields for isomerization are dependent on the structure of the sensitizer when its triplet energy is lower than about $335\ kJ\ mol^{-1}$ ($80\ kcal\ mol^{-1}$), and the Schenck mechanism is involved for at least part of the isomerization.[26] Different sensitizers give rise to different intermediate biradicals by addition to the alkene (8.12), and the biradicals break down to give *cis*- and *trans*-alkene isomers in ratios which vary from one sensitizer to another (low yields of oxetane are also produced with carbonyl sensitizers, see Chapter 7, section 7.7.1).

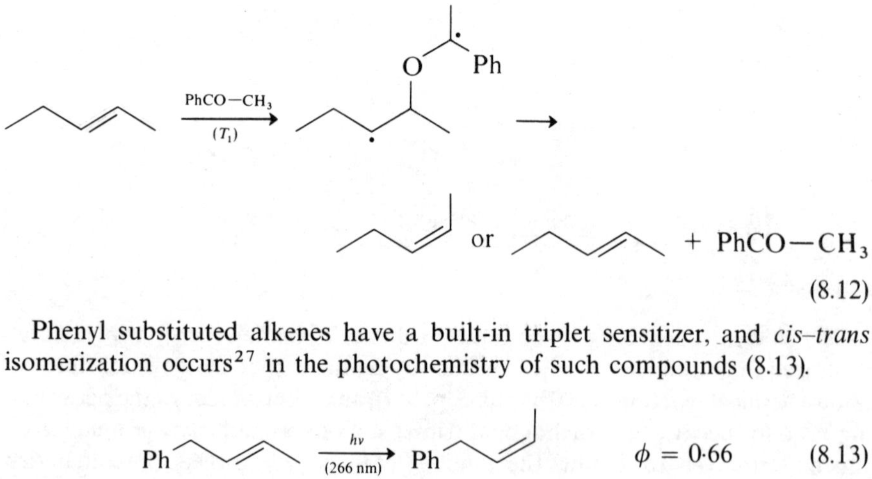

$$\text{(8.12)}$$

Phenyl substituted alkenes have a built-in triplet sensitizer, and *cis–trans* isomerization occurs[27] in the photochemistry of such compounds (8.13).

$$\text{Ph} \quad \xrightarrow[\text{(266 nm)}]{h\nu} \quad \text{Ph} \qquad\qquad \phi = 0.66 \qquad \text{(8.13)}$$

8.2.2   **Cycloalkenes**

Medium and large ring cycloalkenes (C_8 upwards) undergo *cis–trans* isomerization[28] in much the same way as acyclic alkenes (8.14).

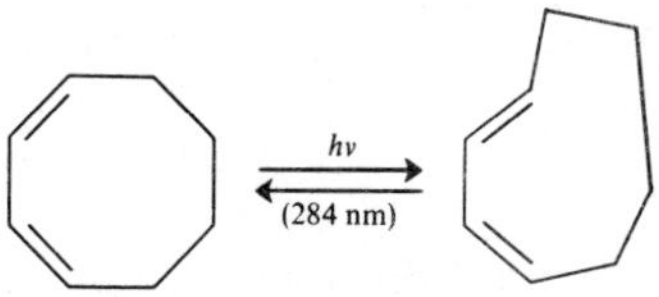

$$\phi_{c,c \to c,tr} = 0.28$$
$$\phi_{c,tr \to c,c} = 0.80$$

(8.14)

The same is thought to occur on direct or sensitized[29] irradiation of cyclohexenes and cycloheptenes, but the highly strained *trans*-cycloalkene cannot be
isolated. It reacts by addition to a ground state molecule of cycloalkene[30] to
produce a dimer (8.15), by Diels–Alder addition to an added diene[31] to give an
adduct with the stereochemistry expected for a *trans*-dienophile (8.16), or by
protonation and addition with a protic solvent[32] such as methanol (8.17) or
acetic acid (8.18).

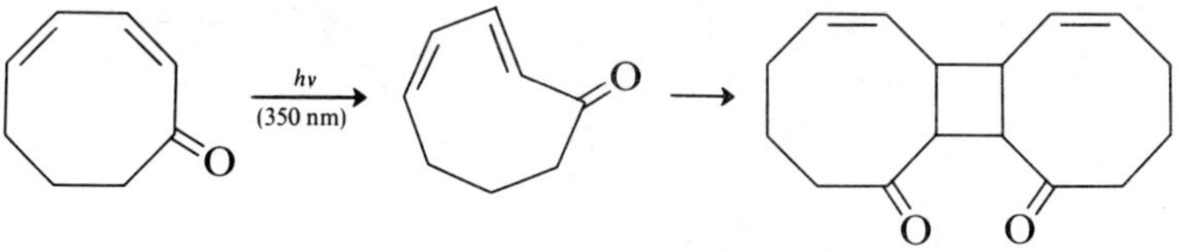

$\phi = 0.28$

85%

(8.15)

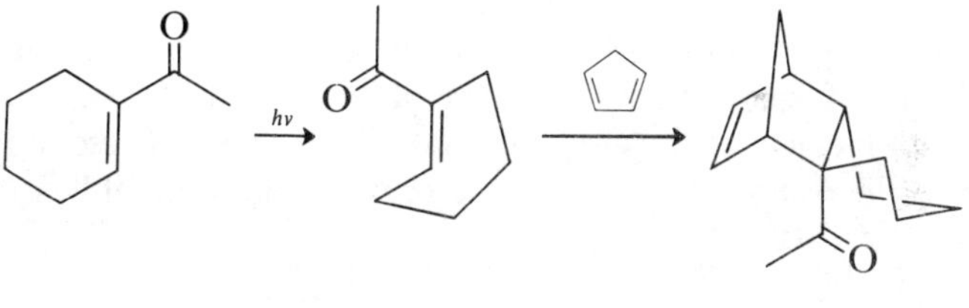

(8.16)

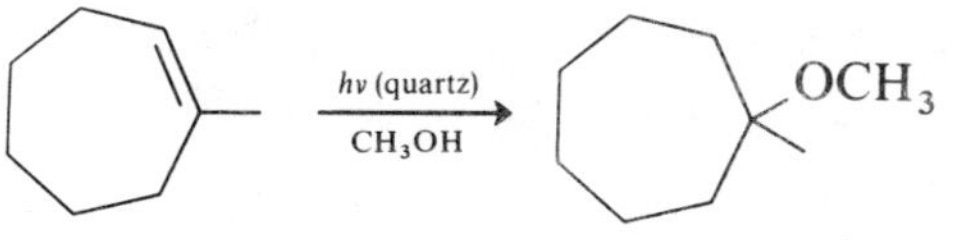

62%

(8.17)

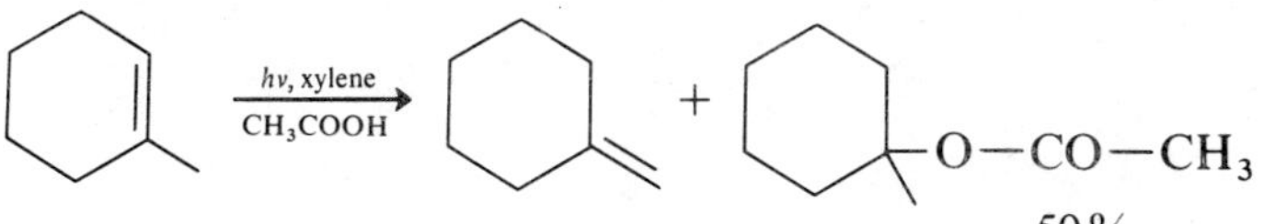

33%   50%

(8.18)

This process allows a selective addition[33] of alcohol to the double bond in
the ring of a compound such as limonene which contains both cyclic and
acyclic double bonds (8.19).

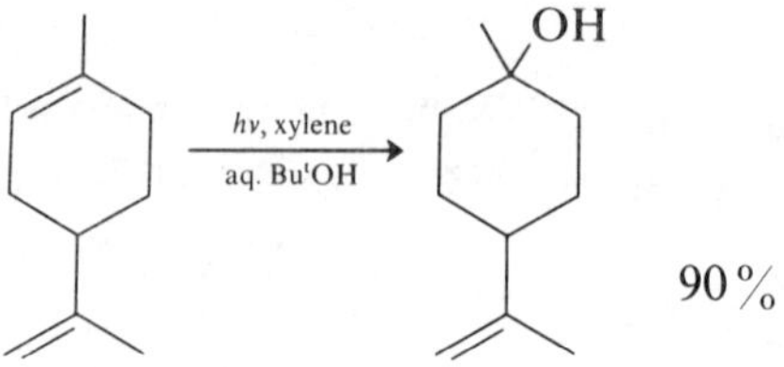

$$(8.19)$$

The mechanism of addition of methanol involving protonation followed by reaction of the carbonium ion so formed is supported by the pattern of deuterium substitution in the products obtained[34] when cyclohexenes are irradiated in the presence of O-deuteriomethanol (8.20).

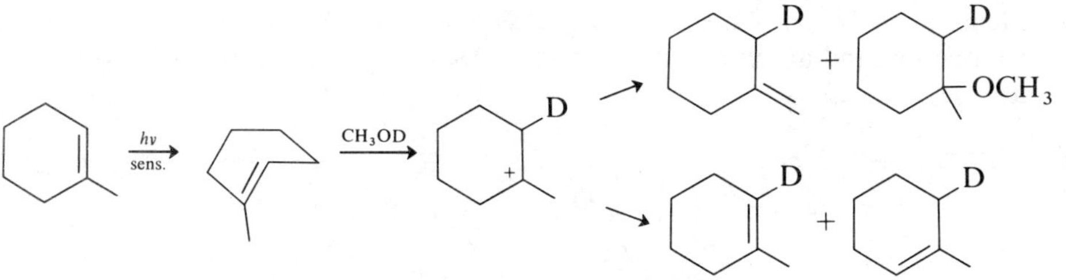

$$(8.20)$$

No deuterium incorporation occurs when cyclopentenes are irradiated in the presence of O-deuteriomethanol, and the observed products of these reactions (which do not include ethers) are explained on the basis of a free radical mechanism (8.21). The *trans*-cyclopentenes are presumably too highly strained to be formed at all.[32]

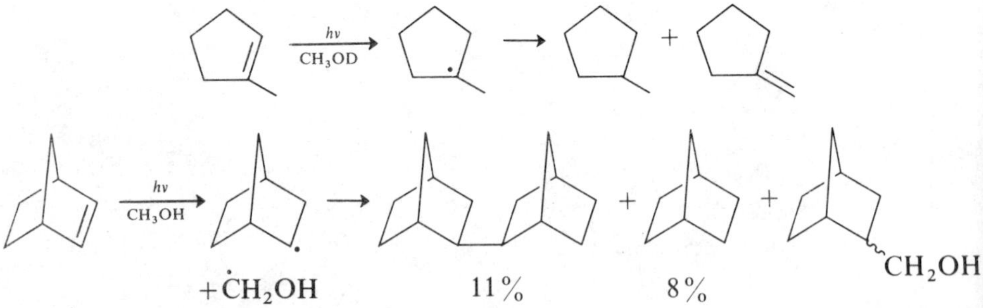

$$(8.21)$$

Cyclic and acyclic tetrasubstituted alkenes behave differently, and when irradiated directly in methanol they give addition products[35] which do not correspond to either of the patterns described above (8.22).

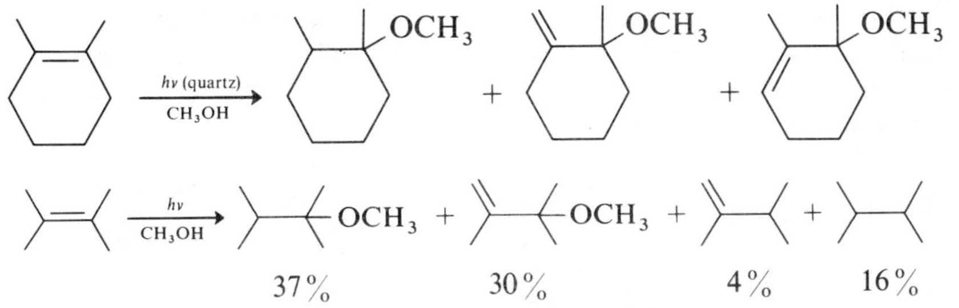

$$(8.22)$$

No deuterium is incorporated in the ethers from O-deuteriomethanol solvent, and it is proposed that the excited state responsible for reaction is the Rydberg $(\pi, 3\sigma)$ singlet state, which may be the lowest energy state for tetrasubstituted alkenes. In this excited state there is a deficiency of electrons in the region of the carbon atoms, and nucleophilic attack by methanol occurs to give a radical species which leads to the observed ether products by disproportionation (8.23). The electrons attack the solvent to give hydrogen atoms which react with alkene to give the observed hydrocarbon products (8.24).

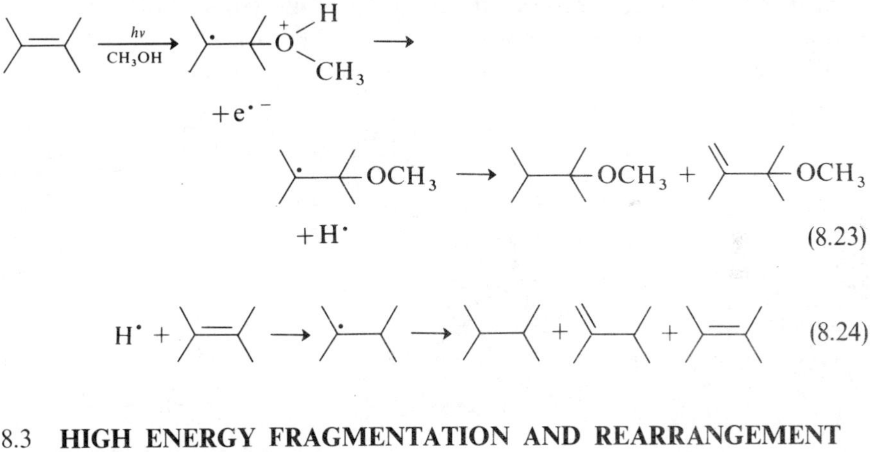

$$(8.23)$$

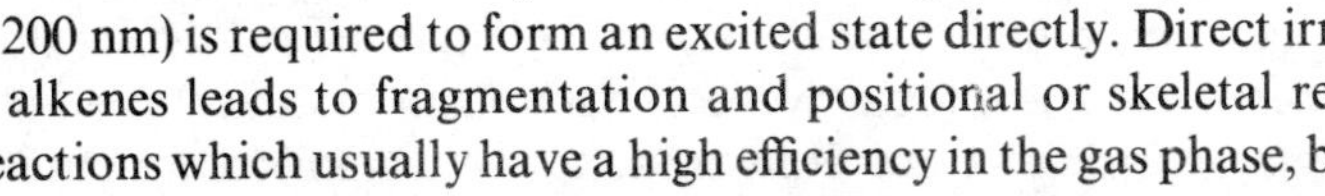

$$(8.24)$$

8.3 HIGH ENERGY FRAGMENTATION AND REARRANGEMENT

Simple mono-alkenes absorb only at short wavelength, and high energy radiation ($\lambda < 200$ nm) is required to form an excited state directly. Direct irradiation of simple alkenes leads to fragmentation and positional or skeletal rearrangement in reactions which usually have a high efficiency in the gas phase, but which are inefficient in solution, where solvent cage effects are important and energy dissipation is very rapid. Ethylene on direct irradiation[36] gives an excited singlet state which undergoes three major primary processes (8.25).

$$
CH_2{=}CH_2 \xrightarrow{\;h\nu\,(185\,nm)\;}
\begin{cases}
CH{\equiv}CH + H_2 & \phi = 0.68 \\
CH{\equiv}CH + H^{\cdot} + H^{\cdot} & \phi = 0.20 \\
CH_2{=}CH^{\cdot} + H^{\cdot} & \phi = 0.16
\end{cases}
\qquad (8.25)
$$

Higher alkenes undergo allylic C—H or C—C cleavage as a major process on irradiation. For propene (8.26) allylic C—H and vinylic C—C cleavage are both important,[37] but for hex-2-ene (8.27) allylic cleavage is the only major primary process other than *cis–trans* isomerization.[38]

$$CH_3CH{=}CH_2 \xrightarrow[\text{gas}]{h\nu \ (185 \text{ nm})} \begin{cases} CH_2{=}CH{-}\dot{C}H_2 + H^{\cdot} & \phi = 0{\cdot}41 \\ CH_2{=}CH^{\cdot} + {}^{\cdot}CH_3 & \phi = 0{\cdot}36 \end{cases} \quad (8.26)$$

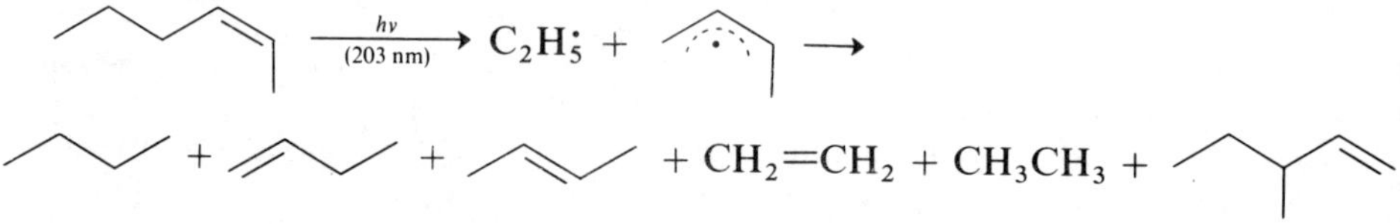

(8.27)

The reactions of the lowest singlet state of 2,3-dimethylbut-2-ene involve 1,2- and 1,3-shifts (8.28), and the intramolecular nature of the rearrangements is indicated by the observations that they occur even in solution and that no deuterium is incorporated in the products from perdeuterio-hydrocarbon solvent.[32] The difference between the reactions of this tetra-substituted alkene and those of less highly substituted compounds may arise because the lowest excited state is a Rydberg state rather than (π, π^*) in nature.

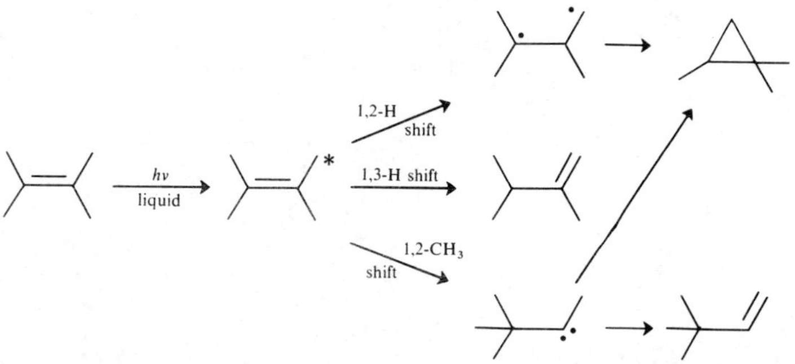

(8.28)

High energy sensitization of alkenes can also provide fragmentation or rearrangement products. A few instances of intermolecular sensitization of simple alkenes using aromatic hydrocarbon sensitizers in solution are known, and the major products are cyclopropanes. Intramolecular sensitization in phenyl-substituted alkenes follows a similar pattern, involving a 1,2-alkyl (8.29)[39] or a 1,2-hydrogen (8.30)[40] shift.

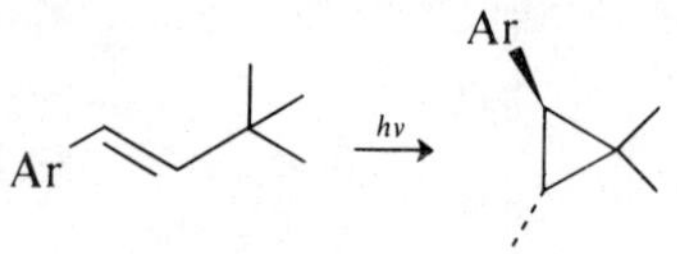

$$\phi = 0{\cdot}0026 \ (Ar = p{-}NC{-}C_6H_4{-})$$

(8.29)

$$\text{Ph} \diagdown\diagup\diagup \xrightarrow[\text{gas}]{h\nu\ (266\ \text{nm})} \triangleright\!-\!\text{Ph} \left(+\ \text{Ph}\diagdown\diagup\diagup +\ \triangleright\!-\!\text{Ph} \right)$$

$$\phi = 0\cdot19 \qquad\qquad\qquad 0\cdot023 \qquad\qquad 0\cdot024$$

$$(8.30)$$

8.3.1 **Mercury Sensitization**

The most extensive investigations of triplet-sensitized reactions of non-conjugated alkenes have been carried out using atomic species as sources of triplet energy, and especially using mercury atoms.[41] A low pressure mercury arc provides intense radiation of wavelengths 184.9 and 253·7 nm (note that in the output from a medium pressure mercury arc the 253·7 nm line is 'reversed' and the centre of the band is missing). If the 184·9 nm radiation is filtered out, the low pressure arc provides an excellent source of monochromatic radiation for specific excitation of mercury atoms to the $6(^3P_1)$ excited state. This triplet excited state (energy 470 kJ mol^{-1}, 112 kcal mol^{-1}) is quenched by added alkene, probably through the intermediate formation of an exciplex (8.31). In the major quenching process the whole of the triplet energy is transferred to the alkene, and Hg $6(^3P_1) \rightarrow$ Hg $6(^1S_0)$. The alkene triplet state is produced initially in a high vibrational level.

$$\text{Hg}(^3P_1) + \text{CH}_2\!=\!\text{CH}_2 \ \rightarrow \ \left(\text{Hg} \leftarrow \begin{array}{c} \text{CH}_2 \\ \| \\ \text{CH}_2 \end{array}\right)^{\!*} \ \rightarrow$$

$$\text{Hg}(^1S_0) + {}^3(\text{CH}_2\!=\!\text{CH}_2)^* \qquad (8.31)$$

Ethylene triplet state reacts to give mainly acetylene and hydrogen. This primary process is intramolecular, since mixtures of $\text{CH}_2\!=\!\text{CH}_2$ and $\text{CD}_2\!=\!\text{CD}_2$ give H_2 and D_2 but very little HD. A vibrationally excited methylcarbene is an intermediate in the process (8.32), and evidence for this[42] comes from the observed similarity in the ratios of H_2, D_2 and HD obtained in the reactions of $\text{CH}_2\!=\!\text{CD}_2$ and $\text{CHD}\!=\!\text{CHD}$ (8.33).

$$^3(\text{CH}_2\!=\!\text{CH}_2)^* \ \nearrow \begin{array}{l} \text{CH}_3\!-\!\ddot{\text{C}}\text{H} \nearrow \begin{array}{l} \text{CH}_2\!=\!\text{CH}_2 \\ \searrow \text{CH}\!\equiv\!\text{CH} + \text{H}_2 \end{array} \\ \searrow \text{CH}_2\!=\!\text{CH}^{\cdot} + \text{H}^{\cdot} \ \ (<3\%) \end{array} \qquad (8.32)$$

$$^3(\text{CH}_2\!=\!\text{CD}_2)^* \rightarrow \text{CH}_2\text{D}\!-\!\ddot{\text{C}}\text{D} \rightarrow \text{H}_2 + \text{HD}$$

$$^3(\text{CHD}\!=\!\text{CHD})^* \rightarrow \text{CHD}_2\!-\!\ddot{\text{C}}\text{H} \rightarrow \text{D}_2 + \text{HD} \qquad (8.33)$$

The major reactions of higher alkenes on triplet mercury sensitization, as on direct irradiation, occur by way of allylic cleavage, and these are illustrated for propene (8.34)[43] and pent-1-ene (8.35).[44]

$$CH_3CH{=}CH_2 \xrightarrow{Hg(^3P_1)} \begin{cases} 90\% & CH_2{=}CH{-}\dot{C}H_2 + H^\cdot \rightarrow \\ & CH_2{=}C{=}CH_2 + H_2 \quad (\text{major}) \\ 10\% & CH_2{=}CH^\cdot + {}^\cdot CH_3 \rightarrow \\ & C_2H_2 + CH_4 + C_2H_6 \end{cases} \quad (8.34)$$

$$\text{\raise2pt\hbox{$\sim\!\!\!\sim$}} \xrightarrow{Hg(^3P_1)} C_2H_5^\cdot + CH_2{=}CH{-}\dot{C}H_2 \quad (\text{major}) \qquad (8.35)$$

In the reaction of pent-1-ene a large number of minor products is formed, and some of these are cyclic hydrocarbons which arise by way of intramolecular hydrogen shifts in the excited state (8.36). These shifts are more important at higher gas pressures (10^0–10^1 mmHg rather than 10^{-3} mmHg), and this may reflect the greater importance of bond cleavage reactions (as opposed to isomerization or rearrangement) in the lower pressure region where dissipation of excess vibrational energy is slow.

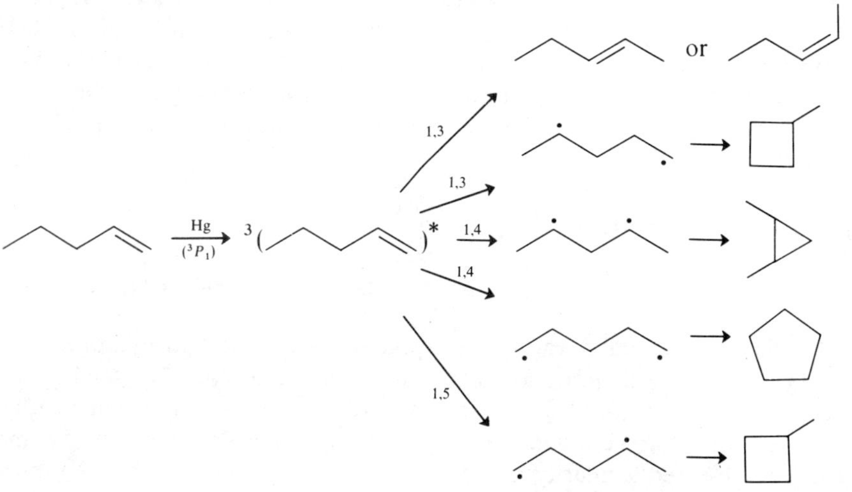

$$(8.36)$$

Cyclic alkenes also give products by way of allylic cleavage in mercury photosensitized reaction (8.37), and these products can be quenched by added radical scavenger such as nitric oxide. Other products are formed which cannot be so quenched,[45] such as vinylcyclopropane from cyclopentene (8.38), and such products may be formed in a concerted ($2\pi + 2\sigma$) reaction through the excited *singlet* state of the alkene produced by intersystem crossing from a high vibrational level of the triplet state. This proposed mechanism is supported[46] by the

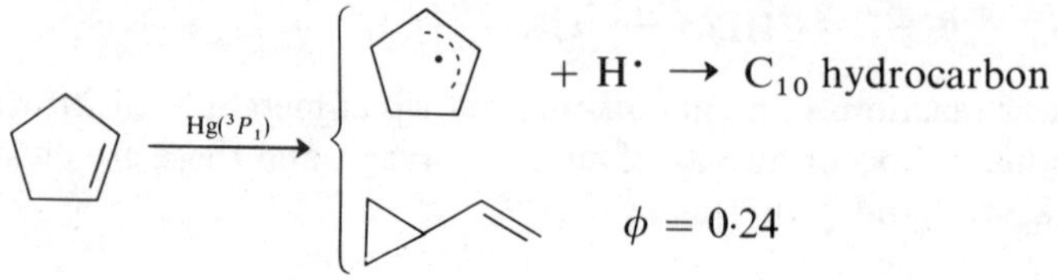

$$+ H^\cdot \rightarrow C_{10} \text{ hydrocarbon} \qquad (8.37)$$

$$\phi = 0.24 \qquad (8.38)$$

occurrence of the reverse reaction (formation of a cyclopentene from a vinyl-cyclopropane) on direct irradiation (8.39).

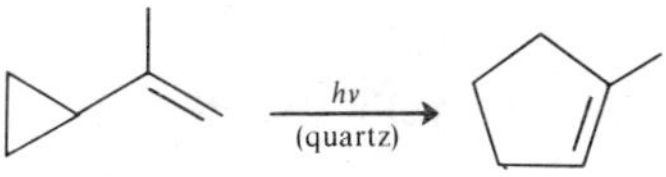

$$55\% \qquad (8.39)$$

Larger ring cycloalkenes give, among other products, bicyclic hydrocarbons which arise by way of intramolecular hydrogen transfer processes, and these are illustrated for cyclo-octene (8.40).[47] The bicyclic products are unaffected by radical scavengers, but this is expected even if radical species are involved, because intramolecular radical reactions are often too fast to be significantly affected by most intermolecular radical trapping agents.

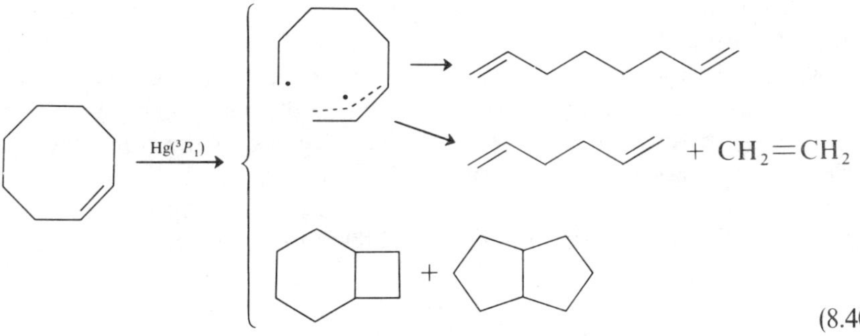

$$(8.40)$$

Conjugated dienes give products arising from 1,3-hydrogen shifts and from dehydrogenation (8.41) under conditions of mercury photosensitization. These products are also formed together with the products of electrocyclic reaction (see Chapter 8, section 8.4.1) on direct irradiation of the diene in the gas phase, and it appears that reaction occurs through a high vibrational level of the ground electronic state of the diene.

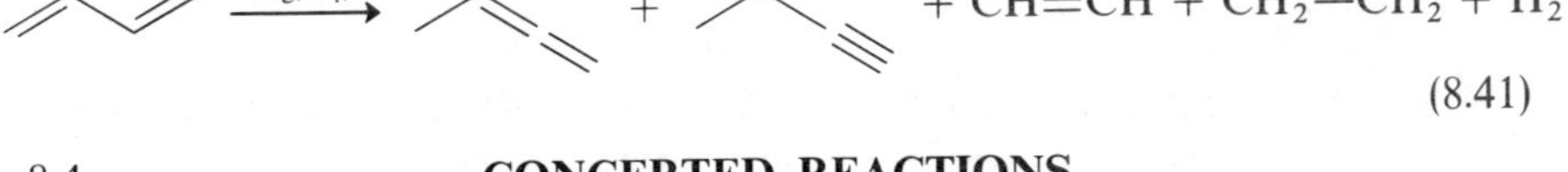

$$+ CH\equiv CH + CH_2=CH_2 + H_2$$

$$(8.41)$$

8.4 CONCERTED REACTIONS

Conjugated dienes and polyenes undergo a wide range of reactions which occur by a concerted mechanism and whose stereochemical course is predicted or rationalized by the Woodward–Hoffmann rules (see Chapter 6, section 6.3.2). These rules are derived theoretically from the concept of the conservation of orbital symmetry, from the concept of the maximum bonding pathway for a concerted process, or on an equivalent basis. The reactions include intramolecular rearrangements involving a reorganization of π and σ electrons accompanied either by ring-closure or ring-opening or by migration of an atom or group, and intermolecular cycloaddition reactions. Concerted reactions occur both thermally and photochemically, and the Woodward–Hoffmann rules predict the preferred stereochemical course of reaction in each case. For a given system the rules do not predict whether the forward or reverse reaction will occur preferentially (the two are not distinguishable by orbital considerations alone), nor which of several 'allowed' processes will predominate if two or more are possible.

Electrocyclic Reactions

The stereochemistry of reactant and product in an electrocyclic reaction must be related by either a conrotatory or a disrotatory mode of ring-closure or ring-opening (see Chapter 6, p. 165). For photochemical processes an electrocyclic reaction occurs predominantly in a disrotatory manner if the number of electrons undergoing major reorganization in the system is a multiple of 4, and in a conrotatory manner if the number of such electrons is not a multiple of 4 (i.e., is $4n + 2$).

Acyclic conjugated dienes can undergo photochemical ring-closure via the first excited singlet state to give cyclobutenes (8.42),[48] and cyclic conjugated dienes behave similarly (8.43).[49] Although the reverse reaction, giving diene from cycloalkene, is allowed photochemically on the basis of the Woodward–Hoffmann rules alone, it is seldom observed because of the transparency of cyclobutenes to normal wavelengths of ultraviolet radiation. This situation is in contrast to that for thermal electrocyclic reactions, for which the position of equilibrium is governed by free energy differences, and acyclic dienes are favoured over cyclobutenes.

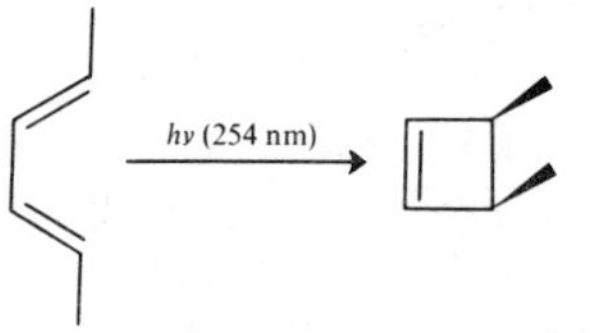

$$\phi \sim 0.01 \qquad\qquad (8.42)$$

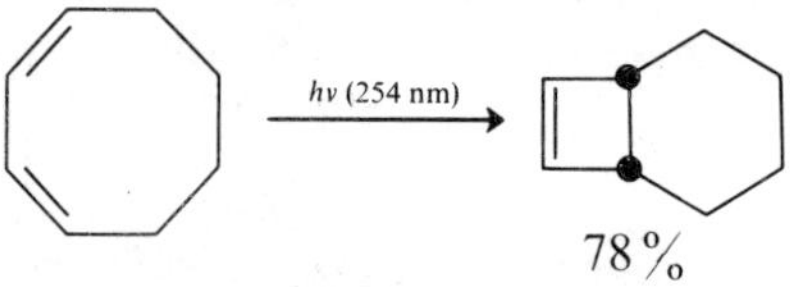

$$\phi = 0.14 \qquad\qquad (8.43)$$

78 %

Another illustration of the influence of factors other than orbital symmetry control on the rate or efficiency of a concerted reaction is seen in the ring-closure of penta-1,3-diene.[50] The *trans*-isomer undergoes moderately efficient ring-closure to 3-methylcyclobutene on direct irradiation, but the *cis*-isomer does not, because steric interactions prevent it from taking up the *s-cis* conformation required for ring-closure (8.44).

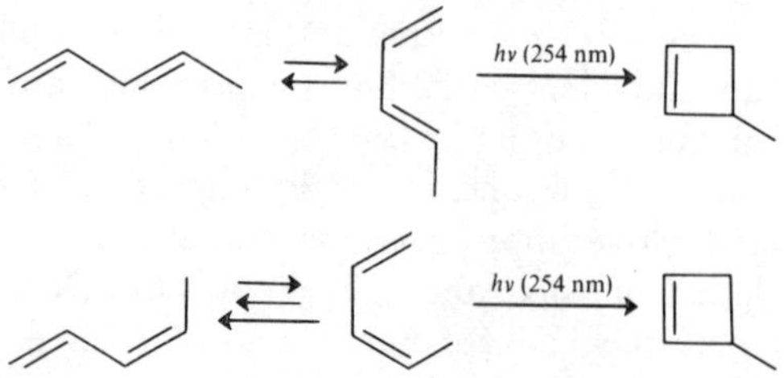

$$\phi = 0.030$$

$$\phi = 0.003 \qquad\qquad (8.44)$$

Acyclic conjugated trienes and cyclohexa-1,3-dienes are interconverted in photochemical electrolytic reactions,[51] although slow side reactions also occur (8.45), and dimerization becomes important at higher concentrations. When substituents are present, the predominant product is seen to be formed in a conrotatory ring-opening or ring-closure process (8.46).

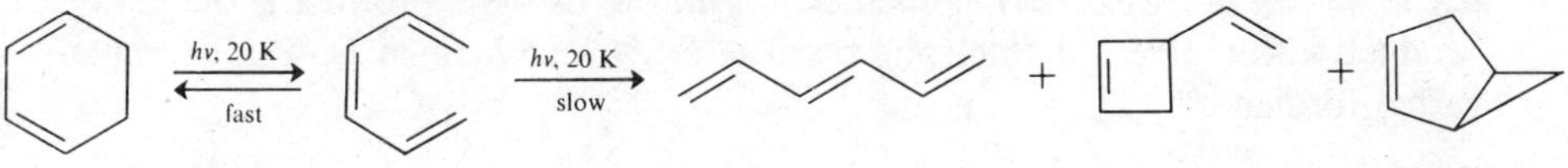

$$\text{(8.45)}$$

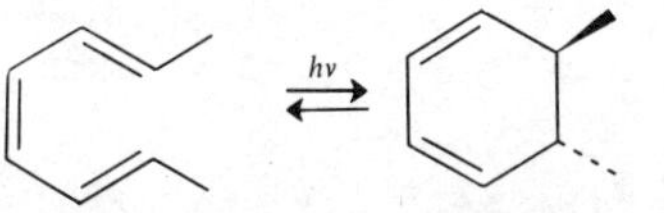

$$\text{(8.46)}$$

The two possible allowed conrotatory modes of ring-opening can lead to different products in some instances, but often one product is formed to the exclusion of the other because of major conformational effects.[52] The cyclohexadiene system (8.47) provides an example of this,[53] and the less stable product is the major one because of the preferred conformation of the diene.

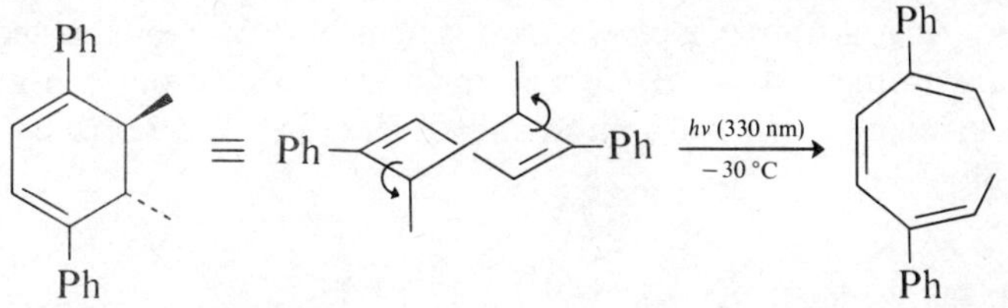

$$\text{(8.47)}$$

Cyclohexa-1,3-dienes often show a preference for $(4n + 2)$ ring-opening, and *trans*-9,10-dihydronaphthalene can be converted[54] to a [10]-annulene (8.48). However, $(4n)$ ring-closure can compete effectively if the ring-opened system is a strongly absorbing species or if the conformation of the diene is not suitable for ring-opening. Such a reaction is employed in the synthesis[55] of bicyclo[2.2.0]-hexa-2,5-diene (Dewar benzene, 8.49).

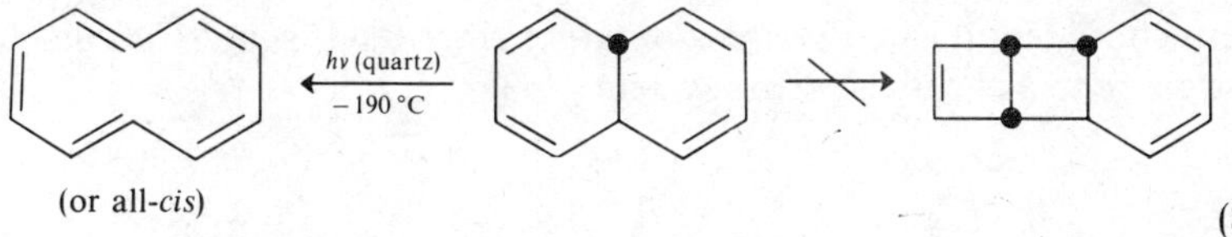

(or all-*cis*)

$$\text{(8.48)}$$

22%

5–20%

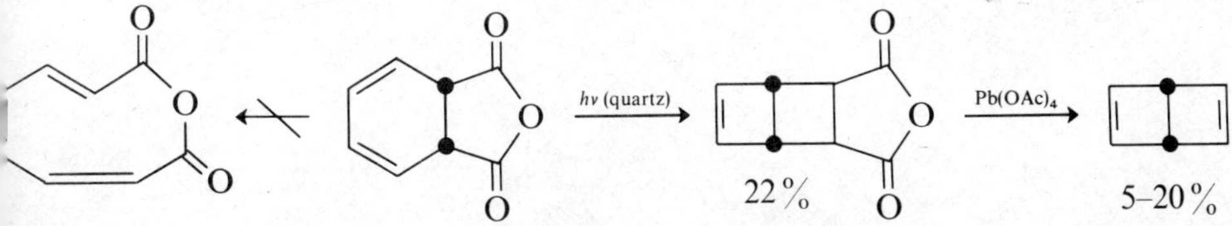

$$\text{(8.49)}$$

For the hydrocarbon analogous to the anhydride used in the Dewar benzene synthesis there is a fine balance of ring-opening and ring-closure (8.50), and the major product of reaction depends on the wavelength of radiation employed. At 254 nm the monocyclic triene is the major product initially because of the high extinction coefficient of the diene at this wavelength, but at 300 nm the triene has a higher extinction coefficient and the tricyclic product is the major product within a reasonable time because any triene formed is rapidly reconverted to diene.[56]

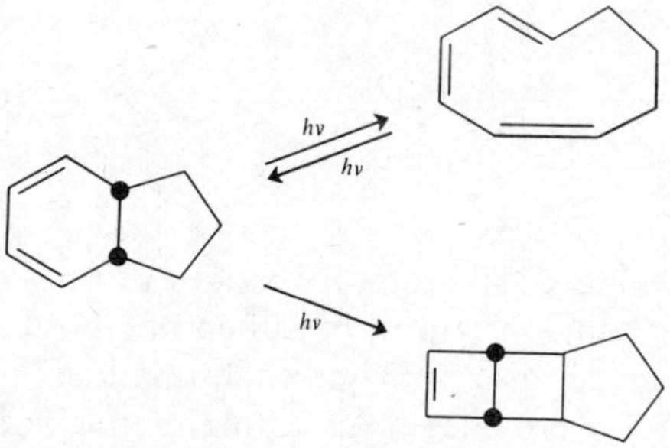

$$(8.50)$$

The ring-closure of *cis*-stilbene to a *trans*-dihydrophenanthrene (8.51) is a conrotatory ring-closure of a 6π-electron system, and this reaction accompanies the *cis-trans* isomerization of stilbene on direct irradiation.[57] Many similar reactions of alkenes with aromatic substituents are known (see Chapter 9, section 9.5).

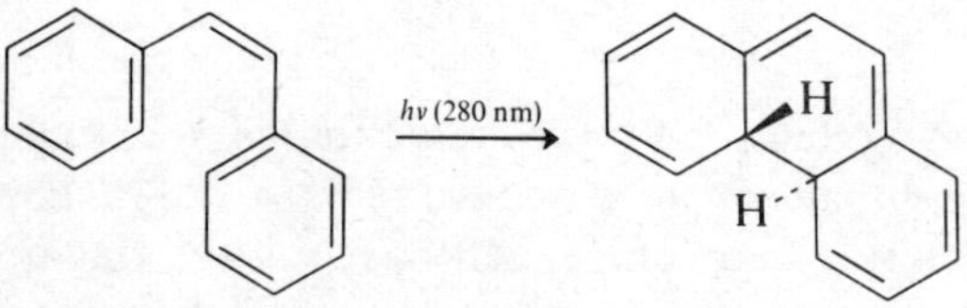

$$hv\ (280\ nm) \qquad 21\% \qquad (8.51)$$

Cycloheptatrienes exhibit a preference for photochemical 4π-electron ring-closure to give a bicycloheptadiene system (8.52).[58] The orientation of ring-closure in a substituted cycloheptatriene is governed by the position of the substituent.[59] Conjugated systems in larger rings tend to undergo 6π-electron ring-closure to a cyclohexadiene system (8.53).[60]

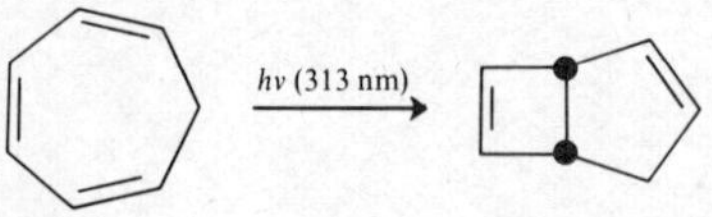

$$hv\ (313\ nm) \qquad \phi = 0.03 \qquad (8.52)$$

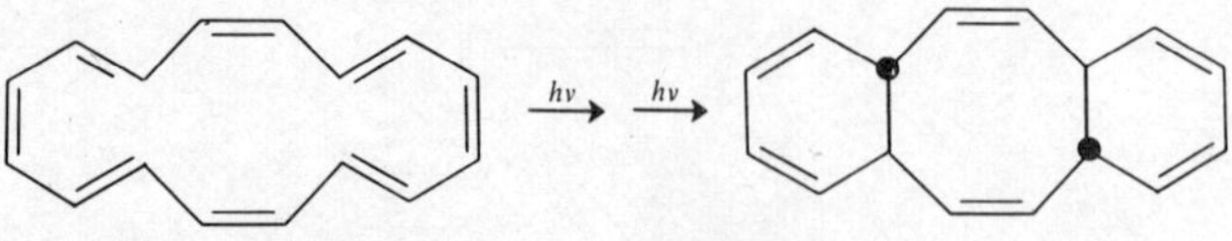

$$hv \longrightarrow hv \qquad (8.53)$$

[16] annulene

Photochemical electrocyclic reactions are nicely illustrated in the sequence of reactions found[61] in the vitamin D series (8.54). Included in the sequence is the synthetically important conversion of ergosterol to pre-vitamin D_2.

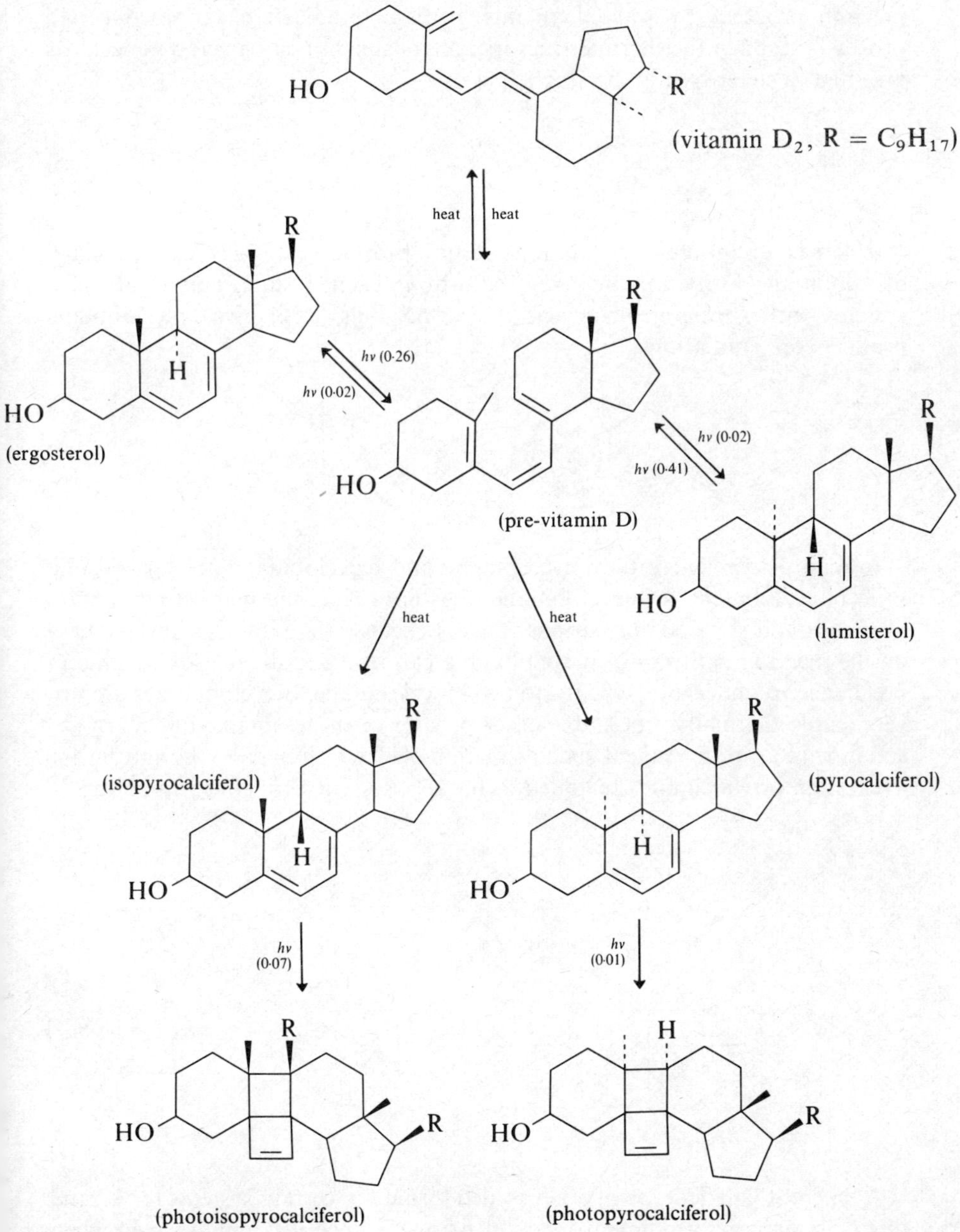

HO
R
(vitamin D₂, R = C₉H₁₇)
heat
heat
(ergosterol)
hν (0·26)
hν (0·02)
(pre-vitamin D)
hν (0·02)
hν (0·41)
(lumisterol)
heat
heat
(isopyrocalciferol)
(pyrocalciferol)
H
hν
(0·07)
hν
(0·01)
(photoisopyrocalciferol)
(photopyrocalciferol)

(8.54)

Bicyclobutane Formation

The electrocyclic ring-closure of conjugated dienes on direct irradiation is accompanied by formation of a bicyclobutane (8.55). Since this product is also formed in the (mercury) triplet sensitized reaction, it is proposed that the reaction proceeds by way of an intermediate biradical in a non-concerted process,[62] though the thermal ring-opening of substituted bicyclobutanes does give highly stereospecific products.

$$\text{(8.55)}$$

Whereas cyclobutene formation occurs through the *s-cis* excited state, bicyclobutane formation involves the *s-trans* excited state, and rigid diene systems with *s-trans* geometry as in the hexalins (8.56) give bicyclobutane products on irradiation.[63]

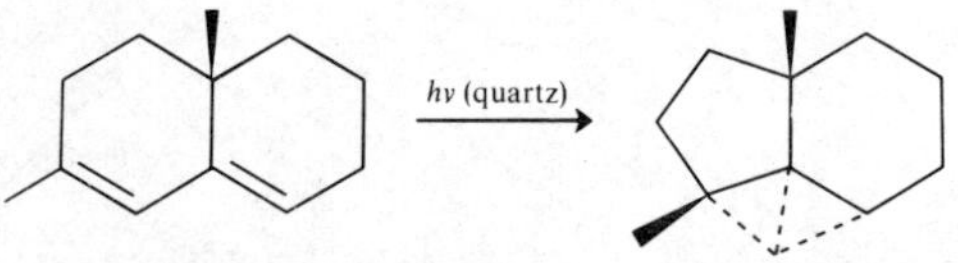

$$\text{(8.56)}$$

In many cases where both cyclobutene and bicyclobutane are formed, the ratio of cyclobutene to bicyclobutane does not reflect the relative population of the ground state conformations. This is because the biradical intermediate on the reaction pathway to bicyclobutane can revert to diene ground state in preference to undergoing cyclization to bicyclobutane. Bicyclobutanes are not very stable thermally, but stable ether products can be obtained (8.57) by reaction with a protic reagent such as methanol,[64] which may be the solvent for irradiation or which may be added to the reaction mixture after irradiation.

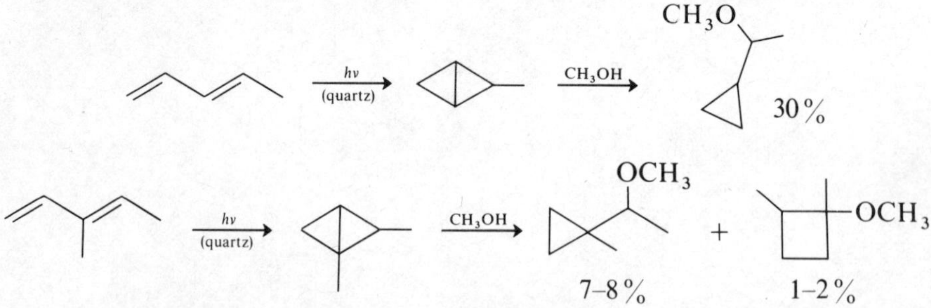

$$\text{(8.57)}$$

A different (allylic) type of ether is also formed in certain systems (8.58), and this must arise by a different reaction pathway since it is only formed when methanol is present in the irradiated mixture, and not if methanol is added after irradiation. The allylic ether must be formed by protonation of the excited state or by way of a vibrationally excited ground electronic state (compare the addition of methanol to mono-alkenes—Chapter 8, p. 237).

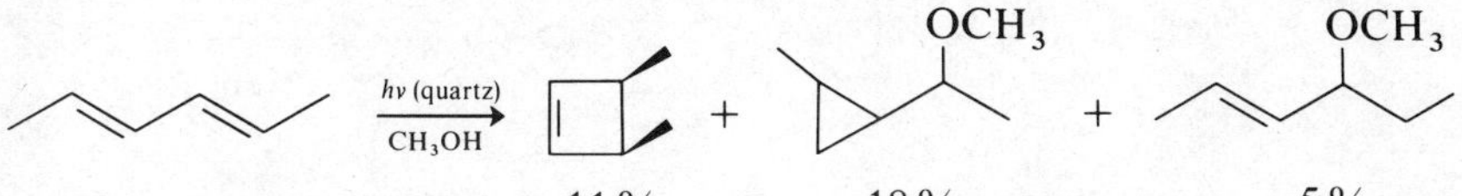

$$\text{(8.58)}$$

In the mercury ($^3P_1$) sensitized reactions of conjugated dienes, cyclopropenes and methylenecyclopropanes are formed (8.59), in addition to the products described earlier (see Chapter 8, p. 243). The same biradical which leads to bicyclobutane might be an intermediate in the formation of these products.[65]

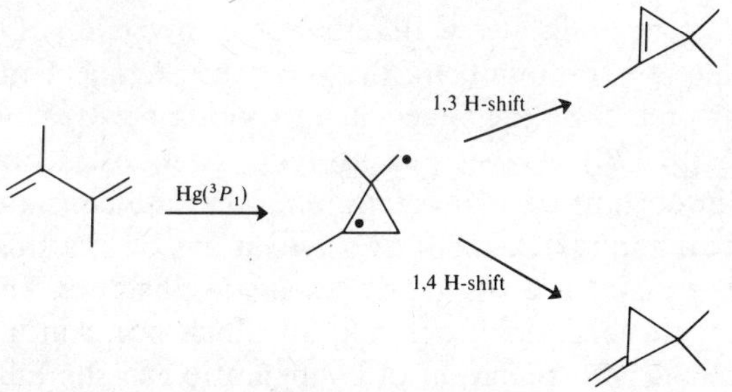

$$20\%, \phi = 0.12 \qquad \text{(8.59)}$$

The biradical intermediate on the pathway to bicyclobutane turns up again in the formation of at least one of the dimers obtained on direct irradiation of a conjugated diene in concentrated solution (8.60). These dimers are not all the same as those obtained in the sensitized reaction (see Chapter 8, p. 260).

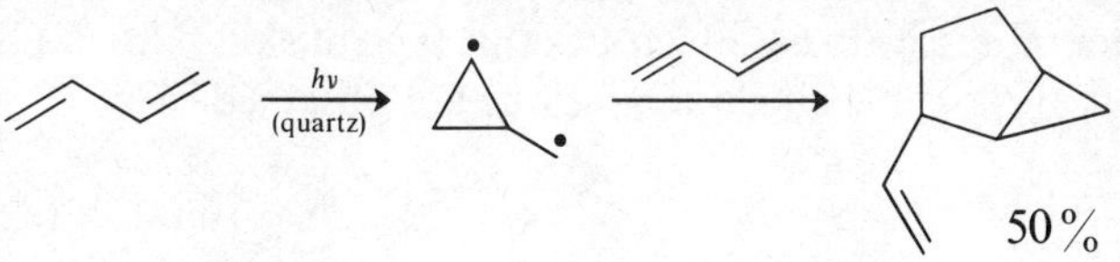

$$+ \text{ other dimers}$$
$$50\% \qquad \text{(8.60)}$$

A related reaction is the ring-closure of conjugated trienes on direct irradiation to give bicyclo[3.1.0]hexenes (8.61). The reaction is 'vinylogous' to bicyclo-

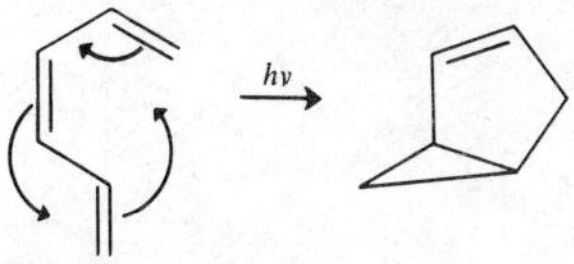

$$\text{(8.61)}$$

butane formation from dienes, though it is stereospecific,[66] at least at one of the centres involved (8.62).

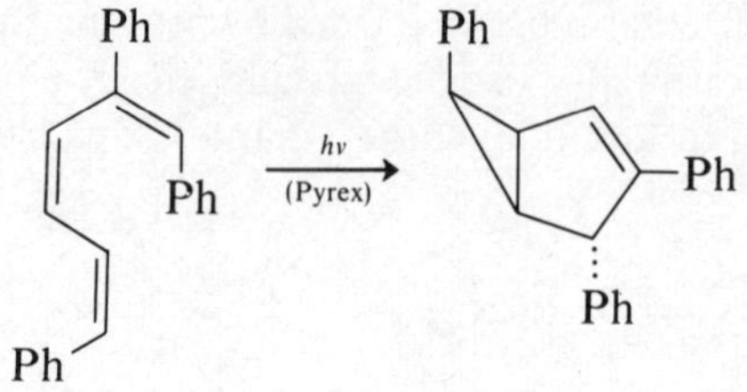

$$\qquad\qquad\qquad (8.62)$$

Sigmatropic Shifts

In an intramolecular migration to a new position of a σ-bond linked to one or more π-systems, the geometry of initial and final compounds can be related according to whether the migrating group is attached to the same face of the π-system after migration as before (suprafacial) or to the opposite face (antarafacial). In addition there is the consideration of whether the migrating group, if it is other than a single atom, migrates with retention or inversion of its own stereochemistry. The rules for rationalizing the observed stereochemistry of sigmatropic shifts, or for predicting the stereochemistry in a new reaction, are described in Chapter 6 (p. 170). As with electrolytic reactions, geometrical restraints also play an important part in determining the rate constant for a particular reaction, and an antarafacial shift over two atoms or in a small ring system is not likely to be a major process in any system, for instance. The most straightforward cases are those shifts of order (m, n) which occur in a suprafacial–suprafacial mode and with retention of configuration in the migrating group. These are thermally allowed if $(m + n)$ is not a multiple of 4, and photochemically allowed if $(m + n)$ is a multiple of 4.

Many photochemical (1,3) sigmatropic shifts are known, especially in hexa-1,5-diene or 4-phenylalkene systems, but also in simpler compounds such as penta-1,3-diene (8.63).[67] Nitrile (—CN) groups are often incorporated in order to bring the ultraviolet absorption of the compound into the accessible range, and the reaction of geranonitrile (8.64) illustrates this type of shift. No 'double' migration occurs, but all possible single migrations are observed.[68]

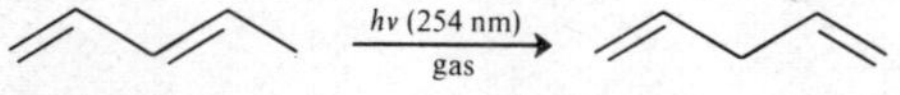

$$\phi = 0{\cdot}007 \qquad (8.63)$$

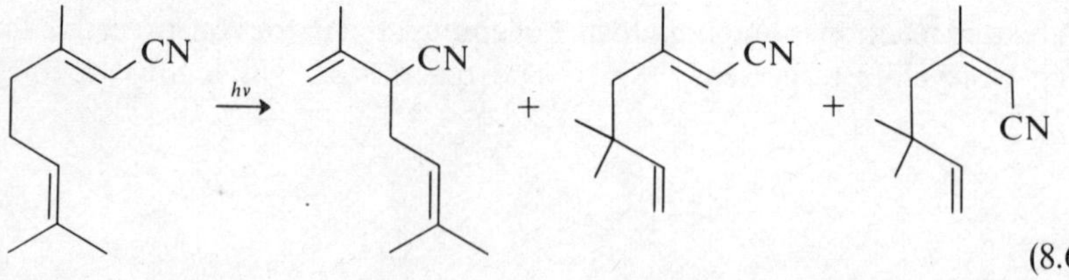

$$\qquad\qquad\qquad (8.64)$$

The reactions occur through the singlet excited state of the alkene and are concerted.[69] The triplet state reactions are different, often leading to the formation of cycloaddition products (8.65).

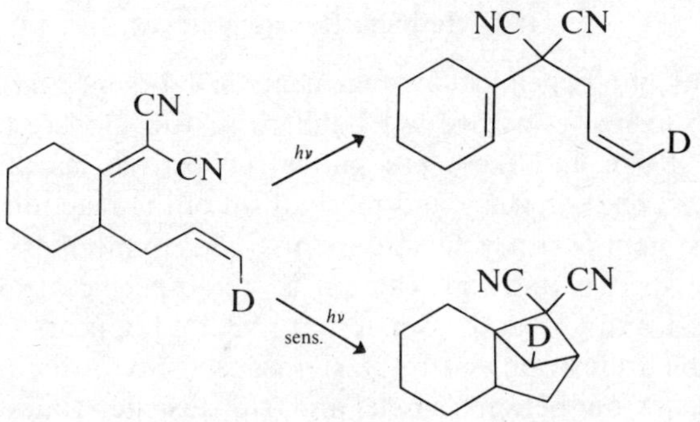

$$(8.65)$$

Photochemical (1,3) sigmatropic shifts are constrained to occur suprafacially by geometrical factors, and retention of configuration at the migrating centre is observed[70] as expected on the basis of the rules (8.66).

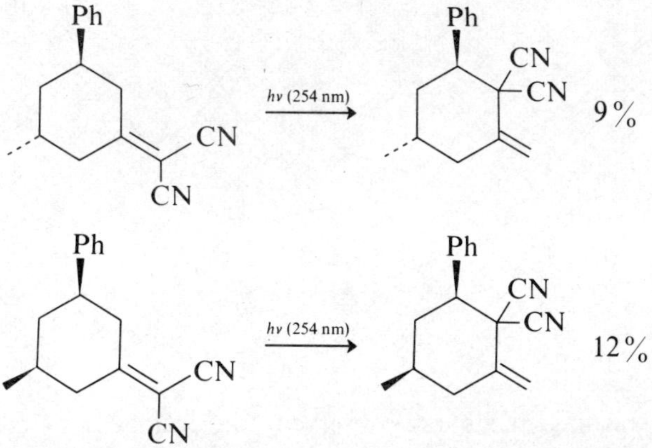

$$(8.66)$$

Photochemical (1,5) sigmatropic hydrogen shifts are known in conjugated diene systems, and where the stereochemical course of the reaction is known it is consistent with an antarafacial migration. An illustration of this is[71] seen in the diene system (8.67). Photochemical (1,7) sigmatropic hydrogen shifts are a major feature of the photochemistry of cycloheptatrienes, and often a sequence of such shifts occurs (8.68).[72]

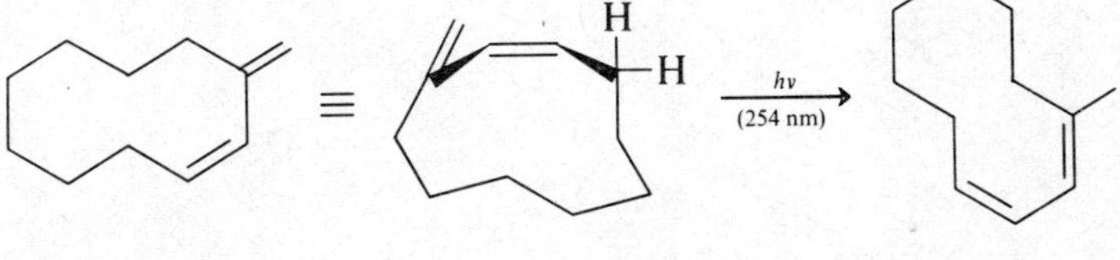

$$(8.67)$$

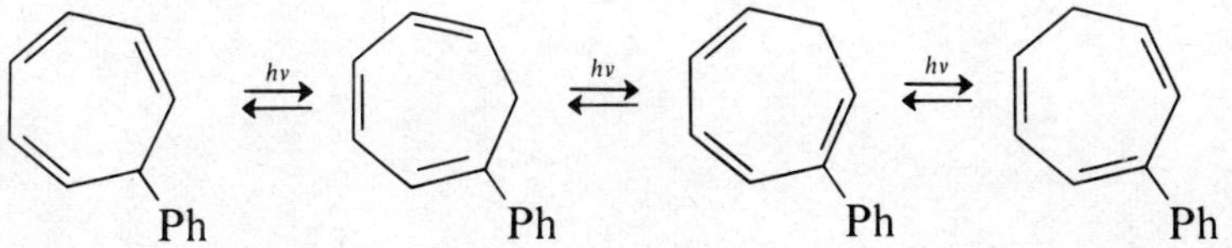

$$(8.68)$$

8.4.4   **Di-π-Methane Rearrangement**

There are many photochemical arrangements in 1,4-diene and 3-phenylalkene systems which formally involve a 1,2-shift with ring-closure to give a vinyl-cyclopropane (8.69), and these are known collectively as di-π-methane re-arrangements.[73] The efficiency and mechanism of the reaction depend on the nature of the system. For acyclic and monocyclic compounds the singlet state is the reactive species, and the reaction is stereospecific at all centres, with retention at C-1 and C-5 and inversion at C-3, and appears to be concerted. A system[74] which illustrates both the stereospecificity of the process and the difference in behaviour between singlet and triplet excited states is the 1,4-diene (8.70). The substituents on the central carbon atom are necessary for reaction to be efficient.[75]

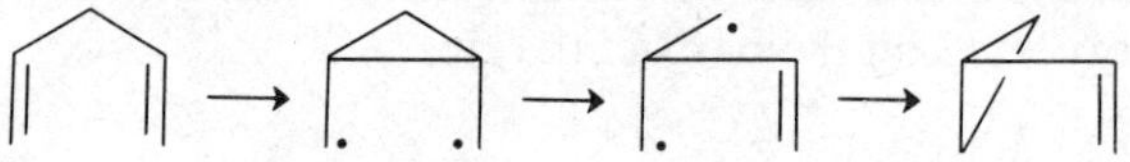

$$(8.69)$$

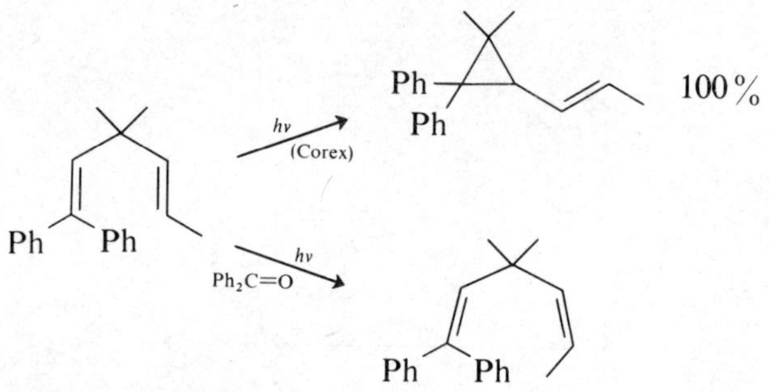

$$(8.70)$$

Similar rearrangements are observed for cyclohexa-1,4-dienes such as (8.71),[76] and for acyclic (8.72)[77] and cyclic (8.73)[78] phenyl substituted alkenes in which the phenyl ring is a part of the 1,4-diene system. The deuterium-labelling in the acyclic system shows that the reaction is a di-π-methane rearrangement and not a simple (1,3) sigmatropic hydrogen shift.

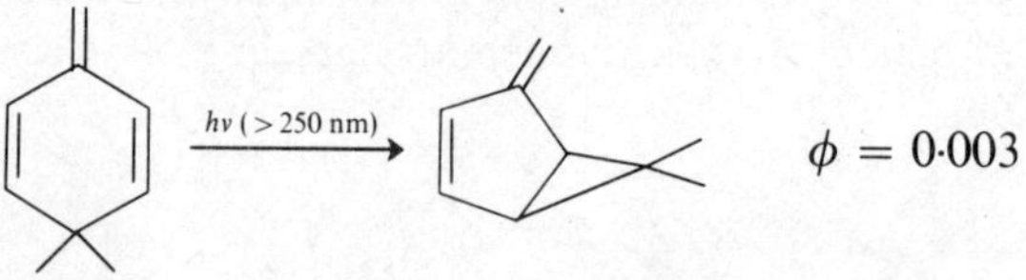

$\phi = 0.003$ (8.71)

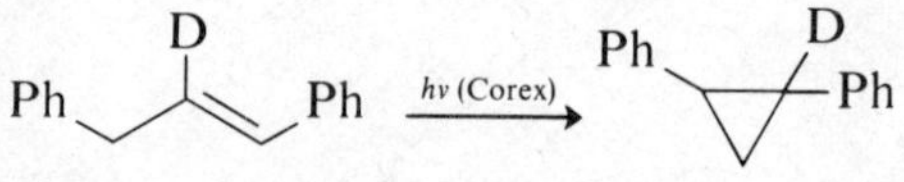

$$(8.72)$$

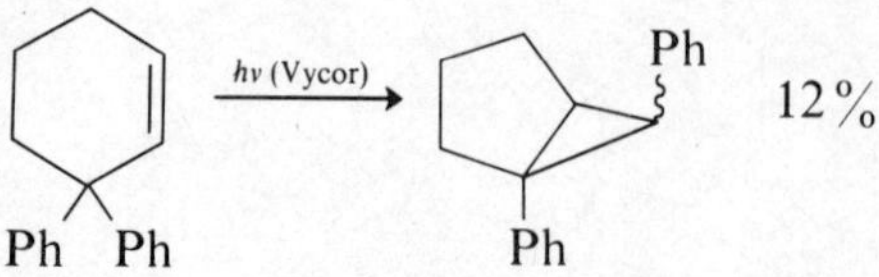

12% (8.73)

The triplet state di-π-methane rearrangement is normally very inefficient in flexible acyclic and monocyclic systems. This is because the 'vertical' triplet state initially formed on sensitization relaxes very rapidly to the more stable orthogonal triplet state and then undergoes rapid intersystem crossing to the ground state. The usual reaction of such triplet states therefore results in overall *cis-trans* isomerization. Exceptions to this generalization are found in systems such as 5,5-diphenylcyclohexa-1,3-diene,[79] for which direct irradiation causes electrocyclic ring-opening, and triplet sensitization leads to a di-π-methane rearrangement (8.74).

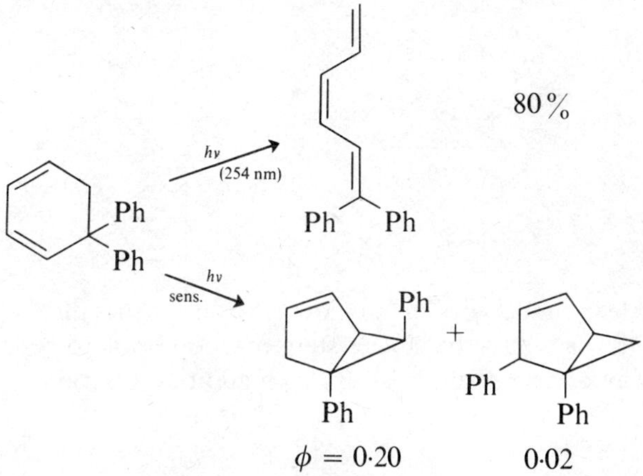

Bicyclic systems behave differently, and on direct irradiation di-π-methane rearrangement is a minor reaction pathway, and other processes predominate. The triplet state, however, does not have a lifetime governed by a very high rate constant for intersystem crossing, since the bicyclic systems are geometrically much more rigid than the monocyclic systems and an orthogonal triplet is not achieved. Such triplet states are able to undergo an efficient di-π-methane rearrangement by way of a non-concerted process, and this is shown in the isomerization of barrelene to semibullvalene (8.75),[80] and of benzonorbornadiene (8.76).[81]

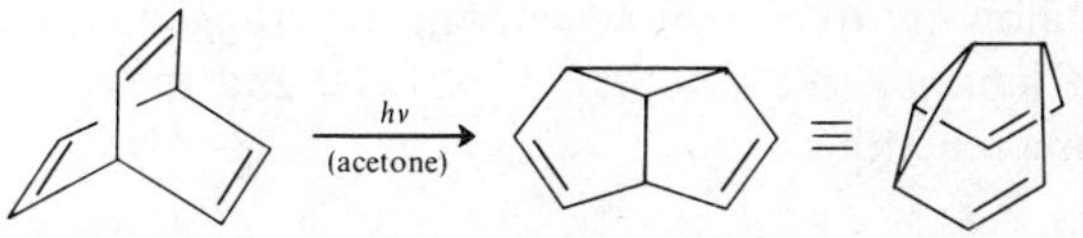

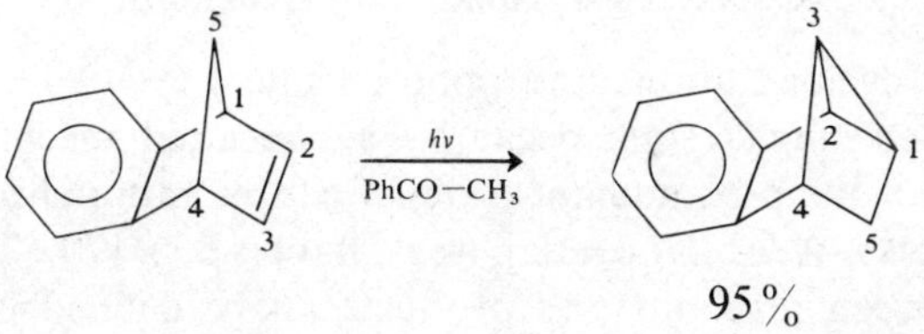

Several reaction pathways can be formulated for the rearrangement, but the correct one can be chosen on the basis of labelling patterns. In the case of the labelled barrelene (8.77, in which the marked carbon atoms are labelled with hydrogen in an otherwise perdeuterated molecule) the observed pattern of hydrogen/deuterium substitution in the product[82] is consistent with the pathway shown, and this is equivalent to the route illustrated (8.69) for a basic 1,4-diene unit.

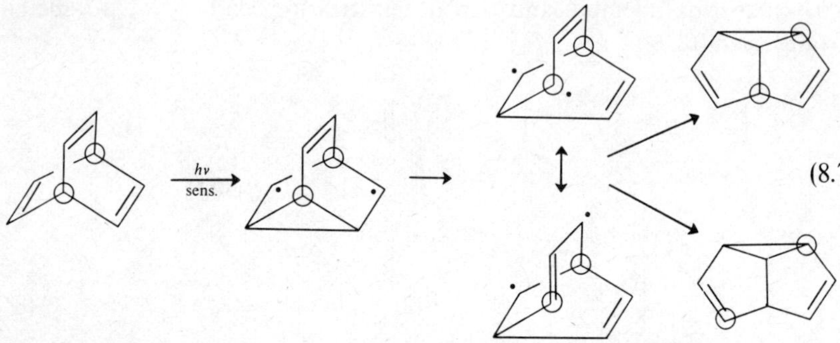

(8.77)

Benzobarrelene undergoes a similar rearrangement through its triplet excited state (8.78), whilst the singlet state reaction leads to benzocyclo-octatetraene by way of an initial $(2\pi + 2\pi)$ cycloaddition reaction.

94%

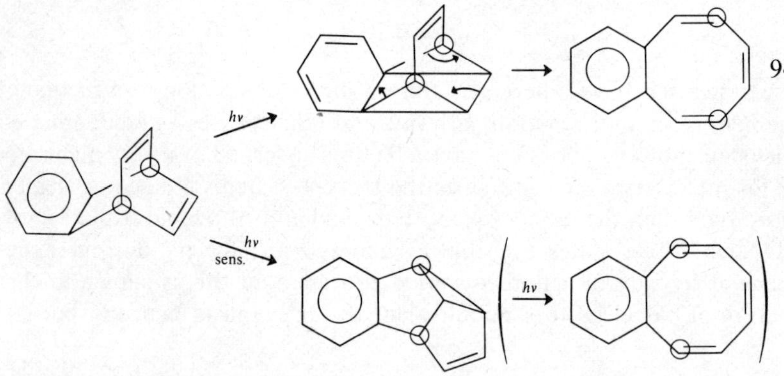

(8.78)

Reactions analogous to the di-π-methane rearrangement can occur with β,γ-unsaturated ketones (see Chapter 7, p. 187), and they are referred to as oxa-di-π-methane reactions.

8.4.5   Concerted Cycloaddition Reactions

Certain thermal cycloadditions and photochemical cycloadditions involving the lowest excited singlet state occur by a concerted mechanism, and the stereochemistry of the predominant product can be rationalized on the basis of orbital symmetry or orbital overlap (see Chapter 6, p. 170). The geometrical relationship between reactants and product can be defined by the mode of addition (suprafacial or antarafacial) to each of the components, and rules can be derived to predict the thermally and photochemically allowed steric course of reaction.

Concerted $(2\pi + 2\pi)$ cycloadditions are allowed photochemically in a suprafacial–suprafacial manner. Although there are a large number of $(2\pi + 2\pi)$ cycloadditions which do occur photochemically, in many cases the stereochemistry is not assigned, and in others the reaction is known to occur by a non-concerted pathway. One example of a concerted cycloaddition for a simple alkene is the dimerization of but-2-ene.[83] Irradiation of either neat cis- or neat trans-but-2-ene gives rise to a mixture of two of the isomers of 1,2,3,4-tetra-methylcyclobutane (8.79). Irradiation of a mixture of cis- and trans-but-2-ene gives, in addition to the adducts formed from the isomers separately, a further adduct which must arise from cycloaddition of cis- to trans-but-2-ene. The observed stereochemistry of the adducts is strong evidence for the concertedness of the process and for the applicability of the derived rules for cycloaddition. These results also imply that in the but-2-ene system the distinct cis and trans excited states react before relaxing to the orthogonal state, which is consistent with the existence of an energy barrier between each vertical excited state and the relaxed orthogonal state (see p. 229, and contrast the stilbenes, p. 232).

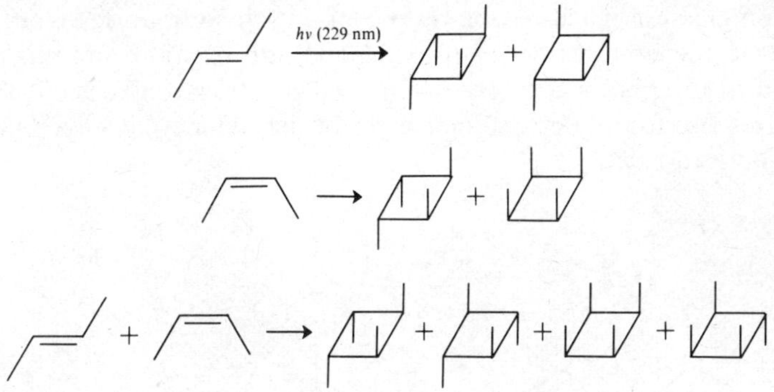

$$\phi = 0.02\text{–}0.03$$

Cycloaddition reactions involving stilbene or substituted stilbenes can occur through the singlet excited state and often lead to one predominant stereoisomer of product. The reactions are not straigthforward concerted cycloadditions because an excimer is formed initially between the singlet excited state of the stilbene and the ground state of the second alkene. The addition process within the excimer appears to have some concerted character, and this is seen in both dimerization (8.80)[84] and cross-addition (8.81)[85] reactions.

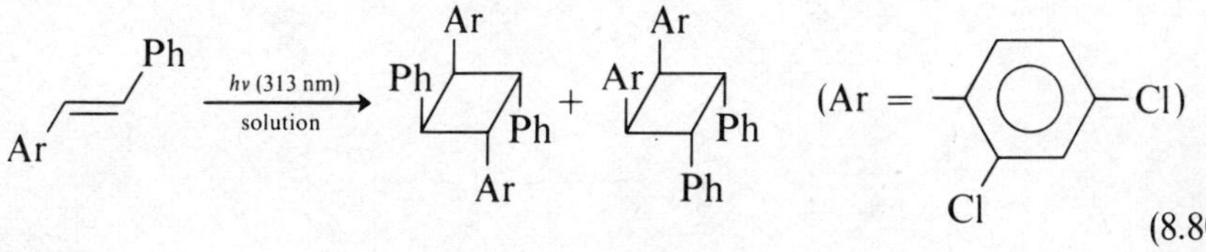

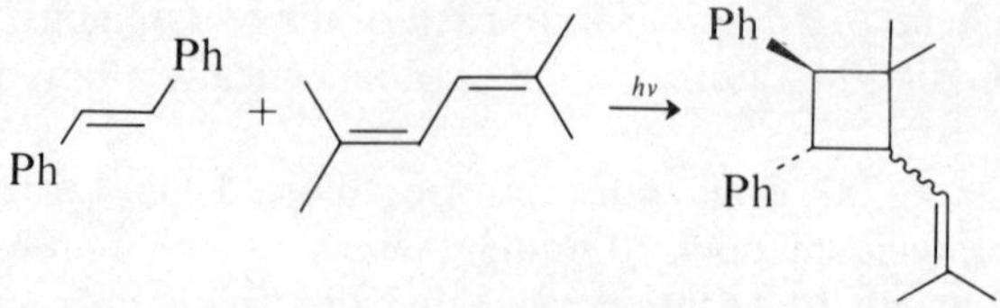

$$\text{Ph} + \text{(diene)} \xrightarrow{h\nu} \text{(product)} \tag{8.81}$$

A related type of reaction is a $(2\pi + 2\sigma)$ cycloaddition of an alkene double bond to a σ-bond of a cyclopropane ring, and this is illustrated by the intra-molecular reaction of the tricyclic alkene (8.82), which occurs through the singlet excited state[86] (triplet sensitization leads to the formation of cyclobutane dimers). The reaction may be concerted, but the geometrical restraints of the ring system prevent the formation of any stereoisomer of the product.

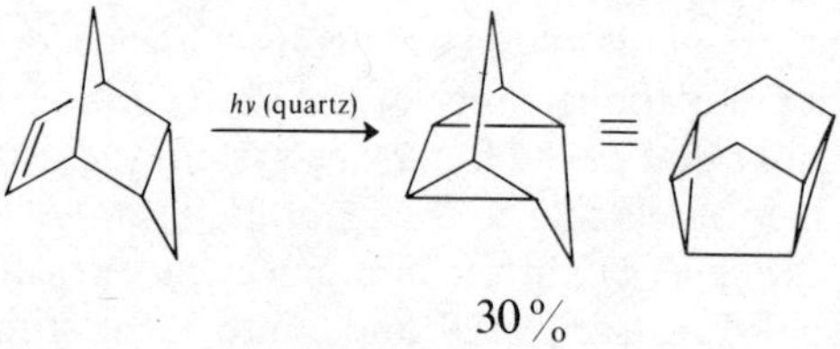

$$\text{(tricyclic alkene)} \xrightarrow{h\nu \text{ (quartz)}} \text{(product)} \equiv \text{(product)} \tag{8.82}$$

$$30\%$$

A cycloregression reaction (the reverse of a cycloaddition reaction) follows the same stereochemical rules as the cycloaddition reaction, and this is nicely illustrated in the photochemistry of the isomeric tricycloalkenes (8.83). Each alkene gives rise to only one stereoisomer of the monocyclic enyne in accord with the derived rules.[87]

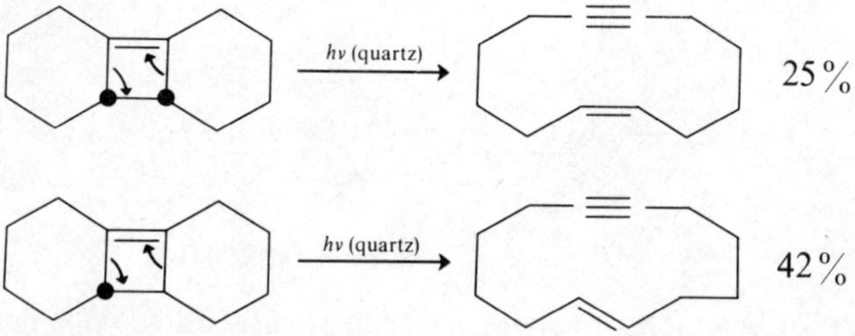

$$\text{(tricycloalkene)} \xrightarrow{h\nu \text{ (quartz)}} \text{(enyne)} \qquad 25\%$$

$$\tag{8.83}$$

$$\text{(tricycloalkene)} \xrightarrow{h\nu \text{ (quartz)}} \text{(enyne)} \qquad 42\%$$

The *cis*-isomer also undergoes electrocyclic ring-opening to give bicyclo-hexenyl (8.84), but the analogous reaction of the *trans*-isomer would lead to *cis*, *trans*-bicyclohexenyl, and this highly strained product is not formed.

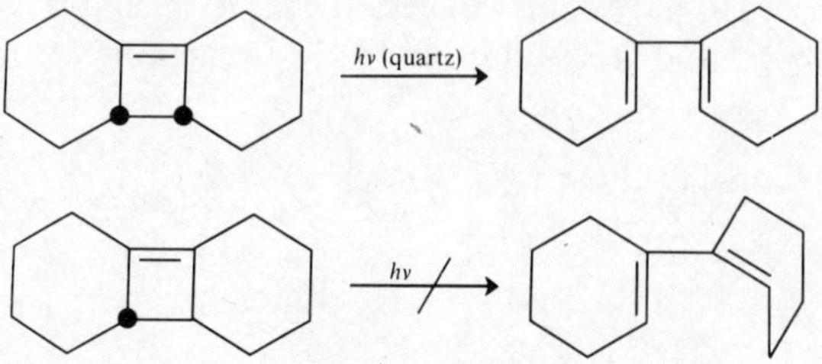

$$\text{(cis-isomer)} \xrightarrow{h\nu \text{ (quartz)}} \text{(bicyclohexenyl)}$$

$$\tag{8.84}$$

$$\text{(trans-isomer)} \xrightarrow{h\nu} \not\longrightarrow \text{(product)}$$

8.5 NON-CONCERTED CYCLOADDITION REACTIONS

Many cycloaddition reactions of alkenes do not follow a concerted pathway, but are two-step reactions involving a biradical intermediate and with non-stereospecific or only partially stereospecific formation of products. Some of these reactions have been described in earlier sections, such as the formation of oxetanes from carbonyl compounds and alkenes (see Chapter 7, section 7.7.1), and the cyclobutane-forming additions to the C=C double bond of conjugated carbonyl compounds (see Chapter 7, section 7.7.2).

Inter- and intra-molecular cycloadditions of alkenes can be brought about by triplet sensitization, and are therefore often accompanied by cis–trans isomerization. Most simple alkenes have too high a triplet energy to be sensitized efficiently by organic sensitizers, and mercury ($^3P_1$) sensitization in many cases causes intramolecular fragmentation or rearrangement rather than intermolecular cycloaddition (see Chapter 8, section 8.3.1). The dimerization of cyclic alkenes with 3-, 4- or 5-membered rings can be sensitized with high energy organic sensitizers such as acetone (but not benzophenone, which leads to formation of oxetanes). The effect of the small ring size may be to lower the triplet energy of the alkene sufficiently for energy transfer from the sensitizer to be efficient, or may be to increase the triplet lifetime sufficiently for intermolecular reaction to be able to compete effectively with intramolecular processes. Larger ring and acyclic alkenes have a very short triplet lifetime because of rapid relaxation to the orthogonal triplet and intersystem crossing to the ground state, but for small ring alkenes the orthogonal state has a higher energy because of geometrical restrictions, and intersystem crossing to ground state is considerably slower. Examples of the dimerization are shown for a cyclopropene (8.85)[88] and for norbornene (8.86).[89] The dimerization of norbornene can also be sensitized by copper(I) compounds, though the yield and ratio of dimers is not the same as in the acetone-sensitized reaction.[90] This observation suggests that one or other (or both) of the sensitized reactions does not involve

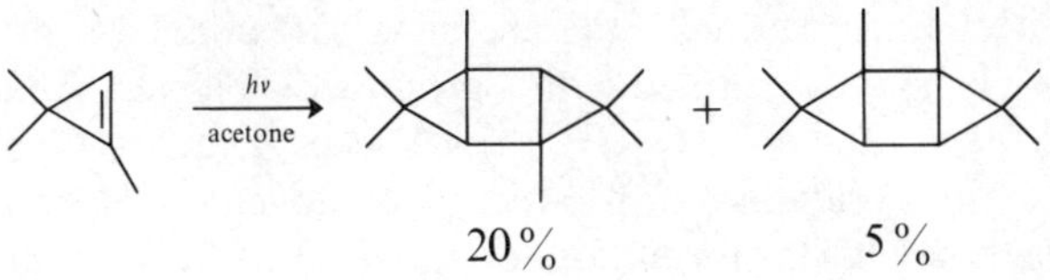

$$\text{(8.85)}$$

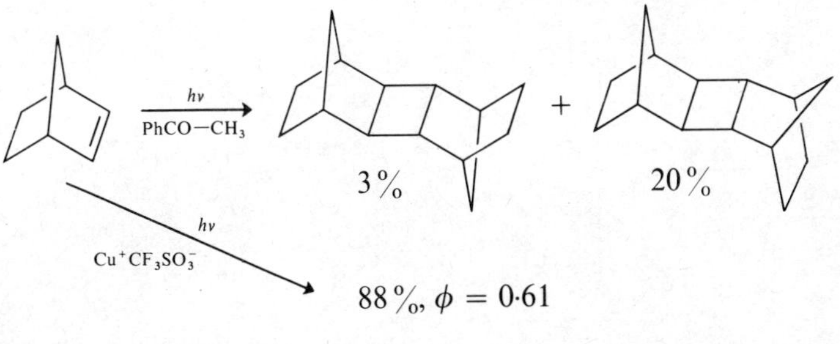

20%   5%

$$\text{(8.86)}$$

3%   20%

88%, $\phi = 0.61$

the free triplet state. It appears that the cuprous-sensitized reaction proceeds by way of a free radical pathway, although initial *cis–trans* isomerization of the double bond has also been suggested as a primary step.

Direct or sensitized intermolecular photocycloaddition reactions of alkenes to form cyclobutanes in a non-concerted process are numerous, and in many cases one of the alkene double bonds forms part of a conjugated carbonyl system (see Chapter 7, section 7.7.2). Two further examples are seen in the addition reaction of butadiene (8.87),[91] and that of stilbene with dihydropyran (8.88).[92]

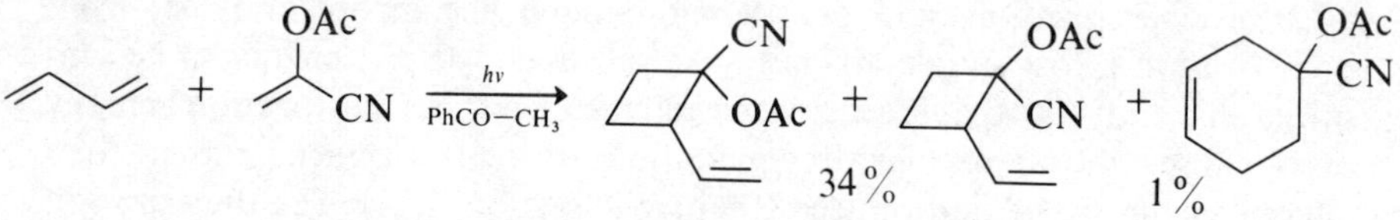

(8.87)

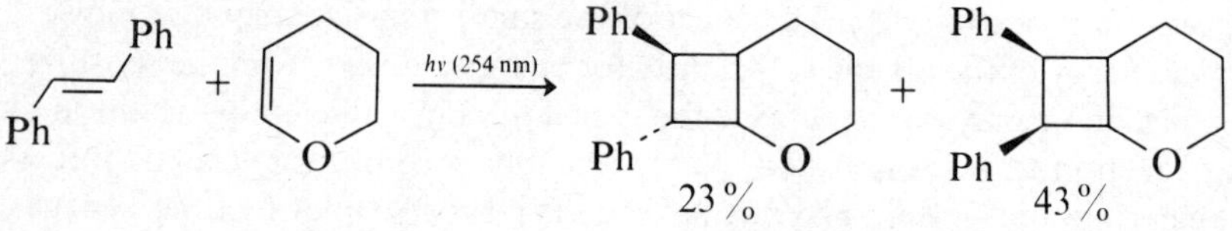

(8.88)

A wide range of sensitized intramolecular cycloaddition reactions of non-conjugated dienes is known, and these reactions can be of synthetic value in the preparation of strained systems. The mode of cyclization varies according to the number of saturated atoms between the double bonds. Where possible, the initial ring-closure leads to a 5-membered ring for formation of the major product, and this is a general feature of cyclization in radical reactions. Mercury ($^3P_1$) sensitization is effective in promoting intramolecular cycloaddition, and organic sensitizers can be used if the alkene is part of a 3-, 4- or 5-membered ring or if the alkene is substituted with a group which lowers the triplet energy of the system (e.g., Ph- or -CO$_2$CH$_3$). 1,4-Dienes give rise to bicyclo[2.1.0]-pentanes as major cycloaddition product (8.89), together with products of the di-π-methane type.[93] The formation of quadricyclane from norbornadiene (8.90) is formally analogous to this.[94]

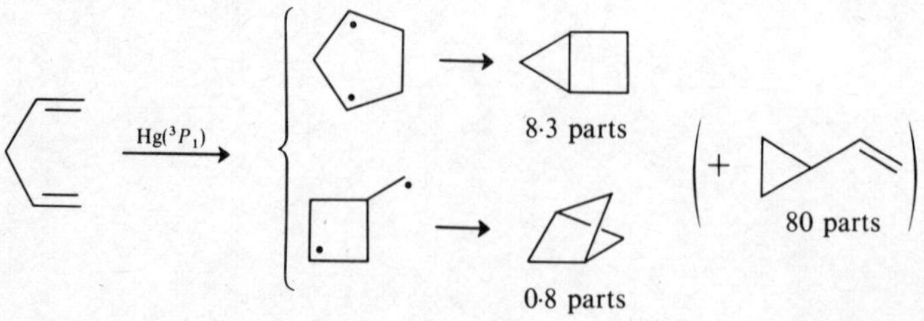

(8.89)

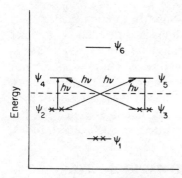

Figure 9.2. The four possible electronic transitions between the pairs of degenerate molecular orbitals in benzene

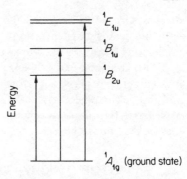

Figure 9.3. The lowest energy transitions in benzene

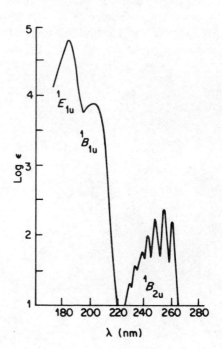

Figure 9.4. The absorption spectrum of benzene (from J. Petruska, *J. Chem. Phys.*, **34**, 1120 (1961), reproduced by permission of the American Institute of Physics)

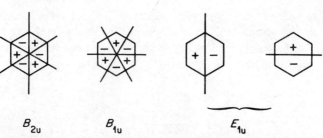

Figure 9.5. Diagrammatic form of the wavefunctions for the lowest excited states of benzene

decreases from 0·18 at 250–280 nm to 0·00 at 242 nm.[4] In solution, ϕ_f decreases with increasing concentration; this self-quenching is attributed to excimer formation[5] since it is accompanied by a new weak diffuse emission at 320 nm.

Although triplet excited benzene has only been observed to phosphoresce in low temperature glasses, it is formed (by intersystem crossing) in all three phases, and in the gas phase[6] $\phi_T = 0.72$ ($\lambda > 254$ nm). Thus in the gas phase at long wavelength the quantum deficit is ~ 0.1. It seems likely that the excited molecules that neither fluoresce nor form triplets are deactivated by ring isomerizations.

9.2   PHOTOISOMERIZATIONS OF AROMATIC SYSTEMS

Monocyclic aromatic systems undergo remarkable photochemical rearrangements. The early observation[7] that liquid benzene, excited into its S_1 state, isomerizes to fulvene (9.1) was followed in the next decade by numerous reports[8,9,10] of 1,2- and 1,3-shifts of alkyl groups around a benzene ring (e.g., 9.2).

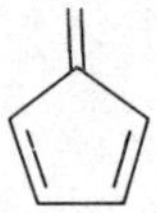

(9.1)

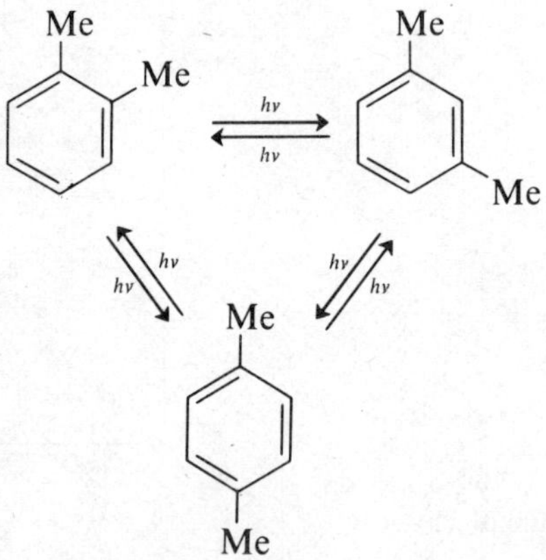

(9.2)

That these shifts involve not photodissociation and recombination of the alkyl residues but scrambling of the atoms of the benzene ring was established by showing that mesitylene labelled with ^{14}C was converted into 1,2,4-trimethylbenzene labelled exclusively in the 1-, 2- and 4-positions (9.3) when

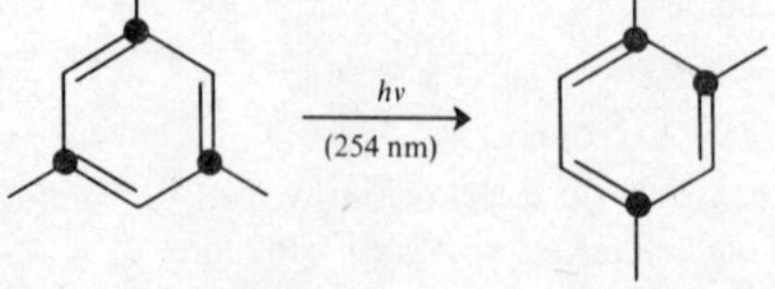

(9.3)

excited with 254 nm light in isohexane solution. Similarly, 1,3,5-trideuterio-benzene is converted in the gas phase into the 1,2,4-isomer (9.4).[11]

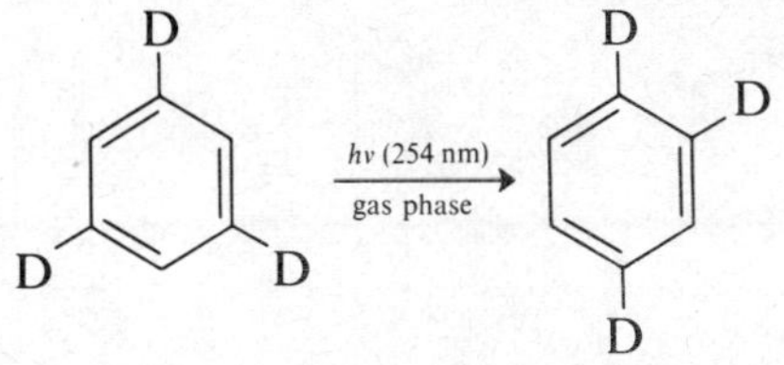

$$hv \text{ (254 nm)} \atop \text{gas phase}$$

(9.4)

The convolutions of the benzene ring clearly implied by these results suggest the intervention of two novel intermediates—benzvalene (9.5) and prismane (9.6), the reversible formation of which give rise to the observed scrambling of the ring carbon atoms.

(9.5)   (9.6)

Benzvalene Intermediate

The formation of benzvalene is formally analogous to the production of bicyclobutane from photoexcited butadiene (see Chapter 8, section 8.4.2). Its reversion to benzene can lead to the interchange of two adjacent carbons and hence gives rise to 1,2-shifts (9.7).

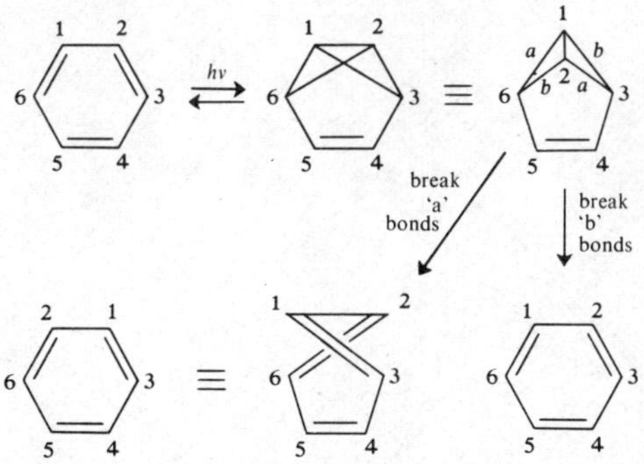

(9.7)

Prismane Intermediate

The production of 'Dewar' benzene (9.8) is analogous to the butadiene–cyclo-butene interconversion (see Chapter 8, section 8.4.1) and its further excitation via a (2 + 2) cycloaddition gives the prismane. If the prismane regresses ther-mally to the three isomeric Dewar benzenes which then revert to benzene isomers (9.9), it can be seen that the overall effect is to equilibrate the three benzene isomers. This would be experimentally manifested as 1,3- and 1,2-shifts of ring carbons and their attached substituents.

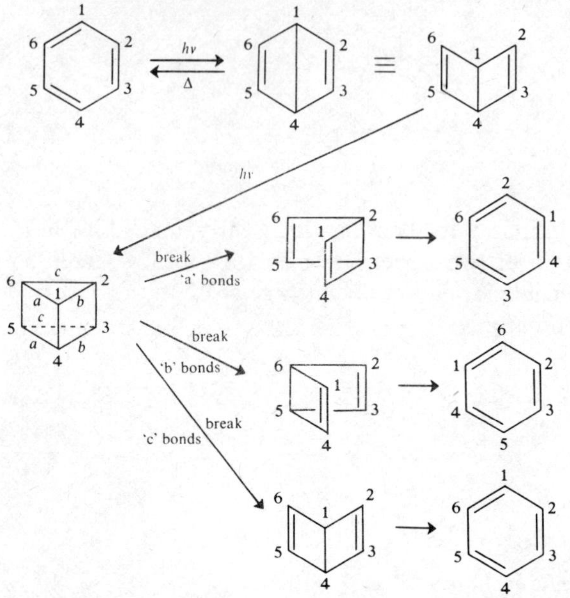

(9.8)

(9.9)

These superficially improbable intermediates have been shown to exist and to participate in photorearrangements. For example,[12,13] the irradiation of 1,3,5- and 1,2,4-tri-t-butylbenzenes leads to the photoequilibrium (9.10), in which the prismane (9.11) is the dominant component.

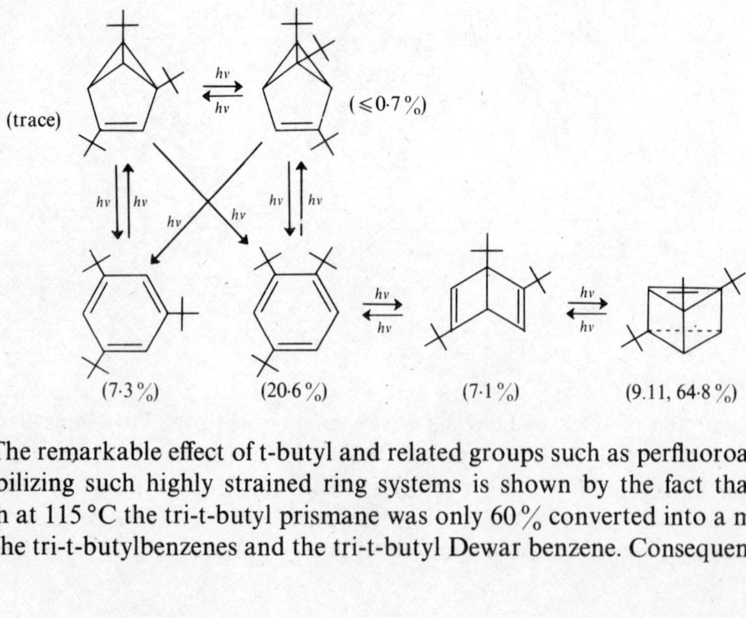

(trace)   ($\leqslant 0.7\%$)

(9.10)

(7·3%)   (20·6%)   (7·1%)   (9.11, 64·8%)

The remarkable effect of t-butyl and related groups such as perfluoroalkyl in stabilizing such highly strained ring systems is shown by the fact that after 18 h at 115 °C the tri-t-butyl prismane was only 60% converted into a mixture of the tri-t-butylbenzenes and the tri-t-butyl Dewar benzene. Consequently the interconversions in this series are by necessity entirely photochemical. In other series, without such substituents, the strained intermediates can revert thermally to the parent and rearranged aromatic systems.

In a similar way,[14] excitation of hexakis(trifluoromethyl)benzene gives isolable valence-bond tautomers (82%) in accordance with the scheme (9.12), and hexakis(pentafluoroethyl)benzene gives the prismane in yields > 96%.

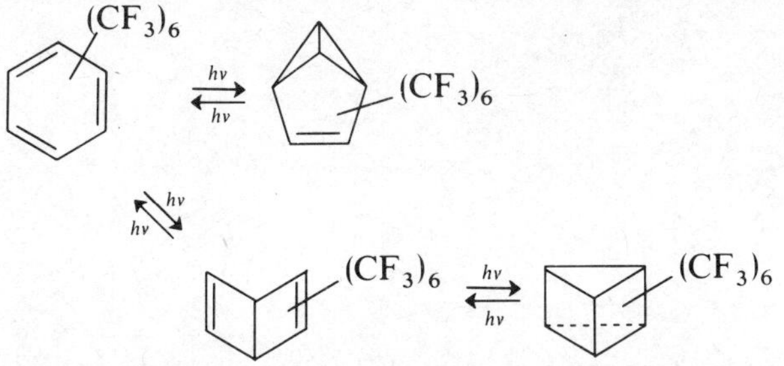

$$(9.12)$$

The effect of multiple substitution in stabilizing the valence-bond tautomers is presumably a reflection of the increased steric strain in the parent aromatic compound. Valence isomerizations and photochemical ring-scrambling processes are also found in heterocyclics (see sections 9.8.2 and 9.8.3).

There is, as yet, no general agreement about which excited states give rise to which valence isomers of benzene. It is known[15,16] that vacuum-ultraviolet irradiation of benzene vapour gives fulvene and cis- and trans-hexadienyne (9.13), and that benzvalene and Dewar benzene can be obtained only in condensed phases (9.14, 9.15), and these results imply that the benzvalene and Dewar benzene are formed with excess vibrational energy which must be rapidly dissipated if they are to survive.

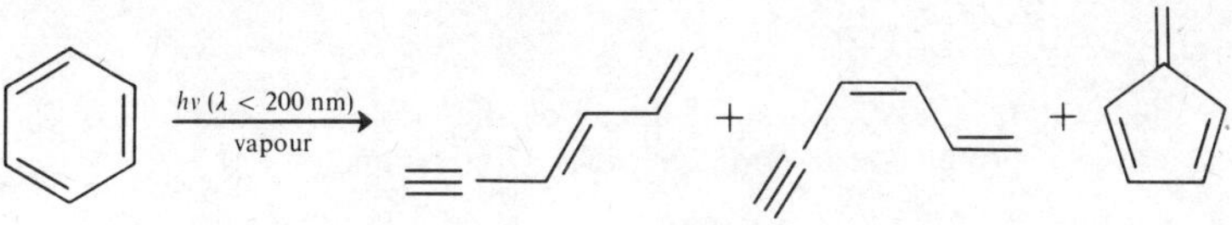

$$(9.13)$$

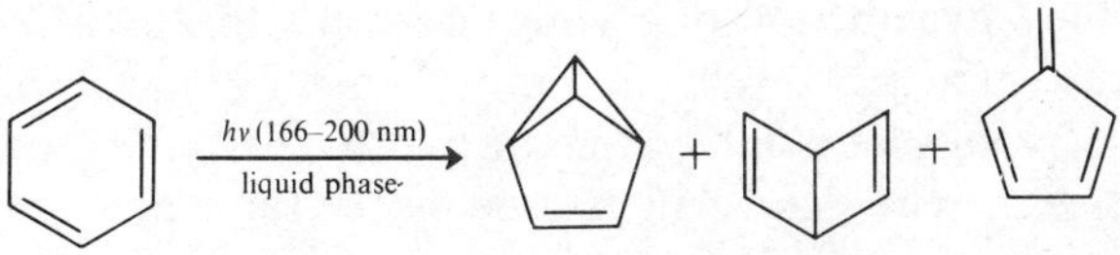

$$(9.14)$$

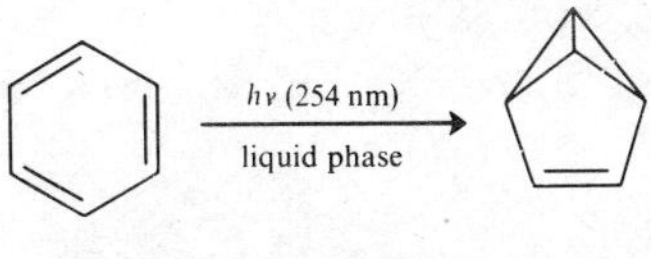

$$(9.15)$$

It is also known that the yield of benzvalene is increased by raising the temperature, implying the possible intervention of vibrationally excited electronic excited states, and the yield is also increased by the presence of triplet quenchers such as but-2-ene, suggesting the probable intervention of singlet excited states in the formation of benzvalene.

9.2.1   Orbital Symmetry Considerations

It has been suggested[17] that these results may be explained by invoking intermediates such as prefulvene (9.16) and 'pre Dewar' benzene (9.17).

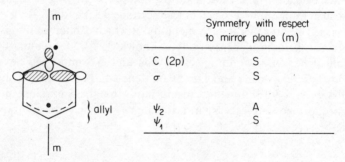

$$(9.16)$$

$$(9.17)$$

The $^1B_{2u}$ state of benzene (Figure 9.5) is antisymmetric to a mirror plane passing through the para-carbon atoms, as is the wavefunction of prefulvene (9.16, see Figure 9.6).

	Symmetry with respect to mirror plane (m)
C (2p)	S
σ	S
ψ_2	A
ψ_1	S

Figure 9.6. Classification of some of the molecular orbitals of prefulvene as symmetric (S) or antisymmetric (A)

The skeletal σ-orbitals, being doubly occupied, can be neglected, and the symmetry of the wavefunction of prefulvene is that of $2p^1 \times \psi_1^2 \times \psi_2^1$. Since ψ_2 alone is antisymmetric to reflection in the mirror plane, it follows that the wavefunction of prefulvene is antisymmetric like that of the $^1B_{2u}$ state of benzene. The two systems therefore correlate and lie on the same potential surface, and the $^1B_{2u}$ state of benzene can be expected to pass adiabatically to the ground state biradical prefulvene, from which benzvalene may be derived by coupling of the radical centres, or fulvene by a carbon–carbon bond cleavage and 1,2-hydrogen shift (9.18).

$$(9.18)$$

Similarly, since the wavefunctions for the $^1B_{1u}$ state of benzene (Figure 9.5) and for the biradical (9.17) are antisymmetric with respect to a mirror plane bisecting the carbon–carbon bonds of the ring, these two states correlate. Thus S_2 benzene ($^1B_{1u}$) and also T_1 benzene ($^3B_{1u}$) should adiabatically transform into the biradical (9.17), which would then be expected to collapse to Dewar benzene, from which prismane may be obtained by further irradiation ($\pi^2s + \pi^2s$ cycloaddition). Some support for these proposals is provided both by the observation that Dewar benzene is formed by irradiating benzene at 200 nm ($S_0 \rightarrow S_2$ transition) but not at 254 nm ($S_0 \rightarrow S_1$), and by the fact that the photoaddition of benzene to alkenes finds a ready interpretation in terms of these biradicals (see section 9.3). However, it should be emphasized that it has not been rigorously proven that the biradicals (9.16) and (9.17) are actual intermediates, nor that, if formed, they play the parts assigned to them, nor that other biradicals such as (9.19), (9.20) or (9.21) are not involved.

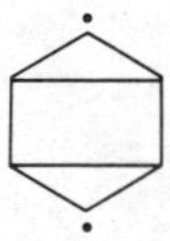

$$(9.19) \qquad\qquad (9.20) \qquad\qquad (9.21)$$

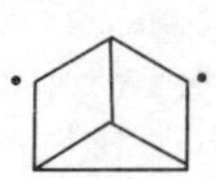

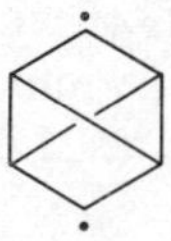

Indeed, inspection of correlation diagrams[18] shows that $^1B_{1u}$ benzene correlates with a singly excited state of Dewar benzene and with the excited state of benzvalene, thus introducing the possibility of a direct photochemical transformation into the photoisomers of benzene.

Finally, note should be taken of the suggestion by Fahrenhorst[19] that photoexcited benzene is converted into the highly strained cis-cis-trans-cyclohexatriene (Möbius-benzene) in which the 2p orbitals have the topology of a Möbius strip. Intramolecular (2 + 2) cycloadditions then give Dewar benzene (9.22) or benzvalene (9.23).

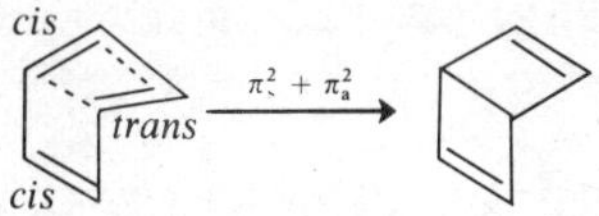

$$(9.22)$$

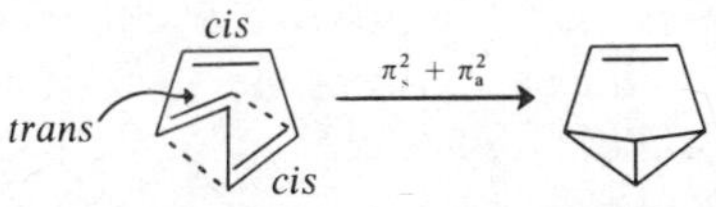

$$(9.23)$$

Clearly, the theoretical position is in a state of flux.

9.3   PHOTOADDITION OF ALKENES TO AROMATIC CARBOCYCLICS

The irradiation of mixtures of benzenoid hydrocarbons and olefinic systems gives rise to a rich diversity of products, formally obtained by adding the olefinic component across the 1,2-, 1,3- or 1,4-positions of the aromatic ring.

This is a difficult and complicated area from the theoretical point of view, because, even if one knows the multiplicity of the reacting species, there are at least four broad mechanistic pathways available:

    (i) Interaction of excited aromatic with ground state alkene;

    (ii) Interaction of excited alkene with ground state aromatic;

    (iii) Formation of biradical intermediate from excited aromatic, and reaction of the biradical with alkene to give product;

    (iv) Involvement of dipolar entities such as (aromatic$\cdot^+$···olefin$\cdot^-$) or (aromatic$\cdot^-$···alkene$\cdot^+$), obtained either by excitation of charge–transfer complexes pre-existing in the ground state or by electron transfer in an exciplex.

These different situations, if concerted, have different orbital symmetry requirements, and these have been analysed by Bryce-Smith.[20] In what follows use will be made of the results of this analysis in interpreting some of the facts. Nevertheless, a note of caution is necessary. The analysis sets out the conditions under which various initially-excited entities correlate with low-lying excited states of the products. For example, the process (9.24) is forbidden for $^1B_{2u}$ benzene but allowed for $^1B_{1u}$ benzene.

$$\left[\bigcirc\right]^* + \begin{array}{c} CH_2 \\ \| \\ CH_2 \end{array} \longrightarrow \left[\bigcirc\!\square\right]^* \qquad (9.24)$$

However, in all of the processes to be described, the product seems to be formed non-adiabatically in the ground state. In this case, the correlation of excited states of starting material with excited states of product is irrelevant unless we are dealing with a situation such as that represented diagrammatically in Figure 6.14 (see discussion of this point in Chapter 6). Other criticisms of the use of correlation diagrams have been made.[21] Until these theoretical difficulties have been resolved, it seems best to continue to use the correlation diagram approach when it is helpful to do so.

9.3.1   **1,2-Photoadditions of Benzenes**

Orbital symmetry considerations[20] impose restrictions upon the concerted *cis-cis*-1,2-photoaddition of alkenes to benzene to produce primary adducts of the bicyclo[4.2.0]octa-2,4-diene type (9.25).

$$\bigcirc + \begin{array}{cc} R_1 & R_2 \\ & \\ R_1 & R_2 \end{array} \xrightarrow{h\nu} \begin{array}{c} R_1 \\ R_2 \\ R_2 \\ R_1 \end{array} \qquad (9.25)$$

The reaction is forbidden for the interaction of $^1B_{2u}$ benzene and ground state alkene, but allowed if there is photoinduced electron transfer between the components as in an exciplex or a charge–transfer complex, or if it is the photo-excited alkene which reacts with ground state benzene. The photoadditions (9.26)[22] and (9.27)[23], if concerted, readily fit into this theoretical scheme because of the donor–acceptor properties of the components.

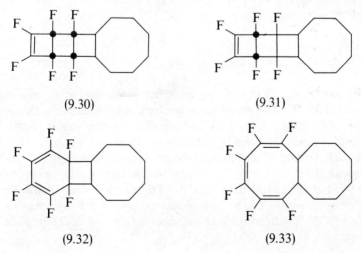

The photoaddition of *cis*- and *trans*-but-2-ene to benzene in the liquid phase[24] gives the 1,2-adducts stereospecifically (9.28 and 9.29), as well as 1,3- and 1,4-adducts (see later).

The stereospecificity of the process implies that it is concerted, and, on orbital symmetry grounds, it is suggested that the cycloaddition occurs in an exciplex in which the orbital symmetry restraints are relaxed. Similar reactions occur with alkyl benzenes or with tetramethylethylene.[25]

The excitation of hexafluorobenzene in the presence of *cis*-cyclo-octene gives 1,3-adducts and also the 1,2-adducts (9.30), (9.31), (9.32) and (9.33).[26]

The compounds (9.32) and (9.33) seem to be thermal isomerization products of the primary adducts (9.30) and (9.31), which are themselves derived, not from the well-known Dewar form of hexafluorobenzene, but from one or more of its precursors such as S_2-hexafluorobenzene or the fluorinated version of the biradical (9.17).

A number of acetylenes, when irradiated with benzene, give cyclo-octa-tetraenes,[27] presumably *via* an intermediate 1,2-photoadduct as in (9.34).

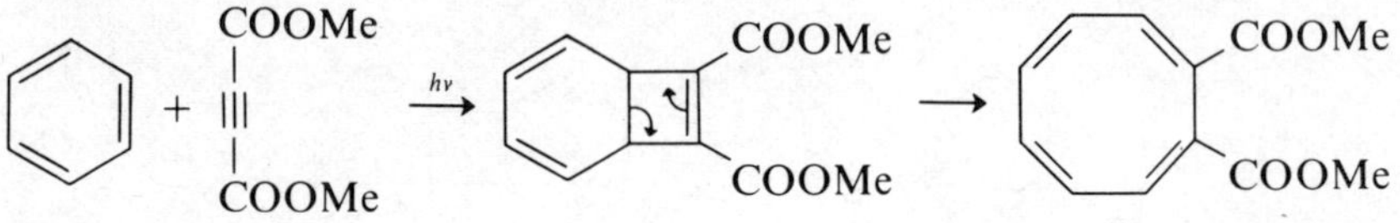

$$(9.34)$$

As an illustration of the complexities of this area, consider now the photo-addition of derivatives of maleic acid to benzene. Irradiation of benzene solutions of maleimides involves excitation of the addend and the primary formation of the 1,2-adduct (9.35), which then undergoes a thermal Diels–Alder reaction with a second molecule of the maleimide giving the 2:1-adduct (9.36). The intermediate (9.35) may be intercepted by tetracyanoethylene, which is a more powerful dienophile than a maleimide.

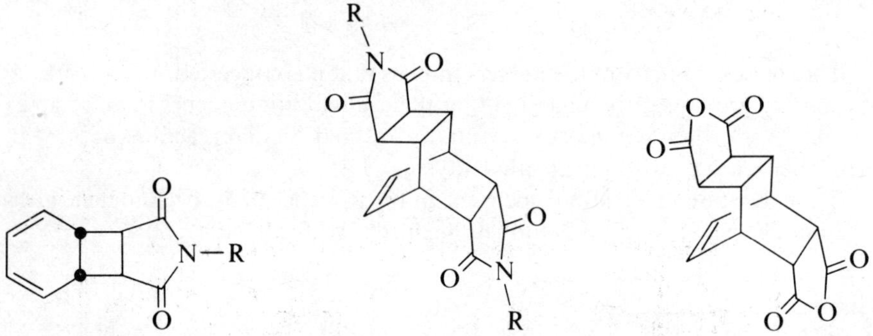

(9.35)   (9.36)   (9.37)

Maleic anhydride, by direct or sensitized excitation, gives an analogous 2:1-adduct (9.37). This is formed, however, by a quite different mechanism,[28,29] probably *via* a zwitterionic intermediate (9.38) which reacts with another molecule of maleic anhydride before closure of the cyclobutane ring. The evidence is that (i) added tetracyanoethylene fails to intercept the 1,2-adduct analogous to (9.35), (ii) it is excitation of the maleic anhydride–benzene charge-transfer band which leads to the formation of the product, and (iii) in the presence of acids the formation of the 2:1-adduct is suppressed and phenylsuccinic anhydride (9.39) is formed instead, by protonation and deprotonation of the zwitterion.

(9.38)   (9.39)

It should be noted that 1,2-photoaddition can also occur[30] in the triplet manifold (9.40).

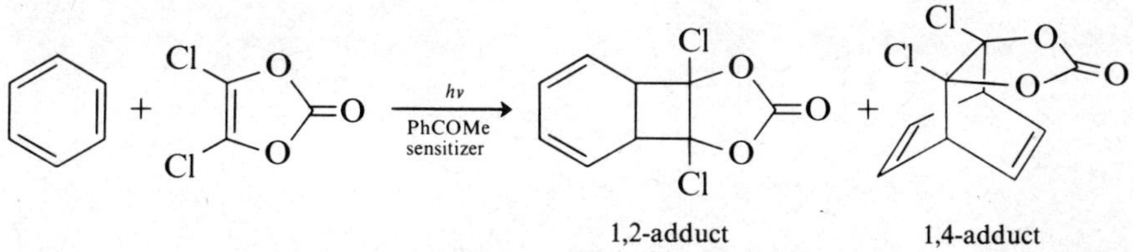

1,2-adduct   1,4-adduct

+2:1-adducts   (9.40)

Chaining the quencher to the chromophore frequently produces interesting results, illustrated[31] in this context by the intramolecular 1,2-photocyclo-addition (9.41).

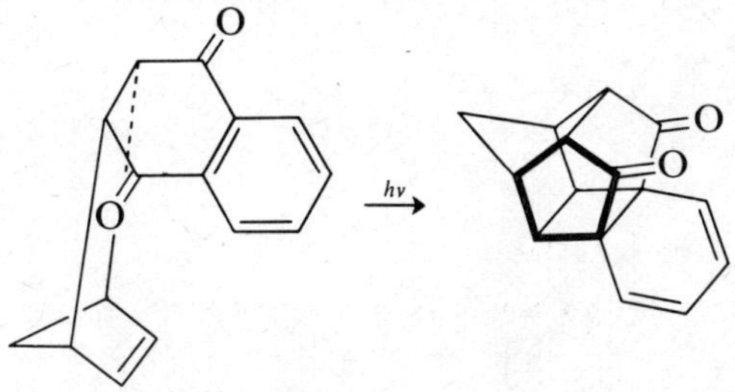

(9.41)

9.3.2   **1,3-Photoadditions of Benzenes**

This remarkable process, discovered[32] in 1966, leads to a tricyclic system in which an olefinic double bond has added across the *meta*-positions of a benzene ring (9.42).

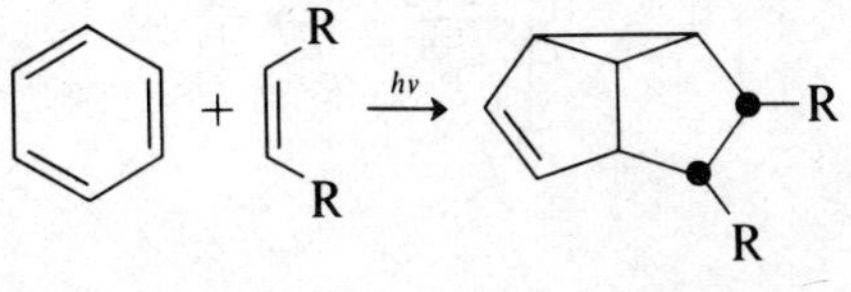

(9.42)

The reaction seems to be restricted to double bonds bearing only alkyl substituents, such as but-2-ene, norbornene,[33] allene[34] and cyclobutene.[35] It occurs with 254 nm light in both the gas phase and liquid phases, it is stereo-specific and it involves singlet excited benzene.[36] It is known that neither fulvene nor benzvalene is a precursor, and it is tempting to suppose that the products are formed from the prefulvene biradical (9.43).

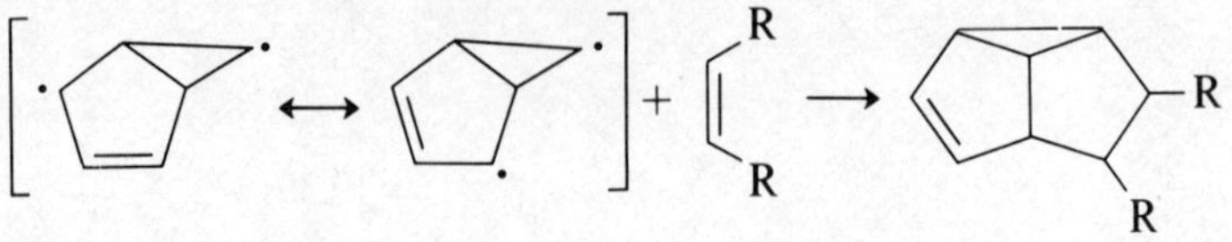

$$(9.43)$$

However, other possibilities exist, for example direct 1,3-photoaddition (9.44), which is allowed by orbital symmetry,[20] or reaction *via* an exciplex.

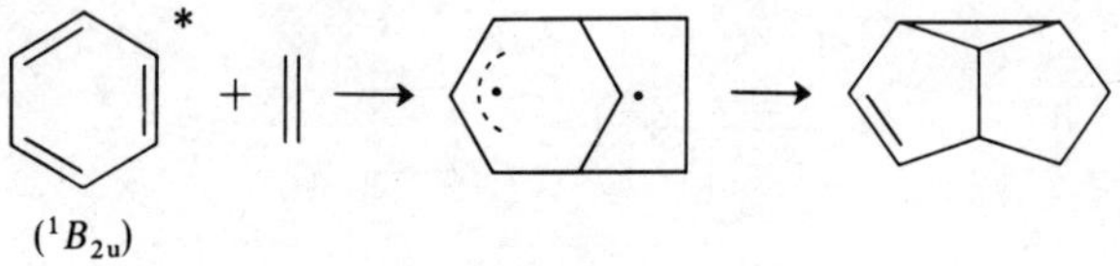

$$(^1B_{2u})$$

$$(9.44)$$

9.3.3   1,4-Photoadditions of Benzenes

Both alkenes (9.45)[24] and dienes (9.46)[37] give 1,4-adducts when irradiated with benzene in the liquid phase. Simple derivatives of benzene undergo similar reactions (9.47).[38]

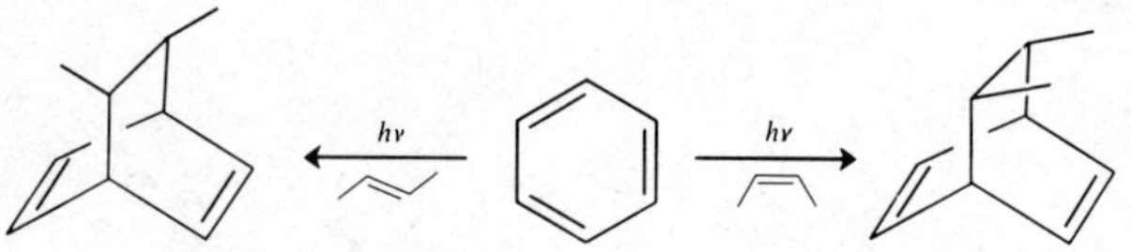

$$(9.45)$$

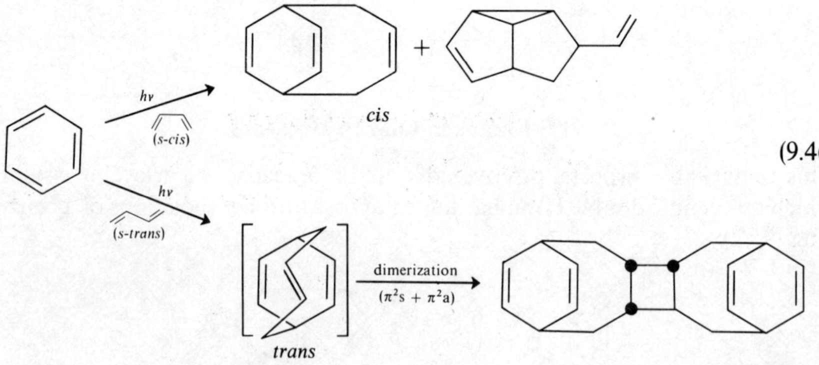

$$(9.46)$$

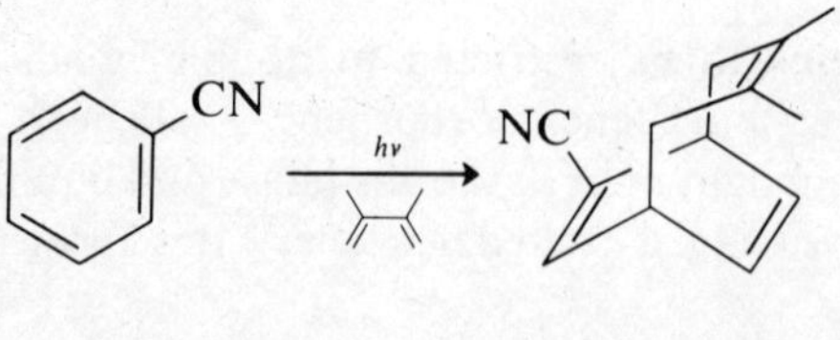

$$(9.47)$$

The 1,4-addition of simple alkenes (like 1,2- and 1,3-photoadditions) is stereo-specific and therefore presumably concerted, and it probably occurs as a re-action of $^1B_{2u}$ benzene with ground state alkene. This process is forbidden under the orbital symmetry rules, and again probably takes place *via* an exciplex.

The photoaddition of olefinic systems is not restricted to benzene and its simple homologues. Analogous processes occur with condensed aromatics such as naphthalene (9.48, 9.49)[39,40] or anthracene (9.50, 9.51).[41,42] In the case of anthracene and dienes, (2 + 4) or (4 + 4) cycloaddition can occur.

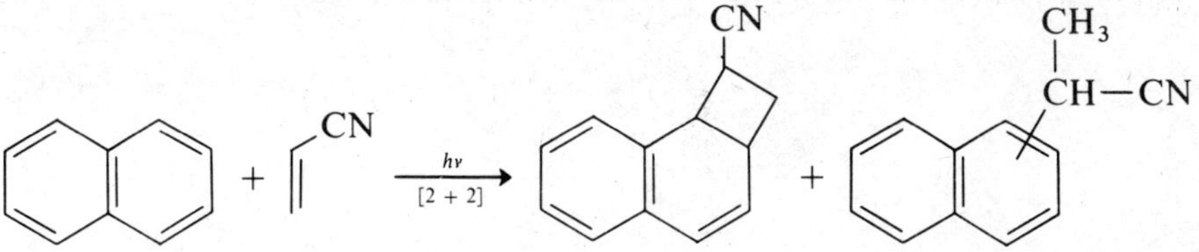

(α- and β-isomers)

(9.48)

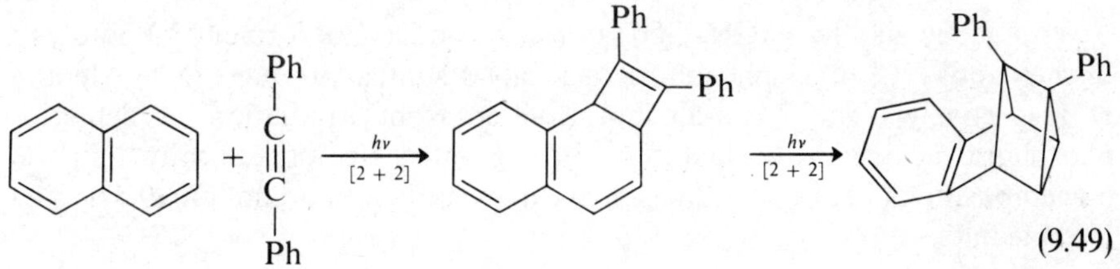

(9.49)

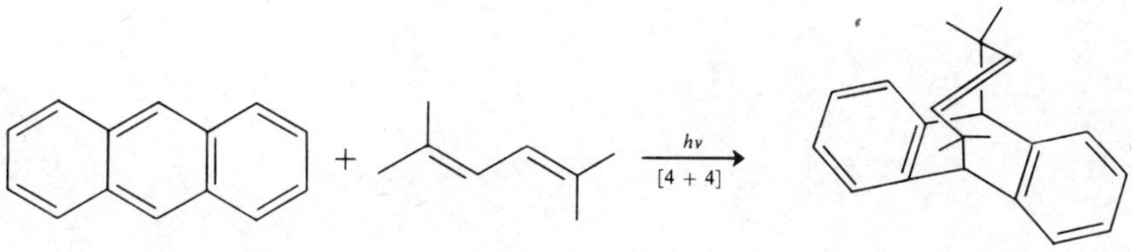

(9.50)

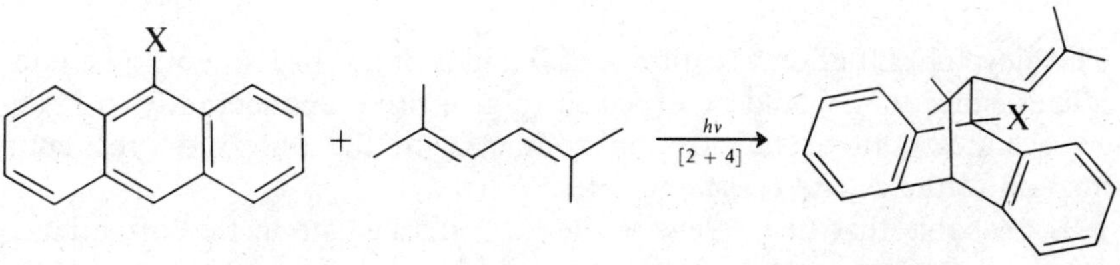

(X = CHO, CN)

(9.51)

Reactions of the type (9.50) and (9.51) are stereospecific, are presumed to be concerted, and are thought to occur *via* singlet exciplexes of the excited aromatic and the diene. Similar exciplexes of the excited aromatic hydrocarbon are invoked in the other examples given.

9.4   **OTHER PHOTOADDITION REACTIONS**

9.4.1   **Addition of Hydroxylic Compounds**

Many hydroxylic compounds, when irradiated with benzene under acidic conditions, afford derivatives of bicyclo[3.1.0]hexene (9.52).

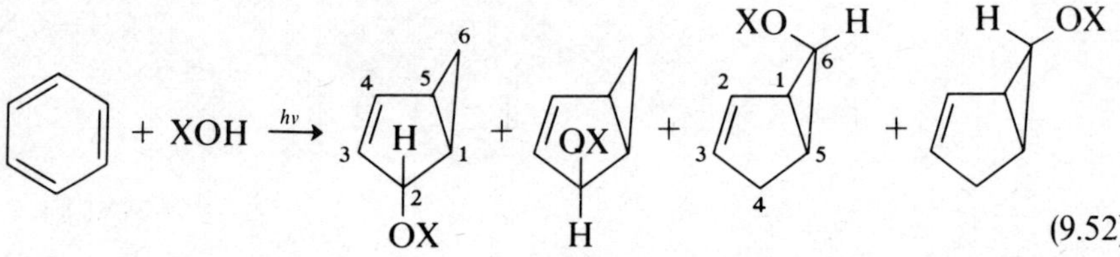

(9.52)

2-*exo*   2-*endo*   6-*endo*   6-*exo*

With acetic acid (X = CH$_3$CO), all four products are formed.[43] Photolysis in methanol/HCl or in phosphoric acid gives similar products (X = Me and H respectively). The 2-*exo*-alcohol isolated from irradiation in deuterio-phosphoric acid has the deuterium incorporated stereospecifically[44] in the 6-*endo*-position (9.53), a result which excludes the benzenonium ion (9.54) as an intermediate.

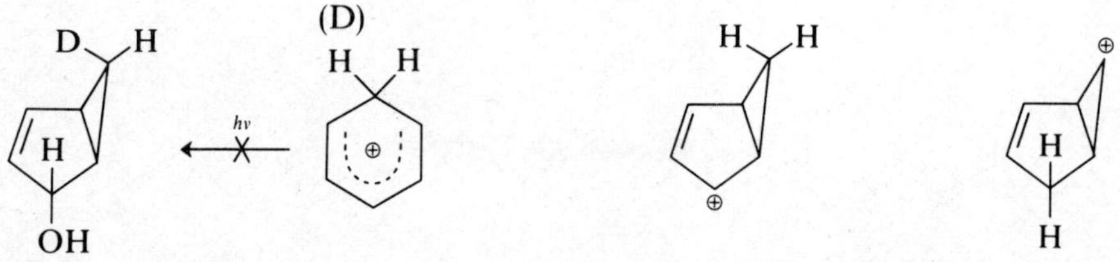

(9.53)   (9.54)   (9.55)   (9.56)

The bicyclo[3.1.0]hexenyl cations (9.55) and (9.56) seem also not to be intermediates, since they would be expected to give rise preponderantly to 2- and 6-*exo*-acetates with acetic acid, in conflict with the observed preferential formation of the 2- and 6-*endo*-acetates.

It is probable that benzvalene is the key intermediate in the formation of these products, because bicyclobutanes are known to undergo acid-catalysed ring-openings in which both the proton and the nucleophile preferentially attack the *endo*-positions. Reaction of benzvalene itself with methanol/HCl gives the same products[45] as are formed by photolysis of benzene in the solvent system.

9.4.2   **Photoreduction**

Many excited aromatics (e.g., benzene,[46] biphenyl, naphthalene[47] and its methoxy derivatives, or anthracene[48]) react with amines to give reduction products and aminated products. The reaction is probably initiated[47,49] by the transfer of an electron from the nitrogen lone-pair orbital to the half-filled highest bonding molecular orbital of the excited aromatic (9.57). The transfer occurs in an exciplex or more probably an encounter complex[50] in which the two components are separated by distances of the order of 7 Å.

$$ArH^* + \diagdown \overset{\diagup}{N:} \rightarrow ArH^{\bullet} + \diagdown \overset{\diagup}{N^{\bullet+}} \tag{9.57}$$

The subsequent steps may be exemplified in the case of the naphthalene/triethylamine system (9.58).

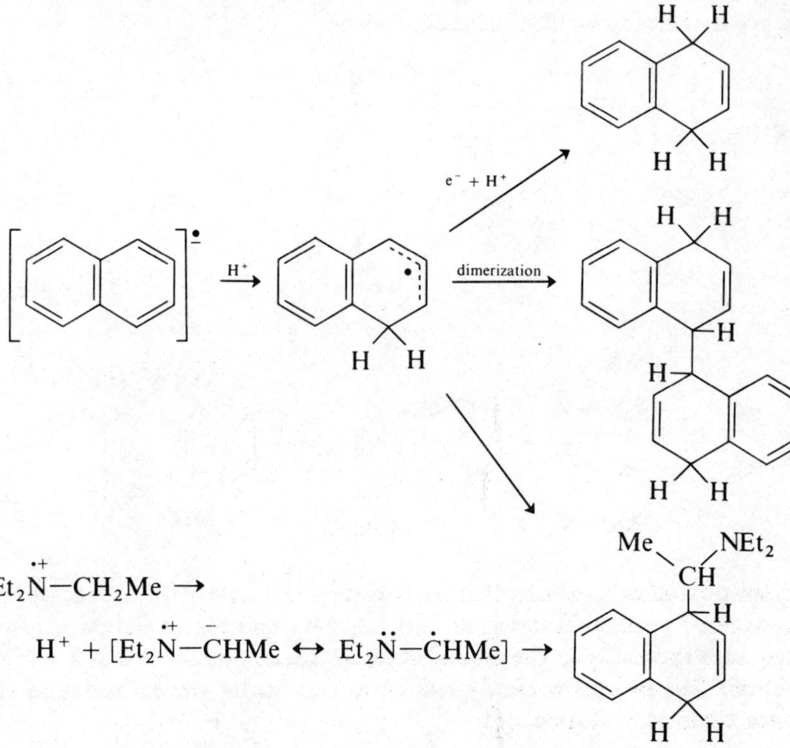

$$(9.58)$$

Et$_2\overset{\bullet+}{N}$—CH$_2$Me $\rightarrow$

H$^+$ + [Et$_2\overset{\bullet+}{N}$—$\overset{-}{C}$HMe $\leftrightarrow$ Et$_2\overset{..}{N}$—$\overset{\bullet}{C}$HMe] $\rightarrow$

Other mechanisms exist for photoreduction (for a review, see ref. 51), notably transfer of hydride ion by NaBH$_4$ (9.59), or hydrogen atom transfer from NaBH$_4$ (9.60).

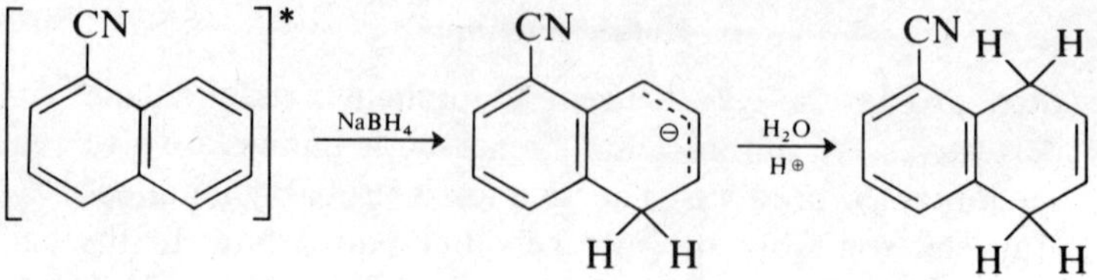

$$(9.59)$$

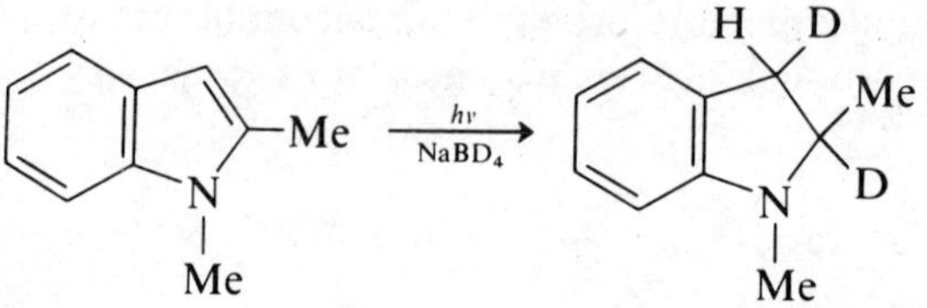

$$(9.60)$$

9.4.3   Addition of Oxygen

Polynuclear aromatic hydrocarbons (notably anthracenes and naphthalenes) form transannular peroxides (9.61, 9.62) when excited in the presence of oxygen, and the point of attack is substituent dependent.[52]

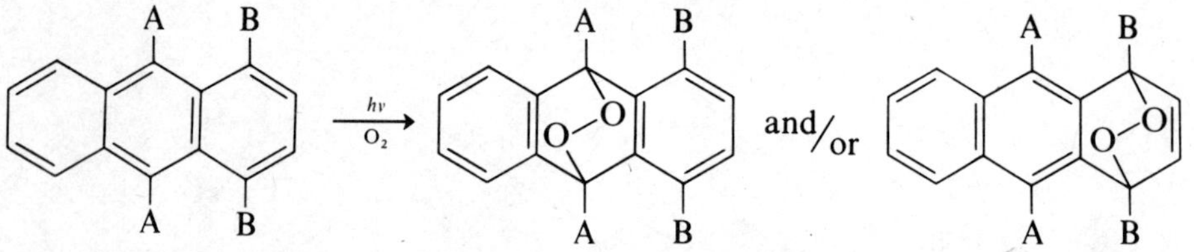

$$(9.61)$$

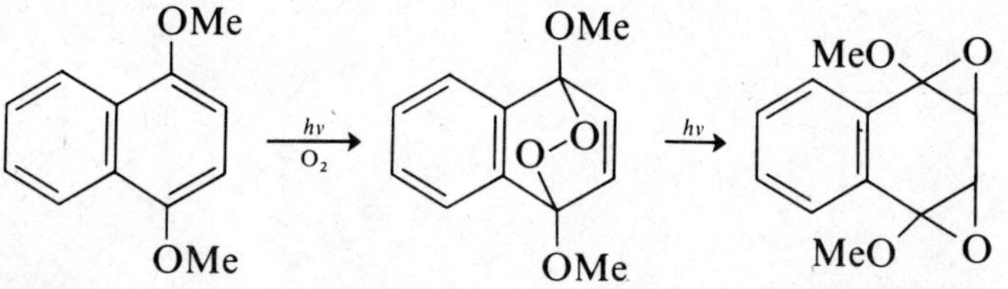

$$(9.62)$$

Mechanistic studies[53] indicate that the triplet excited state of the hydrocarbon excites oxygen, by energy transfer, into its $^1\Delta_g$ state, and the singlet oxygen so produced adds thermally to the ground state hydrocarbon (allowed as a $\pi^4s + \pi^2s$ reaction). The process is clearly related to that for the photo-oxidation of dienes (see Chapter 8, section 8.6).

9.5   INTRAMOLECULAR PHOTOCYCLIZATION[54]

This area is dominated by the photocyclization of stilbene and its derivatives to phenanthrenes (9.63).

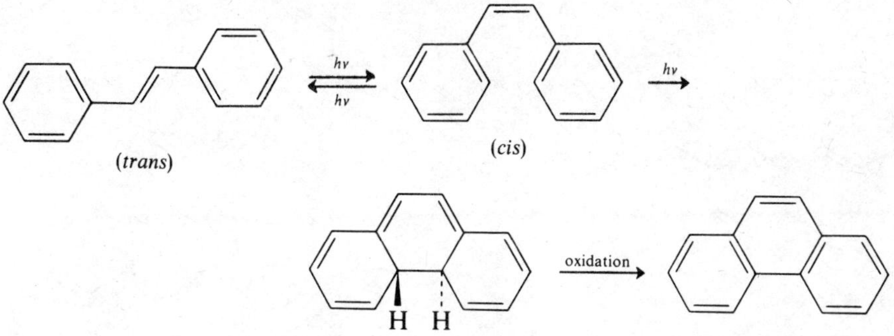

$$(9.63)$$

It seems that the lowest singlet excited state of *cis*-stilbene in the absence of oxygen forms a yellow intermediate which has been shown to be a dihydro-phenanthrene. In the presence of hydrogen acceptors (O_2, I_2, $FeCl_3$ etc.) this is dehydrogenated to phenanthrene. The yellow intermediate eludes isolation, but in related systems it seems that the 'extra hydrogens' are *trans*-oriented,[55] so that one may think of the reaction as the aromatic equivalent of the conrotatory cyclization of *cis*-hexatriene (9.64, see Chapter 6, section 6.4).

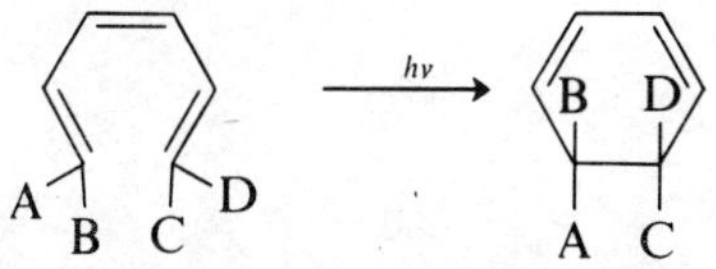

$$(9.64)$$

Although the mechanistic details of the photocyclization are a matter of current controversy, the overall process remains a general and powerful technique for producing in good yields polynuclear systems containing a phenanthrene skeleton. These include carbocyclic systems such as the helicenes (9.65)[56] and related compounds (9.66)[57] or triphenylene (9.67),[58] and heterocyclic systems formed from imines (9.68)[59] or azo-compounds (9.69)[60] or by building on an existing heterocyclic framework (9.70).[61]

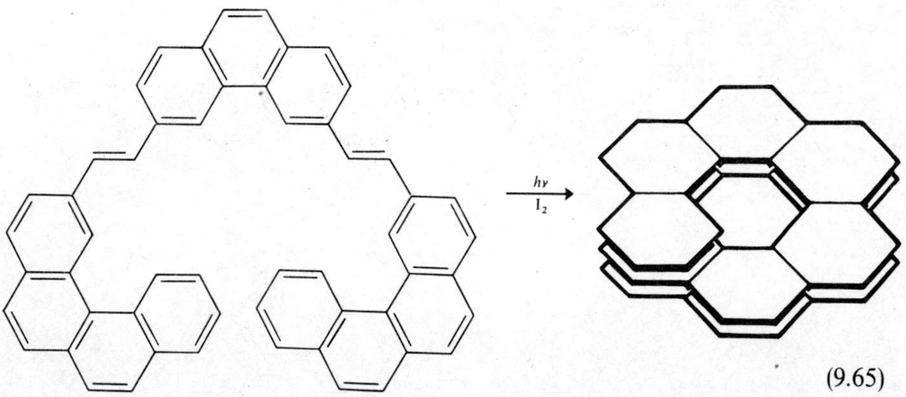

$$(9.65)$$

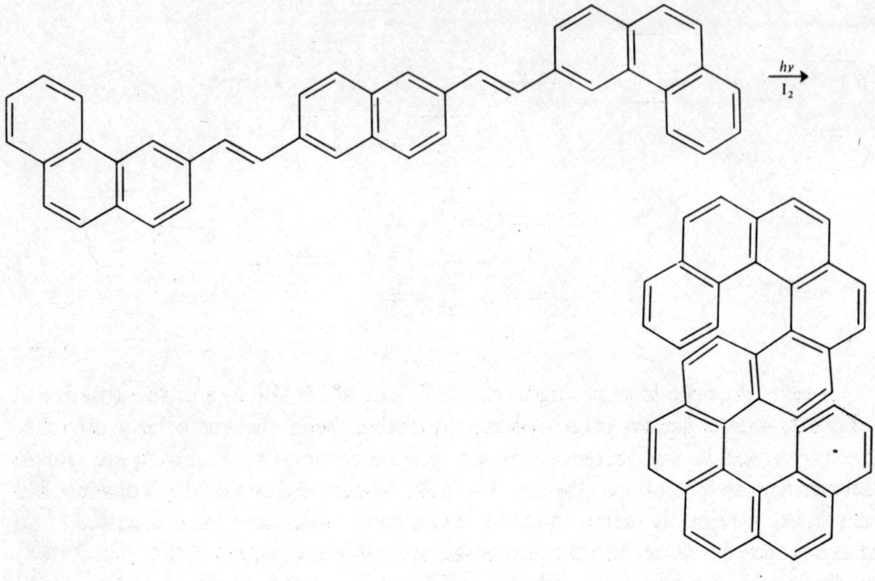

(9.66)

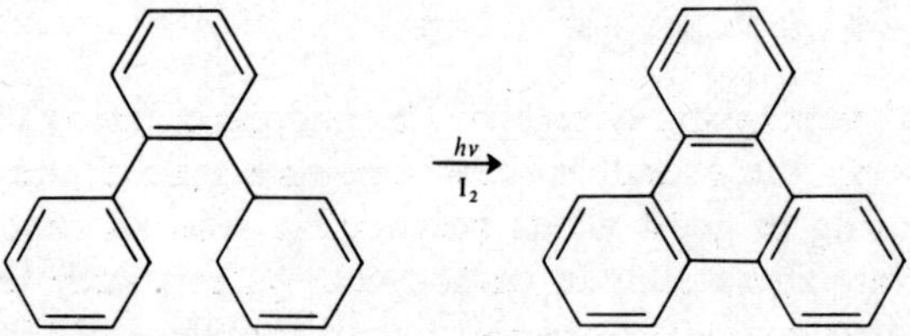

(9.67)

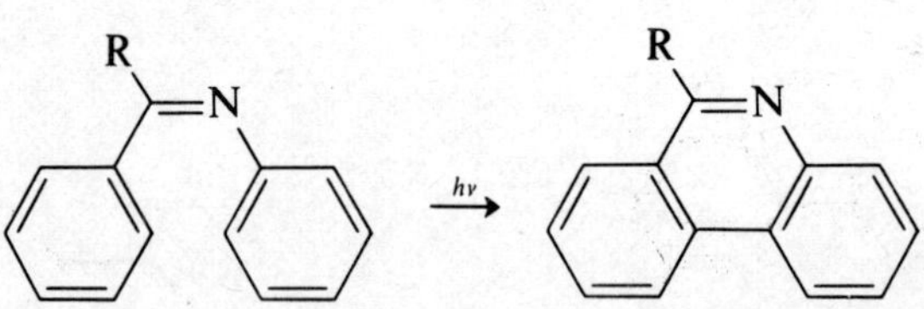

(9.68)

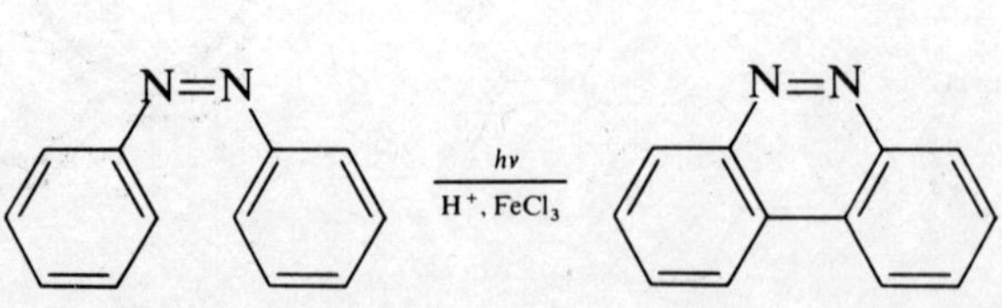

(9.69)

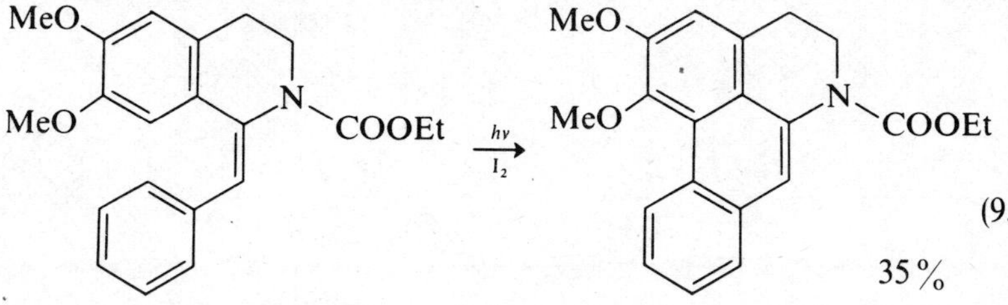

(9.70)

35%

The reaction (9.70) illustrates the use of the photoelectrocyclization of stilbene derivatives in the synthesis of aporphine alkaloids. Other photochemical routes to isoquinoline alkaloids are reviewed in ref. 61.

9.6   AROMATIC PHOTOSUBSTITUTION

Light-induced substitutions in the aromatic series can occur either on a ring carbon atom or on an atom of a substituent, and they can proceed by heterolytic or radical mechanisms. From the theoretical standpoint, the most interesting are the heterolytic substitutions discovered and investigated by Havinga,[62] who found that there were two broad classes of reaction—Type A in which the orientation rules were different from those obtaining in the ground state, and Type B in which the orientation rules were the same as in the ground state but the process was accelerated by light absorption. All these reactions seem to be functions of the lowest singlet excited state, and many proceed in high quantum yield.

9.6.1   Type A Photonucleophilic Substitution

Whereas in the ground state a nitro group activates a benzene ring to nucleophilic substitution in the *ortho* and *para* positions (9.71), in the excited state the nitro group seems to activate the meta position (9.72).

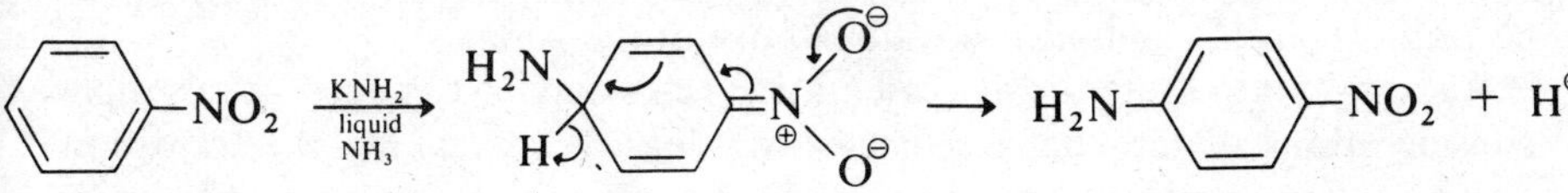

(9.71)

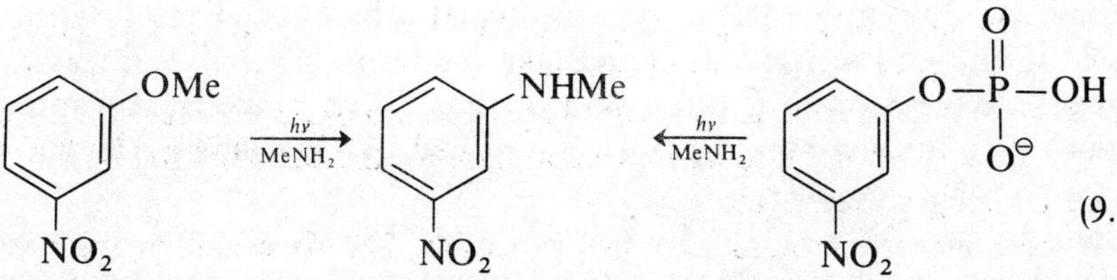

(9.72)

A particularly nice example is seen in the contrasting thermal and photochemical reactions of 3,4-dimethoxynitrobenzene with hydroxide (9.73).

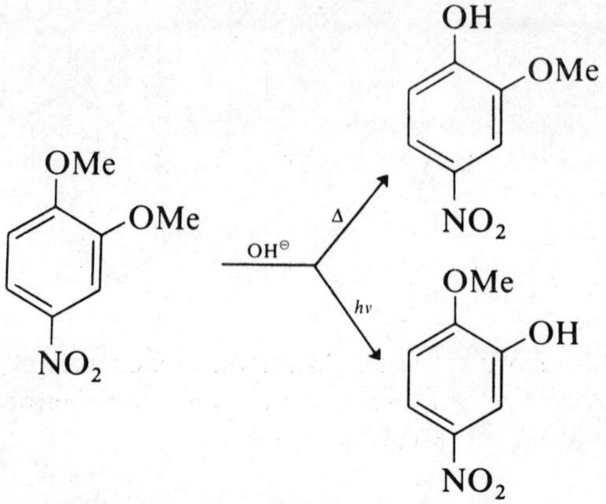

$$(9.73)$$

In many cases the cyano group behaves similarly, and analogous reactions are found with substituted naphthalenes.

9.6.2   **Type A Photoelectrophilic Substitution**

On irradiation in the presence of CF_3CO_2D, toluene gives mainly m-deuterio-toluene (9.74), anisole gives o- and m-deuterioanisoles (9.75, 9.76), and nitrobenzene gives mainly p-deuterionitrobenzene (9.77).

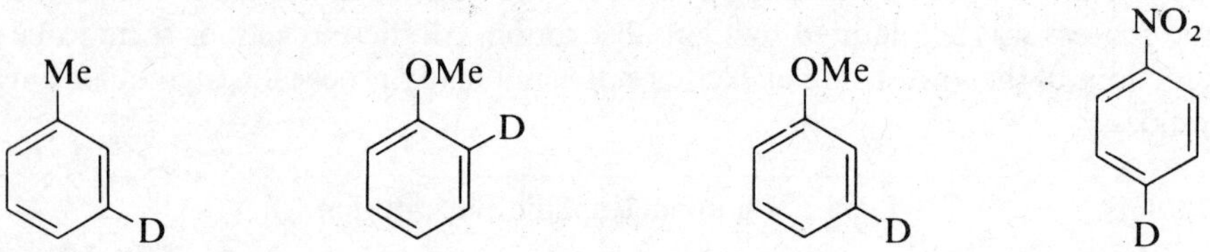

Notice that the observed orientation differs from that expected in the ground state—the NO_2 group activates the *para*-position and the donor methoxy group activates the *ortho* and *meta* positions but not the *para*.

An insight into these curious orientation rules may be obtained from simple considerations of the changes in electron-density on making the transition $S_0 \rightarrow S_1$. For simplicity, take the benzyl cation and the benzyl anion as models for benzenes substituted with electron-withdrawing and electron-donating substituents respectively. The seven π-molecular orbitals of a benzyl system are related to those of benzene itself and take the form of Figure 9.7, where the numbers are the electron densities at the carbon atoms (in the Hückel approximation). ψ_4 is the non-bonding molecular orbital, and the pairs ψ_2, ψ_3 and ψ_4, ψ_5 are no longer degenerate.

Consider first the ground state benzyl cation. The six electrons are accommodated in ψ_1, ψ_2 and ψ_3, and ψ_2 is of lower energy than ψ_3 since the $CH_2^{\oplus}$ substituent can delocalize the electron density at C_1 in the corresponding benzene molecular orbital. Hence the lowest energy optical transition corresponds to a

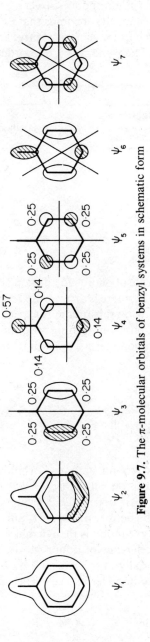

Figure 9.7. The π-molecular orbitals of benzyl systems in schematic form

transition $\psi_3 \rightarrow \psi_4$. In ψ_3 the electron density in the α-position is zero, and in ψ_4 it is 0·57. Thus the transition induces a very large increase in the charge density at the α-position. Similarly, one finds that excitation leads to the o- and particularly the m-position becoming more positive and the p-position becoming somewhat more negative. These changes are recorded in diagrammatic form in Figure 9.8.

Position	Change in electron density on excitation	
α	–	–
o		+
m	+	+
p		–

Figure 9.8. Benzyl cation

Position	Change in electron density on excitation	
α	+	+
o		–
m	–	–
p		+

Figure 9.9. Benzyl anion

If one assumes that the rate of photosubstitution is correlated with the charge density at the point of attack, there is an immediate explanation of the observed data—an electron-withdrawing substituent is expected strongly to decrease the electron density in the m-position, leading to photonucleophilic substitution in this position. In valence-bond terms, this can be thought of as (9.78).

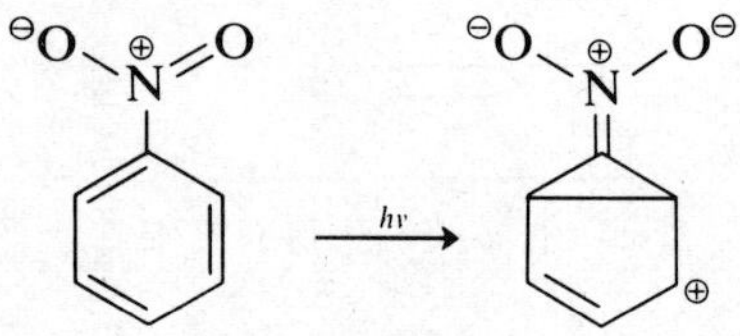

$$(9.78)$$

The benzyl anion, in its ground state, has orbitals ψ_1–ψ_4 doubly occupied, and the lowest energy transition will be $\psi_4 \rightarrow \psi_5$ (ψ_5 will be of lower energy than ψ_6 since the electron pair on the $CH_2^\ominus$ group will be stabilized by inter-action with the zero electron density on C_1 in ψ_5 and destabilized by the charge on C_1 in ψ_6). The effect of excitation is then deduced to be that given in Figure 9.9, which explains the preference of donor-substituted benzenes to undergo photodeuteration preferentially in the m-position and to a lesser extent in the o-position. Again in valence-bond terms, the m-effect can be crudely represented as in (9.79).

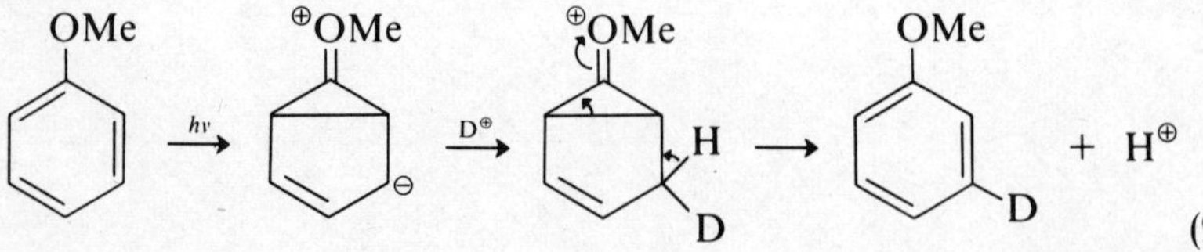

$$(9.79)$$

In summary, it seems that (i) in the excited state, donor and acceptor groups retain their donor and acceptor properties, but operate mainly on the m-position and to a smaller extent on the o-position; (ii) the o-/p- versus m-dichotomy of the ground state is replaced by a o-/m- versus p-distinction in the excited state; (iii) the effect in the p-position is opposite from that in the o-/m-positions.

9.6.3   Type B Substitutions

These reactions occur only on irradiation but have orientation rules characteristic of the ground state. They seem to involve attack by uncharged nucleophiles, as in (9.80), but the detailed mechanism of such processes is not yet clear.

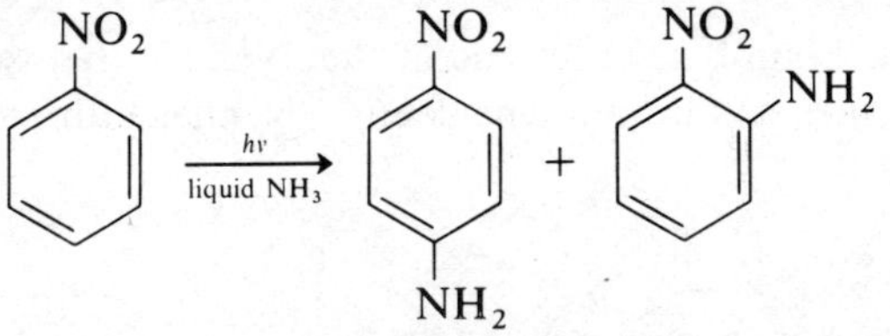

$$\text{(9.80)}$$

9.6.4   Radical Substitution

The effect of irradiating halo-aromatics is to homolyse[63,64] the carbon–halogen bond to produce radicals in their ground states (9.81), which then give rise to products by well-established thermal pathways. However, when chlorobenzene is excited (254 nm) in the liquid phase, the Ph$\cdot$ and Cl$\cdot$ radicals seem to recombine within the solvent cage generating transiently π-chlorobenzene[64] (9.82), an isomer of chlorobenzene in which the chlorine atom forms a π-complex with the phenyl radical.

$$\text{Ar}-\text{X} \xrightarrow{h\nu} \text{Ar}\cdot + \text{X}\cdot \qquad \text{(9.81)}$$

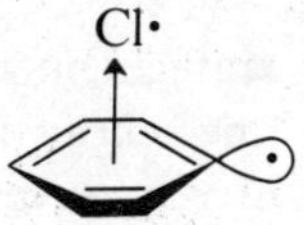

$$\text{(9.82)}$$

π-Chlorobenzene, in which the halogen atom is much more selective than the free chlorine atom in its reactivity, for example with respect to tertiary or primary hydrogen atom abstraction, reverts to ground state chlorobenzene if no other reaction supervenes. Bromobenzene behaves similarly.

The light-induced homolysis of aryl-halogen bonds has important synthetic implications. Irradiation of iodinated aromatics in benzene often gives the phenylated aromatic in high yield[65] (9.83, 9.84).

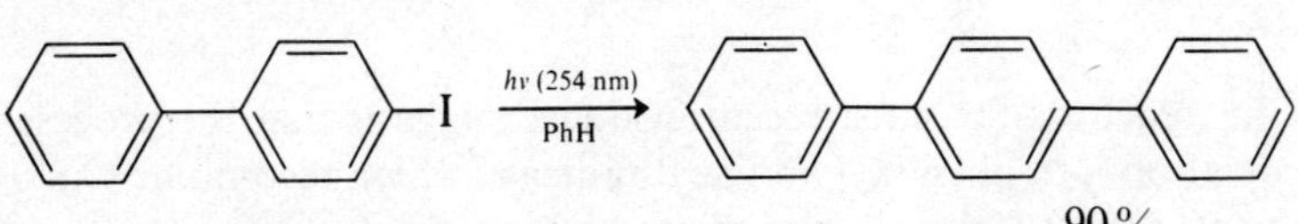

$$\text{(9.83)}$$

90%

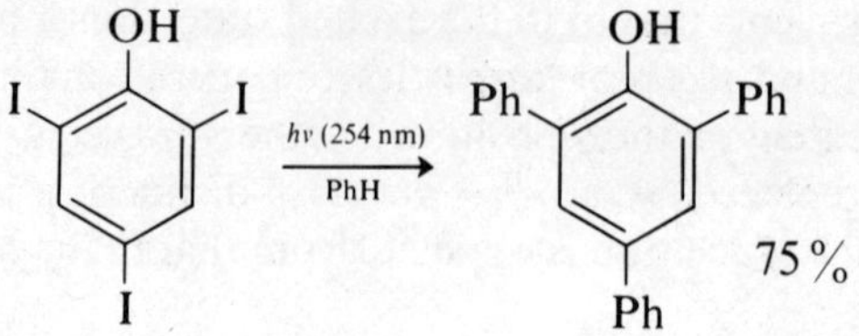

$$\text{(9.84)}$$

75%

A related process (9.85) gives diaryl acetylenes from iodoacetylenes.[66]

$$Ph-C\equiv C-I \xrightarrow[PhH]{h\nu} Ph-C\equiv C-Ph \qquad (9.85)$$

75%

Intramolecular homolytic reactions of this type can be used to form poly-nuclear systems such as fluoranthene (9.86)[67] or phenanthrene (9.87).[68]

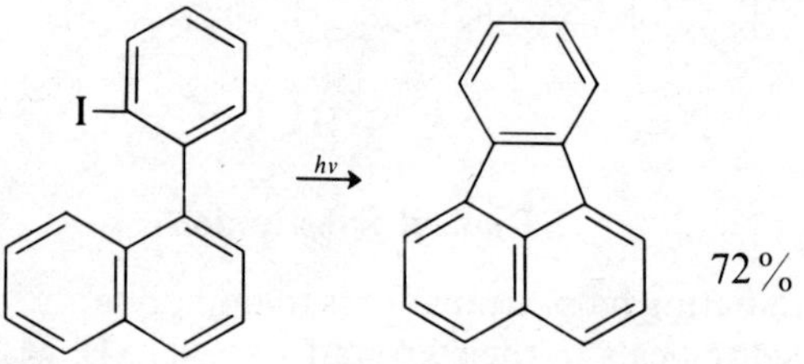

$$\text{(9.86)}$$

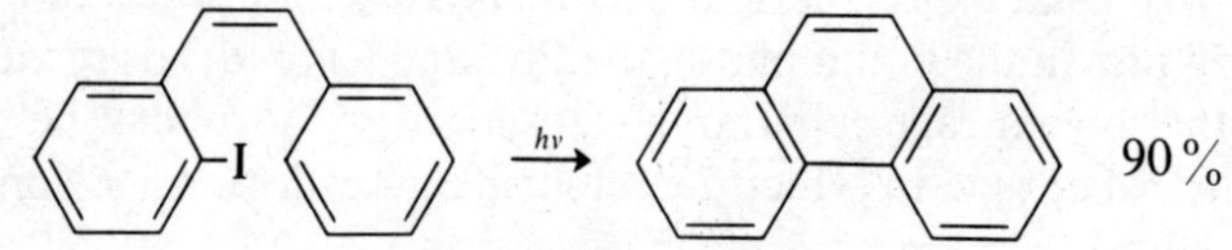

72%

$$\text{(9.87)}$$

90%

9.7   LATERAL–NUCLEAR REARRANGEMENTS[69]

The classical lateral–nuclear rearrangements of 'dark' chemistry have their photochemical counterparts in the photo-Fries, photo-Claisen and other rearrangements. The most extensively investigated is the photo-Fries re-arrangement, for example (9.88).

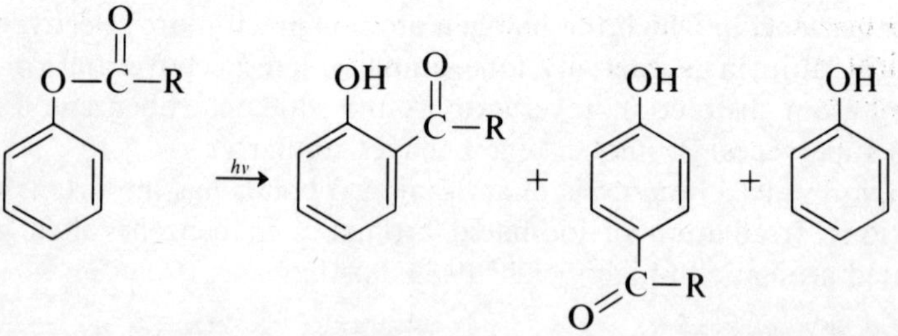

$$\text{(9.88)}$$

For this reaction, the most recent evidence suggests that the process occurs by dissociation into phenoxyl and acyl radicals, which recombine (9.89) within the solvent cage to give intermediates which then enolize to the product hydroxyphenyl ketones. The phenol is produced by phenoxyl radicals which escape from the solvent cage.

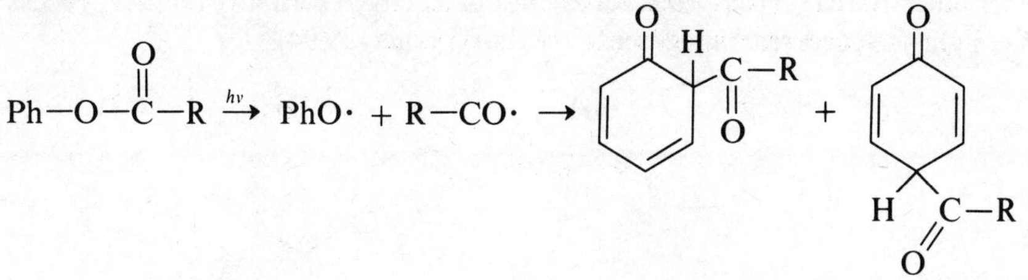

$$(9.89)$$

In support of this mechanism, it has been shown[70] that in the gas phase (no solvent cage) phenyl acetate on irradiation gives large amounts of phenol but no rearranged products. Also, flash photolysis of phenyl acetate in solution gives transients whose absorption spectra correspond to that of the phenoxyl radical and to that of the acyl-substituted cyclohexa-2,4-dienone.

The photo-Fries rearrangement seems to be a rather general process, occurring with a large range of esters of aromatic systems, both carbocyclic and heterocyclic. There are many other rearrangements formally related to the photo-Fries rearrangement, such as those of aryl esters of sulphonic acids (9.90),[71] N-aryl carbamates (9.91),[72] or N-aryl lactams (9.92).[73]

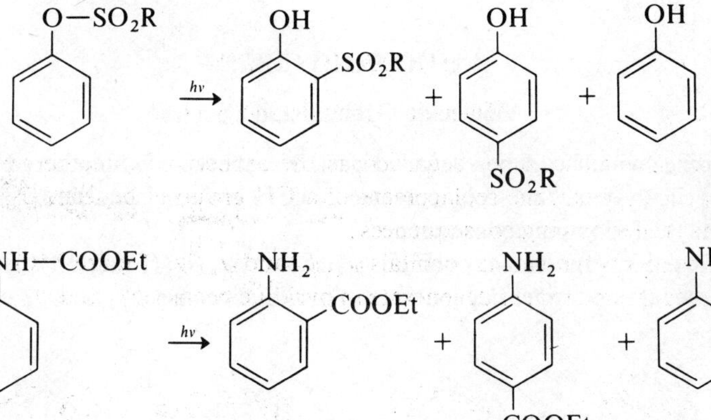

$$(9.90)$$

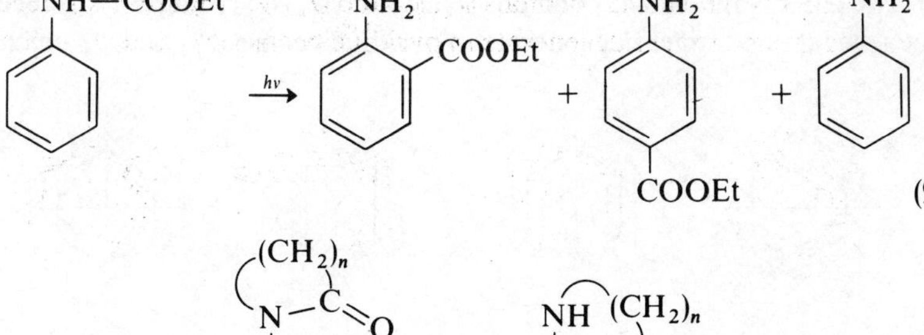

$$(9.91)$$

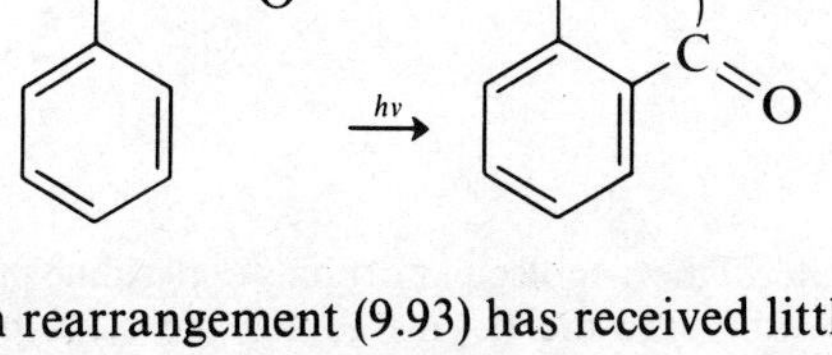

$$(9.92)$$

The photo-Claisen rearrangement (9.93) has received little attention, but the evidence suggests[74] that the rearrangement mechanism is related to that of the

294

photo-Fries reaction and involves dissociation into phenoxyl and allyl fragments, which recombine within the solvent cage. The process seems to be intramolecular in that cross-products are not detected. A formally related process is
the light-induced rearrangement[75] of diaryl ethers (9.94).

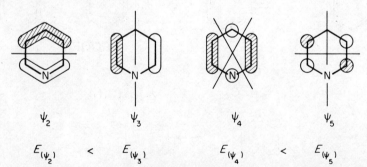

9.8   HETEROCYCLICS

9.8.1   Molecular Orbitals and Spectra

In pyridine, which can be regarded as the archetypal 6-membered aromatic
heterocyclic system, the replacement of a CH group in benzene by a nitrogen
atom has the following consequences:

(i) The pairs of molecular orbitals ψ_2, ψ_3 and ψ_4, ψ_5 (Figure 9.10), degenerate
in benzene, are no longer degenerate in pyridine because ψ_2 and ψ_4, which have

$$\psi_2 \qquad \psi_3 \qquad \psi_4 \qquad \psi_5$$

$$E_{(\psi_2)} \quad < \quad E_{(\psi_3)} \qquad E_{(\psi_4)} \quad < \quad E_{(\psi_5)}$$

Figure 9.10. Four of the π-molecular orbitals of pyridine in schematic form.
The nitrogen atom lifts the degeneracy of the corresponding benzene
orbitals

electron density on the electronegative nitrogen atom, are more stable than ψ_3 and ψ_5, which have not.

(ii) The symmetry of the molecule is reduced from D_{6h} (in C_6H_6) to C_{2v}.

(iii) The nitrogen atom has an unshared pair of electrons.

These changes are manifested in the ultraviolet absorption spectra in two ways (see Figure 9.11). First, pyridine shows a $n \rightarrow \pi^*$ transition, which is the

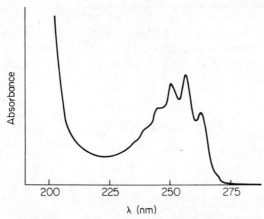

Figure 9.11. Absorption spectrum of pyridine in methanol

absorption at longest wavelength because the n-electrons are at a higher energy than those in ψ_2 or ψ_3. The band is not clearly resolved and appears as a long-wavelength shoulder (near 270 nm) on the more intense $\pi \rightarrow \pi^*$ band. It is a weak band because it is overlap forbidden. Secondly, the $\pi \rightarrow \pi^*$ absorption bands are of increased intensity with respect to those in benzene (λ_{max} 255, 195 nm; ε_{max} 1800, 7500 l mol^{-1} cm^{-1} respectively). The point is that the electronegative nitrogen atom distorts the symmetry of the π-orbitals from that in benzene, but not by very much, so that the $\pi \rightarrow \pi^*$ transitions of pyridine resemble those of benzene but are more allowed.

The introduction of more nitrogen atoms into the ring tends to increase the intensity of the $n \rightarrow \pi^*$ transition and to shift it to longer wavelengths (to ~ 340 nm in pyridazine). This shift can be understood (Figure 9.12) in terms of

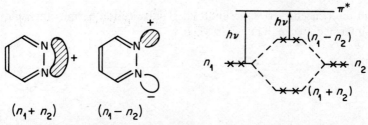

Figure 9.12. Bonding and antibonding molecular orbitals derived from the nitrogen lone pairs of pyridazine, showing the origin of the long-wavelength transition

the interaction of the two unshared pairs of the nitrogen atoms giving rise to molecular orbitals $(n_1 + n_2)$ and $(n_1 - n_2)$.

Within this framework, the long-wavelength absorption is due to a transition to the π^* orbital from the higher energy $(n_1 - n_2)$ orbital.

9.8.1 Ring-scrambling Processes in Six-membered Heterocyclics[76]

The ultraviolet irradiation of aromatic six-membered heterocyclics leads to transpositions of the ring atoms analogous to those encountered in the benzene series. With pyridine and its alkyl derivatives the products are those expected from the intervention of aza-Dewar benzene (9.95)[77] and aza-prismane (9.96, 9.97)[78] intermediates.

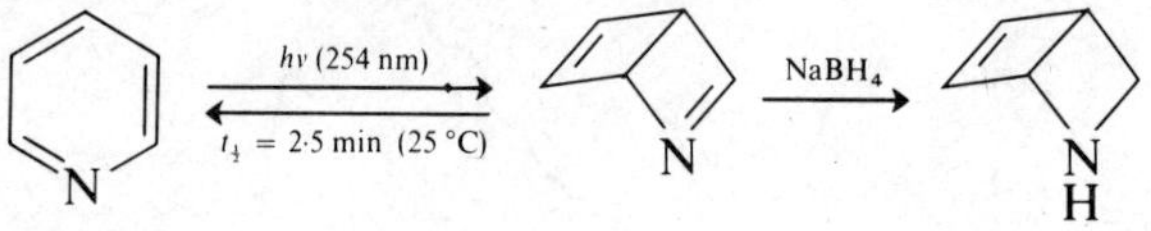

$$(9.95)$$

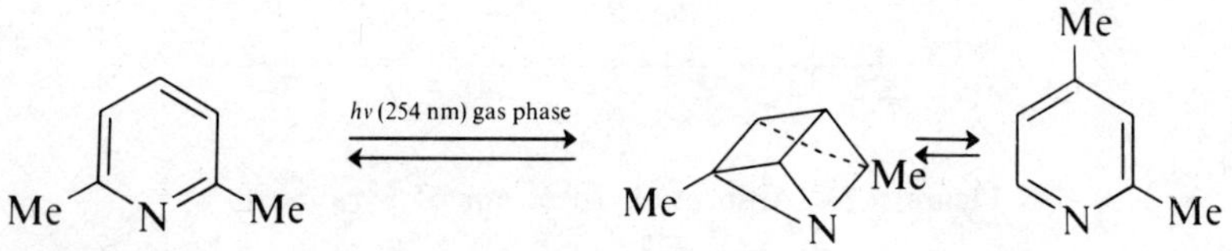

$$(9.96)$$

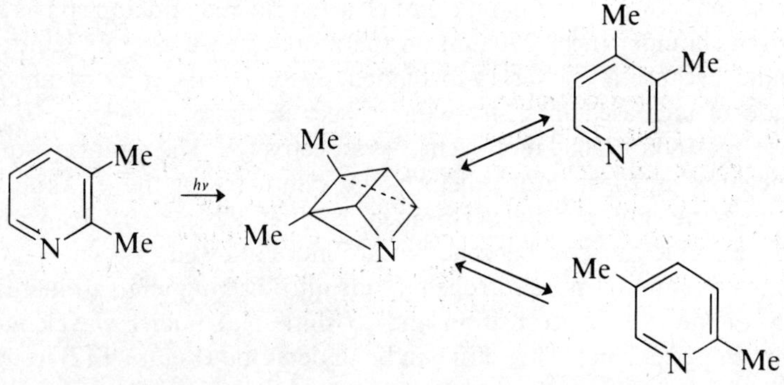

$$(9.97)$$

As in the benzene series, perfluoroalkyl substitution permits the isolation[79] of the intermediates in high yields (9.98).

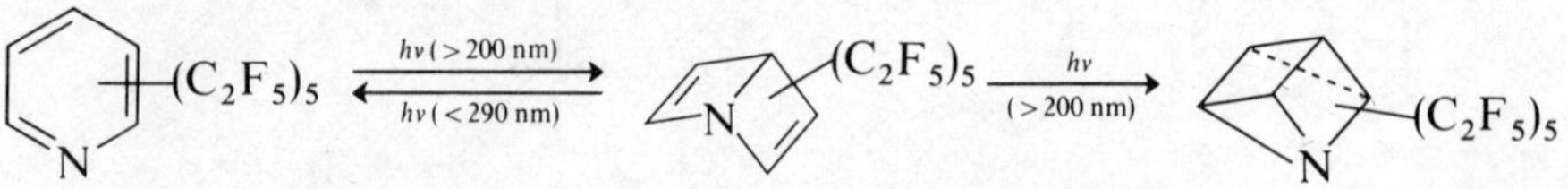

$$(9.98)$$

It seems that these processes may be the consequence of exciting an (n, π^*) state of the pyridine ring, for irradiation of quaternized pyridinium salts,[80] which can only experience $\pi \rightarrow \pi^*$ transitions, gives products (9.99) which seem to derive partly from the nitrogen analogue of benzvalene.

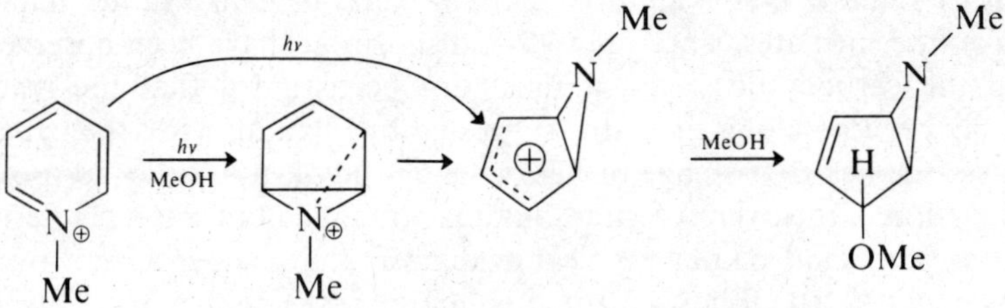

(9.99)

In a similar vein, $\pi \rightarrow \pi^*$ excitation of pyrylium salts[81] leads to 1,2-transposition, possibly *via* a benzvalene type of intermediate (9.100).

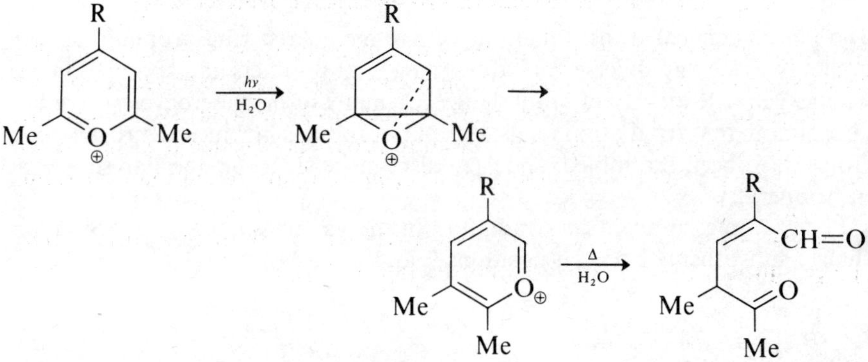

(9.100)

A ring-scrambling process occurring in a heterocyclic compound with at least two heteroatoms[82] may give a new ring system, as with pyridazines (9.101) or pyrazines (9.102, 9.103).

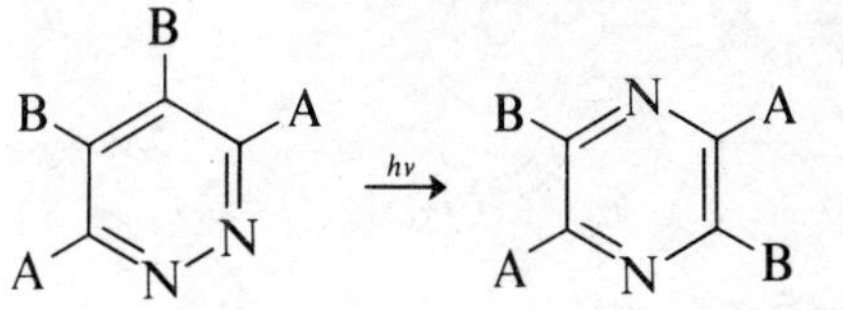

(9.101)

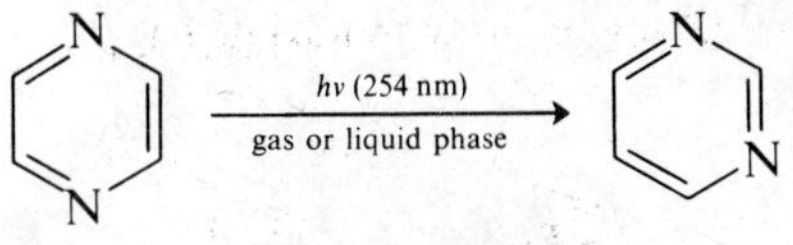

(9.102)

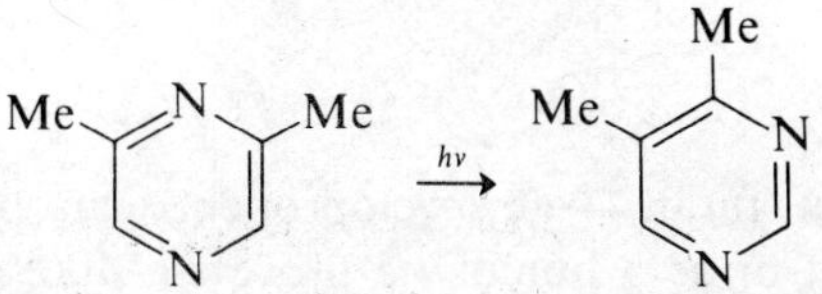

(9.103)

The pyrazine rearrangements are wavelength-dependent. They occur with 254 nm radiation, but not with 313 nm, and they seem to involve the S_2 (π, π^*) state rather than S_1 (n, π^*) and to occur *via* benzvalene-like rather than prismane-like intermediates, since only 1,2-transpositions have been observed.

As in the carbocyclic series, it must not be assumed that the reactions necessarily proceed along the paths suggested by the authors of the papers. In most cases the heteroprismane or benzvalene is invoked in order to provide a rationalization of the overall results, but it is not difficult to think of alternative explanations, and indeed there is clear evidence[82] that the pyridazine–pyrazine rearrangement (9.101) follows a route which involves neither benzvalene nor prismane systems.

9.8.3 Ring-scrambling Reactions in Five-membered Heterocyclics[83]

The photochemical transformations of five-membered ring aromatic hetero-cyclics is a rich and diverse field abounding with products and speculation but with as yet little by way of proof. However, many of the heterocycles have this in common, that irradiation tends to lead to products in which the ring carbon atoms have been scrambled, and here also, more than one mechanism seems to be operative.

For example, furans[84] on irradiation in the gas phase or in solution inter-change substituents between positions 2 and 3 (9.104).

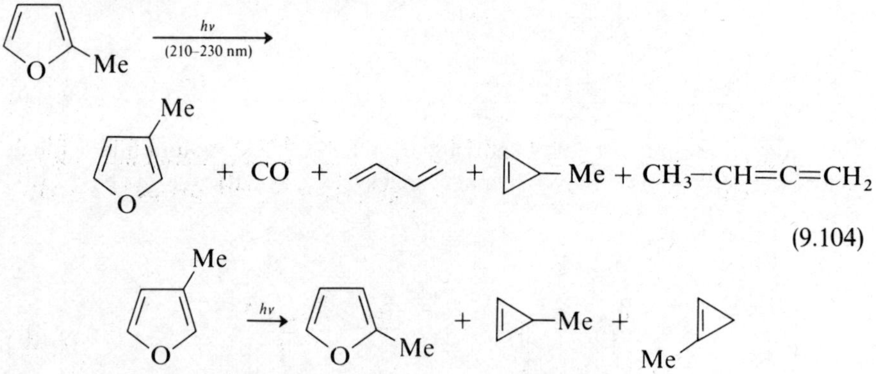

$$(9.104)$$

The currently accepted view is that the overall process probably depends on the reversible formation of acylcyclopropenes (9.105).

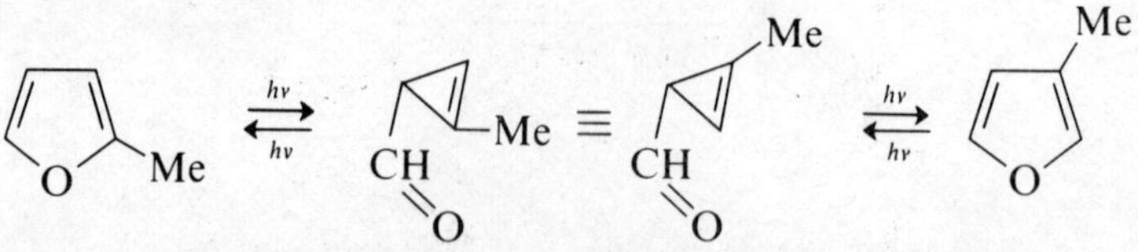

$$(9.105)$$

The rearrangement furan $\rightleftharpoons$ acylcyclopropene can be thought of as a (1,3)-sigmatropic shift or as a homolytic process (9.106), and the subsequent ring-expansion as another (1,3)-sigmatropic shift or as a cycloaddition of the carbonyl group to the σ-bond of the cyclopropene.

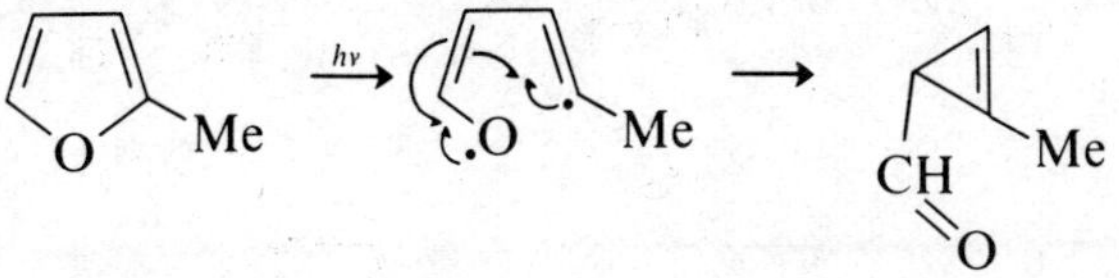

$$\text{(9.106)}$$

Whatever may be the truth about the mechanism of its formation from furan, the acylcyclopropene seems to be a real intermediate because in favourable circumstances it can be isolated,[85] for example from 2,5-di-t-butylfuran (9.107). On further irradiation the cyclopropene gives rise to the 2,5- and 2,4-di-t-butylfurans.

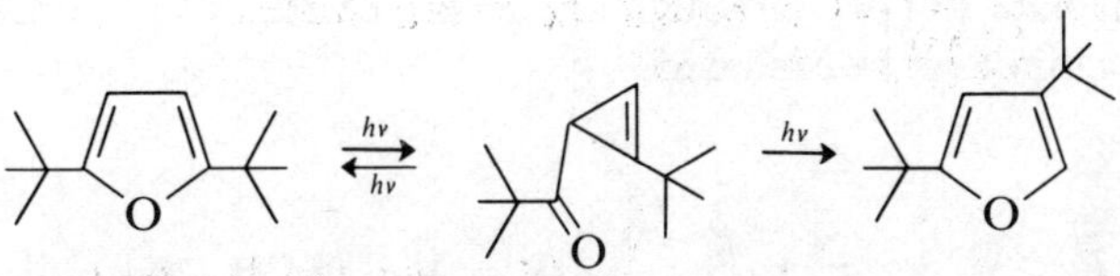

$$\text{(9.107)}$$

The intermediacy of an acylcyclopropene also provides an attractive explanation of the fact that furans irradiated in solutions of a primary amine afford N-alkylpyrroles (9.108).

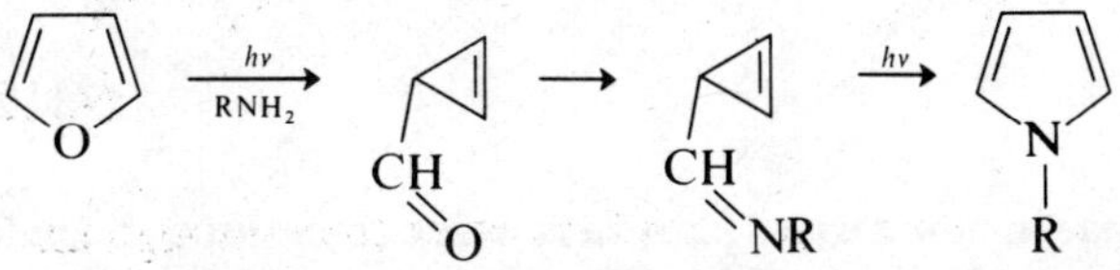

$$\text{(9.108)}$$

Finally, decarbonylation of the acylcyclopropene accounts for the formation of cyclopropenes during the irradiations (9.109, note the double bond shift).

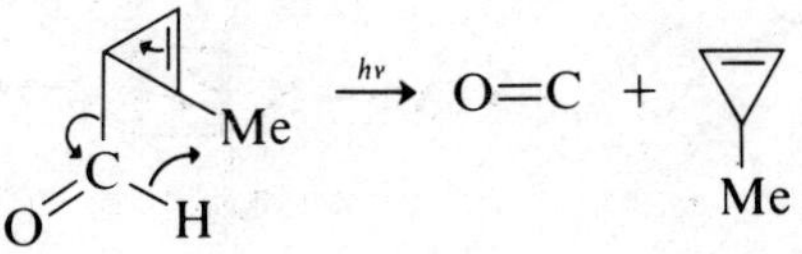

$$\text{(9.109)}$$

Phenomena similar to those just discussed for furans are found for other five-membered heterocycles such as pyrroles (9.110), isoxazoles (9.111) and pyrazoles (9.112).

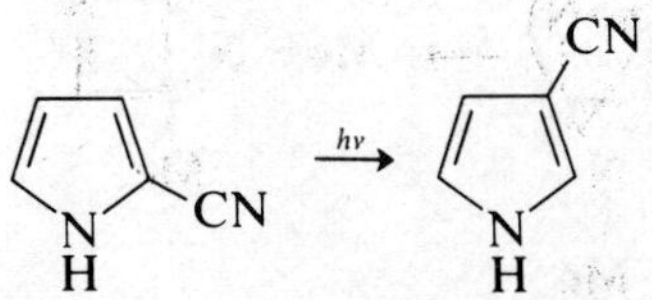

$$\text{(9.110)}$$

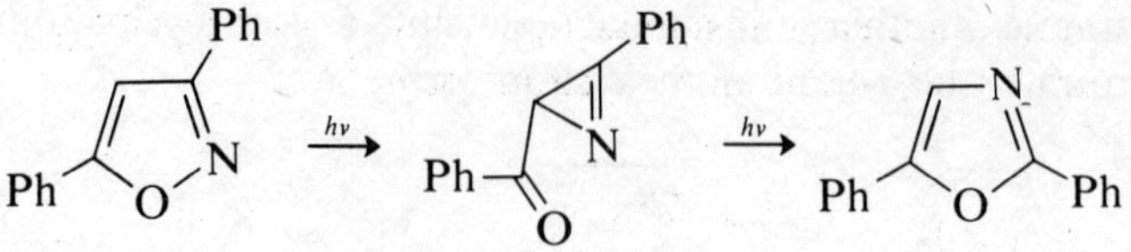

$$(9.111)$$

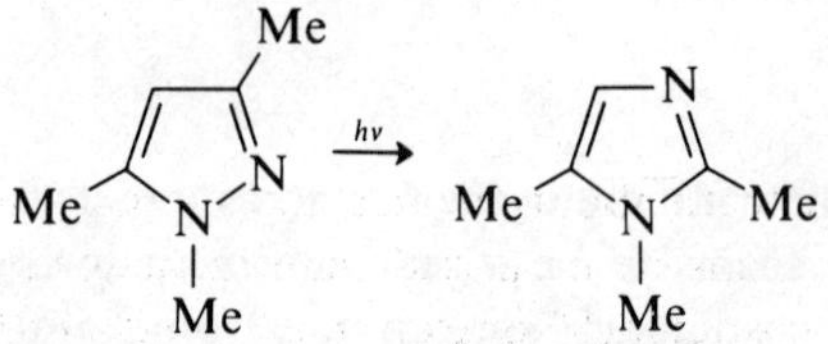

$$(9.112)$$

Ring-scrambling also occurs in thiophens (9.113),[86] and these give pyrroles when excited in the presence of amines, possibly *via* a cyclopropene–thioaldehyde intermediate (9.114), although the intermediate (9.115), analogous to a benzvalene, cannot yet be excluded.

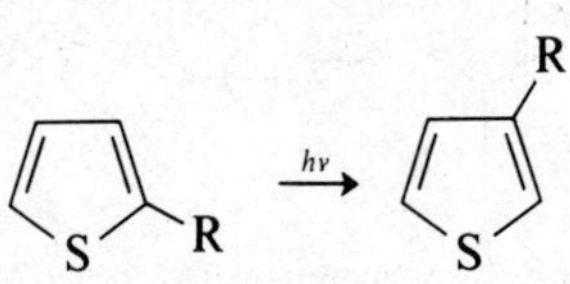

$$(R = Me, PhCH_2, Bu^t) \qquad (9.113)$$

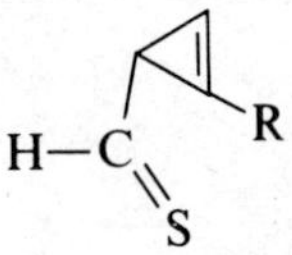

$$(9.114)$$

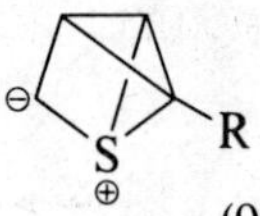

$$(9.115)$$

It is clear that in some cases pathways other than through analogues of the acylcyclopropene exist, as for example in the rearrangement[87] of the imidazole in (9.116), a process which can equally be thought of as a di-π-methane rearrangement (9.117, see Chapter 8, section 8.4.4).

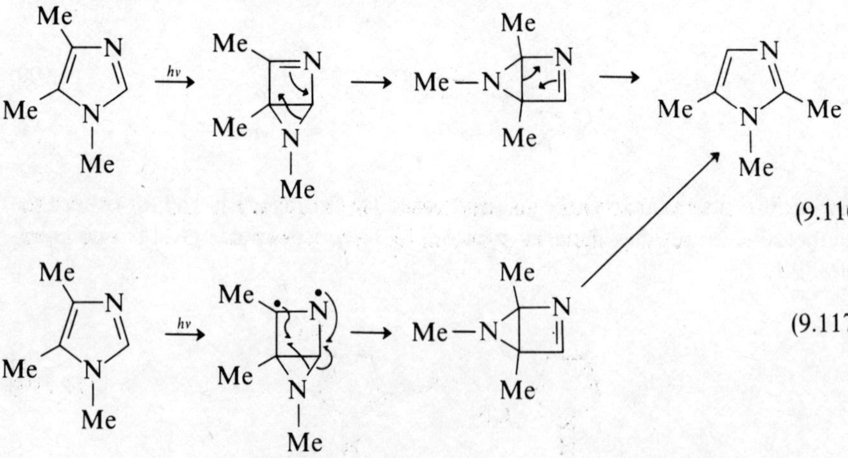

$$(9.116)$$

$$(9.117)$$

The same possibility exists for the ring-transposition of the oxazole in (9.118).

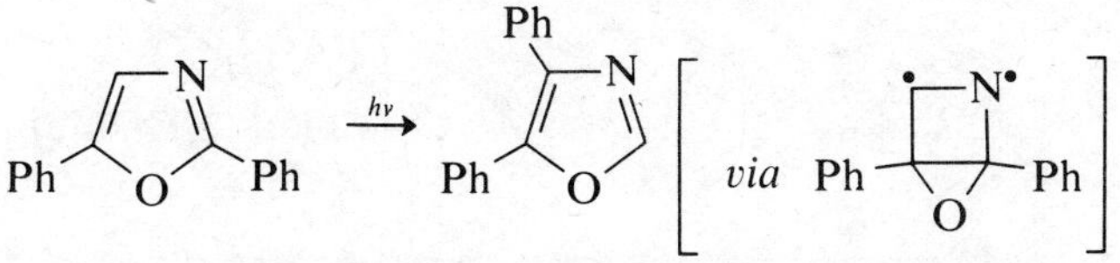

$$\text{(9.118)}$$

9.8.4 Photoadditions to Heteroaromatics

The cycloaddition reactions of heteroaromatics seem to be surprisingly sparse by comparison with those of their carbocyclic analogues. Few mechanistic studies have been performed, the theoretical position is unclear, and we can do little except report some of the results obtained.

It has been found that benzophenone will form oxetanes (see Chapter 7, section 7.7.1) with a number of five-membered heterocycles, for example with furan (9.119).

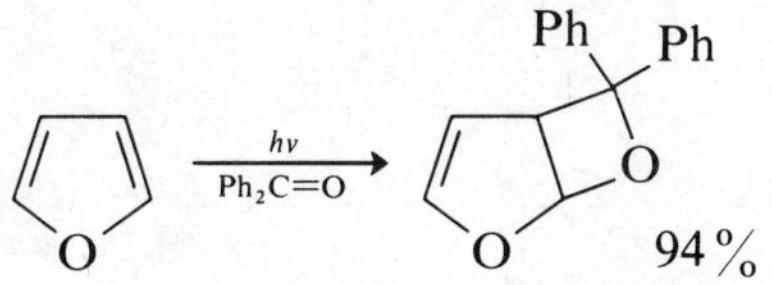

$$\text{(9.119)}$$

Acetylene dicarboxylic esters are versatile addends, giving[88] the 2,5-adducts with furan (9.120) and a 2,3-adduct with pyrrole (9.121).

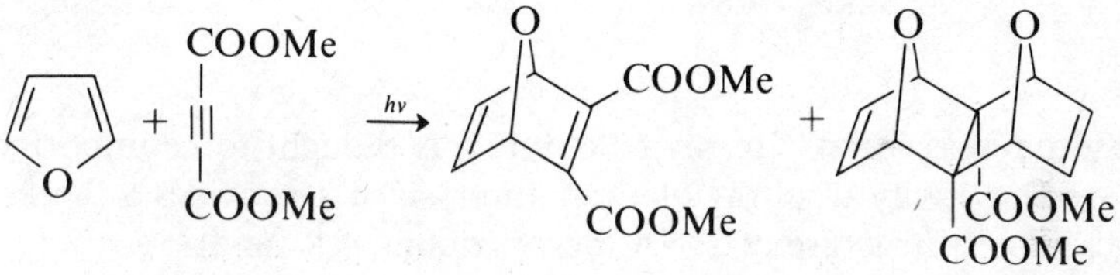

$$\text{(9.120)}$$

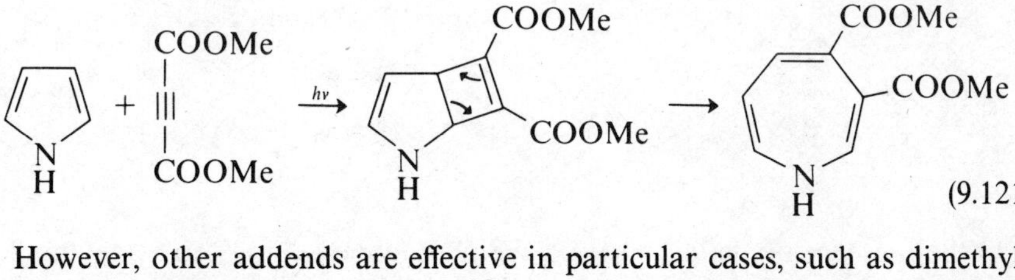

$$\text{(9.121)}$$

However, other addends are effective in particular cases, such as dimethylmaleic anhydride with furan (9.122),[89] and a variety of vinyl compounds with indoles (9.123).[90] In both of these examples the reaction is brought about by a triplet sensitizer.

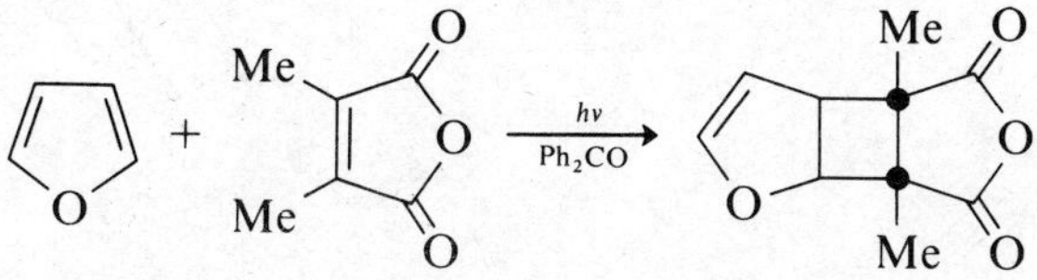

$$\text{(9.122)}$$

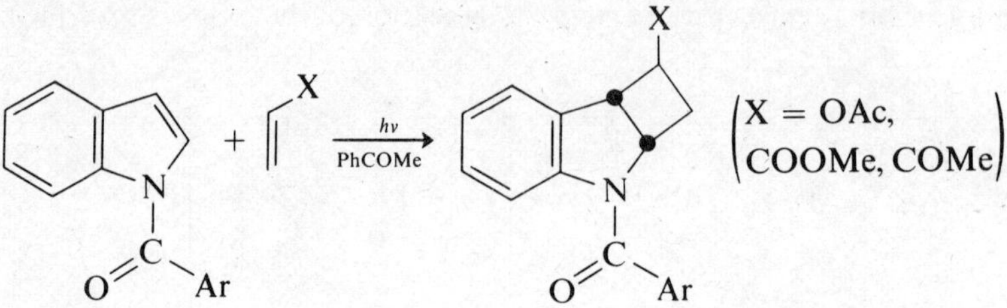

$$(9.123)$$

For the six-membered heterocyclics, cycloadditions seem to be largely restricted to their keto derivatives. These tend to dimerize if irradiated alone,[91] as shown for a pyrone (9.124) and a pyridone (9.125).

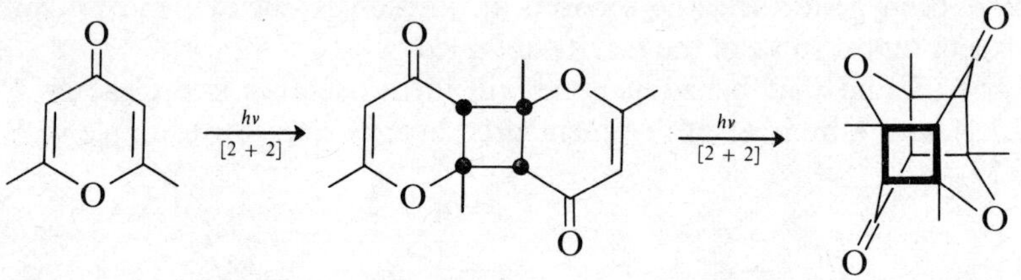

$$(9.124)$$

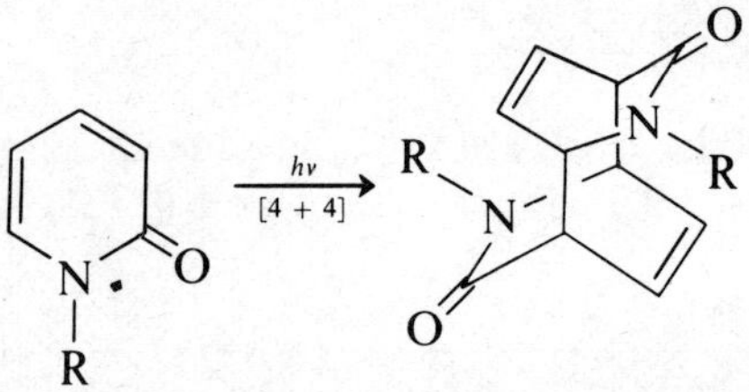

$$(9.125)$$

The photodimerization[92] of thymine (9.126) is thought to be important for the mutagenic activity of ultraviolet radiation, since it provides a mechanism for modifying the structure of DNA (desoxyribonucleic acid).

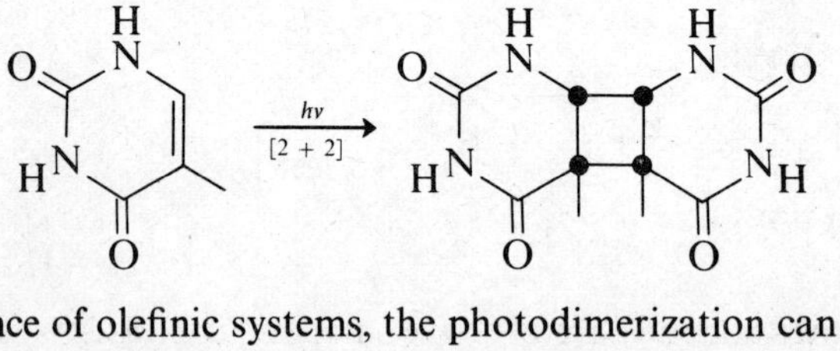

$$(9.126)$$

In the presence of olefinic systems, the photodimerization can sometimes be suppressed in favour of the formation of cycloaddition products, as for example with 2-quinolone (9.127),[93] coumarin (9.128),[94] or chromone (9.129).[95]

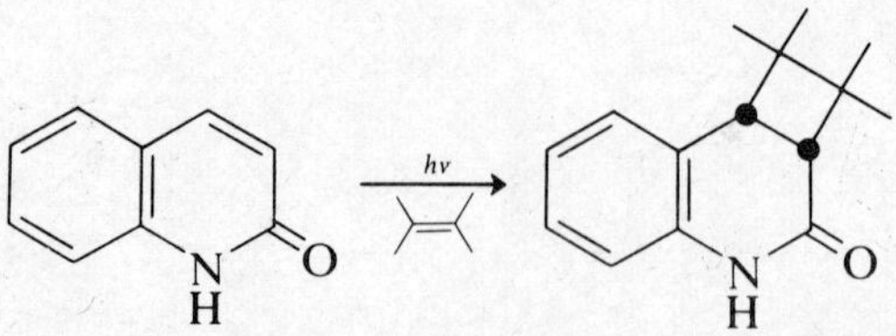

$$(9.127)$$

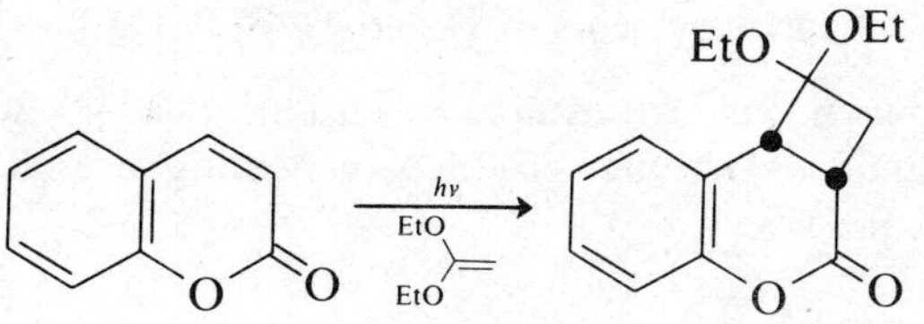

(9.128)

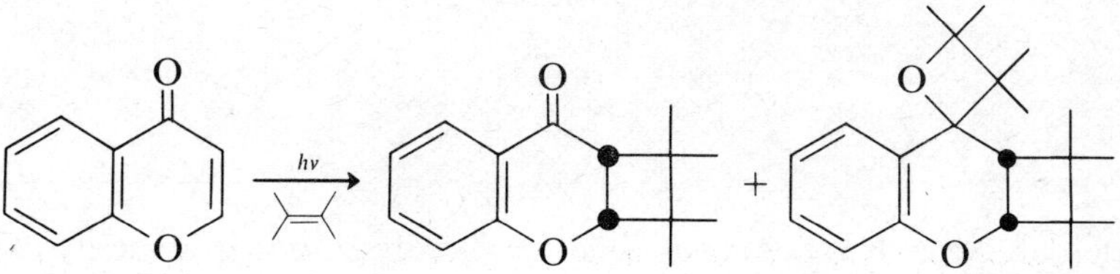

(9.129)

Several photoadditions of water or primary alcohols to six-membered heteroaromatics have been discovered, but in insufficient numbers to permit any useful generalizations to be made. For example, irradiation of pyridine in aqueous solution gives glutaconic dialdehyde derivatives, for the formation of which the scheme (9.130) has been suggested.[96]

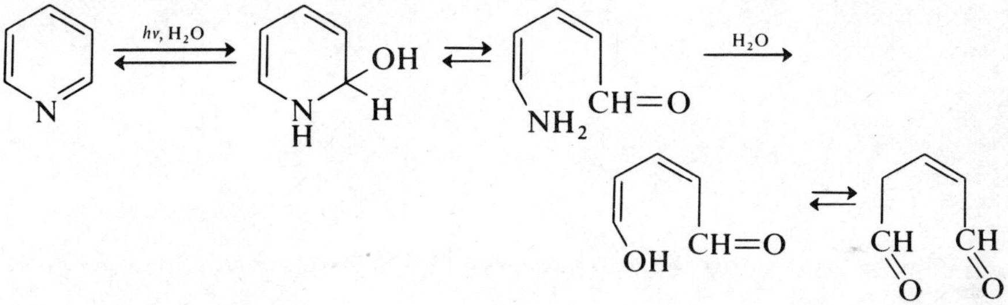

(9.130)

More recently[77] it has been shown that Dewar-pyridine is the critical intermediate (9.131).

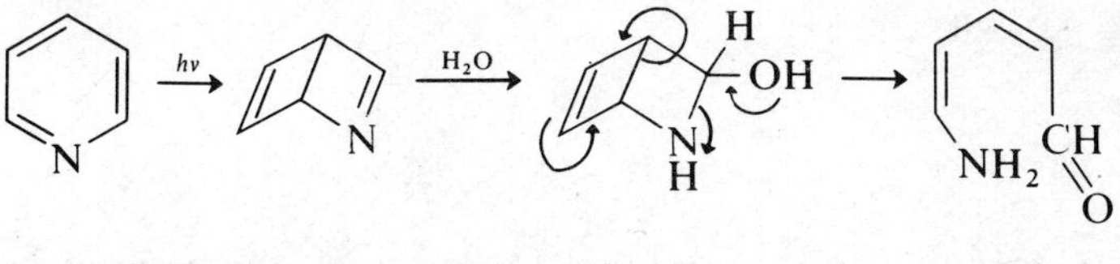

(9.131)

Photohydrations also occur with pyrimidines such as uracil derivatives (9.132).

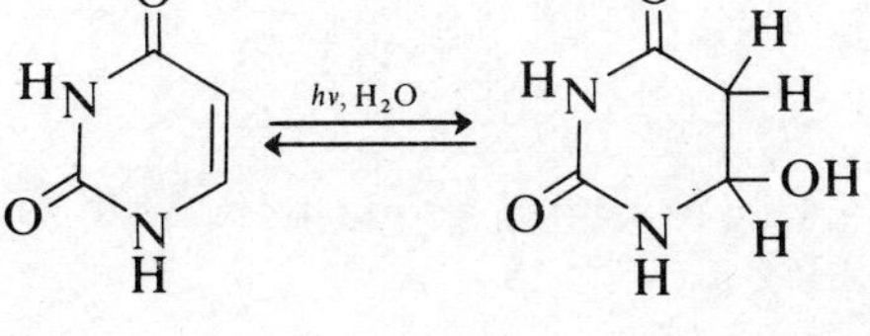

(9.132)

9.8.5   **Rearrangements of Heterocyclic *N*-Oxides**

Irradiation of nitrones gives unstable oxaziridines (9.133), which are readily converted into amides and other products on heating or on further irradiation (see Chapter 10, p. 317).

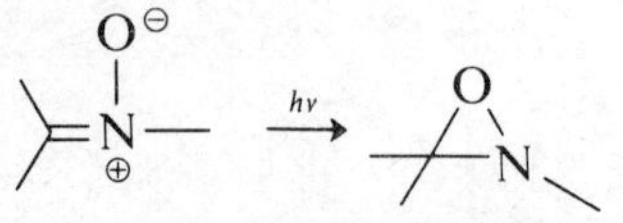

(9.133)

Heteroaromatic *N*-oxides also undergo photolysis to give a variety of products. The nature of these products is a sensitive function of the heterocycle, and for their formation an intermediate oxaziridine is usually postulated. The area is very complex, but the following examples may convey the flavour of the subject. Quinoline *N*-oxide gives 2-quinolone (9.134),[98] whereas 2-substituted

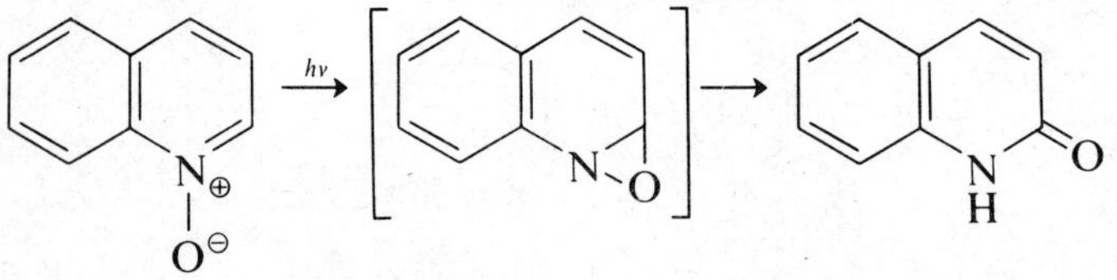

(9.134)

quinoline *N*-oxides give benzoxazepines (9.135).[99] Pyridine *N*-oxide itself gives

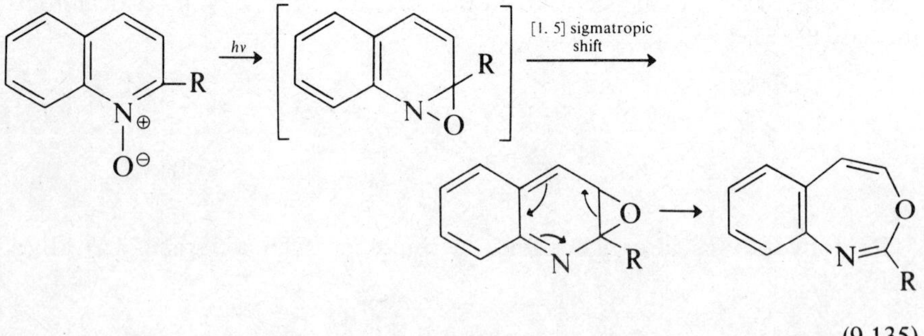

(9.135)

ring-contracted and ring-expanded products as well as 2-pyridone and 3-hydroxypyridine (9.136).

become bonded to the same molecule from which it was produced—that is, no 'cage' effect operates.

Intramolecular hydrogen abstraction in the alkoxy radical predominates[19] over α-cleavage in flexible systems except when the hydrogen atom to be abstracted is a primary C—H. In less flexible systems there may be geometrical constraints on the abstraction of hydrogen through a six-membered transition state, and α-cleavage may predominate as with the bicyclic nitrite in (10.15) above.

It is reported[16] that oxidation to the corresponding nitrate is the major process when 1-octyl nitrite is irradiated in the presence of oxygen (10.18), and the mechanism proposed involves conversion of nitric oxide to nitrogen dioxide in the oxidation step. However, the observation[20] that 1-pentyl nitrite in the presence of oxygen gives 4-nitratopentyl derivatives (10.19) suggests that intramolecular hydrogen abstraction in the alkoxy radical is faster than the oxidation and recombination steps.

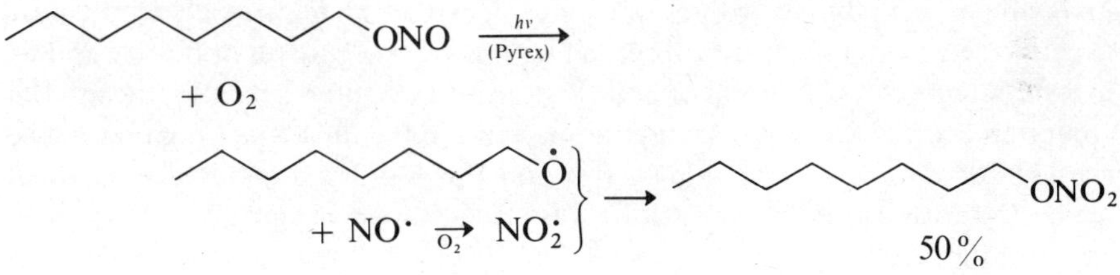

$$\text{CH}_3(\text{CH}_2)_6\text{CH}_2\text{—ONO} \xrightarrow[\text{(Pyrex)}]{hv}$$

$$+ \text{O}_2$$

$$\left. \begin{array}{c} \text{CH}_3(\text{CH}_2)_6\text{CH}_2\text{—}\overset{\displaystyle\cdot}{\text{O}} \\ + \text{NO}^{\cdot} \xrightarrow[\text{O}_2]{} \text{NO}_2^{\cdot} \end{array} \right\} \longrightarrow \quad \text{CH}_3(\text{CH}_2)_6\text{CH}_2\text{—ONO}_2$$

$$50\%$$

$$(10.18)$$

$$\text{CH}_3(\text{CH}_2)_3\text{CH}_2\text{—ONO} \xrightarrow[\text{(Pyrex)}]{hv} \quad \overset{\displaystyle\cdot}{\text{O}} \longrightarrow \quad \overset{\displaystyle\cdot}{}\quad\text{OH}$$

$$+ \text{O}_2$$

$$\xleftarrow{\text{NO}_2^{\cdot}}$$

ONO$_2$ ... OH + ONO$_2$ ... =O + ONO$_2$ ... ONO$_2$

$$13\% \qquad\qquad 6\% \qquad\qquad 10\%$$

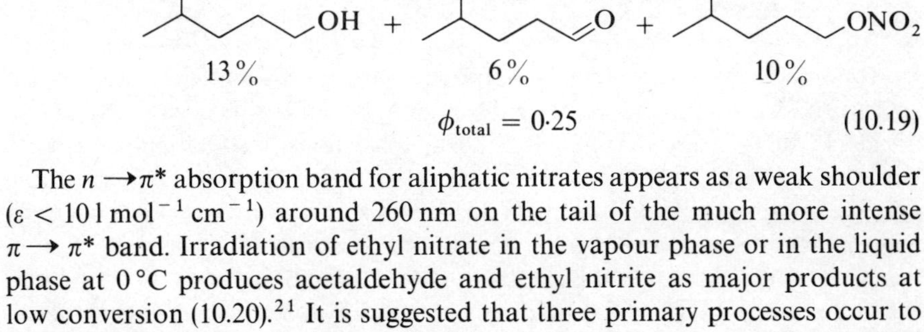

$$\phi_{\text{total}} = 0.25 \qquad\qquad (10.19)$$

The $n \rightarrow \pi^*$ absorption band for aliphatic nitrates appears as a weak shoulder ($\varepsilon < 10 \, l \, mol^{-1} \, cm^{-1}$) around 260 nm on the tail of the much more intense $\pi \rightarrow \pi^*$ band. Irradiation of ethyl nitrate in the vapour phase or in the liquid phase at 0 °C produces acetaldehyde and ethyl nitrite as major products at low conversion (10.20).[21] It is suggested that three primary processes occur to give ethoxy radical and nitrogen dioxide, ethyl nitrite and oxygen atom, and acetaldehyde and nitrous acid respectively. It is possible to account for the products on the basis of radical reactions alone, but the failure of radical scavengers to suppress the reaction completely is taken as evidence for the molecular decomposition pathways. Fluorescence of the ethoxy radical has been observed in this reaction, and this implies that at least some of the radical is generated in an electronically excited state.

$$CH_3CH_2ONO_2 \xrightarrow[\text{gas}]{hv\,(313\,nm)} \begin{cases} CH_3CH_2O^{\cdot} + NO_2^{\cdot} \\ CH_3CH_2ONO + O \qquad \phi_{CH_3CH_2ONO} = 0.14 \\ CH_3CH{=}O + HONO \qquad \phi_{CH_3CHO} = 0.094 \end{cases}$$

$$(10.20)$$

10.3   NITROSO COMPOUNDS

Monomeric nitrosoalkanes absorb weakly ($\varepsilon < 20\,l\,mol^{-1}\,cm^{-1}$) in the visible region, with the absorption maximum around 685 nm. This band accounts for the blue colour of the compounds, and it arises from a $n \longrightarrow \pi^*$ transition involving a lone pair electron on nitrogen. The $n \longrightarrow \pi^*$ nature of the transition is confirmed by the blue shift observed with increasing solvent polarity and by measurement of the direction of polarization (see Chapter 2, p. 24). The nitroso-compounds also have a weak absorption band in the ultraviolet region centred around 280 nm ($\varepsilon \sim 80\,l\,mol^{-1}\,cm^{-1}$), and this band is attributed to another $n \longrightarrow \pi^*$ transition, involving a lone pair electron on oxygen. Much stronger absorption around 220 nm ($\varepsilon \sim 5000\,l\,mol^{-1}\,cm^{-1}$) corresponds to a $\pi \longrightarrow \pi^*$ transition.[22]

Irradiation in the long wavelength band produces an excited state which undergoes efficient ($\phi \sim 1.0$) cleavage of the C—N bond. The major product of reaction for simple nitrosoalkanes[23] is a relatively stable nitroxide radical (10.21). In some instances the nitroxide radical combines with the nitric oxide, and trifluoronitrosomethane forms an N-nitritoamine quantitatively at low conversion (10.22).[24]

$$R{-}NO \xrightarrow{hv\,(visible)} NO^{\cdot} + R^{\cdot} \xrightarrow{RNO} R_2N{-}O^{\cdot} \qquad (10.21)$$

$$CF_3NO \xrightarrow{hv\,(540{-}780\,nm)} NO^{\cdot} + {\cdot}CF_3 \xrightarrow{CF_3NO} \xrightarrow{NO^{\cdot}} (CF_3)_2N{-}ONO \quad (10.22)$$

Monomeric aromatic nitroso-compounds are green, and their absorption spectra show a weak band in the visible region. For nitrosobenzene the maxima are at 745 nm ($\varepsilon \sim 50$), 325 nm ($\varepsilon \sim 5200$) and 280 nm ($\varepsilon \sim 10\,300\,l\,mol^{-1}\,cm^{-1}$). The weak absorption in the visible region is attributed to a $n \longrightarrow \pi^*$ transition, and the other two bands correspond to $\pi \longrightarrow \pi^*$ transitions. In the lowest energy excited state produced by irradiation with visible light no dissociation occurs, and this probably reflects the greater strength of the C—N bond and the lower energy of the excited state as compared with a nitrosoalkane. Irradiation of aromatic nitroso-compounds with ultraviolet radiation does cause dissociation, and as with aliphatic compounds the major product is a nitroxide radical.

Aromatic nitroso-compounds can be photoreduced in the presence of a good hydrogen donor such as propan-2-ol. This reaction occurs when ultraviolet radiation is employed, but not with visible light, and it seems to be a reaction of a higher excited state. For nitrosobenzene the reaction occurs through the monomer rather than the dimer,[25] and the initial product of reaction is phenylhydroxylamine (10.23), which undergoes further reaction with nitrosobenzene.

$$Ph-N{=}O \xrightarrow[\text{(CH}_3)_2\text{CHOH}]{h\nu\,(313\,\text{nm})} Ph-NH-OH \xrightarrow{\text{PhNO}} Ph-N\overset{+}{=}\underset{\underset{O^-}{|}}{N}-Ph$$

$$\phi_{-\text{PhNO}} = 0.047 \qquad (10.23)$$

In the photoreduction of nitrosobenzene there is a resemblance to the chemistry of (n, π^*) excited states of carbonyl compounds, although the quantum efficiency of reaction for the nitroso-compounds is lower.

10.4   NITRO COMPOUNDS

Nitroalkanes have an absorption spectrum[26] which shows a low intensity maximum ($\varepsilon \sim 20\,\text{l mol}^{-1}\,\text{cm}^{-1}$) centred around 275 nm, and much more intense absorption ($\varepsilon > 10\,000\,\text{l mol}^{-1}\,\text{cm}^{-1}$) around 210 nm. The longer wavelength band corresponds to an electronic transition from a non-bonding orbital on oxygen to an anti-bonding π^* orbital, and the position of the band is at shorter wavelength in a more polar solvent. For aromatic nitro-compounds the $n \to \pi^*$ absorption band is observed only as a shoulder because of the more intense aromatic $\pi \to \pi^*$ absorption in the same region of the spectrum.

The (n, π^*) excited state of nitroalkanes undergoes cleavage of the C—N bond to give an alkyl radical and nitrogen dioxide, and direct observation of these species by electron spin resonance spectroscopy has been achieved.[27] The major secondary processes involve recombination of the initially formed radicals to give the original nitro-compound or its nitrito isomer (10.24).[28] Subsequent photolysis of the nitrite may then occur.

$$CH_3NO_2 \xrightarrow[\text{gas}]{h\nu\,(313\,\text{nm})} \cdot CH_3 + NO_2^{\cdot} \to CH_3NO_2 \text{ or } CH_3ONO$$

$$\phi = 0.22 \qquad (10.24)$$

The major reactions of the excited states of aromatic nitro-compounds in the absence of hydrogen donors appear to involve cleavage of a N—O bond and oxygen transfer to another molecule. Thus the vapour phase photolysis of nitrobenzene[29] leads to the formation of nitrobenzene and p-nitrophenol (10.25) as major products. In solution in the presence of cyanide ion the excited

$$(10.25)$$

state of nitrobenzene is reduced to nitrosobenzene[30] whilst the cyanide is oxidized to cyanate (10.26). It is not necessary to postulate cleavage of the N—O bond in the excited state, since a reasonable mechanism can be based on C—N cleavage followed by photochemical cleavage of nitrogen dioxide to nitric oxide and atomic oxygen. The latter process is known to be efficient.

$$PhNO_2 + CN^- \text{ (aq.)} \xrightarrow{h\nu\,(254\ nm)} PhNO + OCN^- \qquad (10.26)$$

An alternative reaction of the (n, π^*) excited state is hydrogen abstraction from solvent or from an added hydrogen donor. In this the (n, π^*) excited state of a nitro-compound resembles the (n, π^*) excited state of a ketone (see Chapter 7, section 7.5.1). In aliphatic systems the initial product of photoreduction is a nitrone, and this can react further to give an amide by way of an oxaziridine (10.27).[31] The photochemical lability of nitrones and oxaziridines has been demonstrated independently[32] in systems such as (10.28), and the oxaziridine formation has been shown[33] to be stereospecific (10.29). An oxaziridine intermediate is also proposed[34] for the photochemical rearrangement of oximes to lactams (10.30).

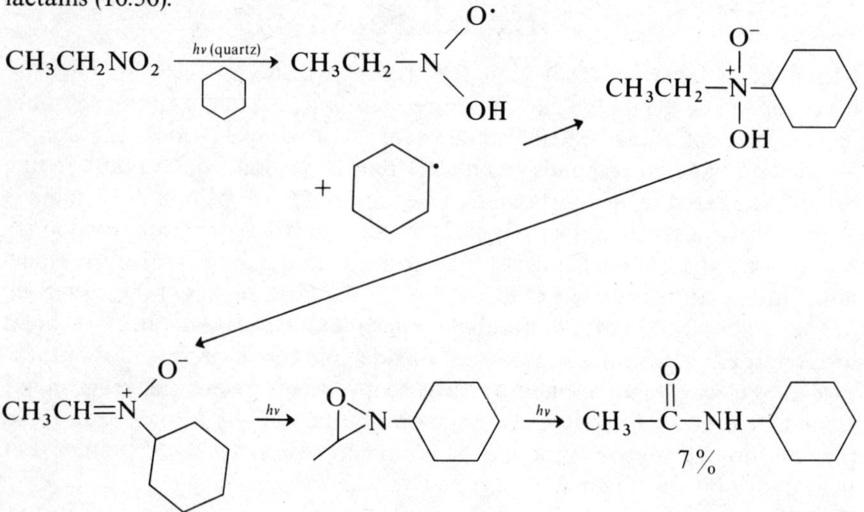

(10.27)

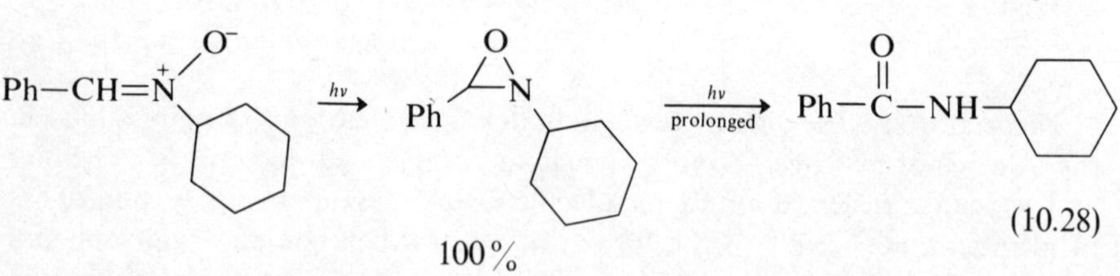

(10.28)

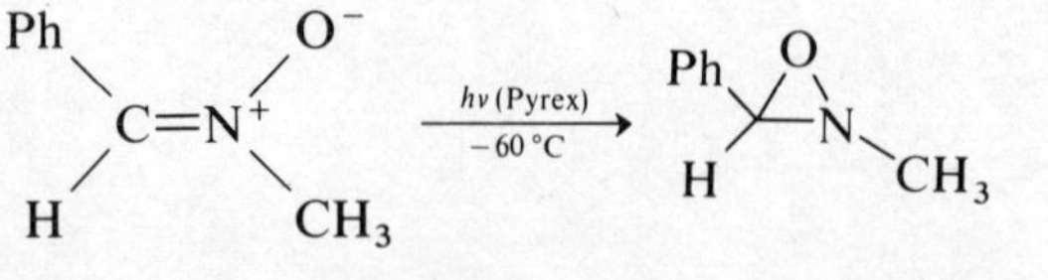

(10.29)

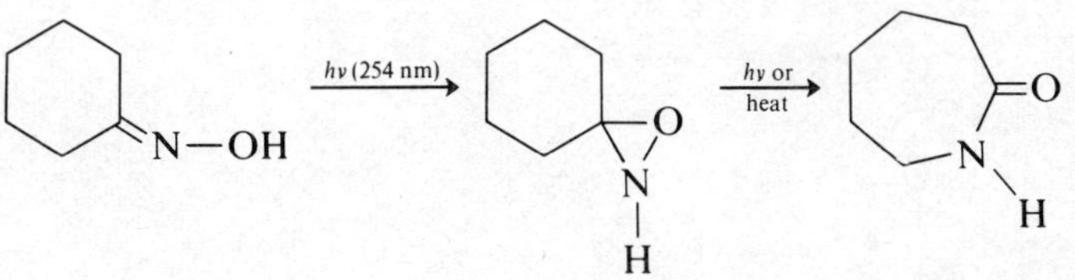

The (n, π^*) excited states of aromatic nitro-compounds also undergo hydrogen abstraction, and the (n, π^*) triplet state of nitrobenzene leads to phenylhydroxylamine as a major product (10.31), possibly by way of nitrosobenzene.[35] Under most conditions the phenylhydroxylamine undergoes extensive further (thermal) reaction. Substituted nitrobenzenes with electron-donating groups in the p-position also form arylhydroxylamines, but electron-withdrawing groups in the m- or p-positions cause the formation of substituted anilines as major products (10.32).[36]

$$Ph-NO_2 \xrightarrow[(CH_3)_2CHOH]{h\nu \, (366 \, nm)} Ph-NH-OH \qquad \phi_{-PhNO_2} = 0.011 \qquad (10.31)$$

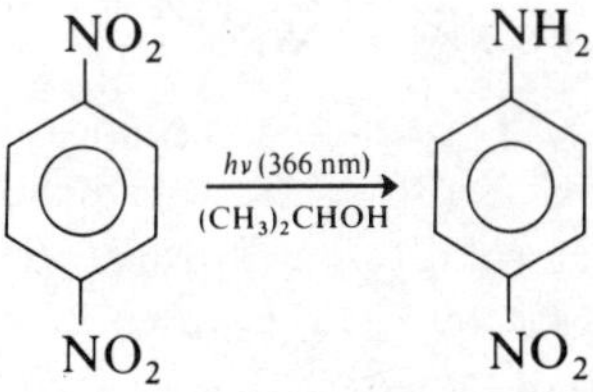

$$\phi = 0.16 \qquad (10.32)$$

As with aromatic carbonyl compounds (see Chapter 7, p. 191) not all nitro-aromatics have lowest excited states which are (n, π^*) in nature. Those which have lowest (π, π^*) excited triplet states, such as 2-nitronaphthalene, are much less reactive in hydrogen abstraction and the quantum yields of photoreduction are lower ($\phi = 0.037$ for 2-nitronaphthalene in propan-2-ol).

Intramolecular hydrogen abstraction is the first step in the efficient photo-chemical isomerization of o-nitrobenzaldehyde to o-nitrosobenzoic acid (10.33),[37] and reactions of this type are responsible for the photochromic behaviour of many o-nitrobenzene derivatives (10.34).[38]

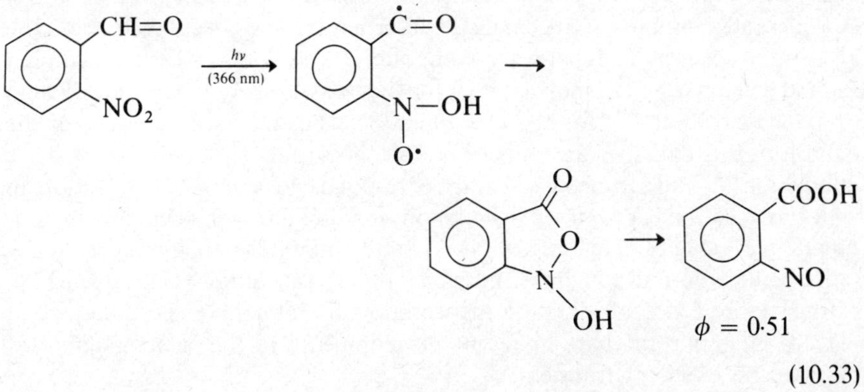

$$\phi = 0.51$$

$$(10.33)$$

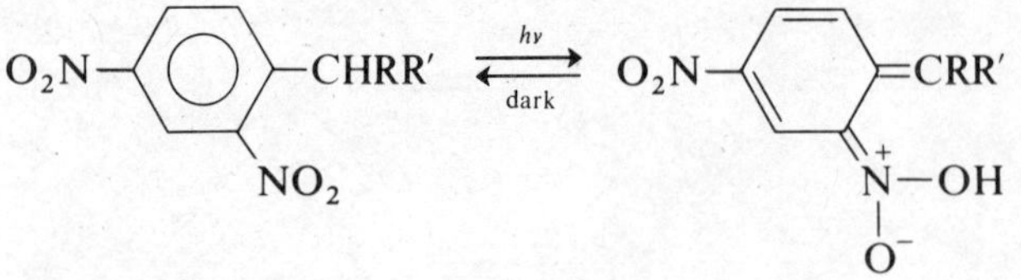

$$O_2N-\underset{NO_2}{\underset{|}{\bigcirc}}-CHRR' \underset{dark}{\overset{h\nu}{\rightleftarrows}} O_2N-\bigcirc=CRR'$$

(R, R' are electron-withdrawing groups)

with $\overset{+}{N}-OH$ and O^- substituent

(10.34)

Aromatic nitro-compounds undergo a non-stereospecific photocycloaddition reaction with alkenes[39] to give dioxazolidines (10.35), and this is a reaction of the (n, π^*) triplet state since benzophenone is an efficient sensitizer of the process and perfluoronaphthalene is an efficient quencher. The formation of the five-membered ring product contrasts with the $C=O$ systems, where oxetanes are produced (see Chapter 7, section 7.7.1).

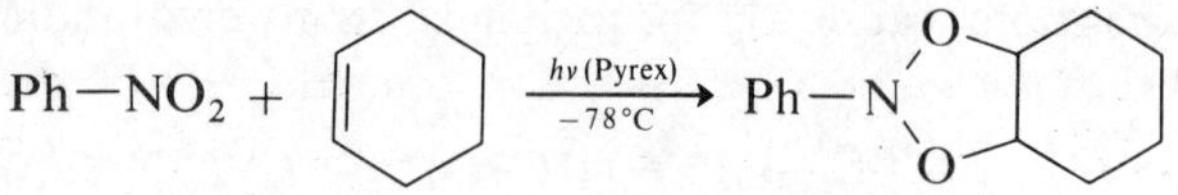

$$Ph-NO_2 + \bigcirc \xrightarrow[-78°C]{h\nu (Pyrex)} Ph-N \underset{O}{\overset{O}{<}}\bigcirc$$

(10.35)

10.5   AZO COMPOUNDS

Aliphatic azo-compounds such as 2,2'-azobis-2-methylpropionitrile have long been used as sources of free radicals (10.36), and the radical-forming reactions can be brought about by light. Normally the radicals are produced for the initiation of radical chain processes, but in the absence of other reagents a reasonable yield of a ketenimine is obtained.[40]

$$\underset{CN}{\overset{CN}{\underset{|}{(CH_3)_2C}}}-N=N-\underset{CN}{\overset{|}{C(CH_3)_2}} \xrightarrow{h\nu (366\,nm)} N_2 + 2(CH_3)_2\overset{\cdot}{C}-CN \rightarrow$$

$$\phi_{N_2} = 0.43$$

$$\underset{CN}{\overset{CN}{\underset{|}{(CH_3)_2C}}}-N=C=C(CH_3)_2$$

(10.36)

Many of the transformations brought about by the photolysis of acyclic and cyclic aliphatic azo-compounds have proved useful in synthesis, and a considerable amount of mechanistic information has been collected. The absorption spectra of aliphatic azo-compounds show a low intensity maximum around 350 nm ($\varepsilon \sim 15\,l\,mol^{-1}\,cm^{-1}$) for the *trans*-isomer, and around 380 nm ($\varepsilon \sim 140\,l\,mol^{-1}\,cm^{-1}$) for the *cis*-isomer. The position and intensity of the band for *cis*-azoalkanes are rather variable, according to the degree of steric interaction.[41] This absorption band corresponds to a $n \rightarrow \pi^*$ transition in which an electron is promoted from a non-bonding nitrogen orbital to an anti-bonding π^* orbital. The singlet (n, π^*) state can return to ground state by radiative emission, although the emission is often weak. Studies of the quenching of this fluorescence of aliphatic azo-compounds by conjugated dienes have played an important part in recent developments in the understanding of mechanisms of energy transfer.

In solution direct irradiation or triplet sensitization of aliphatic azo-compounds leads to isomerization about the N=N bond as a major reaction.[42] This behaviour parallels that of alkenes and is expected for a (π, π^*) excited state, in which the lowest energy conformation has 'orthogonal' geometry (see Chapter 8, section 8.1.1) and is capable of relaxing to either geometrical isomer of the ground state. The reactive excited state for the azo-compounds might be a low energy (π, π^*) state, but there is no spectroscopic evidence for such a state, and the planar (n, π^*) triplet state is expected to be lower in energy than the (π, π^*) state. Inversion of the nitrogen atom in the excited state is a possible alternative route for isomerization.

There is strong evidence that the triplet state of simple acyclic azoalkanes does not undergo dissociation in solution, and that the major chemical reaction is *trans–cis* isomerization about the N=N bond. In many cases the *cis*-azoalkane formed undergoes thermal decomposition[43] to nitrogen and radicals at room temperature (10.37), but at low temperature no nitrogen is evolved and the *cis*-isomer can be characterized. Some *cis*-azoalkanes, such as *cis*-azo-2-methylpropane, are thermally stable at room temperature.

$$\underset{\substack{\\ \text{CH}_3}}{\overset{\text{CH}_3}{\text{N=N}}} \quad \underset{\phi = 0\cdot09}{\overset{h\nu\,(366\,\text{nm})}{\rightleftharpoons}} \quad \overset{\text{CH}_3 \quad \text{CH}_3}{\text{N=N}} \quad \rightarrow \quad 2\,^{\bullet}\text{CH}_3 + \text{N}_2 \qquad (10.37)$$

Bond cleavage accompanies *trans–cis* isomerization on direct irradiation or singlet sensitization of azo-alkanes in the vapour phase,[44] and at low pressures the quantum yield for production of nitrogen approaches unity in the photolysis $(\lambda = 366\,\text{nm})$ of azomethane. At higher pressures the quantum yield for nitrogen production is lower (often $<0\cdot1$), either because of recombination of the initially formed radicals, or because decomposition occurs through upper vibrational levels of the *cis*-isomer and vibrational relaxation is faster at higher pressures.

The extrusion of nitrogen involves a two-step process for alkyl aryl azo-compounds. Evidence for this is that racemization accompanies dissociation on photolysis of an optically active compound (10.38),[45] and that allylic isomerization is observed (10.39).[46] Both of these results are most readily explained on the basis of C—N fission followed by radical recombination.

$$(10.38)$$

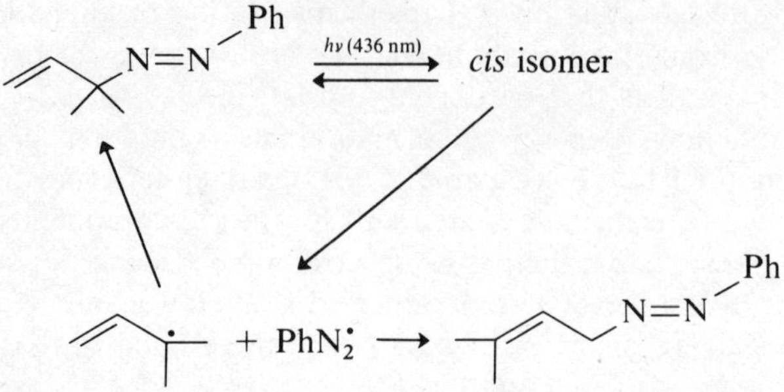

$$\text{(structure) } \quad \xrightarrow[\longleftarrow]{h\nu \ (436 \text{ nm})} \quad \textit{cis} \text{ isomer}$$

(10.39)

$$\text{(structure) } + \text{ PhN}_2^{\cdot} \longrightarrow \text{(structure)}$$

Cyclic aliphatic azo-compounds undergo loss of nitrogen on direct irradiation or triplet sensitization. The dissociation of the triplet state is a result of the inability of the state to dissipate energy by relaxation to a twisted state and thence to the ground state. The reaction is a very useful way of producing strained ring systems, as illustrated in the routes to bicyclo[2.1.0]pentane (10.40)[47] and to a bicyclo[1.1.0]butane (10.41).[48]

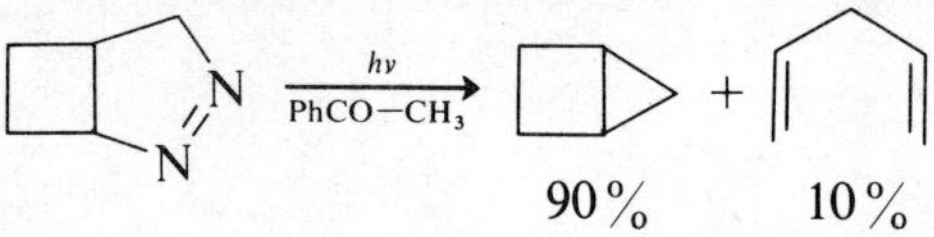

$$\text{(structure)} \xrightarrow[\text{PhCO-CH}_3]{h\nu} \text{(structure)} + \text{(structure)}$$
$$\qquad\qquad\qquad\qquad 90\% \qquad\quad 10\%$$

(10.40)

$$\text{(structure)} \xrightarrow{h\nu} \text{(structure)}$$
$$\text{Ph} \quad \text{Ph} \qquad\qquad 51\%$$

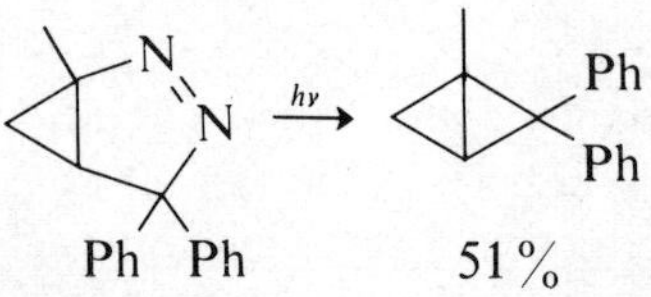

(10.41)

The products of these photochemical reactions contain a much higher proportion of bicyclic compound than do the corresponding thermal reactions, and the best photochemical yields are obtained in solution. This is because the excited state leads to a vibrationally excited state of the bicyclic hydrocarbon, and in solution this rapidly dissipates excess vibrational energy to the environment. In the gas phase other products are formed at the expense of the bicyclic product because collisional deactivation of the 'hot' bicyclic hydrocarbon is slow so that bond cleavage intervenes.

Biradicals are likely intermediates in the photochemical reactions, although the lifetime of those produced on direct irradiation must be short because cyclopropanes are formed stereospecifically[49] from the singlet excited state of pyrazolines (10.42) and similar systems. This is in contrast to the thermal reaction in which there is a considerably lower degree of stereospecificity as well as a lower chemical yield of cyclopropane.

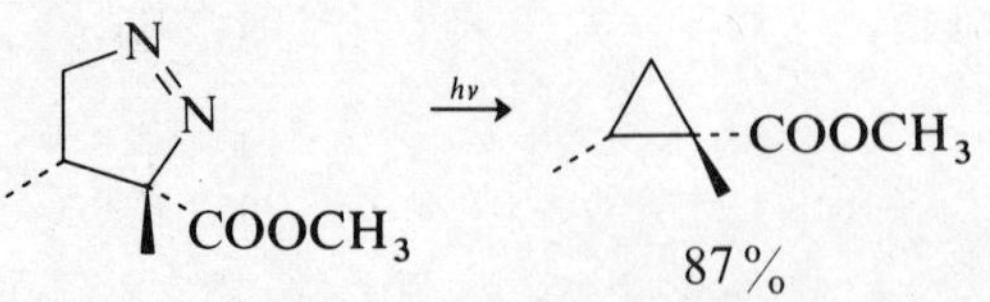

$$\text{(structure) } \text{COOCH}_3 \xrightarrow{h\nu} \text{(structure) } \text{COOCH}_3$$
$$\qquad\qquad\qquad\qquad\qquad\qquad 87\%$$

(10.42)

Triplet-sensitized (benzophenone or Hg $^3P_1$) decomposition of either *cis*- or *trans*-3,4-dimethyl-1-pyrazoline gives the same ratio of cyclopropane products (10.43), and this is indicative of a common intermediate from both isomers, which is probably a long-lived biradical.[50]

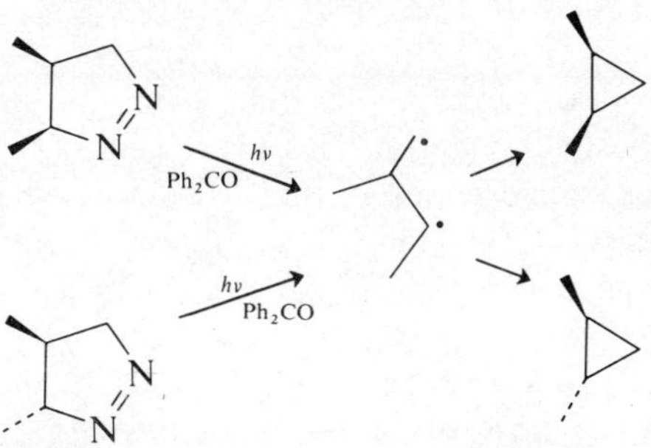

$$98\% \qquad (10.43)$$

Similar results are obtained in six-membered ring systems, and direct irradiation of *meso*- or ($\pm$)-3,6-diethyl-3,6-dimethyl-3,4,5,6-tetrahydropyridazine (10.44) gives rise to cyclobutane products with 96% retention of stereochemistry, whereas the triplet-sensitized reaction is much less stereospecific and gives more alkene.[51]

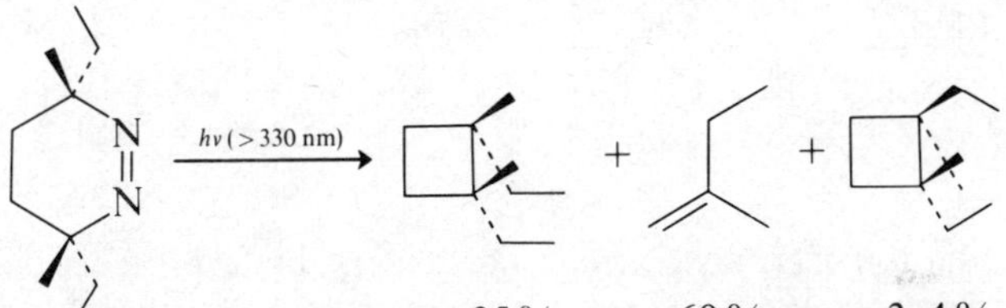

$$35\% \qquad 60\% \qquad 3\text{-}4\% \qquad (10.44)$$

Diazirines, which are cyclic isomers of diazoalkanes (see next section), absorb only in the ultraviolet region, and the weak band ($\varepsilon = 180 \, l \, mol^{-1} \, cm^{-1}$) centred around 310 nm is ascribed to a $n \rightarrow \pi^*$ transition. The (n, π^*) excited state, like that of a diazoalkane, readily loses nitrogen to form a carbene (10.45),[52] and diazirines have the advantages over diazoalkanes as carbene precursors that they have greater thermal stability and greater stability to visible light.

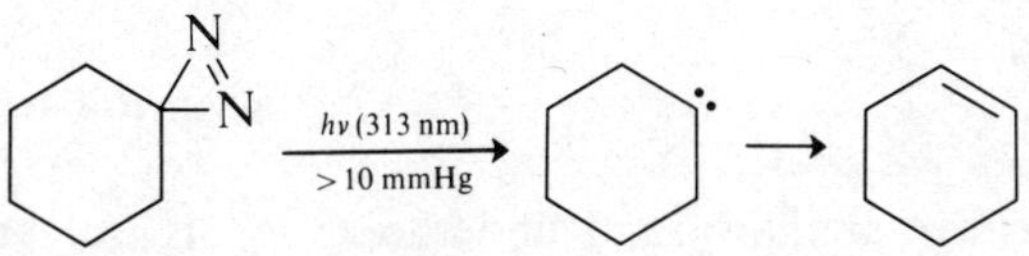

$$97\% \qquad (10.45)$$

The photolysis of 4-methylene-1-pyrazolines provides a route to the theoretically important species of which trimethylenemethyl (10.46) is the parent member.[53]

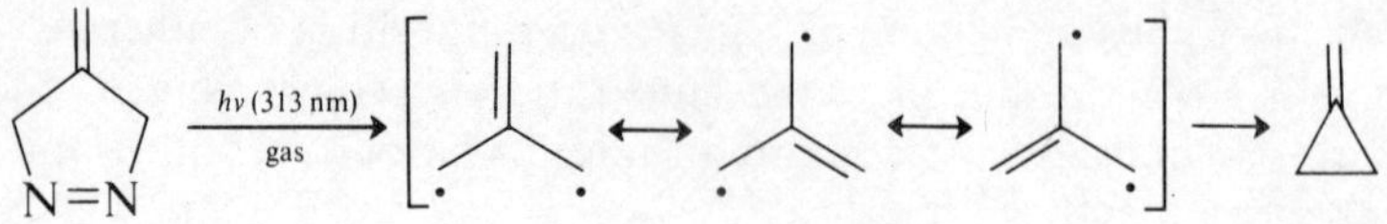

$$\phi = 0.37$$

$$(10.46)$$

The absorption spectra of azoaromatics differ from those of their aliphatic counterparts as a result of conjugation. There is a band of moderate intensity in the visible region (for azobenzene λ_{max} is at 443 nm ($\varepsilon = 500 \, l \, mol^{-1} \, cm^{-1}$) for *trans*, and at 433 nm ($\varepsilon = 1500 \, l \, mol^{-1} \, cm^{-1}$) for *cis*), and a strong band in the near ultraviolet region (for azobenzene λ_{max} is at 320 nm ($\varepsilon = 20\,000 \, l \, mol^{-1} \, cm^{-1}$) for *trans*, and at 281 nm ($\varepsilon = 5200 \, l \, mol^{-1} \, cm^{-1}$) for *cis*). The lower energy band corresponds to a $n \rightarrow \pi^*$ transition, and the higher energy band to a $\pi \rightarrow \pi^*$. The energy of the lowest excited state is too low to effect cleavage of the relatively strong C—N bond in these compounds, and *trans–cis* isomerization (10.47) is therefore the predominant process in both the direct and sensitized photolyses,[54] although some dissociation does occur when shorter wavelength radiation is employed. The isomerization reaction resembles the corresponding reaction of stilbene (see Chapter 8, p. 232), although the situation is more complex because of the existence of (n, π^*) as well as (π, π^*) states.

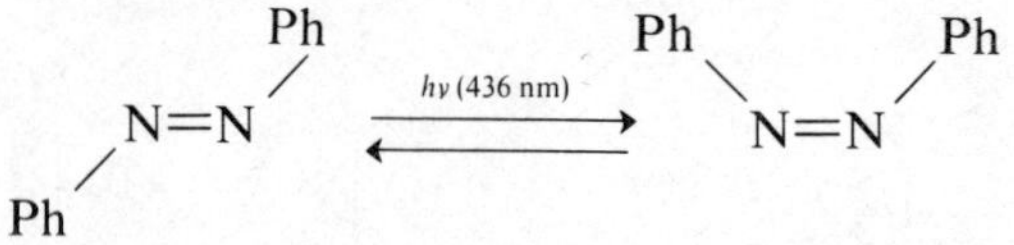

$$\phi_{cis \rightarrow trans} = 0.48$$
$$\phi_{trans \rightarrow cis} = 0.24 \qquad (10.47)$$

Direct irradiation of *cis*-azobenzene, particularly in the presence of a proton or Lewis acid, leads inefficiently to a dihydrodiazaphenanthrene (10.48) via a singlet excited state in a concerted cyclization analogous to the formation of 9,10-dihydrophenanthrene from stilbene (see Chapter 8, p. 246). This 'side reaction' may account for the fact that the sum of the quantum yields $\phi_{cis \rightarrow tr}$ and $\phi_{tr \rightarrow cis}$ for azobenzene is not equal to unity.

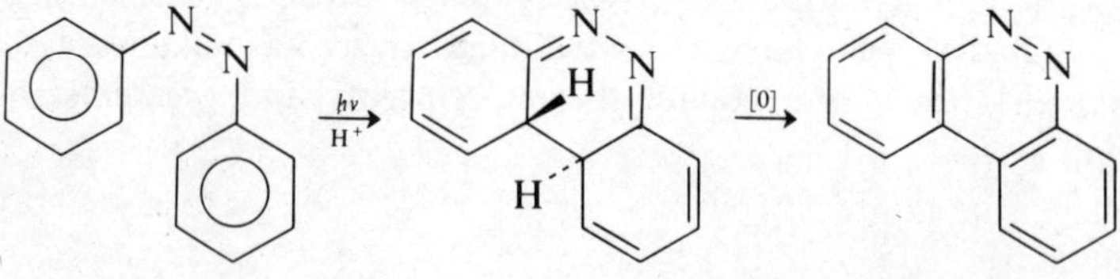

$$\phi \text{ up to } 0.015 \qquad (10.48)$$

Azoxyaromatics behave similarly and undergo *trans–cis* isomerization on irradiation, and in addition transfer of oxygen to the other nitrogen occurs,[55] presumably through an oxadiaziridine intermediate (10.49). The formation of this product from the azoxy-compound is analogous to the formation of an oxaziridine from a nitrone (see Chapter 10, p. 317).

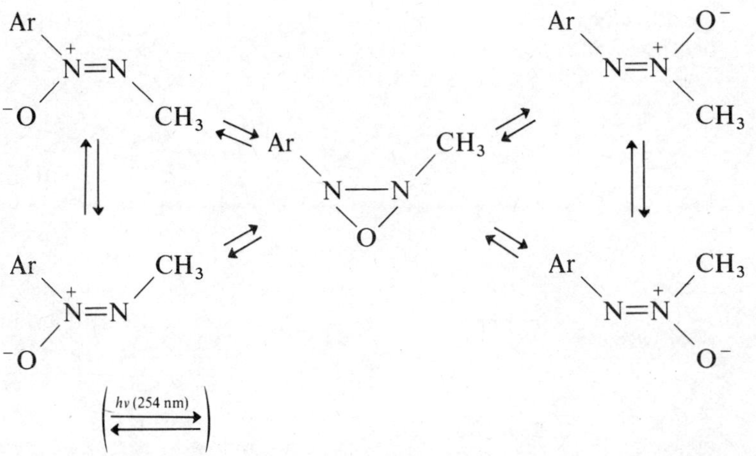

(10.49)

Azobenzene can be photoreduced to hydrazobenzene in the presence of a hydrogen donor such as propan-2-ol (10.50), and this seems to be a reaction which involves direct hydrogen abstraction by an excited state of the azo-compound.[54] The hydrazobenzene can undergo further reaction which involves cleavage of the N—N bond, and overall reductive cleavage of this type is responsible for the photochemical bleaching of azo-dyes.

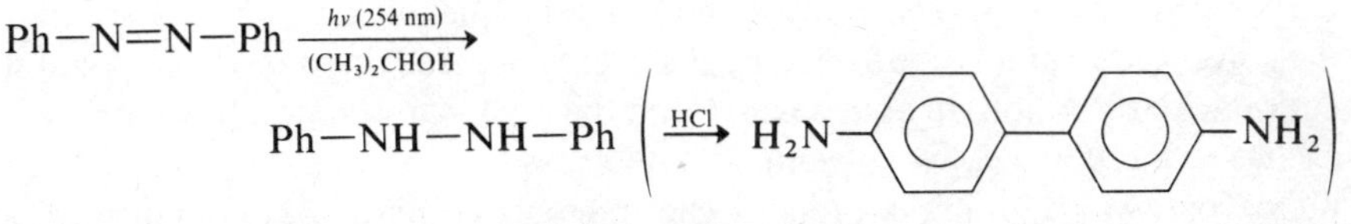

$$Ph-N=N-Ph \xrightarrow[(CH_3)_2CHOH]{hv\,(254\,nm)}$$

$$Ph-NH-NH-Ph \left(\xrightarrow{HCl} H_2N-\bigcirc-\bigcirc-NH_2 \right)$$

72%   (10.50)

10.6   DIAZO AND RELATED COMPOUNDS

Diazoalkanes $[R_2C\overset{+}{=}N\overset{-}{=}N \leftrightarrow R_2\overset{-}{C}-\overset{+}{N}\equiv N \leftrightarrow R_2\overset{+}{C}-N\overset{-}{=}N]$ exhibits a weak absorption band in the visible region (400–500 nm) which is attributed to a $n \rightarrow \pi^*$ transition. The corresponding excited states fragment readily to give molecular nitrogen and a carbene, and this is a widely used method for the generation of these divalent carbon species. Direct irradiation leads initially to a singlet excited state of the diazo-compound and thence to a singlet carbene, although collisional deactivation to give triplet carbene before further reaction occurs may be important for those carbenes whose triplet state is lower in energy than the singlet state, especially at higher pressures in the vapour phase or in solution. Triplet sensitized decomposition of diazoalkanes gives rise directly to triplet carbene by way of the triplet excited state. These processes provide a valuable route to carbenes (10.51),[56] and are employed for intramolecular reaction as in the ring-contraction of cyclic diazoketones (10.52)[57] and in β-lactam formation from α-diazoamides (10.53).[58]

$$Ph_2CN_2 \xrightarrow{h\nu\,(288\,nm)} Ph_2\ddot{C} + N_2 \qquad \phi = 0{\cdot}78 \qquad (10.51)$$

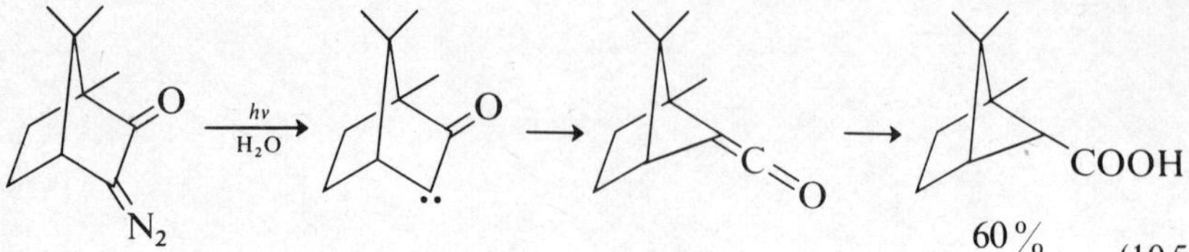

$$60\% \qquad (10.52)$$

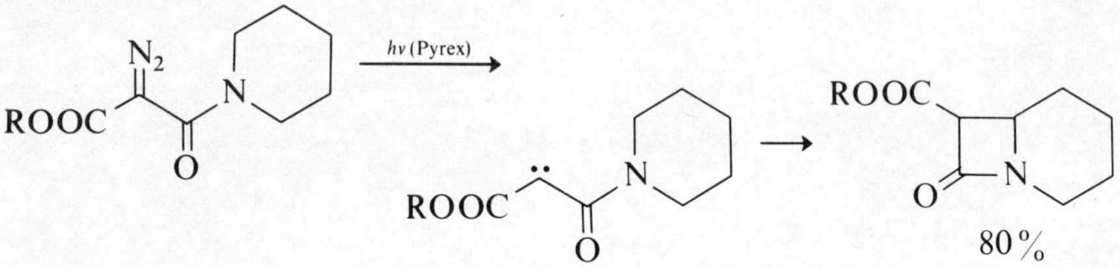

$$80\% \qquad (10.53)$$

10.6.1   Diazonium Salts

Aromatic diazonium salts on photolysis in solution undergo efficient loss of nitrogen in two primary processes with a combined quantum efficiency approaching unity. One of the processes leads to a carbonium ion, and the other to a carbon radical (10.54). The relative importance of the two pathways varies with the solvent used and is controlled by the degree of association of the diazonium cation with the anion. In aqueous solution the ionic pathway is followed almost exclusively, but in alcohol solvents where ion association is more important the radical pathway is the major process. The difference is seen in the photochemistry of *p*-nitrobenzenediazonium chloride (10.55). In aqueous hydrochloric acid the products are largely *p*-nitrophenol and *p*-nitrochlorobenzene formed in an ionic process, but in ethanol as solvent the major product is nitrobenzene formed by a radical mechanism.[59]

$$ArN_2^+ \; X^- \xrightarrow[0\,°C]{h\nu\,(366\,nm)} \begin{cases} Ar^+ + N_2 + X^- \\ Ar^{\cdot} + N_2 + X^{\cdot} \end{cases} \qquad (10.54)$$

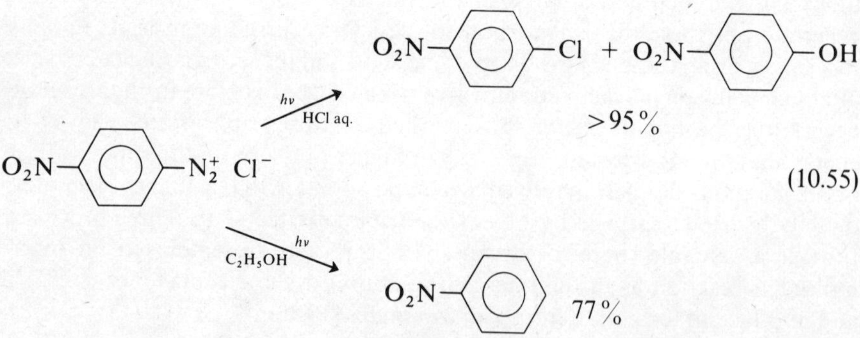

$$(10.55)$$

10.6.2   Azides

Azides are similar to diazoalkanes in that elimination of molecular nitrogen occurs in the excited state, and from the azide a nitrene is formed. The quantum yield for formation of the nitrene on direct irradiation is 0·7–1·0 for aliphatic azides in the gas phase, and 0·7–0·9 for aliphatic or 0·4–1·0 for aromatic azides in solution. Aliphatic nitrenes normally react[60] by a hydrogen shift to give an imine (10.56), or by an alkyl shift if there is no available hydrogen atom. Intermolecular hydrogen abstraction from solvent can also occur to give an amine (10.57). In the presence of added alkene a singlet state nitrene undergoes efficient cycloaddition to give an aziridine (10.58).

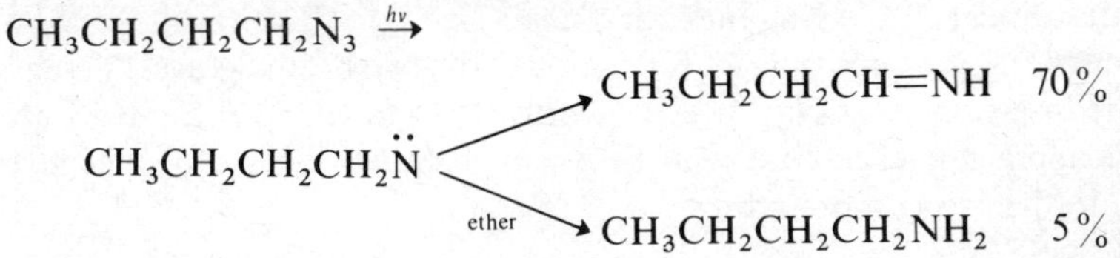

$$CH_3CH_2CH_2CH_2N_3 \xrightarrow{h\nu}$$

$$CH_3CH_2CH_2CH_2\ddot{N} \nearrow CH_3CH_2CH_2CH=NH \quad 70\% \quad (10.56)$$

$$\xrightarrow[\text{ether}]{} CH_3CH_2CH_2CH_2NH_2 \quad 5\% \quad (10.57)$$

$$+ N_3COOC_2H_5 \xrightarrow[\text{(254 nm)}]{h\nu} \quad N-COOC_2H_5 \quad (10.58)$$

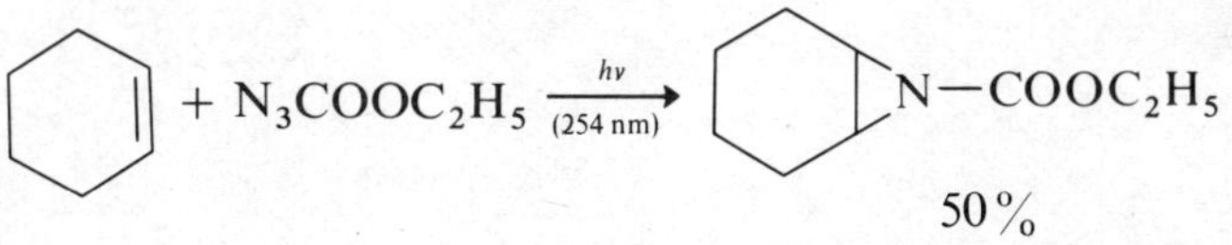

50%

Singlet state vinyl or aryl nitrenes undergo intramolecular cycloaddition to form a 1-azirine (10.59).[60] In the aromatic systems this can lead to ring-expansion[61] with eventual formation of a 6H-azepine when a nucleophile is present (10.60). In the absence of nucleophile an azo-compound may be formed (10.61).

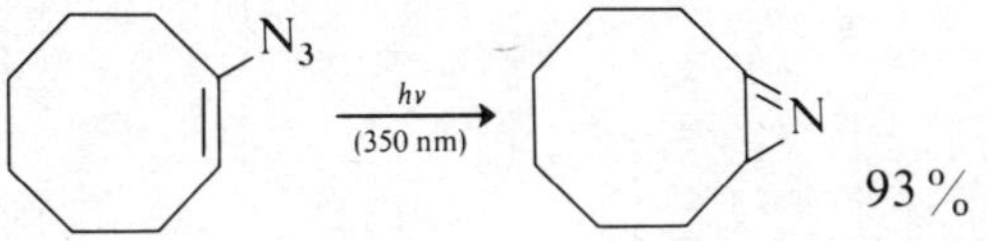

$$\xrightarrow[\text{(350 nm)}]{h\nu} \quad 93\% \quad (10.59)$$

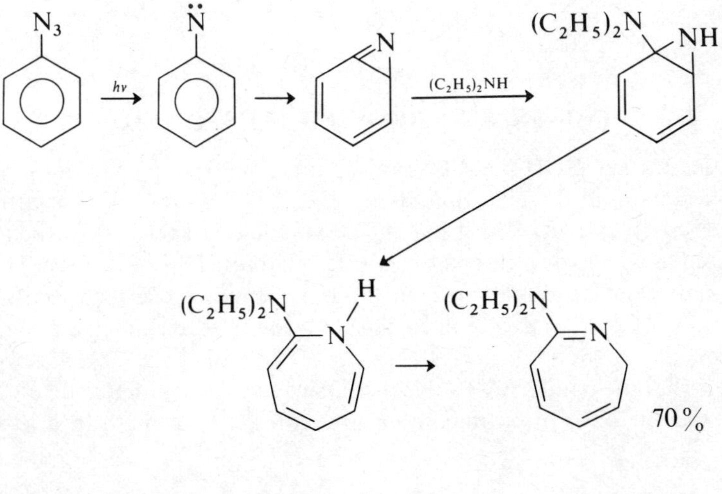

(10.60)

70%

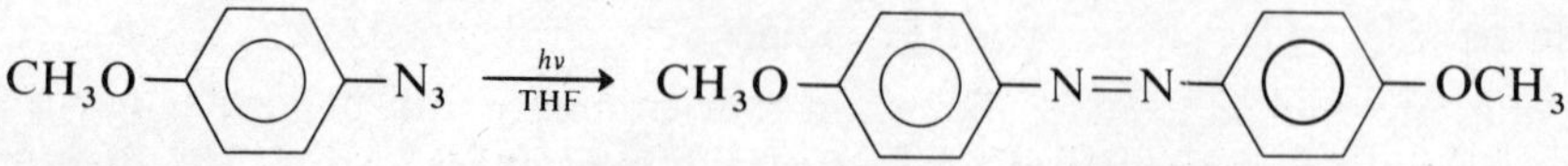

$$CH_3O-\langle\bigcirc\rangle-N_3 \xrightarrow[THF]{h\nu} CH_3O-\langle\bigcirc\rangle-N=N-\langle\bigcirc\rangle-OCH_3$$

$$82\% \qquad\qquad (10.61)$$

Intramolecular C—H insertion by the nitrene obtained on photolysis of 2-azidobiphenyl gives carbazole, whereas intermolecular reaction yields 2,2'-azobiphenyl.[62] On direct irradiation both products are formed (10.62), but in the presence of a conjugated diene the yield of carbazole rises to 89% and that of the azo-compound drops to <1%. On the other hand, when benzophenone is employed as a triplet sensitizer the yield of azobiphenyl rises to 49% and that of carbazole becomes very small (<5%). These results suggest that the excited states responsible for formation of the two products are different, and that the insertion reaction proceeds mainly through a singlet excited state of the azide and a singlet nitrene, whilst the triplet state azide and nitrene give rise largely to the azo-compound.

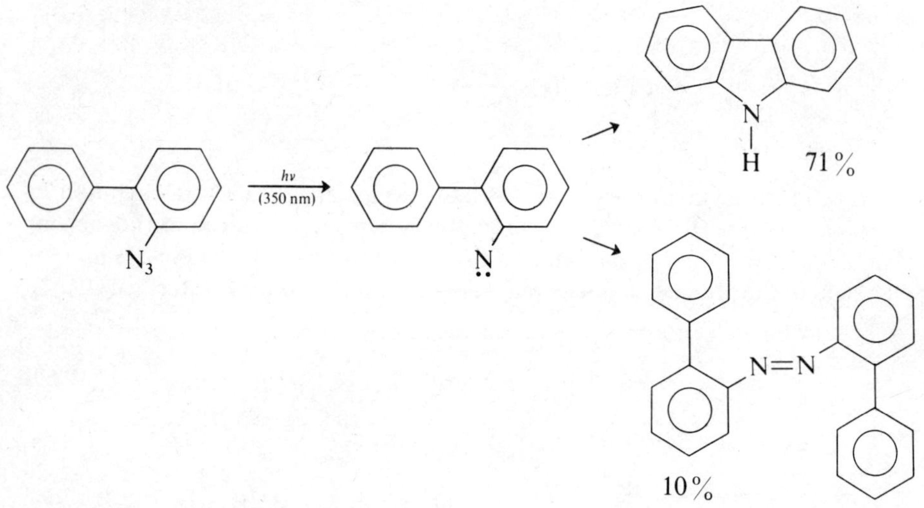

71%

10%

$$(10.62)$$

10.7 IMINES AND RELATED COMPOUNDS

Aliphatic imines (RCH=NR') show a fairly weak ($\varepsilon \sim 100\,l\,mol^{-1}\,cm^{-1}$) absorption band in the ultraviolet region ($\lambda_{max} \sim 235\,nm$), which is attributed to a $n \rightarrow \pi^*$ transition. This is of considerably higher energy than the $n \rightarrow \pi^*$ transition in aldehydes and ketones (see Chapter 7, section 7.1). The long wavelength band of hydrazones, on the other hand, is ascribed to a $\pi \rightarrow \pi^*$ transition because the extinction coefficient is high (for acetaldehyde N-methylhydrazone in ethanol, $\lambda_{max} = 230\,nm$, $\varepsilon = 5000\,l\,mol^{-1}\,cm^{-1}$). The surprising feature is that the band shows solvent shifts which are characteristic of $n \rightarrow \pi^*$ bands, and the absorption maximum is at longer wavelength in a less polar

The species CdH$\cdot$ and HgH$\cdot$ have been detected in the reactions,[13,14] and the action of Hg($^3P_1$) is to abstract a hydrogen atom (11.14),[15] after which the species HgH$\cdot$ can act as a direct or indirect source of hydrogen atom in the secondary processes. Divalent paramagnetic Hg($^3P_1$) is behaving rather like a carbene.[16]

$$C_3H_8 \xrightarrow{\text{Hg}(^3P_1)} C_3H_7^{\cdot} \qquad \phi = 0.57 \pm 0.12 \ (63\ °C) \qquad (11.13)$$

$$C(CH_3)_4 \xrightarrow{\text{Hg}(^3P_1)} (CH_3)_3C-CH_2^{\cdot} + {}^{\cdot}HgH \qquad (11.14)$$

Other species are known to abstract hydrogen atoms from alkanes, and the irradiation of an alkane with oxygen leads to a hydroperoxide in a stepwise radical reaction (11.15).[17] The initial step in this reaction is excitation of a contact charge–transfer complex between the hydrocarbon and oxygen (see Chapter 2, section 2.2.6). Similar reactions occur for ethers, secondary alcohols and other saturated compounds.

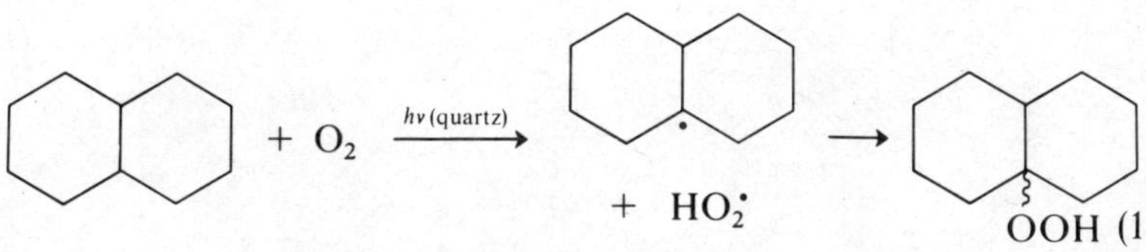

Another process of this type involves the abstraction of a hydrogen atom by a halogen atom (11.16). The most familiar examples involve Cl$\cdot$, in the photochlorination of alkanes with molecular chlorine (11.17),[18] or in the photo-oximation of alkanes with nitrosyl chloride (11.18).[19] Cyclohexanone oxime is manufactured by this method on a large scale for conversion to caprolactam and Nylon-6.

$$R-H + Br^{\cdot}(^2P_{\frac{1}{2}}) \rightarrow R^{\cdot} + HBr \qquad (11.16)$$

$$R-H + Cl_2 \xrightarrow{h\nu} R-Cl + HCl \qquad \phi \sim 10^3\text{--}10^4 \qquad (11.17)$$

$$O=N-Cl \xrightarrow{h\nu} NO^{\cdot} + Cl^{\cdot}$$

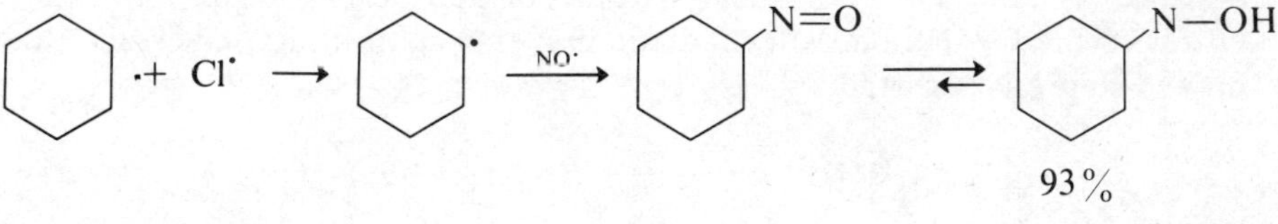

$$93\%$$

$$(11.18)$$

11.2   OXYGEN CHROMOPHORES

11.2.1   Alcohols

Alcohols and ethers differ from alkanes in that they have non-bonding electron pairs on the oxygen atom. The lowest electronically excited states of these compounds are (n, σ^*) or Rydberg states, and the transitions leading to them show maximum absorption around 180–185 nm in the ultraviolet spectra of simple alcohols and ethers. The absorption is only moderately intense ($\varepsilon_{max} \sim 10^2$–$10^3 \, l \, mol^{-1} \, cm^{-1}$) because of the partially forbidden nature of the transition, and alcohols and ethers are used as solvents for 'normal' ultraviolet spectroscopy since there is little tail absorption at wavelengths longer than 200 nm.

Homolytic bond fission processes predominate in the chemistry of the (n, σ^*) excited states, although molecular modes of decomposition are also found. The photolysis of methanol yields hydrogen and ethylene glycol as major products in the liquid phase (11.19),[20] and in both liquid and gas phase the most important primary reaction in the excited state involves O—H bond fission. (The $CH_3O \cdot$ radical abstracts hydrogen from methanol to give $\overset{\cdot}{C}H_2OH$, which dimerizes to give the glycol.) Some molecular elimination of hydrogen occurs, and a small amount of C—O fission (11.20).

$$CH_3OH \xrightarrow[\text{liquid}]{h\nu \, (185 \, nm)}$$

$$H_2 \; + \; HO{-}CH_2CH_2{-}OH \; + \; H_2C{=}O \; + \; CH_4 \qquad (11.19)$$
$$\phi = 0{\cdot}83 \qquad\quad 0{\cdot}78 \qquad\qquad\quad 0{\cdot}06 \qquad\; 0{\cdot}05$$

$$CH_3OH \xrightarrow{h\nu} \begin{cases} CH_3O \cdot \; + \; H \cdot & 79\% \\ H_2C{=}O \; + \; H_2 & 20\% \\ CH_3^- \; + \; \cdot OH & 1\% \end{cases} \qquad (11.20)$$

2-Propanol gives rise to acetone and hydrogen[21] as the major products of photolysis (11.21) with smaller amounts ($\phi < 0{\cdot}05$) of methane, acetaldehyde, pinacol, propane, carbon monoxide and other products. The major primary process is again homolytic O—H bond fission, with molecular processes as substantial minor primary reactions of the excited state (11.22).[22] The formation of acetone rather than pinacol is a reflection of the different chemistry of 2-propoxy and methoxy radicals. The preference for O—H rather than C—O or even C—H cleavage is surprising in view of the higher homolytic bond strength of the O—H bond, and it is probably a result of initial localization of vibrational energy in the O—H bond after radiationless transition from excited state to ground state has occurred.

$$(CH_3)_2CHOH \xrightarrow[\text{gas}]{h\nu \, (185 \, nm)} (CH_3)_2C{=}O \; + \; H_2 \qquad (11.21)$$
$$\phi = 0{\cdot}42 \qquad\quad 0{\cdot}64$$

$$(CH_3)_2CHOH \xrightarrow{h\nu} \begin{cases} (CH_3)_2CH{-}O \cdot \; + \; H \cdot & 61\text{-}9\% \\ (CH_3)_2C{=}O \; + \; H_2 & 21\% \\ CH_3CH{=}O \; + \; CH_4 & 5\% \end{cases} \qquad (11.22)$$

t-Butanol undergoes similar radical and molecular fragmentation reactions in the gas phase, but in the liquid phase radiation of wavelength 185 nm produces a complex mixture because of effectively competing intermolecular reactions. The alcohol $(CH_3)_3C-O-CH_2C(CH_3)_2OH$ is one of 16 products so far identified.[23]

11.2.2 Ethers

In the liquid phase photolysis of diethyl ether[24] the major products are ethanol, ethane and 2-butyl ethyl ether (11.23), and it is estimated that in the excited state 70% of the decomposition involves homolytic C—O bond fission, and 30% molecular fragmentation (11.24). This is in contrast to the behaviour of ether under conditions of $Hg(^3P_1)$ sensitization, when the major products are hydrogen and a radical dimer (11.25).

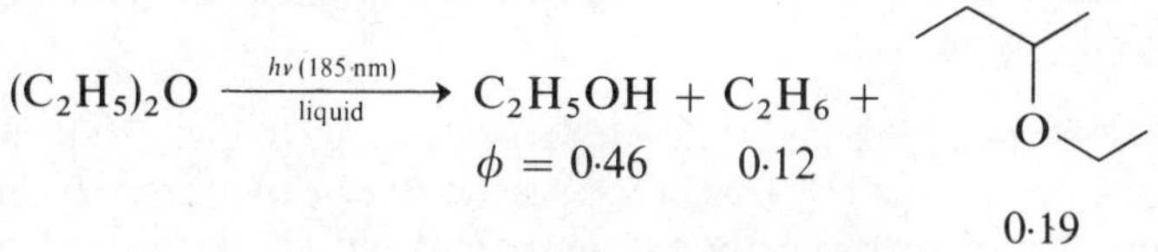

$$(C_2H_5)_2O \xrightarrow[\text{liquid}]{h\nu\,(185\,\text{nm})} C_2H_5OH + C_2H_6 + \qquad\qquad (11.23)$$
$$\phi = 0{\cdot}46 \qquad 0{\cdot}12$$
$$0{\cdot}19$$

$$C_2H_5-O-C_2H_5 \xrightarrow{h\nu} \begin{cases} C_2H_5-O^{\boldsymbol{\cdot}} + C_2H_5^{\boldsymbol{\cdot}} & 70\% \\ \text{molecular products} & 30\% \end{cases} \qquad (11.24)$$

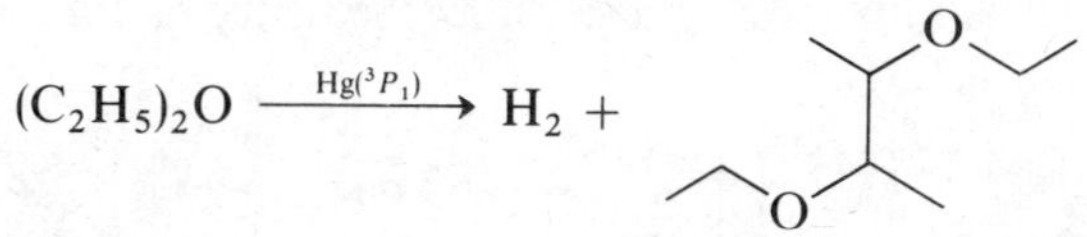

$$(C_2H_5)_2O \xrightarrow{Hg(^3P_1)} H_2 + \qquad\qquad\qquad (11.25)$$

Oxiranes similarly undergo C—O cleavage on irradiation, and the biradical produced leads largely to a ketone or an aldehyde by way of a 1,2-hydrogen shift. Some reduction to a saturated alcohol also occurs, and these reactions are exemplified by methyloxirane (11.26).[25] For some vinyloxiranes the process can be sensitized by acetone (11.27),[26] and it is possible that reaction occurs through a triplet excited state in these systems.

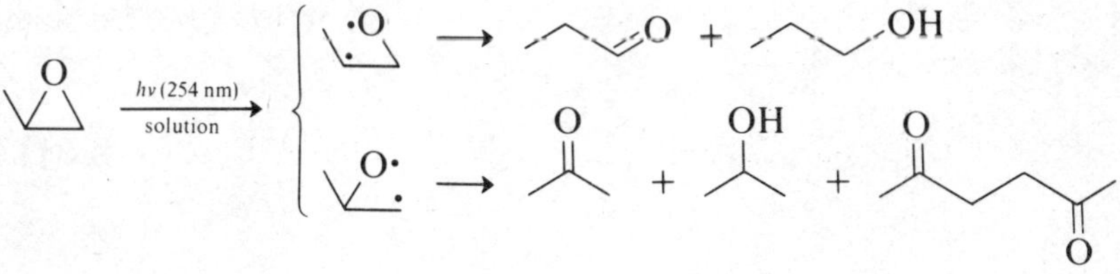

$$\xrightarrow[\text{solution}]{h\nu\,(254\,\text{nm})} \qquad\qquad (11.26)$$

$$\xrightarrow[\text{acetone}]{h\nu\,(300\,\text{nm})} \qquad \phi = 0{\cdot}15 \qquad (11.27)$$
$$60\%$$

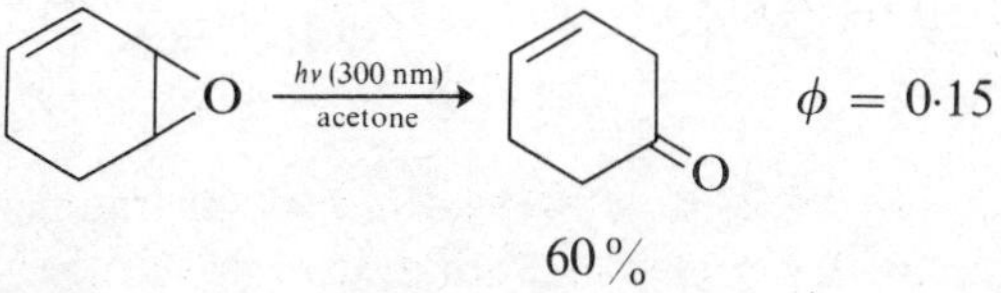

Cleavage of a C—O bond in an oxirane can be accompanied by C—C cleavage to give a carbene and an aldehyde or ketone, and this is especially important for aryl-substituted oxiranes. Low temperature irradiation of the 2,3-diphenyl-oxiranes (11.28) shows that phenylcarbene and benzaldehyde are formed,[27] possibly by way of a coloured ylide. In addition stilbene is produced directly by loss of oxygen from the oxirane without the intermediacy of the carbene.

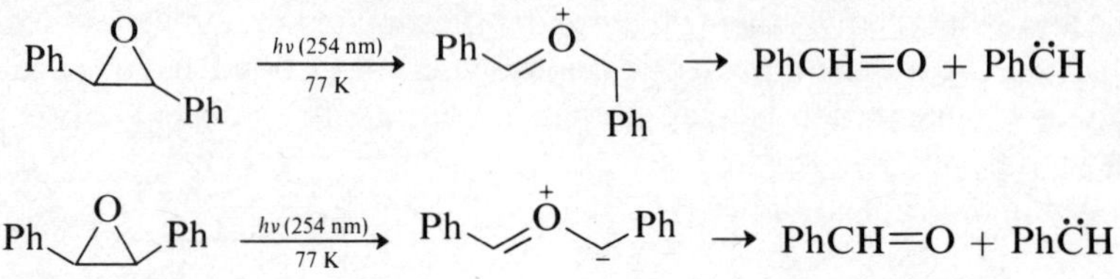

$$\text{Ph} \xrightarrow[\text{77 K}]{hv\,(254\,\text{nm})} \text{Ph} \overset{+}{\text{O}} \underset{\text{Ph}}{\overset{-}{}} \rightarrow \text{PhCH}{=}\text{O} + \text{Ph}\overset{\cdot\cdot}{\text{C}}\text{H}$$

$$\text{Ph} \xrightarrow[\text{77 K}]{hv\,(254\,\text{nm})} \text{Ph} \overset{+}{\text{O}} \text{Ph} \rightarrow \text{PhCH}{=}\text{O} + \text{Ph}\overset{\cdot\cdot}{\text{C}}\text{H}$$

(11.28)

Oxetanes in their (n, σ^*) excited states undergo cleavage to give an alkene and an aldehyde or ketone (11.29).[28] This process is the reverse of oxetane formation by photochemical reaction of a carbonyl compound and an alkene (see Chapter 7, section 7.7.1), although it requires shorter wavelength radiation to cleave the cyclic compound than to form it. The mechanism of the reaction is uncertain, but initial C—O bond cleavage may occur. Larger ring ethers also yield fragmentation products which can formally be accounted for by initial C—O bond cleavage, and this is illustrated in the photochemistry of 1,4-dioxan (11.30),[29] and dihydrofurans (11.31).[30]

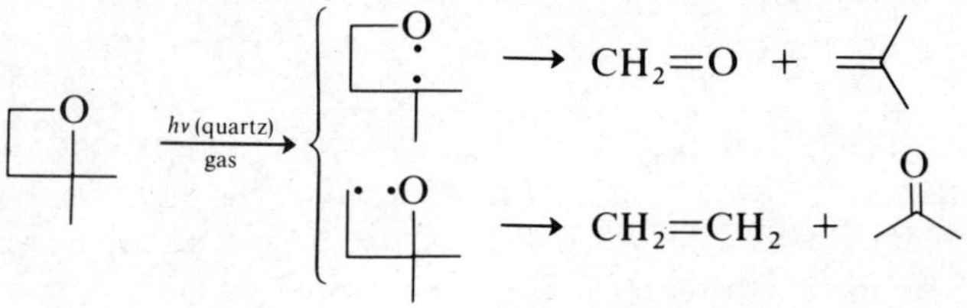

$$\longrightarrow \text{CH}_2{=}\text{O} + \text{}$$

(11.29)

$$\longrightarrow \text{CH}_2{=}\text{CH}_2 + \text{}$$

$$\text{CH}_2{=}\text{O} + \text{CH}_2{=}\text{O} + \text{CH}_2{=}\text{CH}_2$$
$$\phi = 0.75$$

$$\text{CH}_2{=}\text{O} + \text{H}_2 + \text{CO} + \text{CH}_2{=}\text{CH}_2$$
$$\phi = 0.10$$

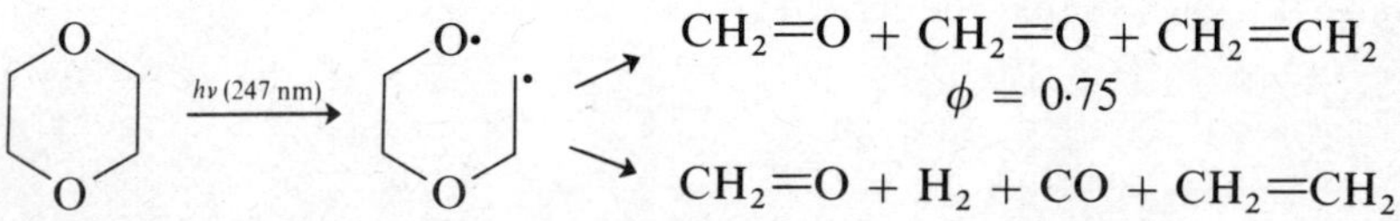

(11.30)

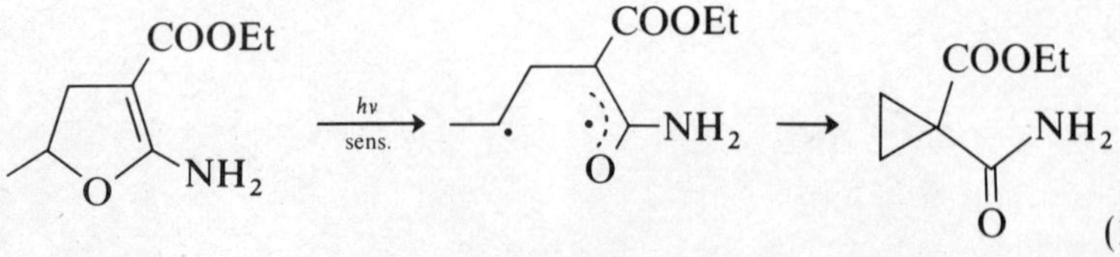

(11.31)

53%

11.2.3 **Peroxides**

Alkyl hydroperoxides and peroxides exhibit ultraviolet absorption at longer wavelength than alkyl alcohols or ethers. Absorption becomes appreciable at wavelengths shorter than about 350 nm, and this is attributed to interaction between the non-bonding electron pairs on the two oxygen atoms. The effect is particularly marked with cyclic peroxides, in which the non-bonding electrons are constrained in a geometry suitable for strong interaction. In the (n, σ^*) excited states of the peroxides the predominant mode of chemical reaction involves O—O cleavage to give vibrationally excited alkoxy radicals (11.32).[31,32] At high energies of radiation ($\lambda < 230$ nm) C—O bond fission is also a significant primary process.

$$C_4H_9-O-O-H \xrightarrow[175\,K]{hv\,(254\,nm)}$$
$$HO^\cdot + C_4H_9-O^\cdot \rightarrow C_4H_9-O-O-C_4H_9$$
$$Bu^t-O-O-Bu^t \xrightarrow[gas]{hv\,(254\,nm)} 2Bu^t-O^\cdot \qquad \phi = 0.95\text{–}1.2$$

$$(11.32)$$

Cyclic peroxides give biradicals on irradiation. Those from 1,2-dioxetanes give rise to aldehydes and ketones (11.33),[33] and those from 1,2-dioxolanes give oxiranes as major product (11.34).[34]

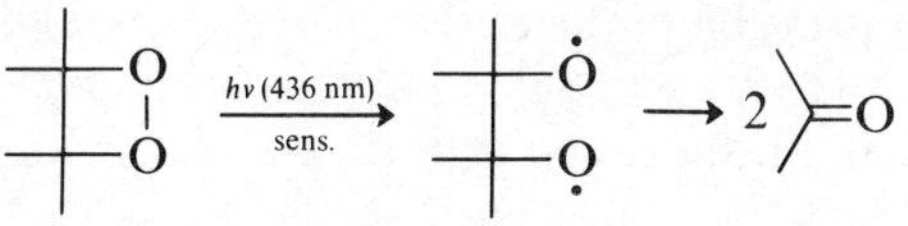

$$(11.33)$$

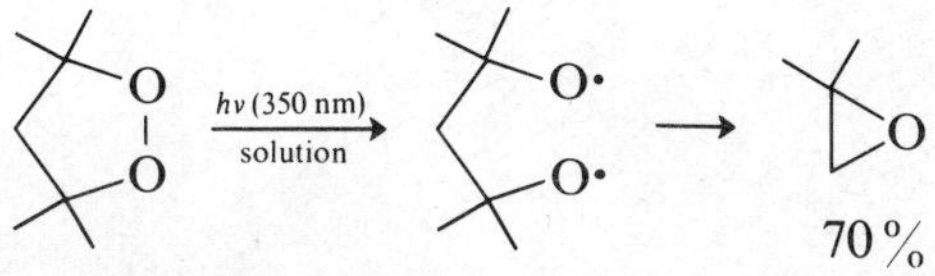

70%

$$(11.34)$$

Similarly, irradiation of acyl peroxides and other derivatives of percarboxylic acids gives acyloxy radicals (11.35).[35] This is a major method for the generation of free radicals to initiate radical chain reactions such as addition polymerization under milder conditions than those required for thermal peroxide initiation. The reactions of hydroperoxides, acyl peroxides and alkyl percarboxylates can be sensitized by ketones with high triplet energy, probably by population of a relaxed triplet state of the peroxy compound. Singlet sensitization by aromatic hydrocarbons is also possible in some instances.

$$Ph-CO-O-O-CO-Ph \xrightarrow[solution]{hv} 2Ph-CO_2^\cdot \rightarrow 2Ph^\cdot + 2CO_2 \quad (11.35)$$

11.3 **SULPHUR CHROMOPHORES**

Formally thiols, sulphides and disulphides are analogous to alcohols, ethers and peroxides respectively, and there are marked similarities in the photochemistry of sulphur and oxygen analogues. However, differences occur in the secondary processes, since alkoxy and thiyl radicals have greatly different chemical properties. Because sulphur has a higher atomic number than oxygen, the transitions are of lower energy in the sulphur compounds, and this can be seen in the ultraviolet absorption data given in Table 11.2.

Table 11.2. Absorption data for oxygen and sulphur compounds

	λ_{max} (nm)	ε_{max} (l mol^{-1} cm^{-1})		λ_{max} (nm)	ε_{max} (l mol^{-1} cm^{-1})
MeOH (vapour)	184	150	MeSH	228	140
Me$_2$O (vapour)	184	2520	Me$_2$S	212	1000
Me$_2$O$_2$ (vapour)	215	<15	Me$_2$S$_2$	255	360
			cyclo-(CH$_2$)$_4$S$_2$	285	350
			cyclo-(CH$_2$)$_3$S$_2$	330	

11.3.1 Thiols

The major primary process in the excited states of alkanethiols produced with radiation of wavelength 254 nm is S—H bond cleavage (11.36).[36] This is in spite of the fact that the C—S bond is considerably weaker than the S—H bond. It implies that dissociation is faster than randomization of excess energy, and it is another illustration of the tendency of many excited states to undergo cleavage of a bond to hydrogen. Subsequent addition of the radicals produced to an alkene in a chain process can provide a useful synthetic route to certain sulphides (11.37).

$$C_2H_5SH \xrightarrow{h\nu \,(254\,nm)} \begin{cases} C_2H_5S^{\cdot} + H^{\cdot} & 90\% \\ C_2H_5^{\cdot} + {}^{\cdot}SH & 9\% \\ C_2H_4 + H_2S & 1\% \end{cases} \qquad (11.36)$$

$$R-SH \xrightarrow{h\nu} H^{\cdot} + {}^{\cdot}SR \xrightarrow{R'CH=CH_2} \rightarrow RS-CH_2-CH_2R' \qquad (11.37)$$

With higher energy radiation C—S bond cleavage is also an important process,[37] and the ratio of C—S to S—H cleavage in methanethiol with 195 nm radiation is 1·0 : 1·7.

11.3.2 Sulphides

For dialkyl sulphides homolytic C—S fission is the major primary process on direct or sensitized irradiation, and the radicals produced undergo secondary reactions of which combination and hydrogen abstraction are the most important. Hg($^3P_1$) Sensitized photolysis of dimethyl sulphide[38] gives methane, ethane, methanethiol and dimethyl disulphide as major products (11.38).

$$(CH_3)_2S \xrightarrow{Hg(^3P_1)} CH_3^{\cdot} + CH_3S^{\cdot} \rightarrow$$
$$CH_3S-SCH_3 + CH_3SH + C_2H_6 + CH_4 \qquad (11.38)$$
$$\phi = 0{\cdot}37 \qquad\quad 0{\cdot}01 \qquad 0{\cdot}47 \qquad 0{\cdot}07$$

D_{6h}	E	$2C_6$	$2C_3$	C_2	$3C_2'$	$3C_2''$	i	$2S_3$	$2S_6$	σ_h	$3\sigma_d$	$3\sigma_v$	Co-ordinates and rotations	d-orbitals
A_{1g}	1	1	1	1	1	1	1	1	1	1	1	1		x^2+y^2, z^2
A_{2g}	1	1	1	1	-1	-1	1	1	1	1	-1	-1	R_z	
B_{1g}	1	-1	1	-1	1	-1	1	-1	1	-1	1	-1		
B_{2g}	1	-1	1	-1	-1	1	1	-1	1	-1	-1	1		
E_{1g}	2	1	-1	-2	0	0	2	1	-1	-2	0	0	(R_x, R_y)	(xz, yz)
E_{2g}	2	-1	-1	2	0	0	2	-1	-1	2	0	0		(x^2-y^2, xy)
A_{1u}	1	1	1	1	1	1	-1	-1	-1	-1	-1	-1		
A_{2u}	1	1	1	1	-1	-1	-1	-1	-1	-1	1	1	z	
B_{1u}	1	-1	1	-1	1	-1	-1	1	-1	1	-1	1		
B_{2u}	1	-1	1	-1	-1	1	-1	1	-1	1	1	-1		
E_{1u}	2	1	-1	-2	0	0	-2	-1	1	2	0	0	(x, y)	
E_{2u}	2	-1	-1	2	0	0	-2	1	1	-2	0	0		

D_{6h} point group because it contains the symmetry elements of that group (A.3). Thus a molecule can be assigned to a point group by writing down the symmetry elements or operations appropriate to the molecular framework and matching these against those displayed on the top line of the various character tables.

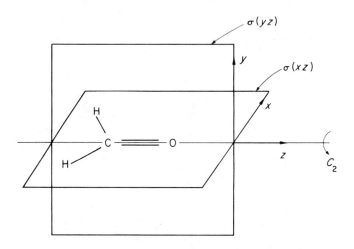

Figure A.1. Formaldehyde belongs to the C_{2v} point group

In the body of the table, the numbers 0, 1, 2 etc. are the characters of the irreducible representations of the point group, one representation per line and each labelled with the appropriate symmetry symbol (A_{1g}, B_2 etc.). What is really meant by the 'character of an irreducible representation' is best discovered from a standard text on group theory, but it is sufficient for our purposes to say that a set of matrices can be found, each corresponding to a single symmetry operation of the group, which combine in the same way as the symmetry operations. The simplest matrices of this sort are said to be irreducible representations of the group. The character, or trace, of a matrix is the sum of the elements constituting its main diagonal.

To illustrate this point, take the C_{2v} point group and the co-ordinate system of Figure A.2, and consider the way in which the vector (x_1, y_1, z_1) transforms under the symmetry operations of the group. If, under the symmetry operation $\sigma(xz)$, i.e. reflection in the xz plane, it is transformed into the vector (x_2, y_2, z_2), then the transformation can be formulated in terms of the equations (A.5).

$$x_2 = x_1 + 0y_1 + 0z_1$$

$$y_2 = 0x_1 - y_1 + 0z_1 \qquad (A.5)$$

$$z_2 = 0x_1 + 0y_1 + z_1$$

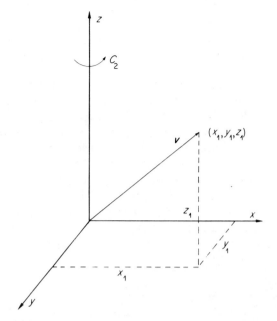

Figure A.2. The vector (x_1, y_1, z_1) in a Cartesian coordinate system

These equations can be written in matrix form (A.6).

$$\begin{pmatrix} x_2 \\ y_2 \\ z_2 \end{pmatrix} = \begin{pmatrix} 1 & 0 & 0 \\ 0 & -1 & 0 \\ 0 & 0 & 1 \end{pmatrix} \begin{pmatrix} x_1 \\ y_1 \\ z_1 \end{pmatrix} \qquad \text{(A.6)}$$

The (3×3) matrix corresponds to the $\sigma(xz)$ symmetry operation; its character is $(1 - 1 + 1) = 1$, and we can write $\chi(\sigma(xz)) = 1$. Similarly, for the identity representation E,

$$\begin{pmatrix} x_2 \\ y_2 \\ z_2 \end{pmatrix} = \begin{pmatrix} 1 & 0 & 0 \\ 0 & 1 & 0 \\ 0 & 0 & 1 \end{pmatrix} \begin{pmatrix} x_1 \\ y_1 \\ z_1 \end{pmatrix} \quad \text{and} \quad \chi(E) = 3 \qquad \text{(A.7)}$$

These matrices are not the simplest possible. By a rotation of the axes, the vector v can be brought into coincidence with, say, the x-axis (see Figure A.2). The matrices, with their characters, are now:
for C_2

$$\begin{pmatrix} x_2 \\ y_2 \\ z_2 \end{pmatrix} = \begin{pmatrix} -1 & 0 & 0 \\ 0 & 0 & 0 \\ 0 & 0 & 0 \end{pmatrix} \begin{pmatrix} x_1 \\ y_1 \\ z_1 \end{pmatrix}$$

or

$$(x_2) = (-1)(x_1) \quad \text{and} \quad \chi(C_2) = -1$$

$$\chi(E) = 1, \qquad \chi(\sigma(xz)) = 1, \qquad \chi(\sigma(yz)) = -1 \qquad \text{(A.8)}$$

These matrices are irreducible, and comparison of their values with the characters of the irreducible representations in the C_{2v} character table shows that a vector along the x co-ordinate transforms according to the B_1 representation. In the same way, a vector along the z-axis belongs to the A_1 representation.

Notice that when the character of the irreducible representation of a symmetry operation has the value 1, it means that the entity is transformed into itself (i.e., is unchanged). Conversely, when the character has the value -1, it implies that the entity is antisymmetric with respect to this symmetry operation and is transformed into its inverse by the operation. For example, consider a $2p_x$ orbital (Figure A.3). Since a C_2 operation about the y- or z-axis inverts the phases or signs of this wavefunction, the character of these symmetry operations will be -1.

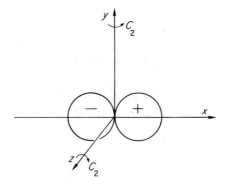

Figure A.3. A $2p_x$ orbital

In some symmetry groups there will be symmetry operations whose characters are 2 or 0.† These correspond to degenerate irreducible representations. To illustrate this point, consider the behaviour of the y co-ordinate axis under the symmetry operations of the D_{6h} point group (Figure A.4).

The y-axis is clearly antisymmetric to C_2 or i and symmetric to σ_h. However, the y-axis and the x-axis are neither symmetric nor antisymmetric to C_3 (rotation through 120°); instead, they are each converted into a linear combination of both in accordance with the equations (A.9).

$$x_2 = x_1 \cos 120° - y_1 \sin 120°$$

$$y_2 = x_1 \sin 120° + y_1 \cos 120° \qquad \text{(A.9)}$$

---

† Some character tables also contain trigonometric functions and complex numbers. These are rarely encountered and will not be discussed here.

Contents

Illustrations

Environmental Action
and the Question of Scale

Graeme Wynn

IN 1972, WHEN RENÉ DUBOS urged people to think globally and act locally, he did more than coin the phrase that became the mantra of the environmental movement. By offering up a slogan that resonated with contemporary circumstances and tapped into a deep well of popular sentiment, he framed a strategy for environmental action and sowed the seeds of a paradox that continues to complicate understandings of environmentalism.[1]

Between Christmas Eve 1968 – when astronauts aboard Apollo 8 entered lunar orbit and photographed Earth rising above the moon – and early December 1972 – when the crew of Apollo 17 transmitted the breathtaking "Blue Marble" view of Earth back to Houston – new Earth-images reshaped perceptions of humankind's planetary home, and made it easier for people to "think globally." Breathtaking pictures of a colourful orb against a dark void seized the public imagination, and gave credence to the idea of "Spaceship Earth," a singular, self-contained, and finite crucible of all life.[2]

To those familiar with the absolute dependence of astronauts upon their capsule and its systems for reclaiming water, recycling air, and scrubbing carbon dioxide, the spaceship metaphor drove home the fragility and interconnectedness of earth systems. For devotees of wilderness, it seemed to buttress John Muir's observation that everything was "hitched to everything else in the Universe." Aldo Leopold's argument (in his land ethic essay) that individuals are "members of a community of interdependent parts" gained new resonance, reflected in the surging popularity of *A Sand County Almanac*. The metaphor also reinforced the message of Rachel Carson's

Silent Spring, about the interrelatedness of Nature's parts, and gave popular purchase to the arguments of such diverse books as Paul Ehrlich's *The Population Bomb* (1968) and Donnella Meadows and her co-authors' *The Limits to Growth* (1972). Indeed, the revealingly titled *Whole Earth Catalogue* – at its most popular during this four-year span – was so named because Stewart Brand, its founder, believed that the image of Earth from space would encourage people to recognize their shared destiny and adopt more environmentally sound modes of living. Thinking globally emphasized the importance of ecological entanglements, even as it minimized cultural differences and disparate histories.[3]

The injunction to act locally brought these entanglements home and encouraged a form of arithmetic calculus in which wise choices replicated across the intricate mosaic of earth-communities would cumulate to shape a better planetary future, and ecological degradation of the earth system would subtract from the quality of local life. At one level, then, Dubos's catchphrase seemed to substantiate the idea, given resonant expression by American essayist and novelist Scott Russell Sanders, that "we can live wisely in our chosen place only if we recognize its connections to the rest of the planet." At another, it gave power to the people by suggesting that planetary-scale concerns could be addressed at the grassroots level by individuals working alone or in small groups. At a third, it implied that local life and local politics would benefit from attention to global issues. There was much of value in all of this.[4]

But the call to local action had other, paradoxical ramifications as it chimed, in diverse ways, with growing contemporary sentiment. In one register, it reified those very differences in culture and history that thinking globally tended to erase, as it encouraged a heightened "sense of place." Expression of the importance and virtues of place attachment seemed to burst forth in these years, possibly as a backlash against pervasive modernist architecture and city planning and the increasingly peripatetic character of contemporary life. Indeed, the geographer Yi-fu Tuan claimed to have coined the word "topophilia" to refer to "the affective bond between people and place or setting" in a 1974 book that at least in part lamented the erosion of that bond.[5]

Certainly the idea that people and place were connected was in the air. In 1977, American environmental philosopher Paul Shepard suggested that it was impossible to know who you were if you did not know where you were. More pointedly, Wendell Berry attributed the mindless destruction of the American landscape to the nomadic character of the American experience, and Scott Russell Sanders proclaimed his desire to "become an inhabitant, one who knows and honors the land," as he lamented the "vagabond wind" that forever drove his restless compatriots to relocate. North of the forty-ninth parallel, journalist Robert Fulford and civic politician John Sewell teamed up on a pamphlet about Toronto with the title *A Sense of Time and Place.* Northrop Frye tracked a more academic but parallel path when he framed "Where is here?" as a fundamental Canadian question, and Neil Evernden did likewise when he concluded that individuals did not exist without context, which was to say as "a component of place, defined by place." Similarly, Canadian historians embraced local difference when they seized upon the concept of "limited identities" as a means to understand the country.[6]

In another sense, the advice to act locally sanctified immediate experience, and entrenched what Ursula Heise and others call an "ethic of proximity" that valued intimate, bodily, sensory engagements with nature over more abstract forms of understanding (that were surely important to thinking on a global scale). In this vein, Norwegian philosopher Arne Naess, founder and figurehead of the deep ecology movement, pronounced that "the nearer has priority over the more remote – in space, time, culture, species." Such sentiments bled easily into broader suspicions, increasingly evident in the 1960s and 1970s, of "things-at-a-distance," of the impersonal character of urban and industrial life, and of modernity in general, even as they supplied oxygen to back-to-the land movements, the counterculture, the embrace of Indigenous ecological wisdom, and enthusiasm for the organization of society on bioregionalist principles. Later they seemed to fit easily with "postmodern" critiques of totalizing grand narratives and claims for the importance of local and "situated" knowledge. In sum, time and circumstances turned a forward-looking, four-word coinage emphasizing the links between everyday actions and global environmental circumstances against itself, to buttress "a general critique of modern socio-political structures" and resuscitate a traditional "sense of place."[7]

Mark Leeming carries us into the heart of this complex, conflicted intellectual terrain with his detailed examination of two decades of environmental activism in Nova Scotia. His astutely yet provocatively titled book, *In Defence of Home Places,* encourages us to think anew about the ways in which the tensions between local and global (and indeed among local, regional, and national) scales, and between traditional and modern perspectives, all seemingly endemic to environmentalism, played out on Canadian ground from the 1960s to the 1980s. From the outset, Leeming makes clear that his story turns on the "recognition and consequences of ideological differences within the environmental movement of a modern country" (p. 3) and that he develops it by considering three major controversies – over nuclear power, the use of chemicals in forestry, and uranium mining – in a single province. He also adumbrates his conclusions in his first paragraph: environmentalists struggled among themselves to find common ground on each of these issues (and by implication the basic questions of their cause). This divided the movement into modernist and non-modernist camps that were readily identifiable, by 1985, as mainstream and fringe players in the continuing drama of local and provincial environmental politics.

Leeming's approach is intriguing. It is focused resolutely on a single province, and more particularly on specific struggles and struggle sites within it. The three issues at the centre of these struggles have been controversial in many jurisdictions and locales, but Nova Scotia is, of course, a place *sui generis*. Relatively small in area (55,284 square kilometres) and population (almost 790,000 in 1971 and 922,000 in 2011), it remained substantially rural through the third quarter of the twentieth century. For thirty years after 1956, successive censuses showed the urban fraction of the province's population fluctuating between 54 and 58 percent, while that of Canada as a whole climbed from 70 to 76 percent. Although Halifax expanded steadily after 1961, most of Nova Scotia has grappled with the consequences of slow economic growth. Fewer and fewer people under the age of eighteen, and increasing longevity, pushed the median age of the provincial population upward.

By the middle years of the twentieth century, when per capita incomes in the Atlantic region were less than two-thirds of the national average, the region had endured a long history of boom-and-bust engagements with resource and industrial development. Nineteenth-century forest and fishing economies were infamously volatile. Later in that century, the economic future of the Maritime provinces seemed bright. With year-round rail access to the interior of the continent and an established presence in seaborne commerce, Nova Scotia and New Brunswick prospered from the development of cotton mills and sugar refineries converting imported raw materials into consumer goods for citizens of the new Dominion.[8] Rich sources of coal and ready access to iron ore supplies also gave rise to a major steel industry, and encouraged celebration of "The Industrial Ascendancy of Nova Scotia" – evidenced, in the words of a 1913 promotional booklet, by the province's "1480 manufacturing establishments, paying out $11,000,000 yearly to 28,000 employees."[9] But prosperity, always unevenly distributed, exacted its toll, as it changed accustomed terms and conditions of labour, and it largely evaporated after 1918 when changing circumstances undercut the industrial base developed in the preceding four decades.[10]

Clearly a laggard rather than a leader in the growth of national manufacturing capacity since the 1920s (or a "have-not" province, in the parlance of Canada's post–Second World War efforts to combat regional disparities), Nova Scotia was the focus – along with its Atlantic provincial neighbours – of several economic development or regional economic expansion initiatives in the 1960s and subsequent decades.[11] Among these were the provincial government's Industrial Estates Limited (IEL), given a mandate in 1957 to build industrial parks and attract new investment, and the federal Department of Regional Economic Expansion (DREE), established in 1969. Reflecting a commitment to the idea that Atlantic Canadians should have opportunities to prosper in their home provinces, the DREE tailored its programs to encourage entrepreneurs and industrialists to locate new operations in centres designated as "growth poles" within the region.[12]

Bold as many of them were, these initiatives had mixed success. Some (such as the Come By Chance oil refinery in Newfoundland and the Bricklin car factory in Saint John, New Brunswick) were notoriously ill-fated.[13] Others such as resettlement schemes in Newfoundland and the Comprehensive Development Plan calling for a massive transformation of agriculture, the diversification of manufacturing, and the development of tourism in Prince Edward Island were disruptive of older ways and hugely controversial.[14] Yet gains were made, at least in the terms defined by governments: IEL brought Swedish forestry giant Stora Kopparberg to Cape Breton, Swedish car manufacturer Volvo to Dartmouth, and French tire manufacturer Michelin to Granton, Bridgeport, and Waterville in Nova Scotia. Redistribution payments and various forms of federal government intervention in the economy helped to bring per capita incomes in the region closer to national standards (they now stand at about four-fifths of the latter mark). By definition, however, efforts to modernize regional economies spell change for regional communities, and such transformations are rarely straightforward.

This, in a nutshell, is where Leeming's story begins. In his telling, government- and industry-led efforts to plant the seeds of a modern economy in areas well known and valued by long-term and essentially conservative rural residents were the catalysts of environmental activism in the province. Because these new development projects threatened to remake particular landscapes in one way or another, and because they also laid siege, more or less, to traditional lifestyles and livelihoods, they were

almost bound to spawn opposition. Following the pattern set in Pictou
County, after the provincial government lured a kraft pulp mill to
Abercrombie Point – creating conditions for Scott Paper to flourish, but
threatening the lobster fishery and killing Boat Harbour, into which the
mill discharged effluent – the sense of personal harm created by the damming of rivers, the pollution of the ocean, or the despoliation of local
ecologies was fanned time and again by industry denials and dismissive
government responses, to produce an upsurge of environmental activism
across the province. Feeling vulnerable in the face of change, Nova Scotia
activists found both economic and emotional reasons to rise in defence of
local places. Yet, it is important to recognize, neither they nor their views
were entirely parochial: television, radio, newspapers, books, immigrants,
and personal connections brought the wider tide of rising concern about
environmental issues into the province and the region. Many heard the
message of *Silent Spring;* CBC television alerted viewers across the network
to the perils of pollution with "The Air of Death," broadcast in October
1967; the environment ministers of the three Maritime provinces sponsored anti-litter and anti-pollution campaigns in the late 1960s and early
1970s; and the government of Prince Edward Island stepped back from
the Comprehensive Development Plan under the influence of ideas developed by E.F. Schumacher and George McRobie.[15]

Diffuse, scattered groups with a range of local environmental concerns
came together in the early 1970s when it appeared that the Nova Scotia
government was negotiating the development of a large nuclear power
generating plant in the far south of the province to supply markets in the
northeastern United States. Opposition aligned on several axes: nuclear
power was linked to nuclear war; Nova Scotians would bear the risks and
Americans reap the benefits; soft energy solutions were better than the
"hard [nuclear] energy path"; the capitalist consumer development that
nuclear generators would facilitate should be abandoned in favour of a
"conserver society"; the government was being less than clear and transparent about its intentions. Tracing these debates in some detail, Leeming
finds that arguments for smaller-scale, non-nuclear solutions and local
economic autonomy were advanced most forcefully by citizens concerned
about the personal risks associated with living near proposed nuclear plants,
and that their commitments were scaled upward as small rural activist
groups joined forces. In time, opposition in Nova Scotia coalesced with
similar hostility to nuclear development in New Brunswick under the
banner of the Maritime Energy Coalition (deliberately named to echo the
acronym of the proposed regional nuclear power utility, the Maritime

Energy Corporation). But as political commitments to nuclear power development weakened after the Three Mile Island disaster in Pennsylvania,
opposition arguments also shifted. Safety concerns and the hope of developing economically viable alternative energy systems trumped calls to
remake society.

Carrying this story forward through convoluted battles over the environmental effects of insecticide and herbicide spraying and uranium mining
in the decade or so after 1976, Leeming discerns growing resistance to environmentalist pressures from government and industry, and increasing
fragmentation among environmental campaigners. To gain a broad understanding of the politics and protests of this period, one might usefully
think of the spraying and mining episodes as engagements in a war fought
on many fronts, by coalitions of interest deploying several units using
different forms of ordnance and operating at a range of scales.

Beleaguered environmentalists sought (if we might borrow and adapt a
phrase from an earlier environmental protest elsewhere) to "secure their
nook of Nova Scotia ground from rash assault."[16] "Industry" – an amorphous coalition of interests – and their allies in government generally stood
shoulder to shoulder in support of economic expansion, which focused
their calculus of concern on measures such as jobs and profits. Still, they
sometimes disagreed over how to proceed, as advisers and strategists (for
example, accountants and trained foresters) preferred different forms of
action. Both sides found backing, of one sort or another, from beyond the
provincial battle zone, and engaged in "deliberate, systematic attempt[s]
to shape perceptions, manipulate cognitions, and direct behavior" to garner
assistance and further their cause.[17]

Each side struggled over the question of where and how best to engage
the other. Company executives sometimes favoured threats, bluster, and
action. Some government officials (particularly at the federal level) believed
that environmentalists might be "turned" or brought to accede to government agendas by the promise of influence; others (particularly provincial
actors) wished no truck or trade with their opponents and preferred
confrontation or judicial and quasi-judicial fora (the courts and Royal
Commissions) to resolve differences. Initially, environmentalists were
inclined to deploy the weapons of the weak – protests, resistance, and
the power of public pressure – to sway popular opinion and influence
politicians.[18] Over time, however, some adopted the view that bureaucratic
minds (and policies) were more likely to be changed by collaboration than
by resistance. Later, for reasons that Leeming details and that included a
growing conviction that scientific argument would win the day and legal

precedent would secure the gains, environmental campaigners were drawn into the judicial arena.

Commission hearings and court proceedings are costly enterprises, and they are conducted according to formal rules of evidence and proof. They also clearly serve to legitimize the state, and tend in long-running cases such as the "herbicide trial" to sap the resources of participants.[19] Insofar as the decision to go to court over the spraying of 2,4-D and 2,4,5-T drew Nova Scotia environmentalists into "a battle with the international pesticide industry" (p. 86), it was a tactic fraught with difficulty; the environmentalists' need for financial support brought disparate groups together to raise money but it also diverted their energies from other actions. And insofar as the presiding judge in the trial eschewed the precautionary principle and insisted on a nigh-impossible standard of scientific proof of harm from the use of herbicide, it was a doomed strategy.

When a Royal Commission on Uranium Mining was established in 1981, with the conservative Judge Robert McCleave at its head, his commitment to an increasingly scientific and legalistic three-stage inquiry seemed designed to wear down environmental protest. As things turned out, vigorous anti-uranium presentations emphasizing the "dangers of radiation exposure, the impossibility of safe tailings disposal, and the inseparability of nuclear energy and nuclear weapons" (pp. 134–35) wore down the commissioner during the first stage of hearings; the inquiry was suspended, and did not report until 1985. Then the commissioner proposed a five-year moratorium on uranium mining, even as he repudiated the claims of the "perverse" leftists who had spoken so forcefully before him.

Yet this was a divisive, if not pyrrhic, victory for the environmental movement in Nova Scotia. When the 1985 report was endorsed by environmental groups whose members had remained on the margins through the first phase of the inquiry, avoiding entanglement in what they described in unguarded moments as "side shows" mounted by "sleazy fanatics," it became impossible to paper over a division that had been developing for years. Lamenting that colleagues who chose to participate in and endorse official systems of decision making were more wedded to the process than the cause, grassroots environmentalists disparaged them as politically naive, "establishment bound" people dependent on government and corporate funding. So the lines between rural place-oriented and urban ecomodernist groups deepened. The former hewed to the long-standing environmentalist conviction that "only political pressure from large numbers of citizens could overcome the established unity of political and industrial interests that the official system was designed to uphold" (p. 134). They believed that "playing the game" thwarted the growth of political consciousness by suggesting that problems were well in hand and capable of bureaucratic solution. The latter held "a modern conception of economic development and risk assessment that devalued the personal experience of place and of home, discounted the inequity of metropolitanism, and preserved the privilege of expertise" (p. 136). Through the rest of the century, Nova Scotia environmentalists remained divided between those inclined to marry deep ecological positions with social justice concerns and those who preferred bureaucratically oriented "market" solutions to environmental issues.

"Think globally, act locally." The stories that Mark Leeming details in the pages that follow both complicate this simple injunction and demonstrate that there is no straightforward history of environmentalism in Nova Scotia or, by extension, any place else on the globe. They remind us that, at base, environmentalism is a resistance movement. Broadly, it casts itself in opposition to the despoliation of particular places (be that despoliation the pollution of a lake, the destruction of a view, or the clearcutting of a forested hillslope); the misuse of common property resources such as air and water or endangered species (think of campaigns against acid rain and whaling); and the insatiable expansionary logic of late capitalism and its close ally and political instrument, the modern state.

Engagement on such diverse fronts encourages, even requires, widely different strategies. Blockades and demonstrations in particular places can stop logging trucks or prevent the quarrying of a hillside (at least temporarily); they can also bring local acts of environmental plunder to global attention (think Greenpeace and the anti-sealing campaign). Not all environmental concerns are "end-of-pipe" ills (where harm occurs in a specific locale) or consequences of specific actions; still, protests against pillage of the commons, or contesting the global-scale consequences of particular behaviours, can have value, and they invariably take place someplace (here the best example may be the demonstration organized by 350.org outside the White House in Washington, DC, to oppose construction of the Keystone XL pipeline because of its implications for global climate change). For all that, blockades and protests are not always possible or effective – as environmentalists in British Columbia discovered when logging moved into remote areas of the mid-coast, where the media coverage so essential to garnering attention and support for any opposition was limited by difficulties of access. In such circumstances, other strategies, including "market campaigns" aimed at reducing corporate profits and tarnishing reputations by persuading consumers not to purchase the product, are likely to make more sense and have greater impact.[20]

Faced with the challenges of opposition, environmentalists also divide, legitimately, on how best to achieve their ends. We can think of their choices schematically. At one end of a spectrum of possibilities lies rejection of capitalism and a trenchant, unflinching oppositional stance. At the other, there is acknowledgment of the enormous power of politics and the marketplace and the conviction that gains can be made by regulation or by nudging the market this way or that to ameliorate its environmental effects. Intermediate positions abound, and members of those groups at the centre of Leeming's narrative occupied several of them. Those whom he characterizes as ecomodernists arrayed themselves towards the political-market end of the spectrum, convinced of the value of working within the corridors of legislative power to facilitate negotiations among environmental, industry, and state interests to limit environmental harm through regulation. Others, Leeming's "non-modernists," who resisted both transformation of their lands and lives and threats to their person, found themselves more comfortable towards the other end of this scale.

But to categorize is to generalize, and it is as well to remember that such schematic descriptions usually shrink or eclipse the range of motivations and convictions that drive individual behaviour, even as they tend to limit recognition of the ways in which individual and group commitments shift with changing circumstances. To flesh out this complexity, let us consider briefly three important environmental activists who appear in Leeming's discussions of the Nova Scotia campaigns against nuclear power and uranium mining, encompassing the span of their activist careers beyond the issues dealt with in these pages. Judy Davis (1951–2010) and David Orton (1934–2011) were friends and fellow environmental crusaders with strong attachments to rural Nova Scotia. Davis, said her obituary, died of cancer and "a broken heart, the latter caused by witnessing the continued destruction of the ecology of the planet through human exploitation, economic greed and war."[21] Orton, who habitually signed off his correspondence with "For the Earth," also succumbed to cancer after a brief illness. Susan Holtz, an American who had recently arrived in the province in 1972, was a member of the Halifax Society of Friends (Quakers) monthly meeting and rose to prominence through her engagement with the nuclear power issue.

Judy Davis was a radical activist who "never stopped fighting for environmental and social justice." Born in Pictou County, she lived for almost thirty years in a small log house on a ten-acre plot five miles from the village of Tatamagouche in Colchester County, where she came to understand "the importance of being rooted 'in place' to become an Earth defender."

Through her commitment to helping those less fortunate, and her in-
clination to direct action, she had what many judged to be "a major impact
on environmental, social justice, feminist, and workers' struggles in Nova
Scotia." She was not, though, a "solidarity with Latin America" type of
person," but someone who was firmly committed to "the human and later
non-human, community" of which she was a part. Most of her organizing
was done at the local scale.[22]

David Orton was a voracious reader, a prolific writer, a "deep green"
activist, and an important Canadian thinker who brought together environmental ethics and social justice concerns. His roots took time to strike
in Nova Scotia soil. Born in Britain, he emigrated to Canada in 1957 to
avoid compulsory national military service, after an apprenticeship in the
Portsmouth dockyard, a year in the Naval Architecture program at Durham
University, and discharge from the Royal Army Educational Core after
nine months of service. After taking a BA degree from Sir George Williams
University in Montreal in 1963, Orton completed an MA in Sociology at
the New School for Social Research in New York and was admitted to the
PhD program there. In 1967, his dissertation unfinished, he returned to
teach at Sir George Williams University. Active in the Movement for
Socialist Liberation (in which Andre Gunder Frank, the left Latin American
economist was involved) and the "Internationalists" (precursors of the
Marxist-Leninist faction of the Communist Party of Canada), Orton
ruffled the feathers of colleagues and administrators during a tumultuous
time in the university.[23]

When university administrators chose not to renew Orton's teaching
contract, he threw himself into political organizing, ran unsuccessfully for
federal office, was arrested at a political rally in Toronto and imprisoned,
and worked on the Montreal waterfront. In 1977, he moved with his
family to a temporary job in a salmon cannery on Haida Gwaii, in British
Columbia. There contact with the Haida, engagement with the emerging
issues surrounding the protection of South Moresby Island, and a kayak
trip into that area persuaded him to focus his organizing work on environmental issues rather than social justice politics. This he did for two years
in Victoria, before moving to Halifax in 1979. In Nova Scotia, Orton
engaged a number of environmental issues, not all of them obviously local.
In the 1980s, he spoke out against acid rain, the Atlantic seal hunt, pulpwood forestry practices, and uranium in well water, even as he penned a
blistering indictment of Judge McCleave and the uranium inquiry on
behalf of the Socialist Environmental Protection and Occupational Health
Group in October 1982.[24] Two years later, he moved with his partner and

infant daughter to a "130-acre hill farm at the end of a dirt road through a forest" in Pictou County, where wood was the only source of heat.[25] At much the same time, Orton became aware of Arne Naess's arguments for deep ecology, which convinced him of "the importance of moving beyond the human-centered values (anthropocentrism) of the anarchist, social democratic, communist, and socialist traditions."[26]

Through the rest of his days, Orton strove to meld his social justice and deep ecological commitments. Rejecting Naess's position that the peace, social justice, and ecology movements should remain separate even as they stood beneath a Green movement banner, Orton became a leading proponent of "Left Biocentrism." Articulating this position through collaboration and engagement with like-minded individuals on a website established in 1988, Orton adopted the view "that the Earth belongs to no one and should be a non-privatized Commons." In furtherance of this basic ecocentric commitment, he produced a remarkable series of reflections on both specific environmental conflicts and more general philosophical issues confronting environmentalists. Published on the "Green Web" and in a wide array of publications (including *Canadian Dimension,* to which he contributed twenty-seven articles, reviews, and comments between 1989 and 2009), this impressive oeuvre defies succinct summary.[27]

In broad terms, however, Orton's written contributions clearly put Earth first and "express solidarity with all life"; they also oppose economic growth and consumerism, advocate voluntary simplicity, and espouse a bioregional rather than a global or strictly local focus, even as they call for a global redistribution of wealth. Equally clear is Orton's insistence on the importance of an environmentalism independent of government or corporate support or influence. But theory and praxis are not always easily aligned. Owning property and insisting that the earth should be a commons was one contradiction that Orton addressed by pointing out that capitalist society offered few alternatives beyond the short-term option of using "private property 'laws' to buy one's own place ... [for those who would further] conservation and wildlife preservation." He also saw ownership of his land as a necessary concession for its stewardship, and thus realization of the goal of "being the change you want to see in others and the world." Similarly, Orton refused to join the government-funded Canadian Environmental Network (CEN) or any of its constituent organizations, although he shared concerns with, and worked alongside, many of those who did. In his view, CEN accepted and legitimized industrial capitalist society, just as the Green Party (towards which he inclined for want of an effective alternative) was ultimately "in the business of putting forward fudge and 'market' or soothing eco-capitalist positions, which do not call industrial capitalism into question or bring about the needed fundamental shift in societal consciousness."[28]

Orton recognized that many Nova Scotians involved in environmental issues were rural dwellers responding to "'not in my backyard' situations." Judy Davis, he observed, was always "prepared to take on those in her immediate environment who wanted to spray biocides on nearby forest lands, blow up beaver dams down her gravel road because of someone's "flooding" complaint, or run a snowmobile trail in front of her house." Although she developed, through her association with Orton and others a sense of "fit" with deep ecology, "this self-awareness came from her own life experiences ... [not] from reading philosophy books or some kind of intellectual conversion." She joined protests against racism in all parts of Nova Scotia and supported Aboriginal issues (including the Mohawk in the 1990 Oka dispute) across the country. Although she stood (unsuccessfully) as a candidate for federal and provincial Green parties in the first decade of the millennium, her métier lay in organizing events such as a funeral procession and memorial service for the trees and the non-human inhabitants of the softwood forest being turned to woodchips at Spiddle Hill in Colchester County, and in rallying crowds with her musicianship. Orton admired such efforts and joined several of them. But wedded as he was to deep ecology's fundamental claim "that major ecological problems cannot be resolved within the existing capitalist or socialist industrial economic system," he found these efforts insufficient because they lacked the staying power to produce regime change. Local action could not achieve the required socio-political revolution. Anxious for transformation, he was also and often scathing of those pragmatists who sought improvement by tinkering with the industrial capitalist system. Such approaches he regarded as misguided; those who pursued them were deluded "pollyannas."[29]

Susan Holtz was one of those with whom Orton crossed swords. Asked to provide guidance to the Halifax Friends meeting on the issue of nuclear power development in 1972, she concluded, as Leeming points out in Chapter 2, that it would be an unnecessary spur to wasteful consumption. Taking an intellectual rather than a categorically oppositional stance towards the nuclear challenge, she quickly forged a nationwide coalition of people and groups opposed to nuclear power that only belatedly developed connections with other campaigners in Nova Scotia. Reflecting emerging ideas about the desirability of developing a conserver society and adopting what American environmental scientist Amory Lovins termed a "soft energy path," Holtz and a cluster of like-minded associates joined Nova

Scotia's Ecology Action Centre and operated under that group's banner from 1975 onward. There they worked assiduously to assess the risks of nuclear power and promote research into alternative energy technologies. By the end of the decade, much of this work was being funded by government contracts: in 1980–81, Holtz was working in conjunction with sixteen Nova Scotia municipalities seeking efficiencies in energy management (funded by the Nova Scotia Department of Mines and Energy); updating an earlier study of soft energy paths under a contract with the federal Department of Energy, Mines and Resources (arranged by Friends of the Earth); investigating legal issues surrounding "solar rights"; and assessing the potential for energy generation from waste biomass (the latter two contracts funded by the Law Foundation and the Technical University of Nova Scotia).[30] Holtz saw herself as a conciliator, someone who could serve as a bridge between radical environmentalists and the decision makers who held the levers of political and economic power.[31]

In that vein, and though staunchly opposed to uranium mining, she agreed (as Leeming notes) to comment on draft guidelines for mining exploration prepared by Jack Garnett, assistant deputy minister in the Department of Mines and Energy. When Garnett subsequently claimed that his guidelines had the approval of environmental groups, Orton made clear his opposition to Holtz's actions and resigned from the uranium subcommittee of the Ecology Action Centre. Here, as more generally, he had no time for "shallow ecologists" (Leeming's "moderns") who "worked the system" and accepted its rewards without seriously threatening its legitimacy.

In the end, Davis, Orton, and Holtz were all local activists to some degree. They made environmental concerns in Nova Scotia, or particular places within it, a major focus of their attention. But the scale, purpose, and tenor of their environmental commitments varied, and they differed over the means most likely to be effective in achieving the ends that they desired. Davis perhaps best exemplifies the resolute defender of particular places against specific threats. Holtz was no less antagonistic towards nuclear power and uranium mining than others, and was taken aback by Orton's trenchant criticism of her efforts to deflect energy development away from the hard path it constituted, considering it a gratuitous attack delivered in a "spiteful and unscrupulous" way (p. 114). Orton, for his part, was increasingly critical of the industrial-capitalist system. Although he was deeply attached to his 130 acres of Pictou County soil, he denied the propriety of his ownership, and addressed himself (particularly through the medium of the Green Web started in 1988) to the radical, thorough-going transformation of human-nature relations.

Together, these vignettes illustrate the conundrums facing late-twentieth- and early-twenty-first-century environmentalism. Acting locally has its merits but the calculus of addition rarely translates into global gains. Stand-ing at barricades in defence of home places might save this patch of forest or that mountainside from exploitation, but such local victories cannot stay the juggernaut of economic development and win the battle. Con-frontations at struggle sites translate into a war of attrition. Rebuffed here, industry moves on to pursue its goals there; environmentalists rally anew to blockade again; the battle runs its course through injunctions, arrests, and acrimony, as the case may be; then the cycle is repeated. En-vironmentalists rarely have the time, numbers, and resources to exhaust the companies' drive to continue. Combatting the growth imperatives of industrial-capitalist society requires other tactics. Time and again, environ-mentalists have divided – just as Holtz and Orton did – over what those tactics should be. Almost half a century ago, Greenpeace espoused non-violent environmental action and used both spectacular protest and market boycotts to protect particular species and specific places or ecolo-gies; sometime Greenpeace member Paul Watson came to favour more direct action: "holding up protest signs, taking pictures and 'bearing wit-ness' while whales are getting killed in front of you doesn't achieve anything at all," he said. Meanwhile, others elsewhere in Canada strove to establish regulations to improve air quality, water and waste management, land-use planning, natural resource extraction, and energy policy.[32]

Should environmentalists work within the industrial-capitalist system to limit and ameliorate exploitation and despoliation of the earth, or should they endeavour to overthrow that system entirely? Are these really alterna-tives? Although Susan Holtz and David Orton divided bitterly when she opted for the former course and he espoused the latter, one has to wonder whether these approaches are mutually exclusive. Might urgent short-term actions – whether they be peaceful demonstrations like that organized by Judy Davis on Spiddle Hill or dangerous clashes of whaling and protest vessels on the high seas engineered by the Sea Shepherd Society – help shift policy horizons and achieve long-term goals of regime change, by raising awareness of the issues and persuading more and more people of the need to act? Or are they likely to have the opposite effect, because conflict can be dysfunctional? Is it possible, in our current circumstances, to achieve the kind of dramatic regime change that Orton sought in time to address the environmental challenges we face? Or does our best hope lie in developing a pluralistic decision-making environment in which conflict is only part of a conversation that encompasses the possibility of working to change things from within? The questions remain – and they turn on the imponderable intersection of scale and process. Surely moving the environmental agenda forward is an art, in which public awareness, appropriate regulation, and improved technology are equal parts of the story.

In the second decade of the third millennium, many of our most pressing environmental problems have diffuse and difficult-to-address origins; although we may experience them "at home," they are generated at some remove, often by persons who (or organizations that) reject responsibility for them, as they stand beyond the reach of local action and territorially bound legal and political systems. In other, more formal words, many of these challenges are transgenerational products of "extraterritorial forces." Thinking about the consequences of this in relation to the problem of climate change, Harald Welzer echoes Michael Maniates's ironic, and iconic, formulation – "Plant a Tree, Buy a Bike, Save the World?" – in describing "individualist strategies" as tranquilizers that soothe anxiety but achieve little. He also finds little prospect of effective political responses at the international level, given the time it takes for political processes to run their course. So he argues for action at an intermediate scale, between the local and the global.[33]

All of this might prompt readers to wonder whether the strong place attachment that Leeming identifies among "non-modernists" remains as vital and as valuable to effective environmental commitment as once it was. If the increasing connectedness of modern-day societies is spawning relatively placeless cultures, then – it might seem logical to argue – the strong sense of place-emphasis that led people to act in defence of local nature in the 1970s may consign environmentalism to irrelevance in the twenty-first century. In this vein, Ursula Heise insists that modern-day environmentalism needs to move beyond efforts to recuperate "a sense of place," to "foster an understanding of how a wide variety of both natural and cultural places and processes are connected and shape each other around the world, and how human impact affects and changes this connectedness."[34]

Others have urged the importance of developing a "progressive" sense of place, one that is "not self-enclosing and defensive, but outward-looking," and adapted to the conditions of postmodernity, in which "space appears to shrink to a 'global village' of telecommunications and a 'spaceship earth' of economic and ecological interdependencies ... and time horizons shorten to the point where the present is all there is."[35] After all, even David Orton acknowledged that although "our place is back-to-the-land ... we also have a whole computer network based there, which means we can communicate around the world."[36]

Should we, then, abandon our idea of places "as areas with boundaries around" them, and imagine them, instead, "as articulated moments in networks of social relations and understandings," with most of those relations and understandings "constructed on a far larger scale than ... the place itself"? Geographer Doreen Massey has argued this case, which requires, in Tim Cresswell's words, a shift from thinking about "place vertically – as rooted in time immemorial – to thinking of it horizontally, as produced relationally through its connections."[37] In this recalibration, the forces of globalization have reframed places; they are no longer locales buffeted by remote forces or the locus of romanticized escapism, but dynamic settings shaped by new philosophies, new forms of engagement, and new synergies between ideas and action at local and global scales.

Leeming complicates this argument by demonstrating that the past was less foreign than such narratives imagine. "Local" communities have long been responding to the actions – be they factory closures or toxic spills – of global scale political-economic-financial powers. They have had to assume responsibility for problems they have not created, and they have grappled with the uneven distribution of power and influence. Leeming's home-place defenders thought globally and acted locally, even as they worked at national and global scales to maintain particular conceptions of their localities. Risk assessments differed, and some communities were divided by conflicting opinions about how best to deal with them; in others, local responses helped build trust, engagement, resilience, and power. Peoples' struggles to secure what they held dear were moulded by individuals deeply committed to particular settings and shaped by local circumstances, but they turned, in the end, on emergent global concerns about the environment. Although they invoked the past, they were oriented to the future. By detailing the often bitter struggle between "the place-loving soul" of the environmental movement on the one hand and "the temptations of a placeless modernity" on the other, *In Defence of Home Places* enriches our understanding of politics and society in Nova Scotia, provides a valuable exegesis of the tangle of relations that brought the state, industry, and environmental activists into conflict and collaboration, and illuminates the strength of local environmentalists' resistance to modernity's "strategy of conceptual encompassment" (pp. 149–50).

Acknowledgments

I WOULD LIKE TO express my appreciation and gratitude to Dr. Claire Campbell, for her good humour, patient guidance, and sound advice during the research that led to the publication of this book. I would also like to thank Drs. Robert Summerby-Murray, Jerry Bannister, Alan MacEachern, Dean Bavington, and John Reid, as well as the two anonymous reviewers, for their helpful criticism and friendly advice. Similarly, I want to declare my gratitude to Valerie Peck at the office of the Department of History at Dalhousie University and to Randy Schmidt, Ann Macklem, and Dr. Graeme Wynn at UBC Press. I also wish to acknowledge the support of Dalhousie University and the Social Sciences and Humanities Research Council of Canada. Finally, my wife, Kazue, deserves and has my deepest gratitude for her wholehearted support of my work. She is more responsible than she knows for my being able to complete this book, and it is dedicated to her.

Abbreviations

2,4-D	2,4-dichlorophenoxyacetic acid
2,4,5-T	2,4,5-trichlorophenoxyacetic acid
AECB	Atomic Energy Control Board
AECL	Atomic Energy of Canada Limited
APES	Association for the Preservation of the Eastern Shore
BBPCC	Bedford Basin Pollution Control Committee
Btk	*Bacillus thuringiensis kurstaki*
CANDU	Canadian Deuterium-Uranium reactor
CAP	Coalition Against Pesticides (Cape Breton)
CAPE	Citizen Action to Protect the Environment
CARE	Citizens Against a Radioactive Environment
CAUM	Citizens Against Uranium Mining
CBLAS	Cape Breton Landowners Against the Spray
CCCC	Concerned Citizens of Cumberland County
CCNR	Canadian Coalition for Nuclear Responsibility
CEAC	Canadian Environmental Advisory Council
CEN	Canadian Environmental Network
CEPA	Chaleur Environmental Protection Association
CFS	Canadian Forestry Service
COPE	Communities Organized to Protect the Environment
CSG	Conserver Society Group
DDT	dichlorodiphenyltrichloroethane
DLF	Department of Lands and Forests (Nova Scotia)

DOA	Department of Agriculture (Nova Scotia)
DOE	Department of Environment (Nova Scotia)
EAC	Ecology Action Centre
ECC	Environmental Control Council
EMR	Energy, Mines and Resources
ENGO	environmental non-governmental organization
FACT	Fundy Area Concern for Tomorrow
FOE	Friends of the Earth (Canada)
FON	Friends of Nature
HFN	Halifax Field Naturalists
HFS	Herbicide Fund Society
IEL	Industrial Estates Limited
JAG	Joint Action Group
KASE	Kings Association to Save the Environment
LIP	Local Initiatives Program
LWR	light-water reactor
MEC	Maritime Energy Coalition
MEC	Maritime Energy Corporation
MOVE	Movement for Citizens' Voice and Action
NBEPC	New Brunswick Electric Power Corporation
NIMBY	Not In My Back Yard
NPSG	Nuclear Power Study Group
NSBS	Nova Scotia Bird Society
NSEA	Nova Scotia Environment Alliance
NSEW	North Shore Environment Web
NSFI	Nova Scotia Forest Industries
NSLPC	Nova Scotia Light and Power Company
NSPC	Nova Scotia Power Corporation
NSPCC	Northumberland Strait Pollution Control Committee
NSRC	Nova Scotia Resources Council
NSRF	Nova Scotia Research Foundation
NSSA	Nova Scotia Salmon Association
OFY	Opportunities for Youth
OPEC	Organization of the Petroleum Exporting Countries
PAN	Pesticide Action Network
PCAC	Purcell's Cove Action Committee
PEP	People for Environmental Protection
RESCUE	Residents Enlisted to Save Communities from Uranium Exploration

SCC	Science Council of Canada
SEP	soft energy path
SEPOHG	Socialist Environmental Protection and Occupational Health Group
SKMS	Save Kelly's Mountain Society
SSEPA	South Shore Environmental Protection Association
TCDD	tetrachlorodibenzodioxin
VOW	Voice of Women for Peace

In Defence of Home Places

Introduction

*I*N DEFENCE OF HOME PLACES traces the origins and development
of environmental activism in Nova Scotia from the 1960s through the
1980s. At the core, it is about the recognition and consequences of ideological differences within the environmental movement of a modern
country. This book considers three major controversies that shaped the
environmental movement through the late 1980s and into the present day:
nuclear power, chemical forestry, and uranium mining. Time and again,
environmental activists confronting these issues struggled to negotiate differences of opinion on the basic questions of environmentalism: the causes
of environmental problems, the proper relationship of humanity to the
non-human world, and the actions most likely to achieve that relationship.
By 1985, Nova Scotian environmentalists riven by these differences had
divided their movement into mainstream and fringe environmentalisms,
with quite different answers to the major questions before them.

Nova Scotian environmental activists mirrored their counterparts in
other modern societies in disagreeing fundamentally over the basic structures of those societies. "Ecological modernists" believed that (most) environmental problems were unintended and technologically remediable
side-effects of otherwise desirable industrial processes, and were inclined
to address them through collaboration with the state and industry. They
did not question modern society's commitment to its fundamental principles, including large-scale industry, economic growth, and the acceptance
of science as the sole legitimate form of knowledge. "Non-modernists,"
meanwhile, viewed environmental problems as the necessary and evil

3

consequences of these same commitments. They favoured radical changes in economy and society. Recognizing that neither the state nor industry would willingly subvert the modernist project, they pursued the politics of confrontation rather than consultation. While these groups cooperated often in the early days of activism, by 1985 the ecological modernists had drifted into a position in which they found further cooperation difficult and occasionally impossible.

Thus the history of Nova Scotian environmental activism speaks also to the definition of environmentalism everywhere. One group of researchers has long favoured an exclusive definition, insisting for more than three decades that the "lifestyle" environmentalism of the affluent world in the 1960s – characterized by the pursuit of clean air, clean water, and outdoor recreation as luxury commodities – is a qualitatively new development in the social history of the Western world, uniquely deserving of the label "environmentalism." It is, they say, a product of demographic and economic changes following the Second World War, set apart from contemporary and antecedent movements by its proponents' unprecedented wealth and comfort; this they call "postmaterialism."[1] Others favour a more inclusive definition, ranking such lifestyle environmentalism alongside prior anti-industrial movements and contemporary environment-themed activism in the less wealthy world, all of them motivated by reactions against the undesirable effects of industry, capitalism, and the dominance of scientific thinking, which impose an unfavourable monetary "discount rate" on the sacred sites, home places, and other economically incommensurable values held by poorer people.[2] This book falls into the latter camp.

Non-modern argument was rarely articulated in theoretical terms, and not until the end of the 1970s, partly in reaction to the consolidation of an ecomodernist mainstream, but to recognize its presence at the heart of environmental activism is to acknowledge that all environmentalism is born of a reaction to late modernity, and the "political, social, and scientific consensus that has dominated the last two or three hundred years of public life." Broadly, modernity is associated with the three key tenets already mentioned, plus a progressive theory of history, nation-state polity, increased use of inanimate energy sources, and a social and intellectual "system of objectification" by which uniquely local or personal forms of knowledge are replaced by a more legitimate universal way of thinking and speaking.[3] Ecomodernist reformers operated within the modernist project; non-modern radicals generally contested it, in diverse and inconsistent fashion. Both responded to its ecological impact.

The commitments of both sorts of environmentalists rested on personal and emotional foundations. Personal vulnerability motivated *all* environmental activism. It most often found expression in defence of economic and affective ties to local places, as the intrusion of new industrial developments into rural settings spawned reaction. Individual and collective protests focused on the degradation of known and familiar villages, coves, hamlets, and landscapes, rather than on some distant prospect or "the environment" in the abstract. Some local activists eventually embraced global perspectives, but rarely at the cost of their personal and local attachments. Only a handful of modernist activists who found it politically inexpedient to acknowledge local and individual experience (an illegitimate source of knowledge, in the modernist world view) truly abandoned the defence of home places as a central point of reference for their actions. In sum, the idea of and attachment to home place here outweighed (though they never completely displaced) network formation, resource distribution, class, race, and gender as the driving forces of environmental action.

For all that individual world views mattered, Canadian political culture also shaped environmentalism in Nova Scotia in the 1970s and 1980s. Government influence, both federal and provincial, helped to drive modernist activists towards conciliatory politics, and by unwitting extension helped to solidify the radical critique. In doing so, it also reduced the political impact of the two strands of activism, which often achieved real political influence when working in combination. In the end, this is the story of several groups of agents – governmental factions, activist groups, and various industrialists – negotiating the politics of social movements with varying levels of success.

Environmentalists in Nova Scotia participated in a national and international political conversation about particular environmental hazards. The issues in question in Nova Scotia – especially nuclear power, chemical forestry, and uranium mining – were subjects of major controversy across the country and around the world, where others took careful note of events in the province and attempted to intervene in Nova Scotia to the advantage of whatever party they represented in Ottawa, Washington, Stockholm, or other locales.[4] Nova Scotians joined federal non-governmental groups such as the Canadian Coalition for Nuclear Responsibility (CCNR) and Friends of the Earth Canada (FOE), and they participated in consultative groups, including the Canadian Environmental Advisory Council (CEAC) and Canadian Environmental Network (CEN). But the substance of the national story, the real action, took place within provincial, not federal, arenas. The story of environmentalism in Nova Scotia is thus a particularly clear window into the story of the movement in Canada.

This study highlights the salience of activity at the provincial level in Canadian environmentalism, and illuminates how activism operated on the ground by attending to the intragovernmental squabbles, the industry/government collusion, the small rural activist groups' interaction with their better-funded urban peers, the cooperation between First Nations activists and other environmentalists, and the collaborations with peace activists, conservationists, and various social justice advocates that shaped events. There is a tendency in national studies to focus on the major players, such as Greenpeace or Pollution Probe, or on the federal and provincial government agencies established in the 1970s to deal with environmental problems, but while these make important contributions to the story of environmentalism in Canada, relationships within provinces best define it. The environmental movement, in Nova Scotia and in most parts of Canada and the world, has been firmly rooted in, and found its motive force at, the local level. Relationships between larger mainstream groups and governments were shaped in large part by the smaller, usually rural groups whose stubborn defence of their home places led them to reject (at least in part) the conclusions that the other two offered about the causes of environmental problems, the role of humanity in nature, and the role of environmentalists in the political process.[5]

That there ever was a national Canadian environmental movement at all is a matter of some debate. Whereas groups across Canada often shared common challenges and opportunities through the constitutional division of powers that gave the provinces jurisdiction over natural resources, there was very little formal cooperation at the federal level among Canadian environmental activists until the late 1970s, and even that seemed often tangential to the most vital emerging concerns of provincial and local movements. There was simply far more coalition, cooperation, and common concern among groups within single provinces. For these reasons, this book speaks of a Canadian environmentalism made up of provincial movements rather than a single Canadian environmental movement, an important distinction.

Far from the simple leisure pursuit of an affluent urban society, environmentalism appears a far more complex social phenomenon in its Nova Scotian iteration. It was a reaction to the ecological impacts of federal/provincial development schemes designed to alleviate regional economic disparity, and it frequently reshaped those plans when they conflicted with small communities' visions of a viable, desirable, and just economic future.

Environmentalism arose in Nova Scotia in response to an unusual congregation of environmental events of national and international significance but crucially local impact. In little more than a decade after 1970, the province witnessed Canada's largest marine oil spill, an attempt to organize a regional nuclear-powered electrical utility, the improbable defeat of a Swedish multinational's plan to spray insecticides on Cape Breton Island, an internationally infamous provincial Supreme Court ruling on herbicides, moves by a French company to begin uranium mining in the midst of agricultural settlement, and the beginning of a decades-long fight to clean up the Sydney tar ponds, legacy of the Sydney steel mills and still the most toxic site in the country. Throughout this book, there appear men and women like Murray Prest, a Halifax County mill owner and bitter opponent of large-scale industrial forestry. Prest's economic self-interest in preserving a smaller-scale sawn-lumber industry and his ecological arguments against clearcut chemical forestry seem at first to invite a distinction between values and interests, but his position actually expresses a commitment to a non-modern way of life in which the continuity and meaning of the human (and economic) community depended on a relationship with the forest in which small trees were left for later generations of woods workers, and the forest cover was left intact to preserve the various ecological amenities (food, clean water, and so on) that the community required. Prest was not concerned with the philosophical coherence of his position or with acquiring a label for it; he accepted capitalism, for example, but rejected large-scale industry, political and economic centralization, and the total commodification of the forest. This book is the story of men and women like Prest – diverse in their dissent, united in their defence of home places.

CHAPTER OVERVIEW

Tracing the origins of environmental awareness in Nova Scotia, Chapter 1, "At Home and Abroad," recounts a diverse set of relatively minor controversies before the early 1970s. From Boat Harbour to Mahone Bay to Chedabucto Bay, these early struggles established the lasting pattern of environmental activism in Nova Scotia as a movement primarily motivated by rural communities' reactions to the negative consequences of industrial modernity: the pollution, the centralization of power in the city, the increasing exploitation of the hinterland, and the threatened destruction of traditional economies. These initial actions make it clear that early environmental

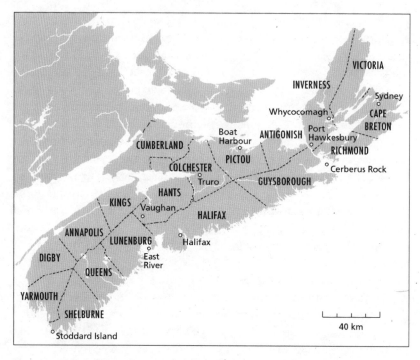

FIGURE 1   Map of Nova Scotian counties and relevant sites. *Cartography by Eric Leinberger*

activists in the province held well-developed and diverse opinions on the nature of their movement. Some were the intellectual brothers of the techno-cratic nineteenth-century conservation movement, others were radically suspicious of government motives. They were a diverse group of people, with starkly different economic and geographic origins, as demonstrated by the map of relevant sites in Figure 1; their number included fishermen, trappers, medical doctors, Mi'kmaq band chiefs, farmers, students, and more. All were aware of the currents of popular ecology in the world at large. Government attempts to manage political radicalism, not limited to environmentalism, helped to create environmental activist organizations such as the Ecology Action Centre (EAC) in Halifax. These groups showed a generalized concern for "the environment" and an early tendency to focus on public education rather than on political pressure. Meanwhile, outside of the city, activists grew increasingly effective at challenging and changing the course of development policies that drew new industrial projects into their communities. By the end of the 1960s, they were mustering scientific arguments to shape public opposition to destructive industries, eventually prompting the government in Halifax to issue its own regulations in an attempt to blunt the critique. In the early 1970s, the rural movement's growing political power culminated in the ignominious defeat of a federal/ provincial plan to create a new national park on the province's eastern shore, and as links among activists of all types and locations grew stronger, they began to take up more ambitious causes.

Chapter 2, "The Two MECs," tracks the development of anti-nuclear activism from 1972 through to the provincial government's decision in 1980 to abandon plans for a single Maritime electric utility. Beginning with a plan to build a large nuclear power plant on the south shore to generate electricity for export to the United States, the narrative tracks a series of external factors responsible for seismic changes in the Nova Scotian movement: the oil price shocks of the early 1970s that ignited debate about the need for nuclear generation in the Maritimes, the limits to growth theory that widely popularized the notion in 1972 that economic growth could not continue forever on a finite planet, and the Three Mile Island nuclear disaster in 1979, which revitalized anti-nuclear politics across many jurisdictions, including the Maritimes. This chapter points to the importance of political jurisdiction in shaping Canadian environmentalism, as the attempt to unite the energy agencies of Nova Scotia, New Brunswick, and Prince Edward Island produced an activist backlash that strengthened and drew together the movements in all three provinces but collapsed with the end of the regional utilities unification plan. In Nova Scotia, organization against nuclear energy brought the first long-lasting and multi-issue activist groups to wide public notice. It also led many activists to realize the possibility of making their appeals directly to government, remaining within the parameters of the existing modernist energy policy rather than continuing on a course of total rejection of large industrial electricity generation. A politically conservative (though still strongly reformist) group of religious peace activists who eventually formed the energy committee of the Halifax-based Ecology Action Centre were crucial and early actors in this shift towards modernism and government consultation, and remained central figures in the movement for many years thereafter. The sense of personal vulnerability of farming and fishing communities remained a key motivating factor and sustained the defence of home places as a theme in environmental argument, but it was joined by the less geographically fixed contributions of immigrant American Quakers. Another central theme of this and subsequent chapters, mistrust of the very secretive provincial government, was a source of disagreement between the more urban ecomodernist activists and their rural peers. Nevertheless, both remained committed to a vision of the movement as a grassroots coalition of local groups with a common, province-wide purpose, so far as anti-nuclear activism was concerned; in neither the nascent ecomodernist mainstream nor the non-modernist groups was the notoriously self-centred NIMBY (Not in My Back Yard) activist ever to be found.

In Chapter 3, "Power from the People," the focus shifts to a hard-fought battle over the use of chemical pesticides in forestry, beginning in 1976 with a famous (and successful) campaign to keep Nova Scotia from following New Brunswick's lead in spraying insecticides from the air in a futile bid to reduce the numbers of spruce budworm on Cape Breton Island. The events recounted in this chapter demonstrate the growing determination of government and industry to resist environmentalist pressure, and the equally strong determination of some activists to change the minds of their opponents in the bureaucracy without resorting to radical tactics or rhetoric. The collision of the two played out in the infamous (and unsuccessful) herbicide trial of 1983, with which the chapter ends. The importance of global trends is central to the narrative of this chapter, with both environmentalists and the forest industry reaching beyond the province's borders for resources in their conflict, as well as for new arenas in which to make use of their respective gains within the province. But the limits of international assistance are fully on display as well in the second half of the chapter, as international connections help the more modernist activists pursue their case in court, where non-modern arguments carried no weight. Several secondary themes rise to prominence here, such as the agency of the news media, the internal politics of the provincial bureaucracy (and its commitment to the preservation of its own power), the continuing vitality of the resource conservation movement, and the rhetorical power of jobs and immigration status to both the advantage and the detriment of activists.

From the air above to the stones below, Chapter 4, "Two Environmentalisms," examines a comparatively brief but intense campaign from 1981 to 1985 aimed at preventing uranium mining in the province. Most of the action discussed in this chapter occurred in and around the hearings of the Nova Scotia Royal Commission on Uranium Mining, during which activists finally acknowledged the full extent of their differences. This is a story of fragmentation – and the fracture in the movement was bitter – but the campaign to halt uranium mining was a surprising success, owing in large part to the persistence of the radicals, who continued fighting desperately to defend their homes. Links with peace and conservation groups are once again on display, along with the urban activists' greater tendency towards modernism and the global links of every set of actors. Among the central themes, the influence (and rejection) of metropolitanism ranks high as a factor in Canadian environmentalism, brought into focus by the vitality of rural activism in the Nova Scotian movement. Indeed, Canadian environmentalism as seen through the prism of the province's experience appears inextricably bound up with the discontents of metropolitanism, between regions of the country as well as between rural and urban locales within provinces. So, too, do the results of the activist schism reflect a national reality: the increased articulation and political consciousness of radical environmental critique, and the growing commitment of the ecomodernist mainstream to consultative processes. Government action around uranium mining in Nova Scotia also illustrates the extent to which judicial and quasi-judicial forms – courts and commissions – serve to reinforce the modernist mode of thought, as well as the extent to which pressure politics continued to win results.

Finally, Chapter 5, "Watermelons and Market Greens," offers a short overview of some major environmental controversies of the 1980s and demonstrates the persistence of the fissure in the province's environmental movement. Important to note is the ability of radical and mainstream activists to cooperate despite their differences, when involved in such issues as the threatened demolition of Kelly's Mountain for a granite quarry. Local grievance and defence of home places still drove environmental activism in the 1980s, as it does today, and the provincial and federal governments still encouraged ecomodernist positions. The provincial movement, however, settled into a new pattern in the 1980s, with radicalism centred geographically on the north and south shores and on Cape Breton Island, and with modernist environmentalism predominant in the city of Halifax and the central mainland.

The process by which the ill-defined activism of the late 1960s became the theoretically and politically articulate alternative environmentalisms of the 1980s, the narrative contained in these chapters, is not a decline-and-fall story. It is a story of intellectual growth and differentiation in a diverse movement. Environmental activists did err at times in pursuing modernist policies too stridently, but they also demonstrated the power of both strands of the movement to effect change in development policy by working, if not in close collaboration, at least on parallel tracks. If there is a lesson to be taken from these pages, it is not that environmentalism in Nova Scotia failed, but that it has always contained the elements of success, and at times has even managed to make them work.

I
At Home and Abroad
The Genesis of Environmentalism

Nova Scotians in the first half of the twentieth century were ambivalent about changes to the provincial landscape. The earliest large hydroelectric dams in the province were remote from settlement and not dramatically larger than the mill dams that had dotted the province's rivers since European settlement began. Even the creation of Lake Rossignol in Queens County by the Mersey Paper Company in the 1920s, which drowned ten natural lakes and over ten thousand acres of land, took place without significant objection.[1] By the end of the 1950s, however, the different biological impacts of small and large dams were better understood, and free-flowing rivers were better appreciated as refugia for Atlantic salmon.[2]

Just as more people were coming to appreciate the value and vulnerability of rivers, the Nova Scotia Light and Power Company (NSLPC) sought permission to dam the Gold River in Lunenburg County for hydroelectric power. The project would have flooded slightly more than two thousand acres from the village of Chester almost to the border of Kings County, creating two long flowage lakes where once there had been a single river channel and eight natural lakes. Worse, two sections of riverbed between dams and powerhouses would be left permanently dry. Had the power company's engineers set out to demonstrate their ability to utterly unmake a river, they could not have designed a better exhibition piece.

There was no question that NSLPC's plan would have ended salmon fishing on the river, and it drew immediate, vocal opposition from sport fishermen in the province, especially in the city of Halifax. In the face of such opposition, the Stanfield government created a Royal Commission in the summer of 1961 to recommend a solution. Led by Halifax lawyer Russell McInnes, the commission operated on a tight schedule. Nonetheless, McInnes promptly expanded the inquiry to investigate dams and fisheries on other rivers. Opponents of the project "consisted almost wholly of persons interested in sport fishing; particularly salmon fishing by rod," but they included, and had the support of, powerful members of the province's economic elite, including Frank Sobey, then head of the province's industrial development agency.[3] They were also supported by the federal Department of Fisheries, whose representative called it "a mistake [to] absolutely destroy salmon and other anadromous fish" on a biologically healthy river merely for "the cheapest possible power."[4]

With remarkable speed, McInnes issued a recommendation before the end of the year, condemning the project and endorsing the conservationist gospel. Since European settlement, he wrote, "the peoples and Governments of this province have used, abused, and for the most part indiscriminately destroyed its natural resources with a prodigality that would be incredible if it were not so patently and painfully obvious."[5] His recommendations included a stay on any and all hydroelectric works anywhere in the province until a long-term planning exercise could be completed and a sport fishermen's licensing system created to support a program of scientific conservation. Significantly, he also adhered to the classic conservationist's elevation of the common good over local preference: the "large majority" of residents near the Gold River had registered support for the project in the hope of new jobs, but McInnes insisted that such jobs would be few and that "the Gold River and every other natural resource and tourist attraction of this province belongs to the people of the province as a whole [and] it should not be available to satisfy the whims or desires of the persons living closest to it."[6] Public opinion carried a weight in his ruling that it did not carry in much conservationist writing – McInnes stopped short of insisting that only scientific experts could determine the common good, and he acknowledged the legitimacy of popular opposition, so long as all citizens of the province enjoyed equal influence – but apart from the threat of public disapproval implicit in the use of newspapers as a forum for elite dissent, genuine popular resistance played very little role in the fate of the Gold River. Unlike the narrowly institutional forest conservation of the provincial Department of Lands and Forests (DLF), however, river conservation in the 1960s reached out to average Nova Scotians with the message that industrial development could be dangerous, and could be stopped.

Among those heard by the McInnes Commission was a group of professional scientific conservationists outside of government who called themselves the Nova Scotia Resources Council (NSRC). Active since 1959, and
dedicated to "ensuring the cultivation and the optimal use of the Province's
resources [via] carefully considered land use and sound management of
plant and animal life,"[7] this private association was no protest group; its
representative at McInnes's inquiry was Harrison F. Lewis, retired chief of
the Canadian Wildlife Service, and its recommendations to the commissioner (largely reproduced in his own recommendations to government)
earned praise for their "reasonableness."[8] But the NSRC represented a
trend in the province that saw natural history clubs seeking incorporation,
hoping to push government to act on a conservation agenda.[9] Wildlife
conservationists in Nova Scotia, whose legislative victories dating back to
1794's Act for the Preservation of Partridge and Blue-Winged Ducks were
dismissed by Lewis as "a sop to an advanced minority and not meant to
be taken seriously," were particularly keen to organize and willing to use
the weight of membership numbers to achieve something more substantial.[10] The Nova Scotia Bird Society (NSBS), founded by Lewis, Robie
Tufts, and other professionally interested naturalists, also witnessed rapid
growth in its public membership over the following months.[11] Later, in
response to the Gold River episode, yet another group of sportsmen and
scientists formed the Nova Scotia Salmon Association (NSSA). Aware of
a need to incorporate pollution abatement into habitat protection, these
new conservation organizations were at once more than simple "wise use"
conservationists and less than the radically skeptical environmentalists of
the late 1960s and early 1970s. They also remained active and collaborated
with various environmental groups in the 1970s and 1980s.

Opposition to the Gold River scheme was further bolstered by an unusual international preservation organization, the Friends of Nature
(FON). The first FON group was created in Maine in 1954 by a young
publisher and shipwright, Martin Rudy Haase, and a small number of
associates. Their initial purpose was to preserve as recreational wilderness
one small island (McGlathery) in Penobscot Bay, then under threat from
a pulp and paper company. Success breeds confidence, however, and the
group's attention soon turned to similarly threatened wilderness areas
elsewhere. Unconcerned with the national loyalties that hemmed in so
many other groups, FON set out to assist in the creation of wilderness
parks in Costa Rica and Tasmania, with uncommon success. In 1967,
Haase and his family relocated to Chester, Nova Scotia, in a bid to keep
his two young sons out of the way of the seemingly interminable Vietnam

War, and FON came with him. More accurately, Haase was FON, and he found in Lunenburg County a new set of associates with whom to continue his wilderness advocacy, in two host nations.[12]

Despite their jaundiced view of America's military adventures and nuclear arsenal, FON was never a politically radical group. Its operations reflected the privileged milieu from which it came: letters and direct appeals to the political elite achieved virtually all of the group's victories. The personal intervention of President John F. Kennedy's adviser and head speechwriter, Ted Sorensen, for example, opened the door to the creation of the Cabo Blanco Nature Reserve in Costa Rica, on FON's advice. In that sense, there was little to differentiate FON from the earlier tradition of both conservation and preservation in North America, in which elite recreationists (hunters, hikers, and mountaineers) and elite technocrats (foresters, biologists, and engineers) pursued novel goals via traditional politics. In another sense, however, their philosophical position linked them much more closely to the new environmentalism of the 1960s. Haase's FON linked concerns about land and habitat to pollution, population, and lifestyle questions. In 1955, seven years before the publication of Rachel Carson's *Silent Spring*, FON began a campaign to encourage a permanent ban on DDT (dichlorodiphenyltrichloroethane) and similar pesticides, which transitioned into a fight against the US defoliation program in Vietnam (and, in the Cold War context, linked FON's anti-chemical work to its anti-nuclear advocacy). The group also frequently insisted that "people must voluntarily turn to a 'lower' standard of living commensurable with available renewable resources," and helped to popularize the ideas of the afforestation advocate Richard St. Barbe Baker and back-to-the-land gurus Helen and Scott Nearing.[13]

Defenders of the Gold River argued not only for a fishing river but for one of the last *healthy* salmon rivers in the province, and their successors in the NSSA, along with Rudy Haase at FON, were well aware of the expansion of clearcut pulpwood forestry in the province and the annual DDT spray program in neighbouring New Brunswick, as well as the fatal consequences of both practices for the insect and fish life of rivers.[14] If their political habits still generally directed them towards technical and scientific expertise and gentlemanly lobbying out of the public eye, there were also hints of a new awareness – in the NSBS's quest for membership numbers, and in the use of the press in defence of the Gold River – that public opinion might now support a reaction against the intrusion of industrial modernity into more and more remote corners of the land. There were even hints of the deeper challenge to come, questioning the need, the value, and the possibility of continued accelerating exploitation of nature. At the end of the 1960s, all of these disparate trends came together, far too quickly for single-issue conservation groups to keep up with, but those groups did not fade away or reject the new political reality. They continued to adapt and to forge alliances with their new and more radical peers, subtly moulding the future of the environmental movement in the province.

"Calamity Howlers": Estuarine Environmentalism

The Gold River episode was a portent of things to come. Threats to bodies of water signalled the beginning of new environmental concerns at the end of the 1960s. In this, as in so much else, Nova Scotia's experience reflected and amplified the pattern in the rest of North America and the world. The provincial government's quest for economic development had changed the face of the province, often for the worse, and the change was not evenly distributed. New industrial projects tended to cluster around harbours for a number of reasons, including ease of access, available workers, clean water supplies, and the availability of the ocean as a sink for industrial waste. By extension, the new activism of the era centred on the same locations, the majority of them rural, as local residents fearing for their traditional lifestyles and livelihoods under new land-use and water-use regimes found the traditional politics of dissent ineffective against polluters working hand-in-hand with government. Fed by direct observation of environmental ills and mistrust of government, as well as by a rising global environmental consciousness, new ideas and patterns of activist behaviour spread across the province from these estuarine enclaves. Environmentalists made increasing use of scientific research, not to convince politicians of their claims as their conservationist forebears had done, but to draw ever greater popular support to their campaigns of political pressure. And with the new style of environmental politics came a new and lasting pattern of participation, with a much greater presence of women, young people, Mi'kmaq, and working-class Nova Scotians.

One of the first instances of the new activism occurred in Pictou County. In 1965, when the provincial government finally enticed the Scott Paper Company to build its newest, state-of-the-art kraft pulp mill at Abercrombie Point, an unusual provision in the agreement had the province rather than the pulp company operating the mill's effluent treatment facility. Seizing on the natural lagoon of nearby Boat Harbour as a cheaper alternative to a purpose-built treatment plant, the Nova Scotia Water Resources Commission put up dams in the lagoon to divide settling and aeration ponds, walled it off from the sea, and constructed a pipeline under the East River of Pictou to carry to the new facility 25 million gallons each day of effluent water, dissolved and suspended bits of wood pulp, and various toxic leftovers from the kraft bleaching process. Economically at least, it was a success story; the Scott mill prospered. Boat Harbour, on the contrary, died. Once a popular site for swimming, boating, and fishing, its waters promptly turned black after the mill opened, as the oxygen demands of decomposing wood pulp left nothing to support life.[15]

Particularly keen to celebrate their sense of belonging to a particular place, and particularly ill treated during the creation of the facility, the Mi'kmaq of Pictou Landing were among the first to react to the environmental downside of development, though even at Pictou Landing they were not alone. From the perspective of the band's negotiators, the destruction of the harbour was not even supposed to have happened. They had been dispatched to meet with federal and provincial officials early in the province's talks with Scott, after the band indicated that it would not accept the conversion of their reserve's beautiful natural harbour into an industrial facility. In 1966, they were taken to a pulp mill in Saint John, New Brunswick, where water issued clear and clean from the outflow pipe, and were reassured that the same conditions would prevail in Pictou. With an offer of $60,000 as compensation for fishing rights on the table, and, according to Pictou delegate Louis Francis, a generous supply of alcohol as well, the band's team agreed to the government's terms. When effluent began flowing into Boat Harbour, they realized their mistake. The Saint John lagoons they had been shown were not even receiving effluent at the time of their visit, and $60,000 was a pittance next to the millions it would cost to build a truly state-of-the-art facility – $4 million for the most modest improvements at Boat Harbour proposed by the optimistic and quite conservative Rust Associates report in 1970.[16]

Members of the Pictou Landing Band had good reason to feel helpless in 1970. "I guess we're beaten," was Chief Raymond Francis's assessment, but they would not give up, and in their fight they had allies as well, willing as never before to challenge the authority of the state.[17] Though environmentalist coalition across the province was not yet common, local solidarity was, and non-Native residents of Pictou Landing felt nearly as deceived as the band. Since 1965 they too had been demanding answers from the water commission, and had received similar assurances that no pollution of water or air would result from the project. As the progressive degeneration of the harbour and its surroundings confirmed their fears, however, more and more residents turned to a local citizens' committee (eventually named the Northumberland Strait Pollution Control Committee, or NSPCC) to press for answers. Municipal councillor and NSPCC member Henry Ferguson wrote for the people of Pictou Landing in 1970:

> With the winds down the harbour we get air pollution from Scott Paper, then with the winds east we again get pollution, this time from Boat Harbour. The fumes are really terrible, almost unbearable. Then we get water pollution coming down the East River from leaks in the pipe across from the Scott Paper Co. to Pictou Landing. Then water pollution from Boat Harbour when the tide is coming up and runs along Lighthouse Beach and into Pictou Harbour.[18]

To that, he added swarms of mosquitoes and gnats, expropriation through flooding of harbourside land without notice and with minimal compensation, and threats to the Northumberland Strait lobster fishery. The last was particularly worrying in communities along the shore, where the Maritime Packers Division of National Sea Products reported a 26.7 percent drop in lobster landings in 1968 and a 42.2 percent drop in 1969.[19] In fact, the threat to the fishery became a major rallying point for activists.

Official response to public outrage at Pictou Landing was muted at best. Accustomed to working without heed to local opinion, E.L.L. Rowe, the chairman of the Water Resources Commission and a former chemical industry employee who had designed the leaking subriver pipeline and had promised minimal disruption to life around Boat Harbour, doubled down on his defence of the facility. He insisted that he personally found the smell of the rotting lagoon and the "rotten egg" hydrogen sulfide fumes from Scott's stacks inoffensive, and that the province could not make funds available for the solution of merely aesthetic problems. He also made it clear that mercury contamination of the mill effluent from the associated Canso Chemicals plant would have to be tolerated, as the development of the plant had "gone too far" and cost too much to be altered.[20] Other officials and politicians holding similar views attracted attention from time to time, including the agriculture minister, Harvey Veniot, who dismissed the affected locals with the oddly poetic epithet, "calamity howlers," or the fisheries experts at the Department of Fisheries in Ottawa who would only repeat that Boat Harbour's effluent had been tested and proven non-toxic to lobster larvae.[21] But it was Rowe who earned the greatest ill will as the man directly in charge of the facility, whose public pronouncements so often failed to hold up to scrutiny, and who had been more than usually candid about his priorities in the use of public funds. More inclined to talk about municipal sewage treatment than the effluent of privately owned industries, he played what Tom Murphy in the *Mysterious East* called "the little verbal game called 'Why-Don't-You-Talk-About-The-East-River-And-Shut-Up-About-Boat-Harbour.'"[22]

Local activists refused to be put off the issue. Unable to secure a hearing and unable to sue the province for nuisance without permission from the government, they turned fully to public opinion as a source of influence. And as a tool for generating public support, they turned to science, and thus to criticism of the facility as a threat to the local fishing economy. The NSPCC commissioned a report from Delaney and Associates that followed the brown film of Boat Harbour effluent twenty kilometres down the shore and calculated that about 185 tons of organic solids spilled into the sea from the harbour each day.[23] D.C. MacLellan at the Marine Studies Centre at McGill University found that the effluent resulted in an unusually high mortality among the plankton at the base of the Northumberland Strait food chain, and Dr. J.G. Ogden at Dalhousie University answered the federal fisheries experts by reminding them that, toxic or not, dark-brown effluent that blocked sunlight from reaching the sea floor would deprive lobster of both food and sheltering seaweeds. "A sheet of opaque glass put over the lawn is not toxic," he said, "but it will kill the grass. The effluent from Boat Harbour is as effective as a sheet of black plastic."[24] So armed with expert authority of their own, the NSPCC members pursued their environmental justice arguments in the press on behalf of the Mi'kmaq and Northumberland Strait fishermen deceived or ignored by the federal agencies designated to safeguard their interests. Nor were their aims narrowly or selfishly defined; one fisherman-activist told Tom Murphy that compensation for losses might not be welcome if it allowed the condition of the strait to continue deteriorating. "We want our environment cleaned up, rather than subsidies for a dirty environment," he said.[25] Dr. J.B. MacDonald, MD, one of the NSPCC's leaders, explained that he had come to the issue out of concern for his own summer home at Pictou Landing, but quickly discovered the medical risks of sulfur dioxide in the air and a stew of unknown chemicals in the harbour's water, as well as the injustice of secretive government deals drawn up without "explanation, justification, or revelation of alternatives" to the public.[26]

In the early 1970s, the NSPCC expanded its range of concerns to include the whole of industrial and municipal pollution affecting the Northumberland Strait, no longer content to entrust the solution of any such problem wholly to the Water Resources Commission or the Department of Fisheries.[27] "Regrettably," noted MacDonald, "one must be hurt to become an active and sustained anti-pollution fighter. This is not as it should be. Intelligent people should not see their environment, their homes, their country being destroyed and sit back placidly and take it."[28]

Boat Harbour was the most bitterly fought of the late 1960s battles, but it was far from the only one. At the same time that Dr. MacDonald was discovering the need for citizen activism, other groups were forming in the province after their own personal experiences with the dark side of development. In Halifax, residents on the shores of the Bedford Basin and Purcell's Cove learned much the same lessons derived from Pictou Landing's experience. Like Pictou, urban Halifax was no stranger to industry, but new developments in the 1960s brought an intensification that seemed to threaten residents' traditional recreational use of the harbour. A growing population in the Halifax area had the city government searching for a site for a sewage treatment system more advanced than the existing one.[29] At the same time, proposals for new industrial developments at previously unused locations, such as a container ship terminal at Navy Island in Wright's Cove on the Dartmouth side of the basin, came up frequently. As in Pictou, the reaction against environmental costs imposed from above came early in the planning process: both the Bedford Basin Pollution Control Committee (BBPCC) and the Purcell's Cove Action Committee (PCAC) began protesting in 1969 that the decision to dedicate greater and greater proportions of the basin's shore to industrial purposes, and the decision to flatten and infill thirty acres in the village and harbour at Purcell's Cove for a massive central sewage treatment plant, were both unjustifiably undemocratic and would subject locals to unanticipated and irreversible environmental dangers. Robert Martin's complaint about the level of public consultation at Purcell's Cove might have come from the pen of any of his peers on the basin or in Pictou County: Should citizens "wait until City Council has done all the thinking they expect to do before expressing their views? Are they to be given a chance to let off steam only after City Council has made its decision?"[30]

Residents of both communities demanded a say in development schemes before plans were finalized, but in Bedford they faced a potential multitude of future projects that could be addressed only with regular public input into the planning process. Accordingly, the BBPCC set to work compiling a major report, which it released in 1970, on the future of industrial and recreational development on the basin, making use of the lawyers, engineers, and planners among its members to argue against unrestrained industrial growth in the city planners' own terms.[31] There was no official response. If the somewhat less wealthy and professionally resourceful defenders of Purcell's Cove, fighting to stop a single project, could be publicly dismissed as ignorant of the inevitability of development, the BBPCC required a different tack on the part of the city and province: silence in public, while the ever-present and tight-fisted Mr. Rowe at the Water Resources Commission circulated among his colleagues a dismissive assessment of the activists as impractical "academic" people with a "lack of appreciation of priorities in the expenditure of public moneys," selfish, unscientific, and not to be taken seriously in their "public relations" exercise.[32]

In the face of such official intransigence, the genuine public relations efforts of Purcell's Cove met with greater success. The PCAC's hectoring methods, undeterred by the condescension of experts or the hostility of bureaucrats, eventually prompted the city council to promise that the cove would not be infilled, scuttling the previous odds-on favourite site and sending the search for a sewage plant location off on its own long and eventful history.[33] The BBPCC, conversely, barely stumbled into 1971 as an active organization, "embittered and in disarray" from its failed attempt to appeal to provincial and city planners.[34] Though certainly possessed of more economic power on average than the cove residents, the BBPCC had no great deal of political clout with which to force their perspective onto the planning agenda, no equivalent to the political power of massed public anger. Successful or not, the lessons learned by both groups echoed the experience in Pictou County, where the NSPCC's public pressure campaign finally resulted in some acknowledgment of the problem at Boat Harbour and some investment in improvements in 1972, though without any corresponding acknowledgment of the activists' role in bringing them about.[35] Government could be moved on single issues, it seemed, but only grudgingly and only under threat of political embarrassment, and not to the extent of giving up its control of development planning.

If Boat Harbour's ruination was not sufficient warning against the urge to "sit back placidly" and let the state's planners have their way, the Canso Strait industrial complex rang in the new decade in 1970 with an infamous debacle, the *Arrow* oil spill. The construction of the Canso Causeway linking Cape Breton to the mainland early in the 1950s turned Port Hawkesbury into a viable deep-water port, and provincial government encouragement brought to the area a pulp mill in 1962, a heavy-water plant in 1970, and an oil refinery in 1971. Port Hawkesbury was to be a "growth pole" for western Cape Breton and the eastern mainland, areas appearing in desperate need of an economic boost.[36] The urgency of industrial growth targets, however, left no room for the assessment of environmental risk, much less for the avoidance of it. Discontent with scantly regulated development existed, but the hundreds of jobs that came with the complex were welcome relief for many, and Dr. MacDonald's assessment held true: people had to be hurt before they could become active pollution fighters.[37]

At 9:30 a.m. on February 4, 1970, a Liberian-registered tanker carrying sixteen thousand tons of thick Venezuelan Bunker C oil struck Cerberus Rock in Chedabucto Bay on the approach to the Canso Strait. Damage to the ship was not great, but the captain was unable to free it from the rock. It languished for eight days, attracting barely any attention from Canadian authorities, before a salvage attempt on February 12 caused the hull to break in half, spilling part of its cargo into the sea and sending the stern to the bottom of the bay, still full of oil. Imperial Oil Limited, owner of the cargo, attempted with some success to burn off and disperse the spilled oil, but neither the company nor anyone else had any experience dealing with spilled oil in such cold waters. For a further eight days, Imperial tried in vain to contain the mess, before the federal government finally stepped in and appointed a three-man task force with instructions simply to "deal with" the oil.[38]

"Operation Oil," as the task force called its work, was a $3 million cleanup that managed to at least prevent a bad situation from becoming worse. Dr. Patrick McTaggart-Cowan, executive director of the Science Council of Canada, Dr. H. Sheffer, vice-chairman of the Defence Research Board, and Captain (N) M.A. Martin, deputy chief of staff at Maritime Command, presided over a three-pronged mission: remove the oil remaining inside the broken *Arrow*, collect as much as possible of the spilled oil at sea, and clean up more than 125 miles of coastline already painted black with crude. Each objective presented novel problems. The bow section of the tanker was largely emptied in the spill, but the oil remaining in the stern section, at rest at the bottom of frigid Chedabucto Bay, had thickened into an immovable mass due to the cold. At sea level, booms could no longer contain the froth of oil and water drifting about, but dispersant chemicals were deemed too toxic to be used in the bay's rich fishing grounds. And on shore, oil clung to rocky cliffs, sandy beaches, and drifts of sea ice, soon to melt and re-oil anything the task force managed to clean.

The three men in charge and the corps of military engineers and civilian researchers under their command had to invent new techniques for each situation; despite the mania for development at Port Hawkesbury and the predictably difficult environment of the Atlantic Ocean in winter, and despite recent high-profile oil spills in California and the English Channel, no one in the provincial or federal governments had ever prepared for the possibility of such a spill.[39]

The cleanup operation took roughly seven weeks. Navy divers managed to tap a new valve into the hull of the sunken stern and inject steam to warm up the congealed oil enough to pump much of it out into a waiting barge, the *Irving Whale*.[40] Their engineer counterparts worked simultaneously at a furious pace to design and build "slick lickers" to pick oil from the surface of the sea, and de-oiling washers to clean fouled fishing gear. On shore, everything from shovels to bulldozers was used to collect oil from the beaches. The task force expressed pride in its achievement; locals expressed frustration at a job half-done. A great deal of oil remained within the sunken hull, occasionally oozing out during stretches of warm water temperatures over subsequent years; as a result, small oil slicks still regularly gummed up beaches as far away as Sable Island. As for the oil-crusted cliffs along Chedabucto Bay, even McTaggart-Cowan had to admit that the oil on rock was "there to stay," and would eventually harden into something like asphalt.[41]

Perhaps surprisingly, the wreck of the *Arrow* did not provoke activist organization on the scale that Boat Harbour or the various developments of Halifax Harbour did, mainly because fishermen already had their own associations through which to urge on the cleanup, and the small number of other activists from communities around the bay seem to have found the issue too large and intractable to deal with. A group called Action Arrow existed briefly, and the author Silver Donald Cameron remembers joining a small group of people who sealed up the corpses of oiled seabirds in plastic bags and mailed them to Ottawa in protest, but in truth Operation Oil had done about as much as was technically possible at the time, and along the way the federal scientists and naval personnel involved had learned a great deal about how to handle oil spills in cold seas. The fate of the bay seems to have been treated less as a winnable fight and more as an object lesson in the consequences of waiting too long to start fighting. "The Gulf Refinery is still there," wrote the *Mysterious East*, "Cerberus Rock is still there. All over the world the big tankers are slipping off the ways into the water. Which one will give Chedabucto Bay its next coating?"[42] What the *Arrow* did for environmental activism in Nova Scotia was confirm the growing suspicion that government planners, the duly appointed experts, could not be trusted to pursue industrial development unsupervised by a concerned public.

The isolated adventures in environmental activism at Boat Harbour, Halifax Harbour, and Chedabucto Bay had a few common elements.[43] One was mistrust of government. Another was the very local focus of concern and the unity of purpose among the more and less wealthy residents of those local places, as well as the roughly equal participation of men and women.[44] The age of men's club conservationism, when sport fishermen from Halifax could save the Gold River with minimal public involvement, was clearly at an end. Finally, the truth of Dr. MacDonald's recipe for activism held in every case: either prospective or concrete, some actual personal environmental harm was required to inspire a pollution fighter and defender of home places. Nova Scotia's leap into industrialization by invitation (its participation in federal/provincial development schemes and its pursuit of foreign direct investment through the Stanfield government's own development agency, Industrial Estates Limited) brought a scale of industry previously unknown to rural corners of the province, and a corresponding surge in the number of affected individuals. Also, once active, no group ever entirely disappeared. The NSPCC carried on its struggle with Boat Harbour and other threats to the strait for years, as did the Pictou Landing First Nation. The dead bird mailing activists of Chedabucto Bay went on to lead a rural recycling campaign in Richmond County and open an environmental education centre in Port Hawkesbury.[45] Even the Bedford Basin and Purcell's Cove committees, both soon nominally defunct, contributed members and experience to successor organizations in the city. The pattern of politically populist and scientifically literate activism crossing lines of race, class, gender, age, and geography became the norm for Nova Scotian environmental activism and provided the basis of further organization in years to come.

ASPECTS OF A GLOBAL WHOLE

Personal experience of environmental harm is not in itself sufficient to explain the rise of environmentalism. The new spirit of activism also had roots in new ideas about the environment, the state, and democratic power that were anything but local. The previous decade had changed the political landscape of the world, especially in Europe and North America. Opposition to the state and industrial society fuelled New Left

and civil rights movements that violently challenged the legitimacy of existing social and political systems. A generation of openly critical and rebel-
lious young people joined forces with labourers and racial minorities to
protest the Vietnam War, the nuclear bomb, the excesses of capitalism,
and the marginalization of the politically weak.[46] Meanwhile, a countercultural movement within the greater uprising encouraged the articulation
of alternative ways of living. It was in this context that an international
environmental consciousness formed. Concerned people made connections
among nuclear warfare, chemical pesticides, economic imperialism, and
the inequity of capitalist industrialism and drew conclusions about the
welfare of the most voiceless of all politically weak constituencies, the
natural world. At the same time, the kind of quality-of-life concerns that
had long vexed urban sanitary reformers – clean water, clean air, and the
like – appeared to have reached a global scale.[47]

The expanding reach of the new activism owed something to the spread
of colour television and the realization among network programmers that
environmental disasters made good dramatic content. More important
was the novel scale of real and potential environmental harm. Worldwide
radioactive fallout from nuclear weapons tests, biologically mobile trans-
border chemical pollution as detailed in Rachel Carson's 1962 bestseller,
Silent Spring, and the disastrous consequences of marine oil spills such as
the infamous *Torrey Canyon* in 1967 were real, and televisual mass media
only magnified their impact as shared experiences. A series of popular
publications in the years following Carson's both fuelled and fed on the
anxiety: Garret Hardin's "Tragedy of the Commons" (1968), Meadows and
colleagues' *The Limits to Growth* (1972), and E.F. Schumacher's *Small Is
Beautiful* (1973), to name but three popular titles, recounted the realities
and offered a diagnosis of dangerously out-of-control modernity.[48]

All of these new ideas about environment, politics, and the value and
possibility of perpetual economic growth were as current in Nova Scotia
as elsewhere. They permeated popular culture and inspired any number
of the small, local anti-pollution fights of the late 1960s. In the city of
Halifax and along the eastern shore, the intellectual inheritance of US
and Central Canadian environmentalism played an especially visible role
in shaping the new movement, where a wave of US draft dodgers and
back-to-the-land immigrants seeking rural ways of life brought their personal experiences to bear. Provincial and municipal leaders were also un-
intentionally helpful in fostering subversive notions in Nova Scotia in
1970, when in the pursuit of their own set of popular ideas about urban
planning, they invited some radical thinkers from the United States, the

United Kingdom, and other parts of Canada to dissect the social problems
of the city of Halifax.

The Progressive Conservative government of Premier G.I. Smith (1967–
70) had great faith in central economic development planning, as had
the Stanfield government before it, and relied extensively on its Cabinet
Committee on Planning and Programs – a group of experts better known
as the secretariat – to direct development spending. Like Premier Smith
himself, the secretariat exercised power only briefly, but they were determined to shake up the structure of power in the Halifax region, where
they believed political stagnation limited politicians' willingness to follow
planners' advice. To that end, just before Smith's ouster, they organized a
weeklong public consultation and investigation of the city by twelve world-
class experts in everything from tourism and labour to public finance
and the structure of government. They called it Encounter on the Urban
Environment. The format was virtually unheard of, having been modelled
on a singular event in Australia two years earlier, and few people outside
of the secretariat seemed to understand what was intended. The Voluntary
Planning Board, which provided funding, expected a major but more
or less conventional development study. The city mayor, Allan O'Brien,
wanted a list of policy proposals. What they got was a structured confrontation, the brainchild of the secretariat's Len Poetschke, in which the visiting experts called on all levels of private and public life in the city to explain
and account for its problems.[49] Open microphones and open minds
brought hundreds of speakers to the panel's daily hearings to expose the
frustrations and social tensions of the city, while detailed television coverage took those hearings to the population at large. Nor did the panelists
sit in silence; they queried, probed, and sometimes criticized those who
came to speak. The churches in Halifax, they said, were out of touch,
newspapers plodding and conservative, and citizens walking in the political darkness with their eyes closed. Discontent was plain to see, but
Haligonians displayed a "reluctance to come to terms with the political
process," according to Dr. Martin Rein of the United Kingdom's Centre
for Environment Studies: all they had was "indignation without information, without tactics and so no strategy."[50]

Len Poetschke was no subversive, much less the rest of the secretariat.
They hoped to harness the energy of public participation coming out of
Encounter as "part of a planning process" in which government (planners)
provided the "framework" for decisions, such as the necessity for industrial
economic development or the central importance of the core business
district of Halifax, and the people provided the political firepower with

which to overcome bureaucratic inertia.[51] Boiled down to essentials, they hoped to manage and direct the power of public opinion the way a road planner directs the flow of traffic. But the secretariat's days were numbered once Gerald Regan's government took office in 1970 – the new premier had his own set of expert advisers. Last-ditch efforts by Poetschke and his colleague Fred Lenarson to create a citizens' arm for the Halifax-Dartmouth Municipal Area Planning Committee (MAPC) in the year following Encounter brought together a coalition of peace, environment, and social justice groups (eventually settling on the name Movement for Citizens' Voice and Action, or MOVE) that frustrated the planners' intentions by challenging MAPC's closed-door planning process. In this they were aided by Encounter panelist Reverend Lucius Walker, executive director of the Interreligious Foundation for Community Development in New York.[52] Walker's participation shows the degree to which Encounter and its consequences represented an accelerated version of trends happening in the rest of the industrial world, in which diverse activists and civil society leaders formed new coalitions and negotiated difficult relationships with politically powerful sponsors. The Nova Scotian version of the story also highlights the participation of environmental activists in those coalitions.

The events of Encounter week and the creation of the MOVE coalition changed the shape of environmental activism in the city, and eventually across the province. The experience of activists during the Purcell's Cove and Bedford Basin controversies, ongoing at the time, was much the same as the experience of the Africville Action Committee or the Heritage Trust, diverse groups united by their confrontations with the stone wall of government.[53] As the *4th Estate* newspaper put it, "our government, politicians, civil service, and government agencies are all somewhat paranoid. They won't answer questions. It is extraordinarily hard to find the real roots of power."[54] But complaints hadn't satisfied the Encounter panel. When a witness at one hearing mentioned a common front of area environmental groups (the "ECO" coalition) dedicated to public participation and education, panelist Edward Logue replied that such groups "merely preserve the survival" of the current political system (an adequate summary of the secretariat's hopes for MOVE); Dr. Rein reminded the audience that there was "no non-political way of effecting changes," and citizen participation without political action was "a grandiose way of preserving the status quo."[55] The critique hit home, and the members of the loose-knit, "amorphous and anonymous" ECO group, still little more than a mailing list put together at the end of 1969, recognized that even though their goal of public education was sound, they could "never be a force in political terms [without] ... achieving form in the eyes of government."[56] In the MOVE coalition they saw a chance to be something more, and as the coalition took shape it therefore included representatives from the fast-fading BBPCC, the PCAC, the Eastern Shore District Pollution Committee, the Cole Harbour Environmental Council, the Halifax Wildlife Association, and a number of other environmentally concerned small groups and individuals.[57] It also included, in the later stages, a new organization called the Ecology Action Centre (EAC).

The EAC, which would go on to have a hand in virtually every environmental controversy in the province for the rest of the century, was born from a bit of experimental pedagogy at Dalhousie University, one of that institution's own responses to the student unrest of the 1960s and demands for more student autonomy. For a brief time, groups of eight or more undergraduates with a faculty sponsor were allowed to design and conduct their own courses. Few applied (do-it-yourself course design was never really a high priority for student radicals), and fewer were approved, but one group did earn approval in the 1970–71 year to run a course on "Ecology and Action," and one of the students who led the effort, Brian Gifford, made it his class project to create an organization in Halifax dedicated to public environmental education (ironically, exactly the insufficiently political goal the ECO group was talking itself out of at the same time). The ecology course and the EAC that came out of it were well aware of the difficulties faced by their activist peers in dealing with the provincial and municipal governments: the Nova Scotia Resources Council's Dr. Ogden, the PCAC's Alan Ruffman, and the BBPCC's John Bentley each conducted classes for the group, and Gifford, in the process of investigating the role his new association might play in the city, took time to consult with Frank Fillmore, activist and editor of the radical paper the *4th Estate*.[58] But the student group, unique among environmental associations in the province, did not find its motivation in personal vulnerability to some specific harm or experience of environmental injustice. As students, middle-class professionals in training, they came nearest of any activist group in the province to the post-materialist definition of environmentalism. They took their inspiration solely from the generalized culture of environmental concern in the industrialized world, and their organizational models from distant student peers like Pollution Probe, not nearby groups familiar with the context of Nova Scotian politics. "Information and education" rather than political action, said Brian Gifford, "are what the Centre is all about."[59]

The earliest concerns of EAC activism were recycling projects, urban air quality, and urban planning: things for which the students could secure federal government funding. Doing so was vitally important. The first two years of the EAC's life, the period in which it established itself as a resource centre for the city, were funded through yet another response to sixties youth unrest, this time on the part of the Trudeau government: the Opportunities for Youth (OFY) and Local Initiatives Programs (LIP), which funnelled hundreds of thousands of dollars to youth groups across the country in the early 1970s. With OFY and LIP funding, the EAC was able to run a demonstration paper recycling project and keep a shifting set of five to eight student staff busy adapting a Pollution Probe environmental education kit for schoolteachers and compiling information kits on forestry, energy, and public transit.[60] Eventually, its persistence earned the group a reputation as a reliable source of information, and people in the city began to contact it for answers to pollution and resource-related questions.[61]

While EAC was, and remained for years, a very local student organization of small influence, its creation (as well as that of the MOVE coalition) demonstrates powerfully the currency of combined environmentalist and social justice ideas in Nova Scotia at the end of the 1960s. Federal spending provided the necessary support, but belief in the importance of the issues provided the people willing to take it up. The EAC did not often share the sense of urgency or political cynicism that drove groups born of single issues to oppose the provincial government's plans so completely. It did not agree with its own associate in Cape Breton, *Arrow* protest veteran Margo Lamont, for example, when she urged it to oppose the entire Canso Strait development plan.[62] But it did share with the rest a genuine concern over environmental harm, and a belief in the imminence of a resource supply crisis according to the projections laid down in the Limits to Growth theory.[63] It also held a firm commitment to public participation, to the point that it was willing to accuse the Regan government of "offending the doctrine of democracy itself" by leaving the mechanisms of citizen input out of its Environmental Protection Act in 1973.[64]

Urban activists, with their focus on education and principle, were joined in the quest for democracy in land-use planning by yet another group of issue-specific rural activists fighting a bitter campaign on the province's eastern shore in 1972 and 1973, a group whose new brand of environmental activism was clearly well informed by the worldwide movement. While almost every environmental controversy at the end of the 1960s and beginning of the 1970s in Nova Scotia centred on a harbour or a river, water woes were not alone in feeding the new activism of the era. Where once land conservation was the purview of fringe preservationists like the FON group, by the late 1960s provincial governments had once again hit on the idea of promoting tourism in the region through the designation of national parks.[65] The ideal and greatest victory of the US wilderness preservation movement was to the government in Halifax simply another form of economic engine with which to raise the province out of relative poverty. Unfortunately for the government, Nova Scotia's coast (to the government's eyes the obvious and only choice for a tourist landscape) offered no convenient unsettled patches. A park, as then understood, had to comprise a natural landscape without human habitation, other than the transient paying tourists. In order to create a park, people living on park lands would therefore have to be removed. That was exactly the plan proposed at the end of the decade for Ship Harbour National Park on the eastern shore, and exactly the occasion for Nova Scotians to demonstrate that the new environmentalism posed a much more radical political threat than its conservationist antecedent.[66]

Plans to create a new national park on the eastern shore existed as early as 1965, but the normal planning process and Premier Stanfield's preoccupation with the leap from provincial to federal politics delayed negotiations for years. By the time federal/provincial talks came to light in 1969, there were still multiple proposals on the table, all of which involved the removal of well over a hundred permanent and seasonal residents' homes between Musquidoboit and Sheet Harbour in Halifax County, and none of which involved inviting those residents to help determine the final shape of the park. Despite some initial resistance from locals who petitioned the federal government to avoid expropriations in the creation of the park, both Premier Regan and Halifax Eastern Shore Member of the Legislative Assembly Garnet Brown were eager to see the plan go ahead, and were little concerned with local sentiment. Negotiations continued behind closed doors, resulting in 1972 in a plan that would remove about 250 homes from newly designated federal lands.[67]

Reaction on the eastern shore was swift and angry. A group of residents came together, calling themselves the Committee to Prevent the Proposed Park (later the Association for the Preservation of the Eastern Shore, or APES), and from the summer of 1972 to the end of 1973 pursued an ever more insistent campaign. The APES comprised both lifelong residents and new arrivals to the province, though their opponents in government found it convenient to focus their criticism on one of the leaders, Gordon Hammond, who had been involved in the recent successful fight to halt construction of the Spadina Expressway in Toronto.[68] The focus on Hammond once more highlights the combination of local grievance and a widely shared culture of populist activism, as well as the determination of the provincial government to undermine and marginalize protest. Hammond became the favourite target of politicians determined to paint him as an outsider, a troublemaker, and "a welfare case who drives a car" (i.e., a possible criminal), and the whole of his organization as a "group of johnny-come-latelys, people on welfare, kooks, hardline Conservatives, and small children," selfishly preventing the development of the eastern shore.[69]

Unfortunately for Regan and Brown (as well as for federal minister of Indian Affairs and Northern Development Jean Chrétien), the kooks went from strength to strength. Four years earlier, they had gathered five hundred names on a petition sent to Chrétien's office; after the controversy broke in earnest in 1972, they gathered nearly three thousand names, carried out a letter-writing campaign to Province House, conducted two marches on the legislature, and won 90 percent support in a *Dartmouth Free Press* poll of the eastern shore.[70] Angered by Brown's talk of kooks and welfare cases, activists struck back, calling the premier (who had stealthily avoided one protest rally on his way to work at Province House) "Rear-Door Regan" and earning press coverage all the way to Toronto with their criticism of his "callous disregard for the existing population of the area, for their ancestors who cleared the land, built their houses and worked so hard that we might enjoy what we have today."[71] Others literally struck back, such as one of Brown's constituents, who launched a punch at the MLA a few days after the "kooks" comment, or threatened to do so on protest rally banners promising "THIS IS YOUR LAST PEACEFUL DEMONSTRA-TION" or in Hammond's slightly more subtle predictions of "civil unrest" if the plan should go ahead.[72] APES tactics included support for much more popular provincial park proposals and for a conception of parkland that included the "existing mutual harmony between man and nature," on the European model of national parks containing human settlements. Members also made time for a frank and factual deconstruction of the province's unsupported assumptions about the job-creating potential of a national park. They knew enough not to be trapped into top-down consultation processes as well, and turned the first of the public meetings held by a federal/provincial advisory committee into a debacle for the park planners, complete with threats of arson and armed resistance.[73]

This was not the only instance of opposition to national park development, but Regan was in a more difficult position than most premiers.[74]

APES activists often referred to an autocratic and bitterly resented eviction
process at New Brunswick's Kouchibouguac National Park, and many
Nova Scotians remembered with ill-will the expropriation of land for Cape
Breton Highlands National Park. On top of that, Regan's government had
acceded to a plan to reconfigure the Cape Breton park's boundary and
allow the conversion of Cheticamp Lake into a hydroelectric reservoir.[75]
With a provincial election expected in 1974, Regan wished to avoid charges
of treating the electorate as "sheep that are driven from pasture to pasture,"
and to the consternation of both Brown and Chrétien, he reversed course
abruptly in December to support a series of provincial parks in lieu of
Ship Harbour National Park.[76] Regan tried to shift blame onto Chrétien's
federal office, but the APES campaign was clearly instrumental in the triumph of grassroots environmentalism over the provincial government.
Earlier decisions – to preserve the free flow of the Gold River or direct a
million dollars or two towards cleanup programs at Boat Harbour, even
the decision to cancel negotiations on a US investor's electricity generation
scheme in 1973 (see Chapter 2) – scarcely dented the development bias of
provincial policy. But the Ship Harbour project championed by the cabinet
in Halifax was turned aside through popular political pressure, by average
citizens who refused to accept the government's modernist assumptions,
by people who were, in APES director Irene Edwards's words, "able to see
beyond the 'riches' promised and hold onto a normal love and desire to
keep that which is ours."[77]

Such attachment to place combined with a growing worldwide awareness of pollution and "the environment" to reframe ways of thinking about
industrial society's intrusion into rural spaces. Buoyed by this new global
politics, activists began in earnest (and in concert) to demand real influence on policy, and to question the promises and premises of modernity.
And they began to work together.

ORGANIZATION AND REACTION

Cooperation among environmental groups in the province seemed far off
at the end of the 1960s. Shared knowledge had resulted in very little shared
action. Local groups conducted their campaigns mainly in isolation. All
of this was changed by an intensely local event affecting a handful of farms
in Annapolis County, an event that aligned with the mainspring of environmental anxiety in contemporary North America: the dangers of bio-
cidal chemicals.[78] Tentative though its gestures towards unity may have

been, the anti-herbicide campaign of 1970–72 epitomized a developing provincial environmentalism that compelled government reaction. The Gerald Regan government (1970–78) dealt with more unexpected environmental campaigns in its first four years in office than any other government in Halifax encountered in a similar period, while holding only minority control of the Legislature, and it took very little time for them to realize that environmentalism held the potential for real power and that they needed to channel activists in a more manageable direction and defuse campaigns like the anti-parks movement that threatened development objectives.

The trouble with herbicides began not with a celebratory announcement of some new industrial development but with the incremental expansion of an existing government program. The use of the herbicidal chemicals 2,4-D (2,4-dichlorophenoxyacetic acid) and 2,4,5-T (2,4,5-trichloro-phenoxyacetic acid) on roadsides by the Department of Highways and at the behest of the Department of Agriculture (DOA) for the control of farm weeds began in the early 1960s and quickly expanded across the province.[79] Chemical weeding was fast, effective, and cheaper than manual clearing – everything the officials in charge could hope for – and the obvious choice when the DOA wished to remove farm weeds from its Crown farmland on the lower Belleisle Marsh in Annapolis County. So it was done, on the third and fourth of June 1969, by a Maritime Air Services plane hired to spray 2,4-D, 2,4,5-T, and Dicamba over the province's 580 acres. But the plane delivered both more and less than promised. Spraying in a strong breeze, the pilot spread his payload over thirteen or more nearby farms, damaging crops two miles from the marsh. The chemicals left grasses unharmed, as intended, but their drift onto vegetable crops was ruinous. It was also undeniable. Local farmers were quick to band together, forming a "disaster group" to press for action and winning a fast response from the DOA, which investigated and awarded about $30,000 in compensation to thirteen farmers, including Robyn Warren, whose 10 acres of cauliflower abutting the government land were utterly destroyed.[80]

An unfortunately common agriculture misadventure then turned into disaster. Believing the DOA's claims that the herbicides were safe for use on grazing land and were harmless to humans and animals, Warren continued to pasture his dairy herd on the marsh and thought little of the direct exposure of himself, his father, and his brother, working in the fields as the spray plane did its work upwind. In the winter of 1970, Warren's cows aborted pregnancies (fourteen times in sixteen months), bore stunted calves, and showed a remarkably high rate of twinning (most of them stillborn). The family, meanwhile, developed abnormal heartbeats and swollen thyroid glands, symptoms consistent with, but not conclusively traceable to, herbicide poisoning. Doctors advised them to stop consuming the produce of their own farm.[81]

Both the manufacturer and the DOA insisted that any herbicides on the ground would have been destroyed by soil bacteria within days. In 1970, Warren replanted his cauliflower and lost the crop, along with his lettuce, celery, and corn. All of them grew quickly to unusual size before withering away. Tissue samples sent to an independent lab returned positive for 2,4,5-T, 2,4-D, and Dicamba – impossible, according to the DOA, which began insisting that the spray plane carried no 2,4,5-T (though it admitted requesting all three chemicals). The department refused to test for 2,4,5-T itself because, as the deputy minister said, "2,4,5-T was not used. Therefore why test for it?"[82] But Warren did continue testing, at his own expense, and continued to receive results showing the chemical in significant quantities in both vegetables and cattle fetuses, well into 1971. The department still refused to acknowledge a link between its own spray and Warren's troubles, but began advising farmers that herbicides might last weeks, months, or longer in cool, salty soil such as that found in the Belleisle Marsh.[83]

Warren's mounting calamities attracted national attention. The *4th Estate* in Halifax and the regional magazine *Mysterious East* investigated, as did the Toronto *Globe and Mail* and the CBC's national Sunday magazine program.[84] By the time the *Mysterious East* reporter arrived in the middle of 1971, half of Warren's apple trees had died – not half of the orchard, but half of each tree, on the side from which the spray had drifted.[85] Doctors who treated the family knew they could never prove conclusively the cause of their ills, but the Medical Society of Nova Scotia had already called on the provincial government a year earlier to review the use of chemical herbicides in roadside and agricultural spray programs; Warren's case offered apparent confirmation of their fears.[86] Environmental activists used the Medical Society's announcement to push for a ban on the use of the offending chemicals. Rosemary Eaton of the Cole Harbour Environment Committee took a particular interest and a leadership role in the campaign after opposing the use of chemical herbicides on the CN railroad passing through the harbour.[87] Beset from all sides, Premier Regan acted to ease the pressure on his government, suspending chemical spraying by government agencies for a year and instructing the Nova Scotia Research Foundation (NSRF) to investigate safety of commonly used sprays. He even convinced CN to impose a moratorium on spraying, and spoke to reporters about chemical persistence, groundwater contamination, and destruction of wildlife.[88]

A year's suspension is little more than a scheduling adjustment, without some further success, and despite the novel common front on a single issue, the campaign against chemical pesticides lost momentum through a year of quiet research by the NSRF. Worse for the activists, despite the premier's use of environmentalist buzzwords, the investigation he initiated adhered religiously to what Bill Templeman in 1971 called the "legal logic" of the Warren case: proof of harm required that a causal chain, A to B to C to D, be traced backward to establish that only A and nothing else could have caused B, then C, then D. It was a standard of proof virtually impossible to meet within the dense web of ecological connections found on an average farm.[89] The NSRF found no evidence that chemical pesticides posed a risk. A year passed, and spraying resumed on roadsides, railways, and power line rights-of- way.

Still, the nascent anti-chemical coalition opened the eyes of the Regan government to the potential power of environmental activism and the need to do something to direct it away from conflict with government objectives. The subsequent serial shocks of pollution fighters, popular democrats, and park opponents ensured that the problem stayed at front of mind. Duplicity was one response. Even as the premier praised activists' concern for a healthy agricultural system, the DOA reacted "as if it were defending Moscow," compiling stacks of public relations material with which to defend and promote the use of herbicides.[90] And the Regan government began to develop an environmental program of its own. "It is essential that the image of a reform be established quickly," wrote the premier's head policy man, Michael J.L. Kirby, to his friend David Watts at the US Department of the Interior, "in order to indicate to the public that we are a reformed [sic] minded government and are serious about solving the problems of pollution."[91] Kirby insisted that the government must take command of the definition of "pollution," not only for the sake of the next election or to counteract the "scare propaganda which is so common today" but to preserve the policy-making prerogative of the provincial level of government.[92] Provincial governments faced a challenge to their jurisdiction over natural resources in the early 1970s, from the new federal Environmental Protection Service. Kirby was deeply anxious that the triple threat of electoral politics, environmentalists' demands, and federal interference could scuttle the already shaky ship of Liberal government.[93] Cost-conscious environmental skeptics in the bureaucracy were equally frightened that feckless, pandering politicians at two levels of government would trap the province into spending promises that it could not afford, or worse, that that federal government would seize jurisdiction and render the province's efforts at crafting industry-friendly regulations moot.[94]

The need to take control of environmental issues gave form to the Nova Scotia Environmental Protection Act. Legislative summaries that Kirby had acquired from his friend in the United States ("so that we can, quite frankly, steal some of them") sped the legislative process along and bolstered the "image of a reform."[95] Above all, however, the act had to co-opt at least a portion of environmental activists and, at the same time, restrict the definition of environmental problems to the kind of sewage-and-smokestacks treatment systems that the Smith government had been pursuing in the late 1960s.[96] Certainly it had to include nothing that might encourage doubts about the manageability of the problems created by modernist economic development or about its sustainability. Activists were invited to comment on draft legislation, and they did, but from the beginning the centrepiece of the act and its primary concession to popular environmentalism was the creation of an Environmental Control Council, including a large number of non-government members, to investigate environmental issues at public invitation and report back to the minister. As a replacement for the Smith government's Environmental Pollution Council, an internal government body widely criticized for its insularity and secrecy, the ECC was supposed to be public, "apolitical, and independent of government."[97] All this it achieved by being also entirely powerless, unable to even initiate hearings or release reports without the minister's permission. For a short time, however, the council was granted freedom to act, and it attracted the interest and participation of some of the more optimistic members of the activist community, Alan Ruffman of the PCAC and MOVE, for example, and Dr. Ogden of the NSRC.

The Regan government's attempt to take the reins of environmental politics met its first test in the early 1970s in Lunenburg County at the Anil Hardboard plant, at the peak of a classic estuarine campaign for the defence of a home place. The controversy in Lunenburg was near enough to Halifax to attract the attention of the EAC and MOVE coalition – another sign of the tendency towards cooperation – but it was almost entirely the work of another small group of affected locals, this time people in and around the village of East River. The leader of the effort, Robert Whiting, was a hunter and trapper living in East River in 1967, when the provincial government proudly announced the opening of a hardboard fabrication plant next to his home. Touted as a triumph of industrial development, the largest plant of its kind in the world, and the first direct North American investment by an Indian firm, the project came complete with promises that there would be no noise and no pollution of the East River or Mahone Bay.[98] It did not take long for Whiting and his neighbours to see that the reality was rather less than promised. The water of the East River ran red and brown after Anil began production, and the stain of dissolved and suspended wood particles spread well out into Mahone Bay. Salmon numbers in the river declined, and not long afterwards shellfishing was prohibited in the bay. Not until much later did the company admit the presence of phenol, formaldehyde, paraffins, and aluminum sulfate in its effluent, in addition to the load of organic matter.[99] Noise and light from round-the-clock operations vexed those living nearby, and a combination of ash and red dust reached somewhat further afield, while a rotting stench rose from the abandoned quarry converted into Anil's effluent treatment facility and from the swamp through which it discharged into the East River.[100]

Residents "grumbled among ourselves," according to one activist, until Whiting began to organize resistance. Having sold his home next to the plant, he could easily have left the problem behind. Instead he wrote reams of letters to the company, the media, and every level of government, and urged his neighbours to do the same. Allies came slowly: Al Chaddock, a newly arrived artist from Halifax, remembered that "when I first met Bob I thought he was crazy, like just about everyone else did at the time."[101]

Government attempts to quiet dissent eventually brought more allies to Whiting's side. In 1969, the highways minister, I.W. Akerley, who served as president of the National Council of Resource Ministers, had told reporters that the problem at Anil had been "cleaned up." At much the same time, E.L.L. Rowe had opined that with "a new industry like Anil establishing [itself], you have to bend over backwards."[102] Three years later, prior claims forgotten, Rowe and a new minister, Glen Bagnell, were celebrating a new cleanup plan and reassuring locals that the plant's effluent, still brown, "does not contain the harmful wood fibres."[103] But residents of East River had long since learned not to take anyone in government at their word. It took long and patient effort at recruitment and public pressure, but when the newly formed ECC decided to make East River's problematic development the subject of its first public hearing, Whiting's "small army" – now organized as the East River branch of the South Shore Environmental Protection Association (SSEPA), with the support of the FON in Chester – was ready for an official confrontation with Rowe and the management of Anil.[104]

The ECC's hearing, held in a school auditorium in Chester in December 1974, was a ringing success for the East River SSEPA and a demonstration of the power inherent in the combination of local grievance and scientific abstraction. Witnesses ranged from local schoolchildren speaking about the loss of their swimming and fishing areas, to artist Al Chaddock, who arrived carrying a thick slab of matted wood fibre pried from the floor of Mahone Bay, to Norman Dale of Dalhousie University's Institute for Environment Studies, who urged the council to examine sea life in the bay for objective evidence of the effluent's effect.[105] Three presiding council members, finally face to face with the problem, were forced to conclude that the Little East River was "grossly polluted" and to recommend a more comprehensive treatment and monitoring system.[106]

In Chester, the ECC appeared to be a viable forum for change, but after the Anil hearing, the council's freedom was reduced and its members soon complained that they were being ignored by the Department of Environment.[107] Their report on Anil failed to satisfy activists, who expressed disgust at the provincial government's plan to dilute the plant's effluent in the wider reaches of Mahone Bay, where it might not provoke broad public reaction. "It's not over yet," Whiting said, "it's no good to build a pipeline if they'll be sending the same stuff offshore."[108] There would never be another ECC event quite like the fiery confrontation in Chester, however.[109] By 1975, it had served its political purpose. The results of the ECC's first hearing had met government expectations, kept the focus on pollution abatement and away from development policy, and convinced some of the most vocal environmentalists to limit their attacks during a time of political weakness for the Liberal government.

CONCLUSION

An efflorescence of environmental activism at the end of the 1960s built, piecemeal, the conditions for a sustained movement, beginning mostly around polluted harbours such as Boat Harbour in Pictou County and Chedabucto Bay in Guysborough and Richmond Counties, and moving from there to other areas and issues. As that movement coalesced, governments, conservationists, pollution fighters, and political radicals each began attempting to mould the emerging environmentalism into something that conformed to their own respective interests. In the process of negotiating a place for environmental activism in the provincial polity, activists proved that they had the power to alter and sometimes entirely frustrate centrally dictated modernist development policy.

Through such means as the Environmental Protection Act and the ECC, the Regan government attempted with some success to mute and redirect radical environmental protest into channels more consistent with the "end of pipe" vision of environmental regulation. In this they were assisted by the fact that their rivals at the federal level were equally intent on preserving the planning prerogative of experts and administrators, and therefore offered no greater opportunities for more profound activist input. The traditional conservationism of the early twentieth century was vastly preferable, for government experts, to a populist non-modern activism that threatened to squeeze governments between an unavoidable obligation to be seen to do something about declining environmental quality and an unchangeable imperative to pursue prosperity through industrial development. Many no doubt hoped that environmental activism was a passing political fad, soon to fade into yet another regulatory commonplace managed by the appropriate experts, and that a few years of temporizing would suffice to negate the trap. Though they acknowledged environmentalism's growing political power in Nova Scotia, few in government appreciated the mutual reinforcement of environmentalist ideas and awareness of industrial blight, or the inevitable continued growth of the former in lockstep with the latter, as more and more rural residents found their homes and livelihoods jeopardized by developments that, from the planner's perspective, promised to be more profitable on balance.

APES's Irene Edwards accused the province's politicians in 1973 of being "in a state of shock," panicking and in search of scapegoats because they lacked "the ability to comprehend that these people ... could see beyond the rosy future painted for them."[110] About their panicked response she could not have been more right, but some in government realized that not all activists rejected the rosy future of modernism, that the link between environmental activism and the radical politics of local place defence might be mutable, and that government might find common ground with a segment of the movement willing to support the prerogatives of science and expertise. The triangular relationship between those perceptive few in power and the modernist and non-modernist sides of the growing environmental movement shaped environmental activism in Nova Scotia over the following decade.

2

The Two MECs
Anti-Nuclear Environmentalism

THE TRIGGERING EVENT THAT brought Nova Scotia's scattered environmental activists together in a lasting way was a surprise to almost everyone. The first indication to the public that the new Regan government might be considering a nuclear project came in June 1972 by way of the Halifax *Chronicle-Herald*. Claiming to have information from a source inside government, the newspaper reported that the premier had met personally and in secret with representatives of a US company, Crossley Enterprises Limited, that wished to build a nuclear plant on tiny Stoddard Island, near the southwestern tip of the province.[1] Details remain scarce, because the project never moved past the informal proposal stage; however, the plan, as it emerged from further leaks and admissions over the rest of the summer and the following winter, was to build ten US-style light-water reactors (LWRs) on Stoddard Island and transmit the electricity generated there directly to New England via undersea cable. Had it been built, the complex would have been the largest generating station in the world, at twelve thousand megawatts, though some immediately doubted that the plan could ever work.[2]

Details about the Stoddard Island project were not available in the summer of 1972, but Nova Scotians were amply supplied with cautionary tales from the province's (and the country's) history of industrial development. Nuclear promotion was, after all, merely another step in a long line of industrial promotions in Nova Scotia, the secrecy of which (and the record of environmental damage) united environmentalists in opposition. After

the election of the Stanfield government in 1956, the province had negotiated secretly with a succession of multinational corporations to bring new
industries to Nova Scotia through an arm's-length Crown corporation
called Industrial Estates Limited (IEL). The notion that a well-chosen set
of incentives and subsidies could overcome the structural disadvantages
of the region within Confederation and reindustrialize its economy ne-
cessitated certain omissions and alterations in the province's industrial
regulations. Unionization was discouraged and transparency in negotiation
eschewed.[3] Environmental concerns were inevitably discounted as well in
IEL negotiations, and environmental activism tended to follow whenever
and wherever the ecological consequences of industrialization threatened
people's homes, farms, or businesses. There was a growing sense in the
province that IEL's secrecy effectively served as an environmental equivalent
to anti-union regulations, keeping any activist backlash from emerging
until it was too late.

Suspicion of IEL combined with a well-founded suspicion that the federal government would sacrifice even more of the public good in its quest
to promote nuclear technology. Atomic Energy of Canada Limited (AECL)
was twenty years old in 1972 and enjoyed an uncommon measure of support from the government of Canada. Originally, nuclear reactor technology had been the domain of the military, part of Canada's contribution
to the Second World War–era Manhattan Project, but over the previous
two decades the federal government had spent billions developing the
Canadian Deuterium-Uranium (CANDU) reactor system in hopes of
becoming a major world supplier of civilian nuclear technology. The high-
technology sectors of reactor design and construction, heavy-water production, and nuclear fuel fabrication drove Canadian ambition, but at every
link of the fuel chain there was a great deal of money to be made in the
business of nuclear energy. The results never did quite live up to the promise, though Canadian CANDU salesmen did at least manage to sell reactors
to India, Pakistan, Argentina, and South Korea.[4] Government enthusiasm
never waned, however. From the days of its infancy during the Second
World War, the Canadian nuclear industry was not so much supported
by government as it was a part of the government, and any challenge to
that relationship would have to be formidable indeed.[5]

Given the context, it was perhaps not wise for members of the Regan
government and the publicly owned Nova Scotia Power Corporation
(NSPC) to initially refuse comment on the leaked Stoddard Island plan
for several days. When the premier finally spoke, he offered only equivocal

denials that any earnest negotiations were afoot.[6] By then, it was too late. The opposition Progressive Conservatives had discovered the issue and happily forced Regan into fresh and ever less credible denials as more information came to light, repeatedly highlighting the government's reluctance to volunteer any facts on new developments.[7] The secrecy surrounding the project united those opposed to the Stoddard Island proposal. For every declaration of disinterest by the federal energy minister ("unless," he said, Canadian CANDU reactors could be used instead of American LWRs), there was a countervailing shock, as when Crossley Enterprises' Canadian holding company was revealed to have purchased Stoddard Island in 1971, or when the man who handled the acquisition, Halifax lawyer Ian MacKeigan, was appointed Nova Scotia's new chief justice in 1973.[8] Through a year of uncertainty, suspicion of the government's intentions was the link that bound environmentalists together.

Unsurprisingly, the earliest reactions from existing environmental non-governmental organizations (ENGOs) focused on the issues of government secrecy and public participation. The Nova Scotia Resources Council (NSRC) was cautiously hopeful of a new attitude of openness to scientific advice on the part of government, embodied in the new Department of Environment and the proposed Environmental Control Council. No doubt any skepticism NSRC members harboured grew stronger after they drew the premier's ire by insisting on a public examination of the risks and benefits of nuclear power – conducted by the appropriate experts, of course – and by wondering aloud, in the case of former NSRC chairman Dr. Donald Dodds, how it squared with the provincial development plan to build light-water reactors in Nova Scotia even as the safety of their design was under investigation in the United States. As the journalist Ralph Surette later reported, the premier was incensed and "had no intention, he said, of scaring off investors by making public the terms of negotiations, no matter what the project was."[9]

Surprisingly even less critical in its public statements was the younger (in terms of both its members and the organization itself) and typically more radical Halifax-based MOVE coalition. The coalition had come together only in 1971 and was still struggling to negotiate funding from the federal government when it was called upon to state its position on nuclear power. The executive, while privately noting its own unanimous opposition to the project, decided to restrict its public statements to a call for open public negotiations.[10] Political calculation may have played a role in the decision, but it is just as likely that the nuclear question represented one of the first of many cases in which the coalition struggled to reconcile the views of its diverse membership with its leaders' desire to use the coalition's higher profile for specific causes. In their different ways, then, both the NSRC and MOVE were reluctant to antagonize the provincial and federal governments. Though they both declared an interest in educating the public about the dangers of nuclear energy, the real work of doing so fell to newer organizations.

Four Acadia University academics calling themselves the Nuclear Power Study Group (NPSG) filled the gap. From October 1972 to the following March, the NPSG produced six bluntly factual and scientifically grounded articles detailing the likely biological, toxicological, and economic effects ten nuclear reactors would have at Stoddard Island, as well as what kinds of effects could not be predicted ahead of time. Chief among the unknowns: what might happen as a result of discharging ten million gallons every minute of twenty-degree-warmer water into the ocean in the richest lobster fishing district in the province, from which a quarter of the annual catch was landed. Radiation dangers, accidents, and employment impacts were also unknowns. The NPSG's stated purpose was to "contribute to informed public discussion." As such, the group explicitly eschewed an "alarmist position," to the point that its writing sometimes drifted into textbook-style scientific exposition.[11] But the four academics were not averse to taking a stand. In summing up the series, they dispensed with scientific reserve long enough to observe that, in contemplating the US connection, including the burden of risks and the disposition of benefits, "the word 'exploitation' comes to mind."[12]

By early 1973, anti-nuclear ENGOs in Nova Scotia were establishing links with each other, and although the NPSG was not heard from again (likely due to Dr. John Brown's departure from Acadia), their essay series was reproduced in the press and by groups across the province who wanted to borrow the scientists' mantle of legitimacy and authority. One group that did so, and also provided some of its own research, was the fledgling Ecology Action Centre (EAC) at Dalhousie University in Halifax, a member of the MOVE coalition and sometime collaborator with the NSRC.[13] Though the EAC was dependent on government funding, its initial membership was, unlike MOVE, wholly environmentally focused and eager to expand beyond its initial preoccupation with urban issues of city planning and recycling. The centre's first publication on nuclear power, in December 1972, openly opposed the Stoddard Island project. Citing routine radiation exposure, accidental releases, thermal pollution, radon from uranium mine tailings, and the unsolved problem of nuclear waste storage, the document concluded that "the electricity, the profits, and even many of the jobs would go to Americans, while Nova Scotians would take the risks." After such a litany of complaints, its closing recommendation for open, public negotiations between Crossley Enterprises and the provincial government seemed comically redundant.[14] In its policy statements and grant proposals throughout 1973, the EAC repeated its emphasis on the injustice of Nova Scotia's assumption of nuclear risks for the benefit of the United States, but it also kept up its agitation for public participation and also showed a strong commitment to the limits to growth thesis. The centre warned that the province would eventually have to deal with its own energy shortage, and by the end of 1973, a year when the international politics of oil pricing bolstered the theory that there were hard ecological limits to the continuation of economic growth, its coordinator was advocating a policy of zero demand growth and zero energy exports.[15]

Another group that leaned on the NPSG's scientific ammunition organized over the winter of 1973. The new South Shore Environmental Protection Association (SSEPA), which would go on to play a central role in the province's environmental movement for a decade, targeted all three levels of government in an attempt to defeat the Stoddard Island proposal politically, rather than merely request public participation or work at public education. Following the lead of the Southwestern Nova Scotia Lobster Fishermen's Association (and sharing members with it), the SSEPA won unequivocal support from the Barrington and Yarmouth municipal councils, Progressive Conservative offshore resources critic and MP for South Shore Lloyd Crouse, and Liberal social services minister and Shelburne County MLA Harold Huskilson, by impressing upon them that, in the words of fishermen's association president Glen Devine, "this whole area [and its voters] depends entirely on fish."[16] Under the leadership of author and activist Hattie Perry, the SSEPA gained its greatest success in October 1973, when Premier Regan attended a public meeting in the tiny village of Barrington Passage, about ten kilometres from Stoddard Island, and found waiting for him hundreds of nearby residents who wanted only one thing. He gave it to them: a clear promise of public consultation on any proposed nuclear plant in Shelburne County, and another that no project would be approved that might harm the fishery.[17]

Political pressure had won a victory for the SSEPA. The assurances given at Barrington Passage, combined with the failure of the proponent to quickly address the suggestion of CANDU reactors from the federal Atomic Energy Control Board (AECB), seemed to spell the end of the Stoddard Island proposal by 1974. There was, however, no corresponding revival of trust in government and no dissolution of the groups that led

the fight. If anything, the continued commitment of the Regan government to two badly functioning heavy-water plants built to supply
the Canadian nuclear industry in the late 1960s at Glace Bay and Port
Hawkesbury suggested a continued interest in nuclear technology.[18] Hattie
Perry continued enthusiastically to lead SSEPA's opposition to nuclear
development schemes through her writing, leaning on the NPSG's research
and adding her own on alternative energy sources and the health effects
of radiation. She drew on an international discussion of nuclear dangers
but always returned to the threat posed to the local fishing economy and
the lack of appreciable local benefit.[19] The SSEPA also showed its continuing concern over the threat of government secrecy at an ECC public hearing
in Yarmouth a month after the Barrington Passage meeting, where, according to the ECC, "the people present cited the example of the apparent
lack of an environmental assessment study for the Strait of Canso [refinery
and shipping complex] as evidence that these kinds of projects and developments can and will go forward without public approval."[20]

Not in Anybody's Backyard

Anti-nuclear activism in Nova Scotia was fully immersed in a worldwide
movement of the same type. It attracted the attention of pacifist immigrants, especially Quakers, veterans of American anti-war protests who
worked to draw attention to the links between nuclear war and nuclear
power, and to introduce other Nova Scotians to a placeless international
peace movement. Anti-nuclear protestors in Nova Scotia were also well
aware of the actions of the Clamshell Alliance later in the 1970s in Maine,
where activists resorted to civil disobedience to fight nuclear development. The strongest and earliest connections outside of the province,
however, were with the anti-nuclear movement in New Brunswick, where
protestors faced an energy project that enjoyed the full support of the
AECB. New Brunswick's activists, like Nova Scotia's, found their strength
in determined defence of local places and economies, especially rural places.
As the two governments (and that of Prince Edward Island) planned new
collaboration, these primarily rural activists demonstrated the emptiness
of the parochial "NIMBY" (Not In My Back Yard) stereotype by reaching
out to each other to form a common regional front.

Premier Regan's clearly reluctant climbdown over Stoddard Island and
continued support for the heavy-water plants were not the only reasons
to fear a resurgence of the nuclear idea in Nova Scotia. Even as word of

the South Shore project spread in 1972, news broke of another nuclear plant proposal across the Bay of Fundy in New Brunswick, which would eventually become the Point Lepreau generating station.[21] The Nova Scotia Power Corporation's general manager immediately denied having any interest in the project, but with the failure of the Stoddard Island plan activists were wary of the provincial government's intentions. Within a year, Regan and the same NSPC official transformed wariness into alarm by promising to decide on whether to buy into New Brunswick's planned second reactor or to build one of their own instead, perhaps somewhere in northeastern Nova Scotia. The success of the Organization of the Petroleum Exporting Countries (OPEC) oil cartel in 1973 gave proponents of atomic technology a second chance to sell their wares to the public, and statements like "we have to get into nuclear power" from the provincial energy agency meant that activists in the two leading organizations had to remain ready to push back.[22]

Whereas the SSEPA transitioned more or less seamlessly to general anti-nuclear activism, the Ecology Action Centre had more trouble. The centre's student staff had better reason than most to appreciate the limits of finite resources in 1973: that was the last full year of federal government funding they would ever see. Across the country, student groups that had enjoyed the largesse of the federal Opportunities for Youth and Local Initiatives Programs during the heyday of student radicalism found the government's commitment to youth empowerment fading along with the threat of youth rebellion. Many groups simply melted away, as MOVE would soon do, but the EAC staff were determined to find the means to keep the organization running. Remarkably, they did, but the transformation from fully funded organization to registered charity – in a sense a liberation from the government purse – had the perverse effect of blunting the centre's radical spirit. With a newly formed board of directors drawn as much from the academic and business communities of Halifax as from the centre's older constituencies of student leaders and environmental activists, the EAC returned to the nuclear power issue in late 1974 at the request of the SSEPA and immediately fell into disagreement over what position to take. Some of the new directors insisted on hedging the centre's opposition to nuclear energy in case of social and economic need, while others argued that such ill-defined hurdles would vitiate any real opposition and urged a categorical rejection of nuclear power. A vote ended with a statement of opposition, but the centre chose to refocus on the less contro-versial issue of public participation as the central matter of energy policy going into 1975, rather than the limits to economic growth or environmental justice arguments that still preoccupied their South Shore peers.[23] The year ended with the centre's survival in doubt and its ability to function severely compromised, leaving a void in the Halifax activist scene that was promptly filled by a group of very different origins but very similar tendency to look first beyond the local context.

During the height of the Stoddard Island controversy, another gathering of Haligonians opposed to the scheme and to nuclear development in general had taken shape from elements of the Voice of Women for Peace (VOW) organization and the Halifax Friends Meeting (Quakers). Both groups had a long association with peace and social justice issues and, in the case of VOW, a history of connecting environmental and anti-nuclear activism, though their work had mainly been in opposition to nuclear weapons rather than nuclear energy. VOW was a national coalition of groups formed in 1960 to oppose Cold War nuclear brinkmanship. Its Halifax leaders, Peggy Hope-Simpson and Muriel Duckworth, pulled together a resourceful group of well-to-do Halifax women and quickly turned the local VOW into the most active branch in the country outside of Toronto, with a keen appreciation for related issues.[24] At the Halifax hearing of the federal Man and Resources program in 1972, the group displayed a keen appreciation of the link between peace, environmental justice, and the limits to growth. In support of "zero economic growth and zero industrial growth" and in opposition to what it called the "consume-exploit-pollute society," VOW explained that

> the Voice of Women has found that we cannot merely espouse an end to war, we must espouse a way of life free from the gearing toward war. Voice of Women cannot advocate a rise in living standards for the rich nations when this rise means a consequent depletion of the resources necessary for the survival of the poor nations.[25]

Both nationally and in Halifax, VOW had close links to the Quakers. From their beginnings as a group of religious dissidents in seventeenth-century England, the Quakers (or Friends, as they call themselves) have maintained a strong commitment to pacifism and to actively opposing violence, warfare, and injustice by moral pressure and "bearing witness." Dorothy Norvell, a recently arrived immigrant from the United States, a Quaker, and a third-generation peace activist, introduced the Halifax Voices to the Quakers' published critique of modern industrial society, *Speak Truth to Power*. Muriel Duckworth, Peggy Hope-Simpson, and Helen Cunningham were also Quakers. Historian Frances Early credits the influence of these Quaker/Voices leaders with the Halifax VOW's "consensual decision-making process."[26]

When Stoddard Island and nuclear power hit the province's front pages in 1972, the Halifax Friends Meeting, lacking any consensus, felt the need for a more complete understanding of nuclear power issues and their relationship to nuclear weapons, war, and social justice. It asked another newly arrived American member named Susan Holtz to do some research for them. As a researcher for the local Friends meeting, Holtz quickly arrived at conclusions in line with the VOW statement at Man and Resources. "Power for what?" she wrote to New Brunswick anti-nuclear activist Doris Calder. "For a way of life for the affluent in North America which is entirely oriented toward disastrously wasteful consumption? For a way of life which cannot be continued if everyone in our global village is to share together our world's resources?"[27] In a matter of months, Holtz became an activist as much as a researcher and deliberately chose to devote herself to the more technical issue of nuclear risk, rather than ecological limits or environmental justice. Those, she told Calder, could be taken care of by other Friends in her group.[28] In reality, the group tends to follow the leader, and Holtz, who brought nearly endless energy to her role, soon became the de facto leader. Fellow activist Alan Ruffman remembers her as a particularly "intellectual" sort (rather than political), and the same might be said of the group she gathered around herself in 1974 to encourage public opposition to nuclear power and discourage government's nuclear ambitions.[29] The group of Quakers, Unitarians, and VOW members, including Norvell and Duckworth, soon began to refer to themselves as the Conserver Society Group (CSG).[30]

The anti-nuclear CSG spent most of 1974 making connections with like-minded activists and organizations across Canada, gathering information and encouraging the formation of a national coalition to take the fight to AECL and Ottawa. Duckworth, Norvell, and Holtz wrote to activists in Vancouver, Calgary, Toronto, and elsewhere, and tried to persuade organizations like the Canadian Wildlife Federation to form a common front. They also cultivated contacts with sympathetic journalists like Ralph Surette and the editors of the *4th Estate,* much as VOW had done in the 1960s.[31] Their approach was decidedly non-local, however, consistent with their peace movement background, and what they did not do was approach the SSEPA, at least not until SSEPA secretary Anne Wickens wrote to Holtz with an invitation to join the group. It was their correspondence that brought the CSG into contact with a growing Maritime coalition.[32]

As the CSG's activities suggest, shared concern over nuclear power tended to draw activists together despite membership in different groups. It also tended to do so at the same level of geographic integration as the program they opposed. The federal jurisdiction of the AECB/AECL provoked some limited national coalition, but provincial jurisdiction and the actions of the three Maritime governments drew more attention to the regional level. Only three months after its apparent success in October 1973, the SSEPA was confronted with Nova Scotia's possible investment in the proposed Point Lepreau nuclear plant in New Brunswick, geographically closer than Halifax to the association's home in Shelburne and Yarmouth Counties. Once again the provincial government stubbornly refused to share information about the negotiations, and New Brunswick's premier certainly could not be influenced in the same manner that Gerald Regan had been. In response, the SSEPA reached out to New Brunswickers who were organizing against the Hatfield government's nuclear plans.[33] It played a major part in drawing together the first New Brunswick anti-nuclear coalition, along with representatives of the Moncton branch of Pollution Probe, the Eel River Bar Mi'kmaq Band Council, and two other local groups.[34] In its first incarnation, as the Chaleur Environmental Protection Association (CEPA), the coalition set out in the spring of 1974 with "no immediate plans to seek meetings with public officials" about Lepreau or about another nuclear plant proposed for New Brunswick's northeast coast. Instead, it said, "we're going to bring experts in to inform the public about the dangers of nuclear plants" and provoke popular opposition in the affected areas, much as local opposition in southwestern Nova Scotia had fed into a provincial and regional movement.[35] By the end of the year, nuclear development plans had provoked a wide enough reaction that a broader coalition became necessary, and CEPA gave way to the Maritime Coalition of Environmental Protection Associations, later rechristened the Maritime Energy Coalition (MEC) for publicity purposes. (It was a shorter name, a clearer statement of the group's interests, and it needled the provincial governments by borrowing the acronym of the proposed regional utility, the Maritime Energy Corporation.) Official statements of support for the MEC came in from Women's Institutes, fishermen's associations, nurses' associations, the NSRC, VOW, and the EAC, along with an application for full membership from the Halifax Meeting of the Religious Society of Friends.[36]

Anti-nuclear organization by small rural activist groups under the MEC banner contributed the greatest vitality to the regional movement during the mid-1970s, and kept questions of limits and local economic autonomy at the forefront. In later years, the alleged self-centredness of local single-issue groups would lead to their being labelled "NIMBYs" and dismissed as unsophisticated reactionaries happy to foist the projects they rejected onto some less fortunate community. Such positions were rarely in evidence. Activists initially motivated by personal risk, living near potential nuclear construction sites, consistently joined forces to oppose nuclear development anywhere in the region. Continuing their focus on public outreach rather than bureaucratic consultation, MEC groups planned alternative energy festivals on New Brunswick's Kingston Peninsula and in Middleton, Annapolis County, in 1976, in an effort to convince the public to work towards local energy independence and the dissolution of central energy agencies like the New Brunswick Electric Power Corporation (NBEPC) and NSPC.[37] In Halifax, meanwhile, the EAC was scarcely active in 1974 and early 1975, uncertain of its position on nuclear power and suffering through the end of federal funding and the steep learning curve of charitable fundraising. By the fall of 1975, however, the CSG had merged into the EAC as its new energy committee, bringing the volunteers and financial support of the Halifax Friends Meeting into the EAC's established structure (and into its free office space at Dalhousie University), and uniting most of Halifax's anti-nuclear activists under the EAC name.[38] If unity was truly the prerequisite for effective action, the stage was set in 1975.

Citizen-Science and Political Power

In the mid-1970s, activists from city to farm shared the opinion of Donald Hamilton of the Northumberland Regional Development Association that "unity is the only way that we can effectively combat the nuclear power menace."[39] Moreover, no single group had achieved a dominant position from which to force upon others a position that might alienate them. Local groups each continued to carry on parallel campaigns without cooperation or sometimes even knowledge of the others, despite nominally being in coalition. This would change, but not before the Halifax activists developed a more scientific environmental rhetoric and a willingness to collaborate with the environmental agencies of the state. In the process of doing so, and especially in the process of organizing publicity for the anti-nuclear coalition, a divergence began to show between the more modernist activists and those who wished to persist in the bottom-up promotion of decentralized, low-technology living.

From the inception of the CSG, the Quaker anti-nuclear activists of Halifax were eager to appear respectable and to appeal to federal and provincial bureaucrats (and corporations) in intellectual terms. Even when asserting limits ideas, as she did in correspondence with the deputy minister of the federal Department of Energy, Mines and Resources (EMR) and with AECL managers, Susan Holtz hedged on implementation, suggesting "an immediate and serious energy conservation policy" instead of a policy of zero economic growth.[40] Nor did the CSG ever share the SSEPA's commitment to argument based on local environmental justice, preferring to argue from the general principle to the specific case rather than vice versa. Holtz remained the public face of the group for years, and was a prolific letter-writer who never forgot her mission to highlight the technical risks of nuclear technology. One of the events that reinforced that determination was the Critical Mass conference in Washington, DC, in November 1974, where US consumer advocate and anti-nuclear leader Ralph Nader brought together citizens' groups from all over North America to share information, and where Holtz learned that "from a political point of view, it is just as effective to put real doubt in people's minds as it is to bring them, 100% committed, to your side – and it's far easier."[41] As she told the SSEPA's Anne Wickens shortly after returning from Washington, the lesson for Nova Scotian activists, unsure about how much weight the premier's anti-nuclear promise would carry, was a pragmatic and practical one: to win over decision makers, "one needn't push one's absolute convictions quite so hard, but one does need lots of facts and credible experts."[42] Facts and experts might be just as convincing to the wider public in order to build a mass political movement, but Holtz, who saw her job as being a bridge between the public and the powerful, immediately began planning a venue where decision makers could be convinced by the intellectual merit of anti-nuclear views.

The Energy and People conference took place at Saint Mary's University in Halifax on September 19–21, 1975. Outside of environmental activist circles, its general context was the global "energy crisis," still in full swing since the OPEC cartel moved to reduce the supply of crude oil to world markets in October 1973.[43] Because of its greater dependence on imported crude for electricity generation, Atlantic Canada had felt the sting of rising prices more severely than any other region of the country. Meanwhile, uncertainty surrounding the plans of the three Maritime governments to replace some of their oil-fired generators with nuclear generating stations fuelled continued speculation and anxiety. Despite the promise extracted from him at the Barrington Passage meeting, Nova Scotia's premier had spent the following year issuing statements alternately promising a major nuclear project and denying that any such project was being entertained. New Brunswickers remained unsure whether or when or where a companion project to the Point Lepreau reactor would be announced, and the question of purchasing electricity from Lepreau still lay unresolved in Prince Edward Island. Given the context, Energy and People could hardly have avoided becoming a forum for nuclear debate, but its organizers hoped for, and achieved, a much wider discussion of related energy and economic issues.[44]

The idea for the conference (and the largest monetary contribution) came in the fall of 1974 from several Quaker members of the Conserver Society Group, who hoped that a well-publicized conference catering to a non-expert lay audience might help advance the notion from which their group drew its name. The term "conserver society" was coined by the Science Council of Canada (SCC) in its 1973 report on *Natural Resource Policy Issues in Canada,* in reference to a society, unlike our present consumer society, in which resource decisions would recognize total costs (account for externalities, in other words), respect the limits of the biosphere, and also bear in mind costs deferred to the future.[45] As a general principle with uncertain implications for economic growth, the phrase "conserver society," like "sustainable development" after it, brought together people with varying interpretations of the meaning of "wastefulness" and the possibility of continued economic growth. Once together, however, activists discovered sharp divergences between their respective interpretations and political intentions. When planning for the Energy and People conference began, in accordance with the consensus principle borrowed from VOW and with committees drawn from several co-sponsoring organizations, those differences fuelled conflict: some wanted an event with strictly public appeal and a strong anti- nuclear and anti-growth message, whereas others wished to use the conference to make connections and build support within federal and provincial bureaucracies, where talk of limits to growth could close rather than open doors. After three months deciding and six preparing, the organizers did manage to achieve the mixed audience and "balanced presentation of controversial issues" they sought, but tensions between alternative visions of the conference only grew sharper as the event got underway.[46]

Limits to growth were a prominent theme at the conference, beginning with the keynote address by Gordon Edwards of the fledgling Canadian

Coalition for Nuclear Responsibility (CCNR). The CCNR represented the fruition of the campaign of Holtz and others for a federal-level lobbying organization made up of ENGOs from across Canada, and in 1975 it was in the process of developing a more or less technical rhetoric of radiation hazards and the hidden costs of nuclear energy, rather than an explicit attack on economic growth or modernist society.[47] Though he stopped short of offering specific answers or policy prescriptions at the conference, Edwards spoke at length on the illusion of scientific analysis of mathematically incalculable values such as wilderness (he cited the current debate over the damming of British Columbia's Skagit River) or of unknown quantities such as accident probabilities at nuclear reactors. For him, the path to a conserver society lay not in the recognition of some narrow technical risks and substitution of lesser risks, but in giving up the obsessive "idea of a machine" and admitting the ecological context of society, complete with limits and unmanageable unintended consequences.[48]

Along with Edwards's warnings about the impossibility of continued exponential growth, others presented radical policy proposals and openly questioned the role of government as an honest regulator and interpreter of policy ideas, particularly where the balance of evidence and argument pointed to conclusions that contradicted economic orthodoxy. Dalhousie University environmental lawyer and EAC board member Ian McDougall argued against continued electricity exports to the United States due to the inevitable exhaustion of Canadian energy sources, a theme echoed by Bill Peden of Ontario's Energy Probe. Challenged by government representatives who claimed that a transition to renewable technologies through a period of rising energy use was the only way to avoid massive increases in unemployment in Canada, MOVE member Don Grady replied for the limits theorists that changing the basic nature of society required "unacceptable" actions, and that in order to avoid such actions, governments would always attend conferences like Energy and People "to lie."[49]

Such outbursts of hostility towards the bureaucracy probably went some way towards inspiring the conference chairperson's retrospective assessment that citizens must do more to "encourage governments to trust groups," but in truth there was at least as much opportunity at the conference for government and corporate attendees to present their views, beginning with their presence on the discussion panels. Compromise and consensus in event planning worked against those who wanted an ideologically consistent presentation. Atomic Energy of Canada Limited, Gulf Oil, Nova Scotia Power Corporation, and the federal Department of Energy, Mines and Resources had displays of promotional material throughout; Robert Stanfield, then leader of the federal opposition Progressive Conservatives, gave an anodyne opening speech; and EMR's direct contribution, an energy primer pamphlet called *See Energy Run,* was distributed to attendees despite organizer Bill Zimmerman's complaints that it was "slanted" and not at all what was promised. Organizers even attempted to prevent the drafting of resolutions for the conference, not wanting to alienate the bureaucratic contributors. This last attempt at achieving the appearance of balance came to naught, however, as attendees assembled on their own to vote and approve a resolution that "this conference rejects nuclear power in its present form as the energy source of the future and calls for a moratorium on the licensing, construction, and sale of nuclear power plants until full national, public, and parliamentary debates are held." While the anti-nuclear resolution certainly highlights the perceived urgency of the nuclear issue in Nova Scotia and New Brunswick (which between them supplied much more than two-thirds of the attendees at the conference), the fate of the limits thesis may better demonstrate the will of the organizers. A proposed resolution to encourage priority in both funding and planning decisions for "small-scale, decentralized" projects was voted down.[50]

The Energy and People conference marked a turning point for the anti-nuclear activist community in Halifax. At the time the conference was planned, the CSG was deliberately embarking on a path of technical argument as a means of attracting the ear of government and bolstering anti-nuclear elements within the bureaucracy. The compromises inherent in consensus decision making put their reasonable and unthreatening modernist approach, which did not question the possibility or desirability of economic growth or the value of centralization and the power of expert decision makers, directly next to the attempt of more radical activists to attract media attention and convince the public of the relevance of self-sufficiency, local autonomy, and ecological limits to energy planning. Given the choice, the media unsurprisingly tended to report more favourably on the former; consequently, the more radical conference planners seem to have abandoned the cooperative effort as a vehicle for political pressure.[51] The explicit insistence on limits and distrust of government were absent from the continuing committee's follow-up report in October 1975, replaced by an emphasis on public education and a national lobbying effort, and a commitment to pure "facts and figures," such as were being compiled at the energy library that Holtz built in the EAC's offices with funding from the Halifax Friends Meeting and the Canadian Friends Service Committee. The continuing committee itself, top-heavy with engineers and professional planners, reflected the near-disappearance of radicals from the Halifax coalition as it settled into its new identity as the EAC's energy committee.[52]

The sort of energy policy advocated by the EAC's committee following Energy and People – a technologically novel but economically orthodox mix of wind power and solar, with some conservation measures added – required financial incentives and investment more than social change, and therefore required more convincing of politicians and bureaucrats than of the public. It also encouraged a national perspective, because activists across the country could direct their combined efforts towards the largest relevant energy technology research body, the federal Department of Energy, Mines and Resources. Cultivating those kinds of connections outside of the region could take two forms: bringing the rest of the country and continent to Nova Scotia, or sending Nova Scotians out to meet them. As one might expect, the most persuasive anti-nuclear spokesmen came from those jurisdictions where the industry had been longest established, particularly in Ontario and the United States, and in 1976 the EAC's energy committee led the effort to bring those speakers to Nova Scotia. Following Gordon Edwards, the first prominent visitor was American energy expert Amory Lovins, whose concept of a "soft energy path" was just beginning to achieve global recognition. The two key characteristics of a soft path are renewability (using solar, wind, or geothermal sources, for example) and appropriate scale and quality (warming houses with on-site heat, for example, rather than with distant, central electric generators). The congruence with conserver society ideas was obvious, and the Science Council of Canada eagerly sponsored a speaking tour. The EAC just as eagerly encouraged Premier Regan to attend. He declined.[53] In fact, despite the energy committee's best efforts to attract them, provincial politicians continued to disappoint.[54] Appeals to corporate self-interest were about as fruitful; when activists sent a five-person delegation to the Atlantic Provinces Economic Council seminar on energy in January 1976, one reported back that the regular delegates would be tempted away from their support for nuclear power only if they could be shown "how to make their money back on the initial cost of alternative energy development and begin to show profit."[55] It was a position to which the committee spent more and more time appealing, but it was hardly a recipe for a radical economic revision of consumer society, and it encouraged a disproportionate focus on the technology of the soft path rather than the issues of scale and ownership.

Outside the province, the leaders of the Halifax anti-nuclear network continued to be heavily involved in the CCNR and found, in Susan Holtz's words, "some openness" to public input in Ottawa, especially from EMR. It was a welcome change from the secrecy and stonewalling back home in the country's "most secretive jurisdiction."[56] Yet the CCNR's strategy itself was essentially conservative. In order to "be taken seriously on technical and policy issues," its arguments inevitably circled the familiar ground of radiation risks and reactor safety, and the constant demand to which all of its public efforts were directed – the nuclear responsibility weeks, rallies, and national petitions in which the EAC participated – was for nothing more than a national inquiry into nuclear power.[57] Even so, it was all to no avail, at least in Ottawa. EMR listened, but preferred the advice of AECL vice president A.M. Aikin, who warned against any forum that allowed demagogues to abuse "honest, capable scientists," and who went on in 1977 to co-author EMR's wildly optimistic Hare Report on the management of nuclear waste.[58] The overrepresentation of nuclear science in EMR's research and development budget revealed their preference just as clearly. What the department wanted, and offered, was an energy conservation and renewable generation program that Canadian ENGOs might support, and that might be kept at a safe distance from the work of forming a more fundamental energy policy based on greater use of nuclear energy and fossil fuels. In one telling example, EMR's Office of Renewable Energy, along with the Ministry of State for Science and Technology, sponsored a series of conferences in 1977 on the conserver society concept, the Halifax edition of which focused on renewable energy sources. Local activists did much of the organizational work, but when the EAC's Susan Mayo discussed the preparations with the ministry's Ray Bouchard, her frustration was plain and her assessment blunt: "The fuckers didn't ask us about nuclear power!"[59] Nor was she alone in noticing that, for federal bureaucrats at EMR and Environment Canada, "alternative" energy seemed more often to mean "additional" energy and a reason to avoid discussing the nuclear issue.[60]

Despite its frustrations, the task of working within a coalition of Canadian ENGOs to access the levers of power in Ottawa preoccupied the EAC's energy committee at the end of 1977. Holtz argued to the Canadian Environmental Advisory Council (CEAC), a federally funded panel of experts and activists among whose number she would soon find a seat, that the federal government's power to seize jurisdiction, as it had done over the nuclear industry, meant that ENGOs had to "get beyond local issues" and concentrate on winning over members of the federal bureaucracy.[61] And the perennial dream of recapturing the policy initiative and influencing the most basic assumptions of Canada's energy economy did not fade. When the CEAC invited ENGOs to formal consultations in 1977, so that it might "better advise the minister," activists aware of their creeping marginalization used the opportunity to form a new national coalition, aiming, in the words of its first president, David Brooks, "to raise the issue from particular[s ...] to the question of a conserver society."[62] Their efforts were somewhat confused, however, and the organization that resulted, Friends of the Earth (Canada) (FOE), immediately became a gathering of the more professional ENGOs, who were reluctant to be seen as radicals. The new organization was also supposed to be independent of government funding, reflecting some awareness of the distortions wrought in the conserver society concept by bureaucratic agenda setting, but the need to avoid competing with its own member organizations for donations left FOE no other option but to depend immediately on government research contracts for its operating budget. That FOE still met with a hesitant reception from the EAC's board of directors says something about how cautious the centre had become; some wondered whether the EAC could fully support the more ambitious version of the conserver society notion, or whether it would prove "too political" a goal for the centre's tax-exempt charitable status.[63] Nonetheless, Holtz continued to represent the centre at FOE and to the CEAC's National Steering Committee (which later evolved into the Canadian Environmental Network), and she became part of the effort when the former group acted to put Amory Lovins's soft energy path on the national agenda.

With the financial backing of EMR, the Science Council of Canada, and Petro-Canada, the first Friends of the Earth soft energy path (SEP) study began in late 1978, only four months after the formal creation of the coalition. Though ambitious, it was not without precedent, even in Canada. While leading EMR's Office of Energy Conservation in 1977, David Brooks had drafted a national SEP study with the help of the University of Waterloo's John Robinson, which was itself intended to be a more detailed and complete version of Amory Lovins's SEP paper on Canada, commissioned the previous year by the SCC. Dissatisfied with his work at EMR, Brooks proposed a study that accounted for the full diversity across Canada's provinces. The FOE study, then, was to be yet another refinement of the basic argument Lovins made in 1976: that a future in which Canadians relied primarily on "renewable, decentralized, and ecologically sustainable" energy sources was both feasible and not even particularly difficult to achieve.[64] The influence of Susan Holtz was all over the decision to take the SEP concept to the next (national) level of collaboration. Having sponsored a panel (including Brooks) on Nova Scotia's

SEP in November 1977 for a largely bureaucratic audience, and having pursued provincial funding for a study on the same from early 1978, Holtz's energy committee was clearly eager to advance the concept among the bureaucratic elite.[65]

In comically Canadian fashion, the first set of provincial SEP studies completed and published in the FOE journal *Alternatives* in 1979 and 1980 was written without common metrics and could not be compiled into a single combined set of data. A revised set was undertaken in 1980, this time funded by Environment Canada and EMR, and the results were published in 1983 as *2025: Soft Energy Futures for Canada*. It was by the lead authors' own admission a remarkably conservative document, explicitly disavowing calculation of the social and environmental costs of various technological options in favour of a strictly economic calculus based on projections of future energy demand drawn from Statistics Canada and EMR economic models. Political considerations were absent as well. Amory and Hunter Lovins wrote in the introduction that "this Canadian study's technical assumptions are so conservative that only a concerted drive for *in*efficient energy use and deliberate *suppression* of cost-effective renewable sources could achieve a Canadian energy system needing more fossil or nuclear fuel than is calculated here."[66] Neither they nor the authors suggested what to do if such a policy did in fact prevail.

One of the intriguing practices of the FOE study was called "backcasting," another term borrowed from Lovins's work. Recognizing that typical economic forecasts subsume the categories of what can happen and what should happen into the seemingly objective category of what will probably happen, backcasters pick a desired economic condition at some point in the future and attempt to determine how and whether it can be achieved. The subjective nature of desirability and probability are acknowledged, and feasibility comes to the fore. So do the authors' assumptions about what is politically palatable. In the FOE study, a hypothetical Canadian economy in the year 2025 provided a "desirable" end point from which the "hard" technologies (central, large-scale, non-renewable, and environmentally destructive) were absent. In practice, this meant that the authors ruled out only nuclear energy, Arctic oil, and coal liquefaction as unacceptable.[67] For the sake of a convincing future scenario, it also meant that the appropriate scale of acceptable technologies disappeared in the analysis, along with questions of ownership, whole-system analysis (including non-energy resources), and any doubts about continued economic growth. Considered politically impractical, those ideas fell outside the category of the "pragmatic."[68] The Nova Scotian segment of the study, for example,

Susan Holtz's own work, shared the general premise of "strong economic growth and substantial increases in material standards of living" and assumed an increased industrial output for the province of 1.3 to 1.9 percent per annum, as well as a 50 percent increase in population.[69]

The *2025* feasibility study, addressed in large part to an audience within government, was intended to lay a foundation for policy. Its omissions provide a sense of just what Canada's first national-level, general-interest, nominally independent environmental coalition felt it needed to sacrifice in order to be heard at the end of the 1970s. The EAC and its Canadian allies wished as much as ever for an equitable and clean energy system in the country, but an apolitical feasibility study in which none of the scenarios approaches the goal of a conserver society – not even one so conservatively defined as in the description Holtz and Susan Mayo of the EAC gave in 1979: a "higher efficiency" society with less material throughput[70] – and that carefully avoids any hint of lifestyle changes was a far cry from the kind of populist, limits-aware environmentalism that initially responded to the threat of a nuclear facility on Stoddard Island in 1973. By demonstrating the supposed ease of the transition to a soft energy path, the study's authors hoped to force bureaucrats to acknowledge its viability and justify any different choice of energy policy; in reality, they removed the need for anyone to seriously address the SEP. Presenting the soft path as not only possible but also profitable encouraged, if anything, a feeling that the usual mechanisms of a market economy should produce the desired society with little or no public encouragement.

1979: "IT CAN HAPPEN HERE"

While the Energy and People conference gave the Halifax-based anti-nuclear activist network a chance to set the agenda in discussions of energy policy, the luxury of doing so proved short-lived. The years 1976 and 1977 saw the resurgence of participation in the Maritime Energy Corporation as a viable policy option for Nova Scotia and the beginning of talks to that end between the Nova Scotia Power Corporation, New Brunswick Electric Power Corporation, and federal Department of Energy, Mines and Resources.[71] Word that Ottawa would exempt AECL projects from the new federal environmental assessment program – and news that NSPC might have already been searching for a nuclear plant site in the central or northeastern areas of the province since 1974 – compounded the renewed sense of urgency.[72] Richard Hatfield's government had committed New

Brunswick to a policy of electricity exports to the US Northeast, and EMR's newest "energy strategy," according to Susan Holtz's assessment, "blatantly exclud[ed] political, social, and environmental considerations," as well as other forms of energy, in favour of continually expanding the supply of electric power produced in Canada.[73] With most of the decision-making power between them, it seemed that the NBEPC and EMR might turn the proposed regional utility into little more than an eastern subsidiary of AECL, and potentially remove the ability of the provincial government in Halifax to reject nuclear energy proposals. Facing the possibility of losing what small influence they had already gained over the province's energy policy, the Halifax network increasingly turned to their federal connections in hope of ending Ottawa's love affair with one of its most protected and promoted agencies. In the rest of the province, however, a new popular activism began to take root, and it provided the basis of the burst of political pressure that scuttled the regional utility proposal for good after 1979.

Part of the explanation for the new intensity of local activism lies in the same federal connection sought in Halifax. Offers of funding from Ottawa to small local groups willing to focus on renewable energy and conservation tended to distract activists from anti-nuclear activism while reinforcing their local character. New Brunswickers opposed to the Point Lepreau project suggested that the federal government was using targeted funding to "defuse strong polarized resistance to nuclear power in particular and large scale energy projects in general."[74] If so, it was a successful strategy: during most of the late 1970s, new energy activist groups in Nova Scotia struggled to put public pressure on elected officials, to imitate the success of forestry activists in Cape Breton, and to lift their political profile beyond basic energy conservation advocacy. This was true even when some of their founding members were keen to make a more direct attack on the political scene.[75] Both the Annapolis Valley Energy Centre and the Amherst Energy Conservation Information Centre, new to activism in 1978, looked to the well-established EAC for guidance, learning from the its staff how to access federal funds and how to undermine the nuclear sales pitch with talk of alternative technologies and energy conservation. Even veteran SSEPA activists began to pursue the funded information centre model during those years.[76]

Unexpected circumstances turned this low-profile wave of local energy conservation activism into just what EMR seemed eager to avoid: a renaissance of anti-nuclear political pressure. The Stoddard Island proposal was long dead at the beginning of 1979, but another nuclear island was about to make world headlines: the accident at Three Mile Island generating station in Pennsylvania on March 28, 1979, reinvigorated popular opposition to nuclear energy everywhere and a political style of energy activism in Nova Scotia. In the aftermath of Three Mile Island and with construction on New Brunswick's first Point Lepreau reactor at a peak of activity (along with speculation on the expected second reactor), new groups took on a more explicitly political and anti-nuclear character. One of the first and most active of these groups began in the Annapolis Valley, much of which lies geographically closer to Point Lepreau than to Halifax and, as the new group liked to point out, downwind of Lepreau as well. The Fundy Area Concern for Tomorrow (FACT) group began at a meeting on nuclear energy organized by the provincial New Democratic Party, held, by odd coincidence, on the very day of the meltdown in Pennsylvania. Within a few weeks, a non-political follow-up meeting was held to formally launch FACT, which proceeded to bombard provincial and municipal politicians (and the general public) with the message that nuclear power was unnecessary, unsafe, and most of all unethical, especially from the perspective of the farming communities of the Annapolis Valley that would bear the brunt of contamination from any Three Mile Island–type accident at Point Lepreau, without ever having had a chance to comment on the project in New Brunswick.[77]

The remainder of the year passed in similar fashion, with mass movement anti-nuclear activism at every turn. Concerned Citizens of Caledonia, Opponents of Nuclear Energy, Citizens Against a Radioactive Environment (CARE), and more stand-alone groups arrived, as well as anti-nuclear committees from the Black United Front Yarmouth, Yarmouth Women's Institute, Cape Breton District Labour Council, North Shore Organic Growers Association, and more.[78] The local character of activism remained unchanged, but the drive to cooperate and to take the fight to Halifax and Ottawa surged. The basic critique of the modern consumer economy – once relegated to the nearly invisible local energy centres – returned to visible public presence, though now in the shadow of a more centralist, technocratic national alternative energy movement. The Cape Breton Alternate Energy Society declared its philosophy to be "one of relative deprivation and changing ways of living," and set about to convince the island population to adopt it,[79] while EAC member Ron Loucks wrote to the premier with a lengthy exposition of consumer society's degrading effect on human happiness and enclosed a copy of Dennis Meadows's *Alternatives to Growth*.[80] Loucks did not represent the public view of the energy committee, but in other respects the EAC followed along with the new popular movement. The dominant narrative that all shared was a return to public political pressure based on fear of nuclear disaster and a defence of one's immediate community, one's place. The title of one widely distributed leaflet told readers that "It Can Happen Here," while the SSEPA put words into action by releasing balloons at Point Lepreau, several of which were soon picked up in the Annapolis Valley.[81]

The climactic event in a year of anti-nuclear protest occurred at the Maritime premiers' conference in Brudenell, Prince Edward Island, on June 4, 1979. Having assembled to discuss privately the details of a regional utility, the three premiers were reluctant to accommodate requests for a meeting with ENGO leaders. As a result, nine days before the event, the collected anti-nuclear groups of the three provinces, acting under the name of the Maritime Energy Coalition, called for a rally outside the resort in Brudenell. Six hundred to eight hundred people responded.[82] Unable to ignore the crowd, Hatfield, John Buchanan, and Angus MacLean made a very brief appearance on the steps of the hall and spent ten minutes accepting a four-point statement from a quartet of rally leaders, including Susan Holtz and Elizabeth May, two EAC board members representing Nova Scotia and the Island of Cape Breton, respectively.[83] With the rally coming barely two months after the accident at Three Mile Island, the number of participants is not surprising. Nor is the fact that the three premiers wished to be seen to appreciate their concerns (even if New Brunswick's Hatfield was "clearly upset," according to the EAC's post-rally report).[84] What the event demonstrates best is the extent to which conservative, bureaucratically agreeable, and scientifically optimistic statements had become the default rhetoric of the anti-nuclear coalition's leadership, to the exclusion of the limits to growth. As one of Susan Holtz's collaborators in Nova Scotia put it in a letter to her about a month before the rally, ideas about changing the capitalist consumer society "in their raw state ... are probably unsalable" and "better introduced by the back door" of safety arguments and alternative energy research as a paying investment.[85] Such ideas, already in accord with the kind of action the EAC, FOE, and CEAC were seeking from Ottawa, prevailed in the correspondence of the EAC energy committee in early 1979, correspondence in which Holtz dismissed as unreal "the allegedly widely different assumptions held by proponents/opponents about lifestyle and economic growth," and in which Elizabeth May advertised the Brudenell rally in support of "conservation and the development of renewable alternative sources."[86] The four points delivered to the premiers at Brudenell, and the petition of thousands collected to support them, followed the same script, stressing the safety concerns and asking for a series of public inquiries into nuclear energy projects. "Realistic" alternative energy policies, the brief noted, are "ones which take into account general social goals such as economic growth, jobs, and environmental protection."[87]

Following the Brudenell rally, the resurgence of MEC anti-nuclear activism in the summer of 1979 included renewed protests against the Point Lepreau generating station, as well as against the nuclear fuel plant in Moncton, and a new front of resistance to heavy-water shipments to Argentina. The rhetoric of place defence intensified, reviving the environmental justice arguments of the SSEPA's Stoddard Island campaign and the association's worldly view of shared responsibility for others' places powerfully enough to obscure the differences between the economically radical and the orthodox positions. And in little more than a year, the anti-nuclear backlash bore fruit: the lack of interest from Prince Edward Island and Nova Scotia finally laid to rest any plans for a unified Maritime Energy Corporation (at least until the twenty-first century). A meeting between the premiers and the federal energy minister in October 1980 offered New Brunswick's Hatfield one last chance to convince the others to share the burden of Point Lepreau, but that was the end.[88]

CONCLUSION

Though confirmation was longer in coming, in the estimation of the EAC's energy committee a week after the Brudenell rally success was already at hand. "It seems," the committee wrote, "that the Maritime Energy Corporation as a facilitator of nuclear energy is dead."[89] In fact, the success of the Maritime Energy Coalition in its last real appearance as a genuine regional coalition was even more complete than that. Strong public reaction following Three Mile Island was a crucial factor in the abandonment of plans for a second reactor at Point Lepreau.[90] In the glow of success, however, it is hard to believe that some of the long-standing activists did not feel a profound disappointment at what had been left behind in the race to Ottawa and Brudenell.

The limits thesis never disappeared from the province's environmental scene in the 1970s, but despite occasional warnings that "the ethical and moral problems are overlooked in favour of technological problems," its proponents slowly and steadily receded into the shadow of a dominant, economically orthodox citizen-scientist mode of operations.[91] In the early days of anti-nuclear activism, the South Shore Environmental Protection

Association's fight for environmental justice for local fishermen and Hattie Perry's adept use of public pressure against the premier sat next to the Nuclear Power Study Group's highly technical exposé of nuclear technology and the profound ethical convictions of the Halifax VOW that a high-energy consumer society could never be an equitable society. The closer a group was to the scene of environmental risk, the more radical its activism tended to be, but all participated alike in pressure politics. By 1979 the sense of common purpose had dulled. A temporary resurgence of anti-nuclear political pressure based on feelings of personal vulnerability succeeded in finally killing the Maritime Energy Corporation in 1980 (and with it the common front of Maritime activists), but the environmental movement as a whole was far different from what it had been when the same sort of action crushed the Stoddard Island proposal in Barrington Passage in 1973. Still primarily a common front of small local volunteer organizations, it now faced a division (still unrecognized as such) between a nascent mainstream that eschewed the radicalism of the counterculture or the uncomfortable predictions of the limits theory, and those who clung to both. "The dramatic language [of political radicalism and limits] which raised ecology-consciousness 10 years ago is missing," wrote Amy Zierler, who claimed to see a brief resurgence of it in the nuclear debate. "The laws and procedures we drew up to institutionalize our concern swallowed up much of its spirit, without eliminating its causes."[92]

Led by the Halifax Friends Meeting and the Voice of Women, and especially by the small group of research-inclined activists behind Susan Holtz who eventually formed the core of the EAC's energy committee, the network of anti-nuclear environmentalists based in the city embraced the laws and procedures of the bureaucratic system and turned towards rigorous analysis of nuclear safety and promotion of research into alternative energy technologies. They began to downplay, then conceal, and finally denigrate as "extreme" and "unacceptable" any arguments that focused on the unsustainability of an expanding capitalist economy and consumer society, and especially any that cast government in the role of villain.[93] Apart from the generous continuing support of the Halifax Friends Meeting and their national co-religionists, more and more of the group's funding came from research contracts with the provincial and federal governments. Very much reminiscent of the Quaker notion that everyone has the potential to access divine truth no matter how committed to a wrong course of action they may be at the moment, this science-inclined network sought relentlessly to access and influence the ranks of the provincial and federal bureaucracies and corporate and political decision makers. By mid-decade the constant refrain of "respectability" began to appear in the documents of the EAC, a group derided as "ecology freaks" by the premier himself only a few years earlier. Susan Holtz described the change herself in 1977:

> I've come to feel that one of my most important roles is acting as a human link between people and ideas usually opposed or not in contact, notably environmentalists and decision-makers in the civil service, the Power Corporation, or politicians. I feel that to change, at an emotional level, a situation from one in which people feel they're opponents to one in which they feel they're working on the same problem but with differing views is to change radically the possible outcome not only of that situation but of all future situations.[94]

In September 1978, Holtz travelled to Arlington, Virginia, to attend a Solar Energy Research Institute symposium, one of very few Canadian attendees. There she heard Lee Schipper of California warn the assembly to "honestly follow through the consequences of some of today's attitudes and policies (right or wrong) that our friends are advocating, lest we find that our friends have led us to our enemies."[95] Back home, eight months later, she attended an EMR workshop on Energy Conservation and Community Economic Development in Truro and heard Dan McInnis turn the story of the Antigonish Cooperative Movement into an even more explicit cautionary tale for advocates of a conserver society. The Antigonish movement, he said, was "ultimately crushed by government in its response to profit-oriented business."[96] Both views of activism were utterly different from her own; as far as Holtz was concerned, and as far as the EAC and the Halifax network of energy activists were concerned, environmentalists had no enemies. Within a few years, however, even she would find some.

3

Power from the People
The Anti-Chemical Campaigns

THE SPRUCE BUDWORM MIGHT have seemed an unlikely trigger for Nova Scotia's greatest environmental controversy in the mid-1970s. The endemic insect had eaten itself into obscurity on Cape Breton Island in the 1950s, exhausting its food supply of aged spruce and fir trees and returning to the status of a minor forest species, repeating its population cycle yet again. Although many, perhaps most, residents recognized the budworm as a New Brunswick problem, there was not an immediate recognition of the incipient irruption in the summer of 1975, when fir trees on the highlands began to turn rust red.[1] Foresters working for Nova Scotia Forest Industries (NSFI), the Canadian corporate identity of Swedish pulp and paper giant Stora Forest Industries, certainly recognized the signs, however, and before the season ended, asked the provincial Department of Lands and Forests (DLF) for permission to begin aerial spraying over 100,000 acres with the same chemical insecticide, fenitro-thion, in use in New Brunswick.

Initial resistance to NSFI's plans centred on Halifax and a swiftly formed coalition of new environmentalists and older wildlife and resource conservation groups. During the winter of 1976, members of the Ecology Action Centre (EAC) joined representatives from five conservation groups to organize a debate-style symposium on the budworm issue, where speakers in favour of a chemical spray program essayed their arguments against the anti-spray position. It was as even-handed a format as one could imagine.[2] The chief organizer from the EAC, Jim Reid, may have absorbed the lessons of the previous year's Energy and People conference; no resolutions from the floor would mar the objectivity of this symposium. Certainly a wide range of opinions was apparent among the six groups, who admitted in their joint statement that "the problem exists [and] some of us became convinced that aerial spraying of insecticides may at present be the only recourse in some circumstances where there are extreme social and economic imperatives."[3] Whatever the reason for its equivocal presentation, the symposium made desperately poor political theatre. Not long afterward, the Department of Lands and Forests registered its own opposition to NSFI's plan, using much the same data and language it had advanced in the 1950s, and much less equivocally than the activists. The budworm, the department argued, was a natural component of the forest ecosystem, controllable through strategic cutting but not really controllable through chemical means. No one credited the activists with any influence over the department's position, nor did it make much difference; within a few weeks, the provincial Cabinet met and overruled the department's foresters: NSFI would be permitted to spray fenitrothion on the Cape Breton highlands.[4]

Scott Cunningham, the man who drafted the joint statement on behalf of the six Halifax groups, plainly did not think the chemical option was necessary or even viable, and appreciated a more antagonistic approach towards government. He also had access to information about the health risks of aerial pesticide sprays that he decided was best used to generate some bad publicity for the spray effort and Cabinet's decision to grant NSFI's request. As a doctoral student in biochemistry at Dalhousie University and a member of the EAC, Cunningham was aware of the research of Dalhousie's Dr. John Crocker into the incidence of Reye's syndrome among children exposed to pesticide sprays in New Brunswick. Rare, difficult to diagnose, and often swiftly fatal, the syndrome had appeared in several children living near spray areas, prompting Crocker to investigate the cluster of cases. Later research would reveal more about the exact mechanism of the disease, but by the winter of 1976, he had already determined that exposure to the petrochemical emulsifiers used to dilute the insecticide and make it stick to trees could enhance the virulence of otherwise common and mostly innocuous viruses. The research was incomplete – Crocker himself was not comfortable discussing it publicly at first – but it was advanced enough to be powerfully persuasive in the public arena, and Cunningham began searching for someone in the press to break the story.[5] He found Parker Donham at the *Cape Breton Post*.

In Parker Donham and Scott Cunningham, the Cape Breton and Halifax movements to oppose a spray program met, though they had otherwise little contact through the first year of the controversy. From the beginning, the Cape Bretoners had a more directly political approach to activism than the Halifax groups. Resistance in Cape Breton had not gotten underway until February, when the provincial government approved the spray. Then, working quickly, a loosely organized group of residents from communities around the highlands began circulating a petition asking minister of environment John Hawkins to approve hearings by the Environmental Control Council before allowing NSFI to go ahead, a deliberately moderate demand that would have meant the cancellation of the spray program anyway, due to delay. At the same time, they began collecting information on the spruce budworm, chemical insecticides, and industrial forestry for use in direct appeals to the people of Cape Breton. News of Donham's impending story about Reye's syndrome in the *Cape Breton Post* soon reached the group, which decided to coordinate its use of public pressure with his headline. On March 31, 1976, the front-page headline demonstrated Donham's own understanding of the power of public outrage. It read: "Fatal To Children."[6]

The pressure tactics of the newly christened activist group, Cape Bretoners Against the Spray, bore fruit immediately. Overwhelmed by a barrage of telephone protests, Premier Regan and his minister of public health, Cape Breton lawyer Allan Sullivan, emerged from an emergency Cabinet debate the day after the *Post*'s bombshell to announce the cancellation of the spray program. From the initial relief of the DLF's declared opposition, through Cabinet's approval and subsequent reversal, the whole fight – the first of the "budworm battles" – had taken only two months.

The two centres of opposition did not have time to collaborate that winter, but apart from their different tactics, the Halifax and Cape Breton activists were not so different. Both were essentially local movements with few outside connections, composed of both long-time Nova Scotians and a conspicuous number of more recent arrivals, which prompted both groups to attempt to forestall the "outsider" label they knew would be applied to them. In Halifax, the attempt contributed to the retreat into scientific respectability, itself a form of localism in the province's administrative centre. In Cape Breton, it led to the avoidance of publicity by the most recent immigrants. Elizabeth May, for example, who would eventually become the face of the anti-spray movement, admitted a few years later that she initially stayed in the background, aware of her problematic image as a young American woman only two years in Canada.[7] In both areas as well, activists drew on links with local wildlife conservation organizations and with ongoing campaigns against the use of biocides on farms, roadsides, and city streets and parks. As 1976 dragged on and more of the province had time to notice and discuss the flurry of activity during the previous winter, more associations cropped up, and more half-forgotten chemical controversies gained new vigour, especially the Warren farm episode.

The Warren family's frightening story had only become more so in the intervening years, as illnesses persisted or worsened. (Robyn Warren's brother would die of a rare soft-tissue sarcoma in 1978.) Along with their tale came a reminder of the failure of activists in the early 1970s to adequately defend the "ecological logic" of epidemiology. Ian McLaren of the Nova Scotia Resources Council (NSRC) began explaining in April, in terms similar to those John Crocker would soon use to explain his own work, that "nothing short of taking several dozen children and exposing them to the same procedure would ever resolve the question of 'proof.'"[8] The government's role in scattering the first anti-spray movement and stymieing Robyn Warren's efforts to learn even the names of the chemicals contaminating his farm was also remembered, especially among the network of organic farmers who participated in the initial petition campaign in Cape Breton. Over the years to come, how far to trust apparent allies in government would become one of the greatest fault lines of the anti–industrial forestry campaign.

"The General Public's Over-reaction"[9]

Within the Nova Scotia Department of Environment (DOE) and the Department of Lands and Forests during the 1970s, battles were fought analogous to the battle between NSFI and the local activists in Cape Breton and Halifax, though they usually went unnoticed outside of government. In 1976, Lloyd Hawboldt, assistant to the deputy minister of lands and forests, led a faction of foresters and entomologists opposed to chemical sprays against the efforts of his department's director of forest planning, R.M. Bulmer, and a pro-spray group.[10] Both groups attempted to win over the minister and deputy minister. The anti-spray faction relied on Hawboldt's decades-old argument against the "vain attempt to offset a natural trend," and pointedly reminded their colleagues that the company could have prevented a budworm irruption by following the advice of DLF foresters rather than clearcutting in easily accessible lowland woodlots and neglecting the aging highland forest.[11] The pro-spray side, conversely, pressed the advantage of its alliance with industry and with natural resource departments in Maine, New Brunswick, and Quebec, where chemical spraying was already taking place.[12] Though the anti-spray position enjoyed greater success, it was a heavily qualified opposition. All of the department's foresters retained their professional commitment to the ideals of environmental management and control and to the preservation of their status as experts; no one was willing to advocate that the spray option be removed from the table for good. Hawboldt himself worked to restrain his allies' more strident statements in order not to "sabotage the department if it chooses to spray in 1977" or to allow activists in New Brunswick to "use us for their own particular purposes."[13] Despite being so completely submerged in the politics of forestry, and strongly opposed to a spray program, the department's foresters remained thoroughly committed to modernist thinking and could not countenance what they saw as the "politicization" of the expert's natural right to decide.

Despite the minister's decision to oppose NSFI's request, he never curtailed the activities of the pro-spray faction within the DLF. Their cooperation with industry began to bear fruit in the fall of 1976, as NSFI took the initiative against the activists ahead of the winter decision-making season. Its first approach took advantage of yet another jurisdictional rivalry and regulatory power struggle, this time between the province's forest agency and its pro-spray federal counterpart. In late September, the Canadian Forestry Service released its predictions for the following year, forecasting a large increase in the budworm population. The director of the Maritime Forest Research Centre promptly recommended a preventative chemical spray.[14] A public meeting in Truro organized by the DLF's Bulmer spread the news to the largest possible audience, presenting it as a disastrous consequence of the government's refusal to act sooner, rather than the expected progress of an insect population bloom.[15] A month later, another Forest Planning Division man, Ed Cloney, attended a regional meeting in Fredericton on "Forest Protection Against Spruce Budworm," where forest experts concluded – somewhat contradictorily but again as publicly as possible – that the harmful environmental effects of chemical insecticides "are insignificant, short-lived, and sensationalized by the general public's over-reaction – but that environmental monitoring is essential in detecting any long range effects which may arise."[16]

The real offensive push came at the beginning of December. A report issued on the first of the month by the Budworm Committee of the Voluntary Planning Board recommended spraying, after an investigation led by John Dickey, a Halifax lawyer and president of NSFI's Nova Scotian holding company, and including as well the ever-present Mr. Bulmer as DLF representative. Finally, the day after the release of the Voluntary

Planning Board report, the president of Stora Kopparberg, Erik Sunbladt, arrived in Nova Scotia from Sweden with a bombshell announcement: without a campaign of chemical control against the budworm menace, he would surely be forced to shut down the paper mill in Port Hawkesbury within the next five years. To back up Sunbladt's claim, his employees at the mill had produced a sixty-five-page report of their own detailing the future of pulpwood supply on Cape Breton and concluding with a demand for insecticide sprays over not 100,000 acres this time but two million acres – the entire island of Cape Breton.[17]

Environmental activists had not been idle since their unexpected victory in April, though most of the action during the summer was on the Halifax side of the still-disconnected Halifax–Cape Breton axis. What activity there was through the quieter months suggested that the urban activists had finally accepted that the provincial Cabinet's decision on spraying would "be an almost purely political one."[18] The EAC, Halifax Field Naturalists (HFN), and other groups that had worked together during the previous winter turned their attention to the press, and thus to the public, or worked to recruit other interested groups into a strong public lobby. As a result, when NSFI and its allies reignited the controversy in the fall, a host of newly interested parties reacted to the expanded spray proposal, and the Halifax and Cape Breton campaigns finally coalesced into a truly province-wide collaborative effort, with statements of support issued across the map, from the Municipal Council of Shelburne to the Women's Institute of Marion Bridge.[19]

If Halifax provided the organizational push, the most effective *action* in response to Sunbladt's ultimatum again originated from Cape Breton, where Cape Bretoners Against the Spray met to plot strategy two days after his visit. At the suggestion of Ron Caplan, editor of *Cape Breton's Magazine,* the group was renamed Cape Breton Landowners Against the Spray (CBLAS), to emphasize its status as a group of locals facing off against a foreign multinational. Its official response to NSFI's pro-spray blitz included a careful deconstruction of the mill's wood supply predictions by former insurance accountant John May, and a report on "Alternatives to Spraying" written with the aid of Frank Reid, a leader of the anti-spray forces from the Victoria County Woodlot Owners and Operators Association. Reid, who had grown up in industrial Cape Breton, understood well the economic reality of the forest industry, where NSFI's large Crown leases enabled the company to push down the price paid to small woodlot owners. Spraying in the highlands, he insisted, was mainly an attempt to preserve the unused "standing inventory" that allowed the company to control the pulpwood market.[20] The CBLAS analysis met a favourable reception on Cape Breton and earned endorsements from primary producers' associations across the island, but achieving recognition in the Halifax media proved more difficult.[21]

The anti-spray movement's break in Halifax came on January 6, 1977, with a major press conference downtown. NSFI's move in late December to request a permit for the insecticide Sevin (carbaryl) in 1977, rather than fenitrothion, left environmentalists struggling to learn about the new chemical and share the information over the Christmas holidays. In Cape Breton, they posted handbills in restaurants and post offices, but it was at the Halifax press conference that they put into practice what they had learned from watching the pulp company's attack on multiple fronts in the fall.[22] A new petition effort and announcement of support from the provincial Medical Society on the conference day helped draw and keep the media's attention long enough for activists to state their case. And cognizant that they could not rely on fear of Reye's syndrome alone twice in a row, especially with the New Brunswick government working hard to discredit Dr. Crocker's findings, environmentalists struck all of NSFI's rhetorical weak points at once. Two woodlot association presidents, Frank Reid for Victoria County and Dan Alfred MacDonald for Inverness, presented their associations' positions on a softwood market dominated by three large paper mills interested in spraying only to maintain their dominance, while John May spoke as a tourist operator. Biologist Ian McLaren joined Lloyd Hawboldt to offer a scientific perspective on forestry, and Dr. Earle Reid, Chief of Medicine at the Halifax Infirmary, discussed Reye's syndrome and the other dangers posed by human exposure to pesticide chemicals. Together they argued that NSFI's proposal was unnecessary, iniquitous, uncertain of success, potentially counterproductive, and still possibly dangerous. Any pretense of objectivity was abandoned, and the event was a great public relations success.[23]

The province's much-delayed decision on a 1977 spray program came almost four weeks after the anti-spray coalition's appeal to the Halifax media, and not before Premier Regan walked into an unexpected face-to-face meeting with activists, reminiscent of his visit to Barrington Passage in 1973. The Port Hawkesbury Chamber of Commerce might have been a refuge for NSFI's side of the fight – certainly the mill turned out a large number of its employees for the chamber's public meeting – but CBLAS turned out more of its own supporters. Nor were the millworkers necessarily the bulwark of support that the pulp company hoped for. Many followed the lead of Ken Calder, NSFI chemical engineer and CBLAS's best source of information from within the corporation.[24] If Reid's contribution to the CBLAS position can be taken as indicative, forest industry workers on Cape Breton also understood that economic desire, not ecological necessity, drove the company's push for a spray program. The combination of strong attendance by anti-sprayers and lukewarm opposition from millworkers turned the event into an environmentalist set piece, with the premier himself as guest of honour. Not missing a chance to press the advantage, CBLAS released its phone poll results for the island of Cape Breton a day later, showing 75 percent of residents opposed to the spray. Results were swift, only this time it was not the health minister, or even the new environment/lands and forests minister, Vince MacLean, but the premier himself who stepped up to announce the decision to refuse NSFI's request for a second time.[25]

The years 1977 and 1978 saw the refinement of environmentalists' use of the power of popular pressure. Appeals to the public along all possible lines of argument obviated any need for consistency among arguments: the economic case for full industrial exploitation of the forest but without chemicals might sway one segment of the population, while the ecological case against all pulpwood-focused forestry practices could sway another group. So long as the objective was to convince the public rather than the premier or any of his ministers, contradiction was irrelevant. As a tactic for generating pressure on political leaders, it was hugely effective.[26] As NSFI became progressively more antagonistic towards the Regan government, and MacLean in particular, anti-spray activists began to reach beyond provincial borders (imitating their industrial opponents) to recruit expert supporters, and Elizabeth May drifted towards the original Halifax-based coalition's conciliatory methods, out of a desire to support MacLean's position.

The end of 1977 and the beginning of 1978 were Elizabeth May's coming-out season as the public face of the anti-spray movement. Having come to the attention of the media during another anti-spray coalition press conference, May then participated in a pair of televised debates that put her squarely in the public eye, a position she quickly grew to enjoy.[27] Further publicity came from a series of running debates for local audiences across the province, pitting May against Kingsley Brown, the maker of a NSFI-sponsored pro-spray documentary called *Mr. Regan's Choice.* Brown too had a measure of public recognition, at least among those employed in forestry. In the early 1970s, he had been an advocate for small woodlot owners, and despite coming off worse in his broadcast exchange with May, still enjoyed a positive reputation in the province. By the spring of 1978,

as the public awaited a third decision on a chemical spray program, May
had already become a seasoned activist, an eager participant in government-
sponsored consultations, and a convert to the economically and politically
unthreatening style of activism advocated by the EAC. At times, she also
defended the Regan government. Only a few years later, she would remark
on her position during this third protest season: "I believed then, and to
this day, that they [Regan and MacLean] acted out of personal conscience
and informed, rational judgement when they opposed spraying."[28]

On March 17, May's faith appeared to be vindicated when MacLean
once again refused to permit aerial insecticide spraying on Cape Breton
Island. Even better, the following fall produced sound evidence (much to
the distaste of the Canadian Forestry Service [CFS]) that the budworm
irruption in Nova Scotia was finally collapsing on its own. Three years of
effort had met apparently unlikely success; environmentalists had prevented
Nova Scotia from following New Brunswick's lead into a permanent spray
program despite the industry's best efforts. They had won. Celebrating
what they took to be a new policy of ecological forestry, few activists noted
the fine print on the province's five-year, $62 million investment in "silvi-
culture." Far from accommodating the will of natural systems, the new
policy encouraged replanting with "commercially desirable" species, especially white spruce and tamarack, in single-aged, single-species stands
suitable for mechanical cutting, and suitable as well for future insect blooms.
There was also too little recognition that an additional $35 million available
to dry-store budworm-killed wood over the following decade would effectively insulate NSFI from the negotiating power that the budworm had
granted to small woodlot owners, and preserve the corporation's growing
political influence.[29] Finally, almost no one seemed to appreciate that the
fine balance between pro- and anti-spray factions within the Department
of Lands and Forests (as well as around the Cabinet table), which had lent
such decisive power to activists' own orchestration of public pressure, could
shift abruptly, or that the logic of industrial forestry guaranteed a return
to the battlefield.

AN UNCERTAIN RELATIONSHIP WITH POWER

The position of government, either provincial or federal, had never been
to side unequivocally with environmentalists. Rather, it wished to shape
the movement. As early as 1976, the federal Department of Environment
recognized the value of environmental activist groups as a constituency, a

tool with which to turn public opinion in favour of the department's legislative agenda. Like any other legislative constituency, environmentalists required careful management, perhaps more careful than most given the difficulty in separating and discouraging any members who held on to unlegislatable ideas about the unsustainability of industrial society as an enterprise.[30] The federal department's cultivation of the movement began with informal meetings between Environment Canada's Atlantic Regional Board and a select few activists, and progressed quickly to invitations for formal participation in national Canadian Environmental Advisory Council (CEAC) meetings.[31] Funding made available directly from the department brought activists from across the country to Montreal in 1978 for two days of meetings to draft "position papers on environment, wildlife, pesticides, and the conserver society."[32] It was only one of many such meetings convened during the last half of the 1970s in which the logic of consensus could work on squeezing out ideas that might seem too extreme. Federal funding placed only one major restriction on participation: groups receiving funds had to maintain "non-political" registered non-profit status.[33] Those selected soon provided their own additional strictures; after resolving in 1978 to create a permanent committee of environmental groups beneath the CEAC (what would become the Canadian Environmental Network), a steering committee went in search of "broadly based" groups capable of responding quickly to queries from the Advisory Council – in other words, professional groups rather than unincorporated "single-issue" groups composed of local activists.[34] By the following year, much of the advice conveyed by those groups to federal bureaucrats consisted of requests for ever closer integration, and for dedicated funding to the groups who would participate.[35]

Those groups in Nova Scotia that saw federal consultations as a route to real influence sought to enjoy the same sort of relationship with the provincial government, and the budworm spray issue evoked from them a concerted effort to achieve it. The EAC and CBLAS, working together closely by the end of 1977, each attempted to convince MacLean to impose an anti-spray member or two from their ranks on the new Task Force on Wood Allocation.[36] EAC leaders also pushed for the Environmental Control Council (ECC) to work more closely with environmentalists and put funds towards research conducted by professional activist groups.[37] This quest for access and funding soon led to even greater emphasis on those areas where some environmentalists could agree with professional foresters' insistence on environmental management and control. EAC and CBLAS propaganda, often jointly produced with conservation groups like the

Canadian Nature Federation, Halifax Field Naturalists, or Nova Scotia Bird Society, lent support to the provincial government's new large industry-oriented silviculture plans, which Minister MacLean insisted would obviate the need for budworm control and also "increase the productivity of our forests by two and one-half to three times its present allowable cut."[38] They also enthusiastically endorsed non-chemical control measures meant to achieve the same effect as chemical insecticides, apparently willing to undermine their and their allies' ecological argument that the budworm belongs in the forest and must be allowed to run its course.[39] Elizabeth May was particularly keen to encourage aerial applications of the biological control agent *Bacillus thuringiensis kurstaki* (*Btk*) as an alternative to insecticides.[40]

Even if some environmentalists were willing to shift their positions towards the Regan government's ideal of industrial promotion, or take on an "advisory arm" made up of businessmen to boost their orthodox economic credentials (as the EAC did in 1978), government proved less flexible. Meeting in the middle works only if both sides move, and the province was never willing to match the federal Department of Environment's efforts to court activist participation. Bureaucrats in Halifax favoured secrecy over manipulation as a tactic for the management of public dissent, and federal/provincial rivalries did not spur the search for legislative constituencies in Nova Scotia as powerfully as interdepartmental rivalries did in Ottawa. MacLean's response to requests for environmentalist representation on the new task force was a casual "no," and his rationale for refusing to order an ECC public hearing was transparently inadequate: he claimed that the council had never requested authorization.[41]

Doubts about the wisdom of professionalization and of accepting the economic growth assumptions of the province's industrial development policy were never far away. With decision-making power within the organization growing ever more remote from the membership, the EAC's leaders defended their role as a "counterweight" to the power of business and government within the existing system, prompting one member, Eve Smith, to reply that the fundraising required by that approach "has become a way of life, and leads at times to modification of approach or policy, or is unequally distributed. It sometimes gives more power to a few who control policy ... this is a money oriented society and we may be corrupted by it, almost without recognizing our own corruption."[42] A few years later, the centre's closest ally on the ECC, the ecologist Dr. Gordon Ogden, insisted that the ECC would not fund action groups, for their own good.[43]

Until the very end of the peak period of forestry conflict, from 1976 until 1983, misgivings about environmentalists seeking professional status or a direct role in government remained at the level of Ogden's advice or Smith's warning, and did not flare up into open recrimination. That could have been due in part to the short time since the EAC's transformation into a professional group, and the even shorter time since Elizabeth May and CBLAS formed their organization amid a direct challenge to government. It was likely also due in part to the incompleteness of both groups' drift towards conciliatory activism: the EAC still regularly debated its own structure and methods during those years, and May, ever more the de facto spokesperson for the Cape Breton group after 1977, remained equivocal for years, alternately defending government actions or participating in direct consultations and then insisting that "power is not achieved through government grants, large budgets, or paid full-time coordinators. Power is people."[44] Perhaps because of such statements, environmentalists making firmly ecological and anti-managerial arguments, even the few still making anti–economic growth arguments, could still see a measure of common ground with the ecomodernists leading the publicity drive. More important, so long as the tactics of the anti-spray campaign encouraged environmentalists to reach politicians through the intermediary of public pressure, all lines of argument – health, economic, ecological, social justice, emotional, or epistemological – and all types of activist could coexist. Adopting the opposing side's terminology and assumptions in a battle of ideas does have consequences, however, and the longer the conflict over industrial forestry dragged on, the more those consequences began to reshape the face of the environmental movement, driving a wedge between the ecomodernists and those who remained skeptical of government's willingness to change.

THE OTHER SPRAY

The budworm spray controversy of the late 1970s and the herbicide spray controversy of the early 1980s tend to be remembered as quite separate events. In reality, they were two sides of the same policy coin: both the government/industry axis and the environmentalists carried forward into the new decade much of what they had learned since 1976, including what they had learned from each other. The ensuing controversy revealed that the political context of environmental activism mattered a great deal to its effectiveness, as a change in government brought the full power of the state into action opposing the environmentalists' anti-spray campaign.

Herbicide sprays were even more familiar in Nova Scotia than insecticides in 1979, when environmental activists took note of their increasing use in forestry. The connection with the brief anti-herbicide campaign of the early 1970s was certainly more direct; one of the first actions by the EAC in 1979 was an attempt once again to stop Nova Scotia Power from using chemical herbicides on its power line rights-of-way.[45] Local agitation against roadside spraying of farm weeds by the Department of Agriculture also continued apace, feeding into the province-wide movement, and organic farmers had for years maintained their case against the use of herbicides, even more than insecticides.[46] In each case, the chemicals in question were old and well-studied agents, most often 2,4-dichlorophenoxyacetic acid (2,4-D) and 2,4,5-trichlorophenoxyacetic acid (2,4,5-T) – albeit often used under the trade names "Tordon" or "Brushkiller."[47] But if environmentalists could arm themselves with more and better information about such long-used chemicals, they also faced even more entrenched opposition.

Arguably dangerous chemicals on their own, 2,4-D and 2,4,5-T saw their first military application in Vietnam in 1962, where a 1:1 mixture of 2,4-D and 2,4,5-T went by the code-name Agent Orange.[48] Unfortunately for the Vietnamese, and for the American and Canadian soldiers and civilians made unwitting test subjects for the wartime defoliation program, the mixtures used during the war were heavily contaminated with 2,3,7,8-tetrachlorodibenzodioxin (TCDD), a particularly toxic form of dioxin. The consequences – death, disease, and deformity – among those exposed in Vietnam or in North America were horrific, but it took years for the US and Canadian governments to grudgingly admit that the chemicals were dangerous. By 1979, neither the US nor Canadian governments had yet repudiated the chemical manufacturers' claim that preventable TCDD contamination was responsible, and not the 2,4-D/2,4,5-T mixture itself, despite the accumulation of studies suggesting just the opposite. Critics of both governments' actions met a stonewall of regulatory indifference and bureaucratic secrecy.[49]

Around the world, chemical disasters seemed to pile upon disasters in the late 1970s, profoundly shaking public confidence that the balance of benefits in fact outweighed the costs of the postwar chemical revolution.[50] Infamous names like Love Canal and Times Beach entered the public narrative in the United States, alongside Agent Orange, and in 1979 the largest-ever class action suit in American legal history pitted veterans exposed to toxic herbicides against their own government.[51] In Canada, however, the first major challenge to the continued use of the same chemicals came not from veterans but from environmental forestry activists in Nova Scotia.

In the forest industry, a single application of chemicals like 2,4-D could kill birch, oak, ash, and the like and leave behind a pure softwood stand for easy mechanical harvest, or wipe out the raspberries and low shrubs and forbs that spring up in a recent clearcut, leaving replanted softwood seedlings with full access to sun, soil, and water. The latter operation, called "conifer release," accelerates the growth of seedlings but leaves the soil impoverished; without the first stage of succession to restore nutrients, or the deep roots of hardwoods to stabilize the regrowing trees, exhaustion and erosion threaten (usually answered with the promise of chemical fertilization, and of insecticides to protect the weakened trees).[52]

Herbicide treatments were well-practised techniques by 1979, yet it took some time for environmental activists in Nova Scotia to take note of the province's nascent forestry herbicide program. Concerned with insecticides, they were slow to shift attention to herbicides, while industry pursued both as part of the same policy of ecological control. In 1978, the NSRC alerted the ECC to a series of experimental applications conducted by the DLF. The ECC attempted to investigate further, but the new minister of environment, Roger Bacon, showed little interest in approving new hearings, in reviving the council's public relevance, or in any action that might draw attention to the issue.[53]

The political context of the fight over industrial forestry had changed sharply with the election of John Buchanan's Progressive Conservative government in September 1978. While some environmentalists continued to praise what they saw as a turn away from industrial conifer monocultures, the pro-spray faction within Lands and Forests continued to promote the industrial model from a new position of strength. The new deputy minister, former forest industry association head Don Eldridge, had supported NSFI's position during the spruce budworm debates, and with the retirement from public service of anti-spray stalwarts like Lloyd Hawboldt and A.M. Wiksten, there was little to impede Eldridge's promotion of industrial forestry.[54] Certainly the new minister of lands and forests would be no obstacle; George Henley, the incumbent from 1978 to 1983, proved an equally staunch supporter of the forest industry's chemical ambitions. In 1979, his department undertook further spray tests with 2,4-D, glyphosate (Roundup®), and fosamine ammonium (Krenite®), including a large plantation site near Trafalgar, Guysborough County, adjacent to the headwaters of both the Saint Mary's River and the East River of Pictou. Challenged by some Pictou County residents, Henley explained that the failure of the prior government to control the spruce budworm meant that reforestation was urgently required, and that once planted, softwood trees required chemical "protection" to speed their growth and forestall a disastrous wood shortage.[55] In general, the Buchanan government pursued a much closer relationship with the pulp and paper industry than had the Regan government, ensured that herbicide spraying on forest land got started with minimal public attention, and encouraged the industry to proselytize its view of forests as crops rather than living systems.[56]

One of the most effective means by which industry and government could promote the strictly economic agro-forestry perspective was to do exactly as environmentalists had demanded since 1976: not use chemical insecticides against the spruce budworm. The DLF eagerly seized upon every chance to employ the relatively new bacterial agent *Btk,* endorsed by CBLAS and the EAC as an organic alternative to synthetic chemicals, even after the collapse of the budworm population in 1978. The DLF sprayed 20,600 acres of forest with *Btk* in 1979 and 70,260 acres the following year, and continued the program indefinitely thereafter.[57] Dubious necessity paled to irrelevance beside the opportunity to advance the idea and practice of industrial forestry – to insist on the need for "control," by one means or another – with a method already declared safe and acceptable by the leading voices against chemical sprays. Thus environmental activists began their anti-herbicide campaign at a severe disadvantage, fighting a practice that had already begun, and one that could be presented by government as a mere extension of the long-established use of the same chemicals on farms and roadsides, all while having undermined their own ability to use ecological arguments against the managerial logic of control.

The move towards annual use of *Btk* did arouse some activist resistance in Nova Scotia, despite the refusal of leading anti-spray activists like Elizabeth May to cease defending it. Biological though it may have been, *Btk* tests by Environment Canada included additional anti-evaporants and adhesives (plus a chitinase enzyme), and most often measured the effect on budworm populations and defoliation rates to the exclusion of wider ecosystem monitoring.[58] That sort of tunnel vision alarmed those who saw in it echoes of chemical insecticide testing. Opposition in Cape Breton led some members of CBLAS to form their own organization in response to the "unacceptable compromise" made by its leaders. The Crowdis Coalition, named after the primary test site at Crowdis Mountain near Baddeck, accused May of "selling out" the ecological argument in favour of a purely scientific approach to chemical safety and a managerial idea of forestry. At a time when the anti-herbicide campaign was warning the public how "in the fast world of chemical discovery and analysis 'safe' is a risky word," the new critics wondered why a broadcast biological agent should be held to a lower standard.[59]

The EAC attracted some criticism on similar lines after sponsoring an equivocal public lecture on agricultural chemicals in February 1980, but both critiques failed to blunt the attempt to achieve respectability and expert status.[60] Convinced more than ever that their victories in the "budworm battles" had rested on sound scientific argument first and public pressure second, Susan Holtz, Susan Mayo, and others at the EAC, along with May (who joined them on the centre's board of directors in 1980 while continuing to represent a loosely organized group of Cape Bretoners[61]), poured ever more of their attention and energy into winning the battle of facts. Wholly in the citizen-scientist mode, they carefully avoided being seen to "blindly oppose any action which may affect the environment."[62] Too much faith in scientific fact, though, can just as easily blind a person to the power of politics, a desperate liability when called upon to formulate new tactics against an unsympathetic government.

Unlike the proposed massive insecticide sprays of the previous decade, herbicide applications in the early 1980s were individual and relatively small operations carried out by various forestry companies, subject to DOE and DLF approval. Accordingly, the campaign against herbicide sprays gained momentum in a piecemeal fashion, jumping from one local battle to another until coming together in a province-wide effort, something industry and government tried very hard to prevent. Henley and Eldridge at the DFL bore responsibility for regulations around public notification, and they took every opportunity to limit the time between notice and spray. Whereas environmentalists had enjoyed months of foreknowledge even for the first 1976 budworm spray proposal, herbicide spray applications became closely guarded secrets, received, reviewed, and approved behind the closed doors of the DLF and announced with only weeks to spare. The DOE signed off on any spray application using chemicals approved by Environment Canada and showed even less concern for public awareness. Secrecy, in short, was policy. An opposition motion in the Legislature in the spring of 1982 would have forced the DLF to publish applications upon receipt; the government voted it down.[63]

But if secrecy made activism difficult, it also sometimes provoked it. In the summer of 1980, NSFI received government permission to use 2,4-D on five hundred acres of leased Crown land at Big Pond, Cape Breton

County. When news of the impending spray reached the community two weeks before the event, residents were incensed. Public notification rules did not specify who should be directly notified, and the pulp company made minimal effort. As one farmer later told *Atlantic Insight* magazine, "it's the goddamn way they were going to do it. We have a priest, fire chief, school principal, and Community Council in this town and NSFI said they didn't know who to contact." DOE and DLF officials reacted to the controversy with a redoubled commitment to secrecy, but Big Pond residents took their case to the Nova Scotia Supreme Court and won an injunction against the spray.[64]

Within months of the Big Pond settlement, another of NSFI's plans faced local opposition in the village of Lochaber, in Antigonish County. The area approved for herbicide treatment was only about 150 acres this time, but the chemical would be a mixture of 2,4-D and 2,4,5-T – the infamous Agent Orange. Once again a community reacted en masse, and this time, with the assistance of Antigonish MLA Bill Gillis, pressured the DOE into a "reinspection" of the site – and NSFI into abandoning the plan – without resort to the courts.[65] In Lochaber, as in Big Pond, it was local activists who began the agitation against herbicide spraying. Most of them had never before been involved in the province's environmental movement, demonstrating once again the centrality of place-defence in generating environmental activism. All of the newcomers reached out to the established groups, the EAC and CBLAS, for advice and help in acquiring the factual ammunition with which to fight, but decided which course to take on their own, with two results. First, this new generation of activists discovered a provincial network in need of new energy. Rather than settle their own problems and retire from the field, they pushed for a united front. Vicki Palmer of Lochaber was at the forefront of the effort. "Although the issue is resolved for the moment in Lochaber," she wrote, "it is plain that the time has come for the government to take a good hard look at their policy on herbicides. Local skirmishes are just symptoms of a greater problem."[66] In the New Glasgow *Evening News,* she was quoted calling for the creation of a provincial organization to carry the fight beyond one small controversy at a time.[67] The second result of the new activists' work, however, was a stark choice of tactics that made close cooperation difficult. Big Pond had enjoyed enviable success in convincing one judge of the potential risks of chemical sprays, while Lochaber did equally well in cowing the DOE. Obviously, an extended campaign would have to be mainly a political effort, but it is a rare environmental campaign that takes its impetus from anything but an imminent threat, and the question of whether to respond to the immediate threat in the legal or political arena would have to be answered. Complicating the question, the pro-spray forces had drawn their own lessons from the season of failure in 1981 and would not so easily succumb to either tactic again.

The new united front of forestry activists took the form of a committee of the EAC during the winter of 1982, a choice of venue that immediately put the group's focus on information dissemination and attempts to directly influence government. On the first account, the Forestry Committee can only be counted a success. Vicki Palmer from the Antigonish Environmental Coalition and Dan MacGillvray from the Big Pond Environmental Association brought the previous year's experience to a group comprising scientists, foresters, and activists.[68] As Don Eldridge feared, the dual controversies of 1981 had given "the advocates of no spray a platform,"[69] and they refused to relinquish it, sending their most persuasive speakers around the province to high schools and community halls to explain once again the dangers of chemical sprays and argue for alternatives.[70]

Unfortunately for the activists, their second principle of action – to directly influence government – turned much of their proselytizing into wasted effort. The Buchanan government readily obliged their desire for formal consultation, but lacking the established infrastructure of Environment Canada's public participation process, reached instead for a time-tested tool and declared a Royal Commission on Forestry. Public inquiries have long functioned to legitimate "the idea of the state" when an organized citizenry attacks government policy.[71] They are also effective distractions from the usual political process. Alexa McDonough, leader of the provincial NDP, wrote the EAC in April to caution against placing too much faith in a process "really only intended to be a further delaying tactic and a ruse for the industry interests."[72] Three weeks later, just as the Forestry Committee noticed a "drying up of information sources regarding the granting of spraying permits," the government announced its official policy that permits and applications would be kept secret from the public.[73] Previously faint hope that the province might prohibit all spraying during the Royal Commission's deliberations faded entirely, only to be replaced by anger when the continued advance notice of approved sprays promised by the environment minister turned out at times to be a week or less.[74] Nevertheless, most activists persisted in asking the public to turn their attention to the three Royal Commissioners, in the hope of convincing them to recommend against continued spraying. No hope was ever so vain. Though the records of the Royal Commission show a tremendous intellectual variety among anti-spraying advocates, including a strong

showing by witnesses arguing in favour of emotion and experiential knowledge as a counterweight to scientific reductionism, the three commissioners were hostile to any but the managerial and economic perspectives on
forestry, and eventually became open public advocates of industrial forestry.[75] In the meantime, the process provided a screen behind which the
premier and his ministers might hide in a time of crisis.

On June 21, the DOE granted permits to all three large pulp companies
in the province to use a 2,4-D/2,4,5-T mixture ("Esteron 3-3E" to the
department and companies; still "Agent Orange" to activists) on fifteen
thousand acres throughout the province.[76] The DOE proved surprisingly
forthcoming with news of the permits, making them known almost immediately, and activists faced a dilemma: try to pull together an effective
pressure campaign against an intransigent Cabinet (after having invited
the public to direct their letters to the Royal Commission), or attempt to
halt all three spray programs in court.

The first reaction was political and could hardly have been otherwise.
As locals learned about the spray plan, committees formed in communities
near spray sites without any outside encouragement. Public meetings,
petitions, and delegations to Halifax followed, the same tactics that had
proven effective in Lochaber. Under the tutelage of the EAC-based coalition, however, nothing disturbed the respectable, scientific appearance of
the movement, at least not until the first week of July. With time running
out and no hint of compromise from government, the Whycocomagh
Mi'kmaq Band had earlier decided to issue its own deadline: an NSFI spray
site on Skye Mountain, near the community's water supply, outraged Chief
Ryan Googoo, and he vowed to uproot one thousand of the company's
seedlings if the permit was not withdrawn. It was not, and on July 7 the
band turned out en masse to fulfill the promise. As other activists at other
times have learned, nothing spurs a government to action like a challenge
to its legitimacy, and that is what Whycocomagh's tree-pulling day represented. Inverness South MLA Billy Joe MacLean arrived immediately with
a promise of personal intervention, and (with Chief Googoo's threat of a
repeat performance weighing heavily) Cabinet moved the following day
to suspend all aerial spraying permits granted for 1982 and until the report
of the Royal Commission on Forestry.[77]

Victory was short-lived. By the end of July, spraying of 2,4-D/2,4,5-T
from trucks had begun on some mainland sites. It took a few days longer
to learn why, but in the early days of August, the DOE revealed that the
aerial spray permits suspended by Cabinet had been automatically converted to ground spray permits. Activists were stunned and faced once

again the choice of tactics: pressure (demonstrably ineffective, short of vandalism, which Elizabeth May and the EAC called "a defeat for the democratic process") or litigation.[78] And this time they deliberated without the luxury of delay. NSFI published its newspaper notices on August 4, promising to begin spraying on the 11th. Driven to their "last resort," in the words of Vicki Palmer, the activists chose to go to court.[79]

PLAYING BY THE RULES: THE HERBICIDE TRIAL

The rules of the courtroom are never simple but the principles are usually straightforward. The court, according to Justice Merlin Nunn of the Nova Scotia Supreme Court, is the "final and proper forum" for determining the veracity of facts and their adequacy to a certain standard of proof.[80] That far, and no further. To win their temporary injunction, the plaintiffs seeking to stop herbicide spraying in Nova Scotia needed only to prove the possibility, not the probability, of harm to themselves or their property, and to do so to the satisfaction of Justice C. Denne Burchell, who had heard and been swayed by many of the same arguments from the residents of Big Pond a year earlier. In addition, they needed to promise to pay the costs incurred by the defendants if the case ultimately went against them. It had been a relatively quick task in 1981, but the pulp companies were better prepared to argue in 1982. And argue they did, for six days – the longest chambers proceeding in Nova Scotian legal history – bringing experts from across North America to testify. Eventually, Scott Paper and Bowater Mersey were dropped from the injunction application when Burchell ruled that no plaintiff lived close enough to the western spray zones, and NSFI alone faced seventeen plaintiffs from Cape Breton and the eastern mainland who found that they had successfully argued themselves into a neat trap. The pulp companies' procedural intransigence had driven legal costs high enough already that for the plaintiffs to withdraw would invite financial ruin (and allow the spray to go ahead), while a loss at trial could be much worse. Having won their injunction, they had little choice but to go ahead with a protracted courtroom battle against an expanding cast of opponents. NSFI's name may have appeared on the suit, but Scott and Bowater kept their lawyers on the case, and were soon joined by the US multinational Dow Chemical, manufacturer of 2,4-D and 2,4,5-T.[81]

Dow's interest was only one part of the internationalization of Nova Scotia's forest controversies. The anti-herbicide movement in Nova Scotia had indeed "unwittingly stumbled into a battle with the international pesticide industry," as environmentalists themselves quickly realized, and if anyone doubted the costs of defeat, they were swiftly educated by Scott Paper. Scott had no official role in the proceedings after the first four days of hearings in Justice Burchell's chambers, but for those four days the company demanded $23,000 in costs from the plaintiffs. Appeals reduced the amount by about a third, but having effectively already lost their case against Scott (and despite the ongoing case against NSFI), the plaintiffs were held liable and given a week to pay.[82] It was an ominous beginning, but driving home the reality of the threat also helped draw attention and supporters to the plaintiffs' side.

The herbicide trial, as it came to be known, was one of the milestone events in the formation of a provincial environmental movement that was more than a mere congeries of independently operating parts. Groups that had existed before the herbicide issue turned much of their attention and energy to it, and new associations found their raison d'être in support of the plaintiffs. Rudy Haase's Friends of Nature (FON), oldest of the new groups formed since 1960, pledged to raise $10,000 towards the legal costs.[83] At the same time, a new organization in Annapolis County, People for Environmental Protection (PEP), which had been speaking out against chemical biocides only since mid-year, legally incorporated in the fall of 1982 and set about fundraising for the case.[84] The plaintiffs even found a lawyer through the provincial network – Richard Murtha, a member of the Sackville Environmental Protection Association and a Vietnam veteran who attributed his own chloracne affliction to exposure to Agent Orange.[85] To avoid duplication of effort and share news quickly, the leading activists formed a phone tree, and by the time the trial began in May 1983, the list of individuals and group representatives they had compiled constituted the first so-named Nova Scotia Environmental Network.[86]

The mobilization of public upset and activist energy around the trial looked very different from the budworm battles of a half-decade earlier, when multiple lines of argument had converged on the Cabinet table in Halifax. Quite apart from the possibility of having to pay NSFI's court costs, simply fighting the case was a massive financial undertaking for seventeen plaintiffs. Though some expert witnesses could be recruited from supporters within the province, such as the veteran anti–budworm spray activist and surgeon Dr. William Thurlow, others could only come from abroad. Ultimately, nine of the plaintiffs' fourteen witnesses lived outside the province, and each one had to be brought to Cape Breton, housed for the duration of their testimony, and returned home. Sympathetic lawyers could serve at reduced rates, especially with the assistance of Elizabeth May and some of her fellow Dalhousie law students, but reduced is not free. Necessarily, then, most of the work done in support of the plaintiffs was fundraising. A band of Cape Breton supporters quickly organized an association – the Herbicide Fund Society (HFS) – dedicated to collecting money for the trial, and other groups began to direct donors to them.[87]

The environmental movement as a whole had only so much effort to expend, and however necessary the fundraising effort was, it could not help but draw attention away from the work of generating political pressure. PEP, for example, after participating in one of Environment Canada's public consultation meetings, writing a brief for presentation to the Royal Commission on Forestry, and soliciting donations to the HFS, might have laboured for half a year without even once coming to the attention of the provincial Cabinet if not for its work in organizing a phone-in day to Premier Buchanan's office.[88] Even there, however, political efforts suffered the additional handicap of having to penetrate the bubble of shifted responsibility that the trial and Royal Commission had placed around Province House. Whatever could not be dismissed with "it is before the court" could be handily redirected with "tell it to the Royal Commission." For some activists, the chance to prove their case "objectively" was a welcome alternative and much more respectable behaviour than forceful lobbying. For others it was a handicap, tolerated out of the necessity to support the plaintiffs, who had put so much on the line on everyone else's behalf.

One new development that under better circumstances might have proven a potent political weapon was the extent of collaboration with environmental groups outside Nova Scotia. Certainly there had been exchanges of information and nominal association with other Canadian and American groups since the early 1970s, when the infant EAC reached out to Ontario's only slightly older Pollution Probe. And the "budworm battles" had seen more of the same, as well as some genuine influence on the course of a similar campaign in New Brunswick, in which the Reye's syndrome episode in Nova Scotia provoked the first mass popular resistance to insecticide spraying across the border.[89] But the end of 1982 saw the first active collaboration between Nova Scotian environmentalists and their foreign peers, aiming at achieving victory not only in a Sydney courtroom but also in a much larger campaign. Cape Breton's trial became news around the world. As the newsletter of the worldwide Pesticide Action Network (PAN) put it in 1983, "the potential implications of the court case are more far reaching than the plaintiffs imagined in that first day in court. Proving to the court's satisfaction that the use of admittedly dangerous substances in dilute quantities according to government approved procedures constitutes nuisance, actionable at law, would broaden the potential for environmental litigation against other hazardous materials."[90]

By early 1983, the EAC's Forestry Committee was also talking about a ban on 2,4-D/2,4,5-T "across Nova Scotia and Canada."[91] Like the writers at PAN, most out-of-province environmentalists tended to see the widest implications to their cooperation – a chance to use the Sydney proceedings as political ammunition – while the Nova Scotians with whom they connected, mainly in the EAC and CBLAS, remained caught up in the fundraising effort. The EAC quote on national ambitions, for example, came quite typically in the course of a fundraising appeal letter. Speaking tours by herbicide trial plaintiffs to Ottawa, Vancouver, Washington, and points between were also, for the speakers transfixed by their financial peril, mainly fundraising tours.[92] The connections built between Nova Scotian environmental activists and Swedish environmentalists in particular brought profound pressure to bear on Stora Kopparberg's headquarters – which probably forced the company to seek a settlement with the plaintiffs at the end of 1983[93] – but never achieved any noticeable pressure on the government of Nova Scotia. In that sense, it, and most of the out-of-province collaboration built around the herbicide trial, were mostly wasted potential.

The reason Nova Scotians had to reach beyond the provincial border to recruit witnesses was the extreme complexity of the set of ideas they had to argue in court. At trial they would be expected to prove the potential for harm within a narrow definition: human health could only be taken as isolated from environmental health, affected by the latter but never part of it, and a legal/scientific standard of proof would apply, meaning that all other possible causes of a given effect would have to be ruled out before the court would accept any herbicides as being responsible for it. The kind of medical research that might produce such results is expensive and rare, and even bringing the researchers themselves from Sweden or the United States could not prevent NSFI's lawyers from arguing the absence of direct, verifiable links. It was the same argument used by the New Brunswick government to defend its insecticide program during the Reye's syndrome controversy. At that time, it led Dr. John Crocker to complain that, short of experimentation on humans, no researcher could ever produce the kind of proof they were demanding of him.[94] The plaintiffs could only hope to educate the court on the nature of epidemiological evidence as the trial progressed. Even so, they would still have to overcome an insurmountable problem of scientific uncertainty, arguing for the effective prohibition of 2,4-D and 2,4,5-T when even the best research sometimes failed to make clear whether the possible danger stemmed from the named phenoxy compounds themselves, from the products of their breakdown over months in storage or in nature, or from dioxin contaminants that occurred in vastly different concentrations from one chemical or one manufacturer to the next. And all the while they would face the attempts of NSFI's lawyers to prolong the court's confusion, arguing the need to sort out every nice distinction before pinning any blame on the chemicals in question, and insisting on the clearest "smoking gun" evidence of their harmfulness. In effect, the plaintiffs had shouldered the nearly impossible burden of transforming a precautionary case into a deductive one – a "probably" into a "definitely." Justice Burchell had warned them in the summer of 1982 that the issue of chemical safety belonged properly to politicians, not judges, but once he granted his injunction, there was little chance to back away.[95]

Given the clearly poor odds of success, the question "why go to court?" was on many minds in late 1982 and through much of 1983, though the urgency of the process precluded its becoming a matter for much open debate until after the settlement with NSFI.[96] The leading activists' having had so little time to decide on their course of action, and the profound fear of what Agent Orange exposure might do, were certainly major factors. For some activists, injunctions were seen as primarily a delaying tactic anyway. Yet haste alone cannot explain it; the first successful challenge to the budworm spray in the winter of 1976 had hinged upon a Cabinet decision in a matter of weeks using political tactics. One reason for the difference was surely Elizabeth May's entry into law school at Dalhousie University. May enjoyed a well-earned respect among anti-spray forces, especially on Cape Breton, and she had led every legal effort since Big Pond in 1980. Most of all, however, the reason for May and her allies in the EAC and other established Halifax groups to pursue the legal route was their maturing faith in scientific argument and the ability to sway politicians (or judges) with statements of pure fact. Since 1975, the EAC had pursued an image of respectability, based on the understanding that a place at the policy-making table and a chance to state its case would be enough to convince anyone; Susan Holtz's insistence that the environmental movement had no real enemies epitomized that attitude. Its allies in forestry activism, the old-style conservationists, were the heirs of a long tradition of even more strict scientific optimism, not always as keen on public education and democracy as the new environmentalists, but very much committed to the idea that technocratic foresters, free of the corrupting influence of politics and commercial interest, could be counted on to follow the dictates of sound science. May and CBLAS might have begun in 1976 with some antipathy towards government, but the experience with Vince MacLean and with Environment Canada's public consultation had turned some heads, especially hers, and close cooperation with the EAC encouraged the change.

With confidence in their facts and experts, then – shaken only slightly by Justice Merlin Nunn's decision to hear the case alone, without a jury[97] – the plaintiffs put together a set of arguments around four essential points: (1) the physical evidence of spray chemicals' mobility beyond the spray site; (2) the ecological evidence of their damage to the forest system; (3) the economic evidence of the lack of necessity for herbicides in pulp-wood forestry; and (4) the medical evidence of their harmfulness to human health. Their witnesses' combined testimony would present a damning description of NSFI's practices and of the government's oversight. Optimistically, supporters hoped that politicians, bureaucrats, and Royal Commissioners might "hear the crucial evidence that will help them make sane decisions."[98] In reality, none of them were listening, and if they had been they would have heard Justice Nunn dismiss each of the first three approaches summarily as irrelevant to the question of human health and safety. If the chemicals could not be proven harmful, he reasoned, NSFI might waste money, kill forests, or spray water supplies at will, even when instructed not to, and the court should still be satisfied that no criminal trespass existed.[99] Arguments that had won over the public against insecticide spraying fell flat in front of Nunn's insistence on the strict legal standard of proof of harm.

According to Nunn's preference, the majority of testimony from the first day of the trial on May 5 to the last day on June 1 concerned the effects of phenoxy herbicides on human health, and it was here that NSFI came prepared to fight. The plaintiffs' witness John Constable, surgeon at Harvard Medical School and one of the world's foremost epidemiologists regarding the effects of Agent Orange in Vietnam, started the contest by detailing the very recent evidence of birth defects there, and suggested that though newer phenoxies had less dioxin, no one could yet know if it was little enough to be harmless. "It is unreasonable," he said, "to use human subjects to determine the minimum toxic dose of a very deadly chemical."[100] It was exactly the argument that the plaintiffs offered again and again, even as the central point of Elizabeth May's end-of-trial summary: to proceed in the face of scientific uncertainty amounts to human testing. NSFI responded by minimizing both the scientific disagreement and the potential harm, producing experts to contradict each of the plaintiffs' experts, and yet more experts to insist on the relatively higher cancer risk from, variously, vitamin A, aspirin, cigarettes, contraceptives, diet soda, peanut butter, milk, and oxygen.[101] As testimony went on, witnesses attacked other witnesses' work. Lawyers for each side attacked the other side's experts for their experimental procedures, grasp of statistics, or closeness to industry. It became, in Richard Murtha's words, "the world series of scientific evidence."[102]

If the trial was a ball game, the plaintiffs struck out often. They seemed to be playing by rules different from those imposed by the umpire. Mikael Eriksson, a Swedish cancer researcher, objected to NSFI's simplistic numeric assessments of risk. "Risk to whom? Benefit to whom?" he asked. With the herbicides themselves as well as their various dioxins and other contaminants understudied, he added, "we just know for sure about sarcomas and lymphomas." Other cancers might well be caused by the same array of toxins.[103] Susan Daum, an epidemiologist from New Jersey, carefully explained to Justice Nunn that rare cancers and birth defects may be "markers" for more common ones whose causes can't be discerned with enough certainty. If thalidomide had caused a common birth defect, she said, its effects would have disappeared into the statistical background and no one might ever have realized what it was doing.[104] In granting the injunction in August, Justice Burchell had mentioned thalidomide and DDT, saying that people do have a right to fear "approved" substances.[105] Nunn would not have it (or simply could not grasp it).[106] The question of risk inequity was beneath mention, but during the trial as well as in his final ruling, Nunn insisted that any chemical as bad as environmentalists made TCDD dioxin out to be should have obvious effects.[107] Barring such effects, no evidence would satisfy him. When plaintiffs' lawyers pressed one of NSFI's witnesses about full or partial bans on 2,4,5-T in Sweden, Italy, Holland, Norway, Japan, and the Soviet Union, Nunn interjected to dismiss "political" matters: "I have received no evidence of any country that has banned them on the basis of scientific evidence." On another occasion, he vented his frustration with inconclusive studies and suggested that the scientists "scrap them all and start over again," as though one good experiment would settle the question forever. Finally, on the last day of the trial, the plaintiffs attempted to call Robyn Warren to the stand, hoping to use his story of herbicide poisoning to "rebut the theory that nothing happens in real life." Nunn refused to hear him.[108]

Justice Nunn's ruling, when it came on September 15, could hardly have been a great surprise. In insisting on the strictest standard of proof, he had earlier said that the "reasonable fear for safety and health" criterion could shut down almost any industry if consistently applied.[109] Those who had attended the trial might have heard echoes of the startling honesty of Dr. Marshall Johnson, witness for the defence, who said that "if you took everything that causes terata [birth defects] off the market, we wouldn't have anything."[110] Only "smoking gun" evidence could have met Nunn's standard, and that (e.g., "John Doe's cancer was definitely caused by 2,4-D") is exactly the kind of evidence epidemiology cannot produce. There was no shortage of criticism of the ruling during the late months of 1983, but in truth Nunn had carried out his assigned task as well as should have been expected. The court is an inherently conservative institution. Arguments in favour of precautionary principles cannot be expected to move a judge when the law he applies has no space for such considerations. May, Murtha, and Wildsmith crafted a case best suited to the court of public opinion and gambled on its success in a court of law.

Criticism of Nunn's ruling on the merits of the case quickly fell away behind rising criticism of his ruling on costs. The plaintiffs had indeed gambled, and stood to lose everything. Not even pausing to hear arguments from either side, the judge awarded full court costs to NSFI. Though the plaintiffs had acted in the public interest, normally a mitigating factor in costs rulings, the final ruling seemed better suited to a frivolous suit. This, along with Nunn's direct criticism of the plaintiffs' witnesses' "lack of objectivity" caused the greatest anger and worry among environmentalists.[111] Regardless of the outcome of the trial, the feeling that a ruling on costs had been used to punish activists and discourage further opposition to chemical forestry incensed their supporters. Sydney City Council (already upset during the trial when one of NSFI's witnesses suggested that the city's tap water was more carcinogenic than 2,4,5-T) passed a unanimous resolution asking the provincial government to pay the costs on behalf of the plaintiffs, and fundraising efforts increased with the help of the Dalhousie University Student Union, the Nova Scotia Liberal Party, and the provincial Federation of Labour, along with environmental groups in several other provinces. Pollution Probe bought an ad in the *Globe and Mail* asking for donations in support of the seventeen Nova Scotians and their families facing financial ruin; Friends of the Earth Canada did the same in the *Ottawa Citizen*.[112] Nevertheless, the precedent set by Nunn's ruling could scarcely be undone by a simple show of support, no matter how strong. The provincial minister of environment made it clear that he had no intention of interfering in the case in any way, and saw "no need to finance the court battles of environmentalist groups,"[113] and the EAC reported that a group of activists in Calgary had abandoned court action against Amoco Petroleum over a sour gas leak in 1982, for fear of a similar ruling on costs.[114]

In the fall of 1983, the herbicide issue was as vital as it had ever been. If government and industry had hoped to squelch opposition in the wake of a harsh court ruling, they were disappointed. Measured by fundraising effort, press attention, and the geographic reach of both, the protest was, if anything, gaining momentum. Yet the whole effort in Nova Scotia – and by extension the connected efforts beyond – was still tethered to court proceedings, and to the shaken confidence and fading energy of seventeen average Nova Scotians. None of them could have expected in the summer of 1982 that sixteen months later they would be contemplating appeal proceedings that might drag on for months or years longer, and the strain of Justice Nunn's costs ruling was wearing them down. The example of Scott Paper's costs award during the injunction hearing was never far from their minds: if NSFI made a similarly sudden demand for payment, the plaintiffs, despite all of the fundraising, could not hope to raise the amount needed. Some of them also questioned the value of an appeal, when the injunction that had grounded NSFI's planes had expired with the end of the trial. They might go through all of the additional expense of an appeal only to be sprayed before ever meeting another judge. Peter Cumming wrote:

> They had become representatives of all Nova Scotians, personally responsible for reforming Canadian law for all environmental and citizen groups who might take a corporation or government to court over anything, figureheads, martyrs, and torch-bearers for people across Canada, the United States, and Sweden ... sooner or later the question must be asked as to how much any group of people should be expected to bear.[115]

For plaintiff Vicki Palmer, the answer was already clear: "People are burnt out."[116] For almost three months the plaintiffs debated whether or not to proceed with an appeal. In early December, they decided to "get out of a bad situation as cheaply and quickly as possible."[117]

The settlement that the plaintiffs signed with NSFI was a divisive one. The final debate over whether to accept the deal was tense and ran late into the night, and when they finally came to a decision, it was based in part on the advice of a corporate lawyer who enjoyed less than perfect trust from all seventeen plaintiffs. The settlement would mean turning over remaining funds raised to NSFI (not a significant amount) and surrendering the right to appeal, but would also finally end the plaintiffs' financial peril. Ryan Googoo, keen to appeal Nunn's brisk dismissal of the Aboriginal rights argument, and Elizabeth May, who wished to keep the debate in a legal forum, initially refused to sign the settlement, though Googoo eventually agreed to abandon an appeal in order to preserve the settlement for the other plaintiffs' sake.[118] The Herbicide Fund Society, not itself a plaintiff, did not need to sign but was asked to give up its funds or risk scuttling the settlement. After some debate, its members decided to capitulate, and also to dissolve their society and begin anew, hoping to raise funds to repay the costs left out of the settlement and to operate at a greater remove from most of the former plaintiffs.[119] Money questions strained the cohesion of the anti-herbicide forces – HFS's Peter Cumming believed NSFI had drafted its offer with that result in mind[120] – but almost everyone was reluctant to criticize too heavily the plaintiffs who had done so much. Financial disagreements alone did not drive the acrimonious debate of late 1983 and 1984; instead fundamental disagreements about how to wage the campaign were breaking through after months of enforced solidarity.

THE STING OF DEFEAT

For a certain subset of the anti-pesticide movement (well represented in the EAC, and including Elizabeth May), court action presented the ideal vehicle for environmentalism: objective, scientific, and respectable. At a time when environmental activists were being accused of harbouring anti-business feelings – dangerously close to anti-capitalist, with all of its sinister Soviet connotations – any method of argument that could borrow the legitimacy of the judicial system was an advantage. Unfortunately, the tactical choice that was supposed to "balance the image" of the movement did nothing to increase its influence on forest policy or alter the growing frequency of attacks against environmentalists in the press, except perhaps by restraining their ability to respond effectively.[121]

Industry spokesmen rarely missed a chance to criticize environmentalists and indeed went out of their way to do so, for example, in newspaper advertisements like those shown in Figure 2, but predictably so and to muted effect. Attacks from high-ranking members of the Buchanan government were far worse. With the pro-chemical faction in ascendancy within Lands and Forests and the DOE, there were few curbs on comments such as deputy environment minister E.L.L. Rowe's, associating activists with "violence and civil disobedience," or Attorney General Harry How's

FIGURE 2 The backlash against activists included a mostly fruitless attempt to usurp the word "environmentalist" and a much more successful attempt to present them as a meddling leisure class. *Source: Chronicle-Herald,* December 10, 1983, 7 *(top); Chronicle-Herald,* December 16, 1983, 7 *(bottom).*

suggestion that half of the movement in Nova Scotia was made up of Americans.[122] The long-serving and prominent minister of lands and forests George Henley gave the most thorough denunciation of environmentalists, as a small group operating in Nova Scotia with the "support of subversive elements from both within and without the nation by both political philosophy and financial assistance and under the guise of environment." The solution to this pernicious influence, according to Henley (and despite the efforts of the most scientifically inclined environmentalists to appropriate managerial language), was to "base our educational program on facts, on scientific facts so that we may go forward with our forest management despite the preoccupation of those I have mentioned who through distortion, through vandalism, through supposition, through emotion, based without truth would destroy the very principle of forest management."[123]

Activists understood what the critics intended. One remarked that "Mr. Henley ... is well aware that about 50 per cent of all those who heard or read his statement already have implemented in their minds the connection between anti-spray and communism."[124] Yet their response was remarkably confused. Since the end of the budworm fight, the largest environmental coalition in the province had grown lax in its public outreach even as it pursued legal solutions and bureaucratic access. As one measure, the EAC's formal membership declined to about three hundred in 1982, from over six hundred just two years earlier.[125] Members who self-identified as part of the movement were replaced by supporters, who stood outside of it. And while some in the movement fretted over the potential loss of social influence as a consequence of the drift towards elite or expert status, it is not clear that anyone anticipated how much easier the change made it for people like Henley or How to portray them as a clique of subversives and outsiders. Nor does anyone seem to have realized how badly losing in court had undermined their own claims to scientific expertise and credibility, or the power of the propaganda weapon that Justice Nunn's decision would place in the hands of the movement's critics. By impugning the scientific objectivity of the plaintiffs' witnesses in his ruling, Nunn effectively gave the endorsement of the court to the statements of the defence, which Lands and Forests' Eldridge immediately began circulating to forestry agencies across the country. (The PAN cited a particularly egregious example by a witness supplied by Dow Chemical, who described Agent Orange as "a mythical substance which caused tremendous damage in Vietnam in the newspapers, but nowhere else."[126]) Even when environmentalists attempted to buttress their mass-movement credentials, they faced criticism designed to undermine their reputation as honest information brokers. A public opinion poll that showed 61 percent of Nova Scotians opposed to the use of 2,4-D and 2,4,5-T on forests prompted a DOE spokesperson to suggest that the numbers had been "cooked."[127]

Henley's talk of an "educational program" was much more than a rhetorical stick with which to beat environmentalists. It was policy. In an ironic reversal of the trend in the nascent environmental mainstream, the Department of Lands and Forests avidly pursued public outreach starting in 1983 and continuing under Henley's successor, Ken Streatch. Active support of industrial forestry had been growing in the department and in the DOE since shortly after the Progressive Conservative government took power in 1978 and brought Don Eldridge from the Nova Scotia Forest Products Association into the deputy minister's office, but it took the herbicide trial to finally shock the chemical proponents out of the narrow reaches of interdepartmental committees and the Eastern Spruce Budworm Council and into acting on Eldridge's exhortations to publicly "answer [the] anti-spray people."[128] The year 1983 brought active cooperation between pulp and paper companies and government on public woodlot tours to sprayed and unsprayed stands, and on attempts to counter negative publicity by seeking out media attention. As momentum grew for mass public appeals by industry during the winter of 1984, plans for the first Nova Scotia Forestry Exhibition were drawn up and funding was secured from Lands and Forests.[129]

Critics of the industry operating within the reach of government agencies found their positions increasingly untenable in this new era. Some, like the ECC, simply swam with the tide, calling it "encouraging" to see industry "beginning to combat this [environmentalist misinformation] with information and education programs."[130] Others resisted. Minister Streatch personally took over as the chairman of the Forest Practices Improvement Board, which had given a voice during the controversy to small woodlot owners opposed to chemical forestry. The former chairman, Hugh Fairn, complained that Cabinet was stacking the board with "people dedicated to the destruction of the [Forest Practices Improvement] Act," while board member and fierce critic of industrial forestry Murray Prest accused the minister of attempting to "cripple" the board and with it one of the few means for small landowners to resist the pulp companies' market control.[131] But industrial forestry advocates refused to cede either media dominance or control of the board to their critics. Just as environmental activists had learned from their opponents during the budworm controversy, their opponents now borrowed environmentalists' tactics and appealed to aesthetic arguments (contrasting the "terrible looking grey mess of trees". killed by the budworm with the "lovely little green trees" in herbicide- and insecticide-treated plots).[132] They also continued to consolidate their claim on scientific expertise, and began crafting forestry education curricula for elementary students. Thanks to "public education" subsidies included in federal/provincial forestry agreements, Canadians' own money brought the inaugural "People of the Forest" lesson to Pictou County at the end of 1983, from which students learned the value and necessity of chemical treatments.[133]

Environmentalists recognized what was happening. Some could even appreciate the irony of their unintended achievement. In one of the first newsletters from the HFS's successor organization, Citizens Against Pesticides (CAP), Peter Cumming noted on a list of activist accomplishments that, while they had earned support from literally all around the world, "we have also mobilized the forest industry, Lands and Forests, Canadian Forestry Service, Truth in Forestry, etc. etc. into trying to sell the chemical forest to Canadians."[134] It is questionable, though, how many realized that public education was necessary only as a weapon against environmental activists. After all, unlike environmentalists, industry had no need of public pressure to turn policy decisions in its favour. The provincial government had made perfectly clear its intention to continue herbicide spraying even as the legal drama unfolded. While waiting for a decision, the environment minister publicly vowed to treat the upcoming ruling as applying only to the few thousand acres covered by the injunction, and Lands and Forests finally removed the last official requirement for prior public notification before spraying. Aerial application of 2,4-D/2,4,5-T was still off the table thanks to the pressure tactics of the Whycocomagh Band the previous summer, but there was nothing to prevent spraying from the ground. In 1983 as in 1982, Lands and Forests approved ground permits for scattered sites across the province and even conducted its own program on five hundred acres.[135] (See Figure 3.) The secrecy was imperfect – the Scott Paper Company was more open than government, and the information could be extracted from the DOE with enough persistence – but it hardly mattered. Letters of protest from the EAC and the provincial New Democratic Party were no more effective than last-minute protest marches by the residents of Orangedale, Inverness County. Government had simply stopped listening.[136]

In the aftermath of Nunn's ruling and the subsequent settlement, as bureaucrats and elected officials joined industry spokesmen in denouncing "ignorance and irrational fear,"[137] the rain of secretly approved herbicides resumed. New rationales were found for insecticide applications as well, as the gypsy moth moved into southwestern Nova Scotia.[138] Environment-alists, largely cut off from information sources in government and fighting on multiple fronts (now against a revived roadside spray program with 2,4-D in various counties), drew meagre comfort from the fact that 2,4,5-T, suspended from use by the US Environmental Protection Agency in 1979 and finally discontinued by Dow Chemical, was no longer available. The herbicide program in 1984 would see instead the first widespread use of glyphosate on Nova Scotia's forests.[139] But the change was due entirely to outside factors, and the last stockpile of 2,4,5-T still made its way into the forest, albeit by the more laborious method of ground spraying. The herbicide trial, epitome of a provincial movement based on science and the law, had in the end achieved only the peculiar and unintended effect of driving the provincial government to abandon any pretence of impar-tiality. There were now, at least, very clearly opposed camps on the issue of industrial forestry, though not everyone in the movement was willing to acknowledge it.

Conclusion

The campaign to prevent insecticide and herbicide spraying on Cape Breton Island garnered unprecedented public support across Nova Scotia, but the court challenge that had effectively locked non-modern arguments out of the contest for over a year had proven disastrous, and mistrust of government among those who would make such arguments was stronger than ever. It was joined by a growing mistrust of those who had led the movement into court. The failure of the legal effort and the shock of Justice Nunn's costs ruling preoccupied the province's environmental activists, and the eventual settlement even more so, but it also freed them from the enforced solidarity of the courtroom proceedings. Peter Cumming, who had pushed his alternative views fruitlessly before the Royal Commission during the trial preparations, wrote several essays in the aftermath. In "The Stink of Defeat," he called it "amazing ... that our shaky structure has held together as long as it has,"[140] but admitted that the effort had left the entire movement "demoralized, burned out, and fighting among themselves."[141] In "Out of the Courts and Back to the Issue," he suggested that "the legal child [had] gobbled up the larger parent, the herbicide issue itself."[142] In

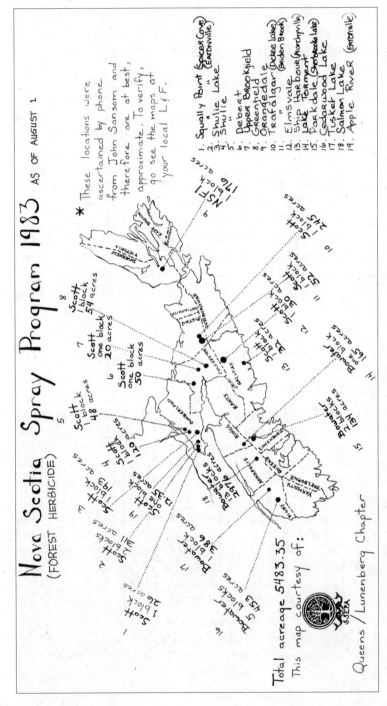

FIGURE 3   Map of spray sites for 1983. *Source: Queens/Lunenburg SSEPA News* 2, no. 2 (1983), Nova Scotia Archives, Betty Peterson fonds, MGI, vol. 3470, no. 6.

both essays, the advice was the same: return in earnest to the political pressure tactics of the budworm era and leave the divisive post-mortem analysis for another time (perhaps never). Too many activists, though, were unwilling to let it go at that. Freed to express their frustration, groups like the Cape Breton Wildlife Association lashed out at a "red herring" legal effort that focused on human health and left no room for defence of wildlife or landscapes,[143] while others – likely influenced by the activist schism around uranium mining (see Chapter 4) – saw behind the question of tactics a deeper division between those who would work within the boundaries set by government and those who would challenge them.[144]

The power of governments to shape social movements has to do with more than simply tempting offers of consultation or research funding. It also involves their withdrawal of the same, as well as of basic information, in order to protect their prerogative of final decision making, and the powerful instruments – such as Royal Commissions' terms of reference, parliamentary procedure, court rules, and bureaucratic compromise – with which they influence the sorts of arguments and ideas deemed legitimate in official proceedings, in the press, and in the public mind. The Queens-Lunenburg South Shore Environmental Protection Association (SSEPA) complained in the summer of 1983 that

> we have been telling the government how we feel about its forestry and public notice policies, especially regarding herbicide spraying. We have followed all the proper channels – letters, phone calls, telegrams, petitions, and even litigation. We have made submissions to the Royal Forestry Commission. Still we are ignored. Still government will not meet with us. Still government issues permits while the Royal Commission on forestry is deliberating; while the Nova Scotia Supreme Court weighs the evidence. Still the government reduces further the feeblest of public notice regulations.[145]

The SSEPA members were feeling the effects of this whole suite of conditions as well as describing the sort of activities that it is designed to evoke. Confronted with such an apparently insuperable challenge, activists were forced to decide between acquiescence and rebellion, but the process of following "all the proper channels" had already set a large proportion of the movement on the path to acquiescence by narrowing the parameters of acceptable argument to suit the institutions of power.

The contest over industrial forestry demonstrates the power of the modern state to shape social movements, but the state never acts alone. These conditions, this exercise of government power, require as well the tacit consent of the governed. It requires the latter's own agency, and in the summer of 1983 a number of environmentalists in Nova Scotia had already demonstrated their determination to remain within the boundaries of respectable discourse, regardless of the cost.

4

Two Environmentalisms
Uranium and Radicalism

I N THE LATE 1970S, Nova Scotians witnessed a boom in uranium prospecting. As prospectors homed in on deposits, moving from aerial
surveys to drilling and trenching, their activities became more noticeable
on the ground, and more and more alarming to Nova Scotian landowners,
some of them already certain that uranium should not be mined, others
still only curious and concerned to find out the risks.[1] The province's anti-
nuclear movement was well prepared to provide information. Uranium
and its dangers had long been recognized as a fruitful avenue of anti-nuclear
argument: granted a hypothetical accident-free nuclear energy system,
uranium mines and tailings piles would present the largest single source
of additional radioisotope releases by the nuclear industry, and the federal and provincial governments had been steadfast in their support of
the industry. The established groups therefore had information ready on
the anti-uranium experience in Saskatchewan, British Columbia, and the
United States.[2]

Organized activism that went beyond merely sharing information
gathered steam slowly during 1979 and the early months of 1980. Prodded
by a growing stream of letters and phone calls from worried landowners,
and enjoying a partial respite from campaigning against the *use* of nuclear
fuel in the region after the Brudenell rally, established groups turned their
attention to cooperation on the issue of its production in the winter of
1980.[3] In January, the Community Planning Association's inveterate coalition builder in Halifax, Joanne Lamey, began assembling interested parties
from around the province into a common front, before quickly deferring

to the Energy Committee of the Ecology Action Centre (EAC).[4] There was a gathering on February 20, but the EAC was initially cautious and conservative. Susan Holtz was determined to cultivate a bridge-building mode of operations, in line with which the EAC made direct calls upon the minister of mines and energy, Ron Barkhouse, to "follow BC's lead," but offered him little reason to comply.[5] The EAC's information-sharing role certainly helped draw more attention to the issue, but it (as well as the Fundy Area Concern for Tomorrow [FACT] group in the Annapolis Valley) generally integrated anti-uranium campaigning into their well-established routine of anti-nuclear activism; there appears to have been little effort to collaborate on a strategy with the growing number of citizens alarmed about uranium mining quite on its own.

The first significant new citizen action against uranium exploration in Nova Scotia came from an unexpected source: the Women's Institutes. Nova Scotia's Women's Institutes were established early in the twentieth century as service clubs for rural women, promoting education, civic engagement, and cultural activities. By the 1970s, however, they were often dismissed as conservative assemblies of older women.[6] Yet they were far from moribund or unresponsive to changing times, and had much in common with the feminist peace groups that were drawn to the earliest anti-nuclear activism in Halifax. The pesticide debates of the late 1970s attracted a great deal of attention in agricultural communities and among institute members, who considered the health of farm families a traditional women's issue. Some institutes were also reinvigorated by the arrival of back-to-the-land families, including women with experience in peace and social justice activism. Early in 1980, several Women's Institutes received information and assistance from the Department of Environment to set up environmental awareness committees; within months, Institutes in Hants and Kings Counties were already at work gathering information on uranium mining.[7] By November, the Women's Institutes of Hants County moved from gathering information into building public support for an anti-uranium movement.[8]

In early 1981, one of the rare Maritime-wide anti-nuclear gatherings under the banner of the fast-fading Maritime Energy Coalition (MEC) united interested parties in demanding a moratorium on and inquiry into uranium mining, but a common set of demands alone made for neither a full-scale movement nor a strategy for organizing one.[9] What was lacking was a triggering event, something personal, something an-alogous to the Genelle blockade in British Columbia, which launched the process of protest, Royal Commission, and eventual moratorium on uranium prospecting in that province. The winter of 1981 provided one, as news spread that one of the companies with claims in the Vaughan/ New Ross area southwest of Windsor was no longer looking for uranium so much as looking at a mineable deposit of it. If any single factor transformed uranium from the obscure preoccupation of a relatively small number of peace activists, anti-nuclear groups, and Women's Institute members into the third major environmental controversy since the 1960s, it was the prospect of an actual uranium mine operating within a few years at a known site. With the encouragement of Women's Institute members who had spent most of the previous year studying the issue, statements of support for a British Columbia–style moratorium on uranium mining and prospecting came from the Hants and Digby Counties Federations of Agriculture, and from the provincial New Democratic Party leader, Alexa McDonough. Most worrying of all, from the industry's perspective, the vote of the West Hants Municipal Council to request a provincial moratorium was the direct result of the work of the Women's Institutes.[10] And the local leaders of the anti-uranium forces from Hants County promised more to come, continuing their drive for support during the winter and spring of 1981, urging more action from other members of the Voice of Women (several "Voices" were already among them) and from other groups in the province.[11]

As opponents were lining up against uranium mining, Aquitaine, the French mining company that had made the discovery near Vaughan, and its allies in the provincial government began a counter-effort. At a public meeting at the Windsor exhibition grounds in April, Aquitaine's senior vice president of explorations and special projects, Mike Hriskevich, joined environmental consultant Leo Lowe and the company's regional office head, Don Pollock, on a mission to defuse local opposition. In the coordinator's chair sat Jack Garnet, Nova Scotia's director of mineral resources and a champion of uranium mining. One of the province's up-and-coming anti-uranium leaders, writer and English professor Donna Smyth, later described the meeting as decidedly hostile towards the presenters. Certainly the men from Aquitaine gained few allies by underestimating the store of knowledge built up by local activists over the previous year; claims that yellowcake uranium was "not radioactive," or that Ontario's Elliot Lake mines (with their long, dead river) represented safe and successful industrial practices, were instantly rebutted and were featured in anti-uranium propaganda for the next two years.[12] A similar performance soon followed at the West Hants Municipal Council, where Pollock gave a rebuttal to the Women's Institute presentation a month earlier and insisted that there was no additional radiation measurable in the town of Elliot Lake (an arguably true claim, but quite beside the point of radioisotope contamination of food and water). The council declined to withdraw its moratorium request.[13] Pollock was one of two leaders of the mining industry's propaganda effort, but his extreme statements and accidentally revealing gaffes consistently provided his opponents with ammunition.[14] The provincial government, on the other hand, apart from Jack Garnet, tried at first to evade rather than confront public fears by appointing an inactive Select Committee of the Legislature to investigate.[15]

As the campaign turned into a contest, the growing activist group in Hants County decided to organize formally as Citizen Action to Protect the Environment (CAPE). Its open letter to Premier Buchanan at the end of June displayed all the hallmarks of the group's activism: requests put to the premier served more to inform other readers of alarming facts, such as that the ministers of health and environment had not been chosen to sit on the Select Committee on uranium mining, or that the Department of Mines and Energy was both regulator and promoter of mining, or that exploration continued while the Select Committee supposedly sat to determine its dangers.[16] The members of CAPE had already decided that the Select Committee was a dead-end process, but writing as though it was not enabled them to point out the de facto support the government gave to mining and the lack of effective regulation in the province. Publicly needling politicians in order to attract the support of people who did not already identify as "environmentalists" was an effective strategy for CAPE in 1981, though it did not always please its allies at the EAC, who persisted in a conciliatory approach.[17] CAPE's direct appeals to the public were less controversial but still populist, such as the much-publicized map that showed areas of the province under claim by uranium prospectors. As geographer Brian Harley remarked, "maps are never value-free images";[18] CAPE's choice to map claims rather than simply focus on the one existing site of potential development encouraged people in broad areas of the province to think of themselves as being personally under threat. (See Figure 4.)

The success of CAPE's more aggressive tactics, along with the steady contribution of the established groups' publicity and public education, drew fresh countermeasures. Aquitaine's Don Pollock travelled to the annual meeting of the Mining Society of Nova Scotia in June with a stark warning and a request: "If these groups [of environmentalists] are successful in creating a seven-year moratorium in Nova Scotia as they have done in

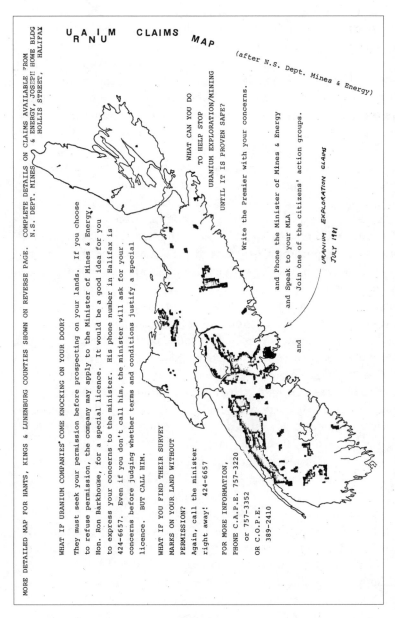

FIGURE 4 CAPE's map of uranium claims in Nova Scotia, 1981. *Source:* Dalhousie University Archives and Special Collections, EAC fonds, MS-11-13, box 30.10.

BC, exploration in the North West Territories could also fall and Canada could wind up with no uranium whatsoever" – a fate preventable only if mining interests could organize political pressure of their own.[19] Whether or not Pollock truly believed that Saskatchewan's entrenched uranium industry was under threat, his advice was sincere. Like the spokesmen for pulp and paper interests in the province, he wanted industry to strike back publicly against environmentalists. By the end of the year, his prodding would result in a new agency, the Chamber of Mineral Resources, to advocate for industry, but Aquitaine's commitment to confrontation was more urgent: on August 28 Nova Scotia's Legislature dissolved for nearly six weeks of election campaigning, during which Alexa McDonough's NDP promised to give the anti-uranium forces a voice that government members could no longer ignore.[20] Under its corporate successor name in Nova Scotia, Kidd Creek Mines, the company brought in its head of environmental affairs, Roy John, to lead a counterattack.

To the members of CAPE and its peers, an election campaign in 1981 was a precious opportunity. Armed with the results of a year and a half of research and self-education, organized, and drawing on the support of agricultural and medical advocacy groups, the activists were better prepared for the campaign trail than some MLAs. In print, in person, and on the airwaves they repeated the facts on radioisotope contamination of water and agricultural produce, reminded their audience of the limits of humanity's technological prowess, and challenged politicians to state a clear position. And they urged more people to join the fight. "The government of Nova Scotia seems in a mad rush to encourage this industry," wrote Muriel Siemers of Centre Burlington in the *Chronicle-Herald*. "Uranium is a relatively common ore. It is mined in places where public opposition is weak, or where the public is not aware of the hazards."[21]

Making the public aware meant appealing to people's personal identification with their home place. The combined claims map did that. So did the constant refrain about the incompatibility of uranium mining and agriculture, an echo of the economic justice arguments made by south shore fishermen during the Stoddard Island nuclear debate. But the greatest coup in the effort to personalize the issue for Nova Scotians was the discovery of a uranium claim overlapping the Pockwock Lake watershed, which supplied drinking water for the entire city of Halifax. When the EAC exposed the claim (by the German company Saarberg) two weeks before polling day, and revealed that the provincial Department of Environment was apparently not even aware of Saarberg's activities in the watershed, it threatened to turn the mainly rural issue into a major urban issue as well, and hugely increase its press coverage.[22] Politicians on all sides jumped on the new information. Alexa McDonough accused Premier Buchanan of attempting to "absorb and silence the protest" with "un-enforceable" voluntary guidelines for prospectors and the unproductive Select Committee.[23] The government, having vested control of exploration licences in Cabinet in August in a bid to reassure the public ahead of the election, found itself directly responsible for the Pockwock claim, and had little choice but to respond. Within three days of the EAC's press release, mines minister Ronald Barkhouse announced a freeze on new exploration permits until the Select Committee submitted its report on exploration. Existing permits would not be renewed, and the last active permit would expire in May 1982.[24] Representatives of Saarberg also came forward immediately to announce that nothing more than aerial surveys had been done near Pockwock Lake, and no more exploration would be conducted there.[25] After nearly two years of silence and misdirection, draft exploration guidelines and second-draft guidelines, never with the force of true regulations – a "talk and dig" policy anticipating British Columbia's infamous "talk and log" – the provincial government had taken its first substantive action on the uranium issue by committing to an investigation that included exploration, just what the activists had aimed at since early in the year. Success might never have come without the support of the NDP during the election, or without the early alliance with established anti-nuclear campaigners, but public pressure is what finally compelled the Buchanan government to act, pressure that could never have been brought to bear without local upstart activists' aggressive publicity.

If Buchanan's government (returned to office after polling day) or Kidd Creek Mines hoped that their desperate concession during the election would quiet opposition or finally channel activists' energy in a less public direction, their hopes were soon crushed. Few in the anti-uranium camp were prepared to trust the government to produce a report truly addressing their concerns. In the four months following the election, into early 1982, activists redoubled their efforts to reach the public and persuade Nova Scotians of the danger and foolishness of uranium mining. The Annapolis Valley Branch of the Nova Scotia Medical Society resolved in November to join the call for a full moratorium, followed two weeks later by the General Council of the provincial medical society.[26] Agricultural groups continued to lend their names to the effort: the Cream Producers Association, the Kings County Federation of Agriculture, and more.[27] And new local anti-uranium groups sprang up like spring grass. Rather than expand the CAPE group geographically, its members helped local activists start their own in Kings County (Kings Association to Save the Environment [KASE]), Vaughan (Residents Enlisted to Save Communities from Uranium Exploration [RESCUE]), New Ross (Communities Organized to Protect the Environment [COPE]), and Chester (Citizens Against Uranium Mining [CAUM]). In Cumberland County and in Colchester County, established anti-nuclear activists launched into anti-uranium campaigns as well, all of them like the southern groups in borrowing heavily from the EAC's collection of videos and documents on radiation health risks.[28]

Local groups borrowed more than just books and pamphlets from their Halifax peers. Elizabeth May became the environmentalist equivalent of Kidd Creek's Roy John, travelling from one local event to the next, where she and he would replay some version of their initial meeting.[29] John, who quite readily admitted having been hired by Aquitaine to combat "the anti-nuclear activist groups," kept up a constant stream of critiques of activist statements and aimed always to bring the debate back from the unscientific realm of nuclear weaponry and technological uncertainties to the firm ground of fact and comparative risk, a battleground where he expected to hold the advantage but where May presented a uniquely capable opponent.[30] While his allies in government employed emotional language of their own (at times verging on red-baiting, such as the insistence of the Department of Health's radiation expert, Ted Dalgleish, that activists "are only trying to gain some personal notoriety or wish to expand upon some personal anti-social dogma"[31]), John insisted on "a rational judgement based on logic and truth rather than emotion and lack of information."[32] And environmentalists replied in kind, with peace activists pointing out the contribution of Canadian uranium to the global nuclear arms race while others, like May, hammered back at John with an equal command of nuclear jargon, exposing the unspoken assumptions beneath his statements (for example, that tailings can be contained for millennia with yet-undiscovered technology, or that radiation exposure can always be expressed as a whole-body average).[33]

Neither side stopped at the provincial border in recruiting new authorities to join the fight. As would soon happen in the lead-up to the herbicide trial, Nova Scotia's environmental controversy became the focus of attention from far beyond. The quartet of Jack Garnett (Mines and Energy), John, Pollock, and Dalgleish (Health) regularly joined forces with men from Atomic Energy Canada Limited (AECL) and Environment Canada who often found themselves addressing points put by Gordon Edwards of the Canadian Coalition for Nuclear Responsibility (CCNR) or Dr. Rosalie Bertell, whose presentations on the "bureaucratic exclusion" of the public from decision making and government's deliberate misrepresentation of health risks at Wolfville and Halifax provoked Dalgleish enough to comment on activists' supposed sinister self-promotion.[34] Dr. Robert Woollard, one of the authors of the British Columbia Medical Association's anti-uranium report, proved a champion of the cause in Nova Scotia as well. Indeed, the battle over uranium mining in Nova Scotia became one of the three great nationally significant challenges to the industry in the years around 1980, alongside Saskatchewan and British Columbia.

The level of cooperation and coordination among anti-uranium groups in 1981 was greater than it had been at any time to that point. By the start of 1982, the new acronyms on the scene – CAPE, KASE, COPE, and so on – could field capable public speakers well versed in the technical jargon of nuclear risk. The EAC and its library had played an important role in education. So did the networks of existing anti-nuclear and social justice groups, drawing on the resources of national and international activist allies.[35] But the vigour in organizing the public into an effective force for political pressure came always from the local level, from residents wanting at first to protect the integrity of their relationship to the natural environment – their farms, gardens, and wells – and discovering by accident that the one issue came hitched to a whole system of political ecology, a system that required them to become true political activists and to fight for the recognition of an ecological perspective and for a forum in which there existed a possibility for meaningful public input into policy decisions, before those decisions were made. Unsurprisingly, there were also differences of opinion among local activists as to how profoundly they should oppose the system; different networks reinforced different views, and acted at times less like neutral repositories of resources and contacts and more like ideological police.

THE CRACKS APPEAR

Differences in activists' approach to government surfaced early in the campaign against uranium in Nova Scotia, well before the election campaign produced such encouraging results. As time passed, those differences seemed more strongly linked to the geography of dissent, with rural activists presenting the most consistent political pressure and stridently non-modern arguments against uranium mining as a threat to local communities and economies, and urban activists trying to convince regulators of the factual merits of the anti-uranium position.

When the FACT group joined the Bear River Board of Trade to organize a public information meeting at Bear River, Digby County, in May 1981, Jack Garnett attended to defend his department's handling of mineral exploration and tout the strength of his draft exploration guidelines. The guidelines, he said, the "most stringent in Canada," had been redrafted after "constructive comments from numerous exploration companies and environmental groups."[36] It was a well-practised line. Garnett, at that point the leading pro-uranium figure in the province, offered much the same defence each time his voluntary, accommodating, and still-unfinished (after five years of exploration) guidelines came under attack.[37] In response, Robbie Bays and David Orton, aghast at how debate had been cut off in Bear River without an effective anti-uranium speaker, issued a press release deriding the guidelines and "strongly repudiat[ing] any claims that environmental groups contributed to or approve of these guidelines, which do not protect the health of Nova Scotians."[38] Garnett was not lying, though, even if he did put the most favourable spin on the truth. He had sought input from at least one source within the environmental movement – Susan Holtz at the EAC – and had received it.[39] For her part, Holtz remained entirely opposed to uranium mining and exploration, and she did not believe that offering comment on Garnett's first-draft guidelines compromised her position. Rather, she saw it as a second front, pushing for the best possible rules and a place at the negotiating table if in fact exploration did go ahead. Unfortunately, the mere fact of her commenting did provide political cover for Garnett, and her refusal to admit as much proved to be an ongoing bone of contention between the EAC and environmentalists who, like Bays and Orton, saw little difference between government and corporate actors.

Differences of opinion on strategy or theory brought tension, but internal confusion at the Ecology Action Center as to its own power structure and role in the anti-uranium fight caused the most significant fracture among allies in 1981. Since the winter of 1980, the EAC's Energy Committee had been the nominal gathering place for anti-uranium activists, but for much of the following year the committee did not meet. Instead, uranium was folded into the activities of a person-to-person network of anti-nuclear activists. After the events of April 1981, a uranium subcommittee began to meet, bringing together CAPE and EAC members with a few other peace activists and environmentalists, mainly from

Halifax.[40] From the start, confusion reigned over the question of their autonomy from the EAC's board of directors.[41] By the end of May, a confrontation over the board's power to veto the group's more overtly anti-government statements provoked the resignation of David Orton and the eventual departure of most of the members from rural areas.[42]

David Orton exemplified the growing ideological diversity of the environmental movement in Nova Scotia since the late 1970s. Moving to the east coast from British Columbia in 1979, he had brought with him long experience as an organizer for various leftist groups in Ontario and Quebec, and a familiarity with direct-action protest.[43] The networks of peace activists and scientific conservationists who had so powerfully shaped the province's environmental scene through the Stoddard Island episode and the "budworm battles" were often politically conservative, and their members within the Ecology Action Center and the peace organizations that initially took up the uranium issue were uncomfortable with overtly political critiques of the government/industry relationship. The new network of politically leftist environmentalists in the province, however, shared a structural view of social and economic power in which government was never neutral and could be moved to act against the interests of its corporate partners only if sufficient numbers of citizens demanded it. This was the thinking behind Orton's proposed coalition statement for the EAC subcommittee, which dismissed the government's Select Committee as a political sedative and insisted that "our [i.e., environmentalists'] main task should be one of convincing and mobilizing the public."[44] Despite agreeing with much of his position, other members of the subcommittee dismissed his draft as "seditious," "one-sided," and "too revealing."[45]

The conflict that began at the Bear River meeting with Jack Garnett's "environmentalist-approved" guidelines festered for months, and was not resolved by Orton's departure from the EAC's subcommittee scarcely a month later. Discontent over the EAC's close relationship with government was openly discussed, and finally erupted into public animosity around the issue of the speakers selected for a public lecture series at the Halifax library, for which EAC planners had scheduled one entirely pro-uranium panel.[46] Three critics offered much the same opinion in September and October, ahead of the November 5 event, but only one, Sherri Cline from the north shore group CARE, offered it without appending further and more profound criticism.[47] On behalf of his new Socialist Environmental Protection and Occupational Health Group (SEPOHG), Orton gave a pointed and public reminder in a letter to the *Dalhousie Gazette* of Susan Holtz's inadvertent hand-up to Garnett (one of the speakers lined up for the library), and even quoted her assessment of his guidelines as "a fairly good first draft."[48] It was the third critic, though, who struck closest to the heart of their various objections. A long-time anti-nuclear ally, the South Shore Environmental Protection Association went so far as to issue a press release instead of direct criticism, decrying the slate of speakers on the simple grounds that "claims by government and mining companies are not reality," and taking the analysis of the guidelines incident one step further: "We [SSEPA] neither solicit nor accept money from corporations or the government to fund our organization, [and] although our association is one of the most prominent environmental groups in the province, we were *not asked to comment* on the guidelines for uranium exploration."[49] In the end, the EAC did accept the judgment of its peers, and on November 5 Elizabeth May was on hand at the library seminar to offer a formal rebuttal.[50] By then, however, Susan Holtz had already moved to defend her methods against what she termed "gratuitous attacks," and to vent lingering ill-will for the "spiteful and unscrupulous" David Orton, whom she accused of trying to "take over the [uranium sub]committee as a political vehicle for his Marxist-Leninist philosophy."[51] As for her commentary on Garnett's exploration guidelines, she insisted on the value of "a precedent for consultation" in case the prospectors could not be stopped, and reassured her critics that, after all, "our two major points were not incorporated into the final version," though how she imagined that this would help her case is unclear.[52]

The two months of strained and bitter correspondence that followed, between Holtz and Anne Wickens at SSEPA or Holtz and Sherri Cline at CARE, was about as public as private correspondence can be, and each page laid bare the mistrust many activists felt towards the EAC, especially those at the greatest remove from the city of Halifax. In the sharpest blow, Wickens explained that the SSEPA had decided on a press release in order to achieve "an element of surprise against the pro-nuclear people," something of which the EAC was obviously not considered capable. She repeated the earlier accusation that, due to its reliance on government funding, the EAC could not be expected to be as aggressive against the uranium industry as "we ... who are being poisoned first and worst."[53] On the same theme, Cline asked two pointed questions: "Do you feel it is better to be conciliatory and let the mining companies/government get away with their misleading and/or false statements? If so, is this because EAC would not jeopardize getting the next corporate contribution or government grant?"[54] Holtz's personal criticism of David Orton prompted a similar rebuttal. Orton's SEPOHG had worked effectively with several other anti-uranium groups in the province after his departure from the EAC's subcommittee, and "as for his Marxist-Leninist views," wrote Anne Wickens, "they may well exist ... [but] SSEPA does not, and never has refused membership to, or assistance from, any person or group because of race, colour, or creed."[55] According to the critics, it was much more likely that his politics had proven incompatible with the EAC's "respectable" image.[56]

Demanding as they did proof of direct opposition to government, in the form of public criticism, as evidence of the EAC's environmentalist credentials, the EAC's critics held a very different view than its staff and directors about what it meant to be an environmentalist. Despite the warning from one of their own in 1979, when former board member Linda Christansen-Ruffman described the organization's development through the 1970s as a turn away from the public and towards experts and policy makers (a "professionalism and legitimacy ... which will undermine EAC's long-standing advocacy of public participation"), the "two Susans"–era EAC remained committed to conciliatory methods and believed that no insurmountable differences divided them from less-established groups.[57] As Holtz insisted on the EAC's behalf in January 1981, "I see little evidence that 'professionalism' has weakened grassroots support."[58] Christansen-Ruffman had the better eye, though. As the winter of 1982 wore on, the war of words between the EAC and its critics petered out. The president of the EAC's board of directors took over the correspondence and closed the file with a bland insistence on its independent voice, regardless of funding.[59] No one, though, would call it a resolution; the major collaborative effort of the winter, a six-group mission to lobby Gerald Regan, included none of the EAC's critics.[60] And though it continued to debate the propriety of government and corporate funding internally from time to time, the EAC remained committed to its respectable path: "We have achieved, over these ten years, high credibility across the province and indeed all over Canada, and have no intention of compromising that."[61]

THE ROYAL COMMISSION ON URANIUM MINING

The Buchanan government returned to office in 1981 in a difficult position regarding uranium. Rather than continue attempting to reassure activists with political attention, the government returned to a policy of distancing itself from the issue and channelling activist energies into a less damaging forum. Reaching once again for the quasi-judicial form that granted government so much power to define the terms of debate, on January 22 the

Buchanan government announced a Royal Commission of Inquiry into Uranium Mining, set to officially begin on February 9 under Judge Robert McCleave.

Anti-uranium activists in Saskatchewan and British Columbia had learned to recognize the potential for inquiries to marginalize opposition or compel participation in a specific discourse (usually scientific or legalistic, but not emotional or ethical), and the lesson was not lost on their Nova Scotian peers. If the Buchanan government's intention was co-optation, however, it could not have chosen a less appropriate commissioner. If ever a man could test a dissident's commitment to cooperation with government, it was Robert McCleave, A sixty-year-old judge of the Provincial Court, former Progressive Conservative MP for Halifax, and current chairman of the Nova Scotia Labour Relations Board, in which position he had helped quash the unionization of the Michelin Tire plant at Granton, McCleave was well known but not well liked among labour activists in the province. His intense attachment to judicial propriety extended to his other public roles, where he felt compelled to defend his own authority from any perceived slight. As an unnamed Halifax lawyer put it to a *Toronto Star* reporter in 1983, McCleave's politics and his judicial role couldn't be separated: "He's firm in his conviction he always knows what is right."[62] In short, he was a democratic authoritarian, suited to the courtroom perhaps, but immediately ill at ease in the commissioner's chair.

In keeping with their year-old strategy, environmentalists were initially pleased at McCleave's appointment. The effective use of public inquiry proceedings in British Columbia had enabled activists there to intensify the pressure on politicians rather than be pushed aside. The first step towards repeating the process in Nova Scotia would be to achieve the widest possible terms of reference for the inquiry. The terms set out by the Order in Council establishing the inquiry eschewed "moral" issues such as the connection between uranium mining and nuclear weapons manufacture, but terms of reference are open to interpretation.[63] An appeal to McCleave himself was clearly in order.

The results of a multi-group meeting of activists in Truro on February 14 offered the first major clue that McCleave might not present the picture of reasonable accommodation that normally draws participants to an inquiry. The nine groups meeting in Truro included SEPOHG and CARE. The EAC's Uranium Committee had a representative there as well, though the centre itself later disavowed involvement. Despite their well-established differences, the activists in Truro agreed to act as if they accepted the inquiry as a genuine venue for public participation in policy making, at least until and unless it became clear that McCleave would not allow them to use the proceedings to generate publicity and political pressure.[64] A CARE press release following the meeting expressed support for the inquiry and asked that McCleave (who had not yet set out any rules of procedure) hold preliminary hearings on the conduct of the inquiry, and that the government consider appointing two additional commissioners and establishing more specific and comprehensive terms of reference.[65] McCleave reacted angrily. Already feeling "insulted"[66] and "pressured"[67] by the advice offered by the EAC and extra advertisements for the inquiry published by CAUM, he lashed out in interviews against "some Colchester groups ... anonymously spreading false information" and "making unprincipled attacks on the inquiry," and he threatened to suspend three unnamed people from the inquiry.[68] Nor did it end with CARE. As the judge's plans for the inquiry slowly became clearer, more reactions from environmentalists appeared – more "attacks," in McCleave's eyes – and he moved beyond threats of suspension.[69] Threats of subpoena and of contempt charges followed, and after McCleave took it upon himself to investigate the fairness of some newspaper reporting, prompted by an editorial in one paper that mildly rebuked him for his overreaction, the *Truro Daily News* briefly refused publication of any inquiry-related material for fear of the same threats.[70]

Reactions to McCleave's unusual behaviour and apparent intention to treat a public process as if it were taking place in his courtroom ranged from total non-cooperation to apology and favour seeking. Activists already skeptical that the judge would allow a truly free debate took his intemperate reaction and his interference with the press as a sign that he would work actively against any attempt to use the inquiry as a vehicle for publicity. The logical next step, then, was to turn the inquiry itself into a public issue. Some continued to participate, like Robert Whiting, the veteran pollution fighter from East River, Lunenburg County, who recognized no distinction between government Royal Commissioner, and who informed McCleave of his intention to pursue charges under the Federal Fisheries Act if the Nova Scotian government allowed uranium mining.[71] His strategy was simply to make a publicity-worthy presentation within the inquiry when his chance came. Others mocked McCleave's overdeveloped sense of propriety, like Judy Davis of CARE, another veteran activist who had attended the Truro meeting and wrote to the judge explaining that she saw his job as deciding "questions of how mining will be done," not if it should be done, and accusing him of scaring the public away from his own inquiry.[72] Others needled McCleave, aiming to draw him further into a debate that could only cast doubt on his political neutrality. Winston Settle made a public response to McCleave's public critiques. In a letter copied to multiple politicians and the press, he refuted the judge's accusations against CARE (that they had corresponded with him anonymously) and called his conduct "paternal, if not threatening." The public "consider themselves as respectable," he wrote, "and do not intend to be talked down to as if they were little children. If this is the type of inquiry we are to have, I for one will have no part of it, and I do reserve the right to write and talk to anyone as is the right of a free citizen of this country."[73] Settle's political tactics and those of several groups from the Truro meeting who wrote to the attorney general to demand McCleave's removal were rewarded by the entry into the debate of provincial Liberal leader Sandy Cameron, who asked for McCleave's removal due to his "contempt for free speech."[74] It was not enough to move Buchanan, but pressure from Cameron and NDP leader Alexa McDonough did inspire McCleave to retreat from his legal threats for a while.

Some on the anti-uranium side persisted in trying to guide the province's mercurial uranium commissioner into a more populist/democratic position. One sent McCleave a copy of the proceedings of the 1979 National Workshop on Public Participation in Environmental Decision-Making, perhaps hoping that he might appreciate its warning against an approach that is "highly discretionary, favouring the rich, the well- informed, and those who have access to the corridors of power, [and which] ... has helped to frustrate many groups, and has tended to make them believe that confrontation, rather than cooperation, is the best way to deal with government."[75] McCleave, however, remained adamant that he would offer no funding to presenters and would treat volunteer organizations no differently from the nuclear industry.[76]

Given McCleave's well-known conservatism and his open resistance to any action that might turn his inquiry into "some form of popularity pressure contest," those attempting to nudge him in the opposite direction faced a daunting task.[77] One of the most common tactics during the initial controversy was to put distance between those who had already drawn the judge's ire, whom he called "the lunatic fringe," and the rest of the movement.[78] Here the activists from Halifax and the Annapolis Valley area had an existing hostility upon which to draw, and in David Orton an ideal figure from the political left upon whom to place the blame. CAPE member Ralph Loomer wrote to McCleave in February to express his support for the judge's criticism of CARE, and to explain why the Truro meeting behaved in so "unprincipled" a fashion:

> The valley group with which I am associated did not send a representative to this meeting because we are aware of an attempt by a declared Marxist-Leninist person to dominate groups in the Colchester area. His tactics and attitude are repugnant to those of us who are members of Ecology Action Centre and our valley groups who oppose the idea of uranium mining here. We rejected this person's attempts to compromise us in his pursuit of class struggle. After he declared his politics and was excluded from further discussions of our committee he became involved with the Colchester and Amherst groups.[79]

Apart from the few early boycotters like Winston Settle and David Orton, the vast majority of the anti-uranium movement decided to participate in McCleave's inquiry. Despite the tight deadline imposed by the judge for people to request time to present to the inquiry, letters poured in from activists as well as those who had never before considered themselves activists. Some still hoped to use the proceedings for publicity, counting on attention from news media that never did muster great interest in the minutiae of inquiry hearings. Many truly believed in the inquiry as a vehicle for public input into policy, while others seemed unsure to what extent they ought to attempt to win over McCleave or to keep their attention on building the movement. One thing immediately apparent was the wisdom of the decision to ask for a public inquiry, at least inasmuch as new groups of activists continued to gather across the province, inevitably forming around a small set of local leaders and extending over a single town or set of villages.[80] The official vehicle of an inquiry also attracted vastly more attention to the issue from church groups of every denomination, whose insistence on addressing the moral issues may have convinced McCleave to give up any attempt to rule the nuclear weapons connection out of bounds.[81] Despite these gains, however, active opponents of uranium mining remained a small minority of the overall population, easy enough to ignore if they ceased to accumulate support. As the winter of 1982 melted away and Nova Scotians prepared for the first hearings of the Royal Commission on Uranium Mining, members of the Buchanan government could have been forgiven for thinking they had finally shunted the uranium issue off to a politically safe venue. Whatever new local opposition arose could be directed towards McCleave, away from the daily conduct of politics in Halifax and out of view of the general public.

HOMELAND DEFENCE

The testimony given at Judge McCleave's forty-four hearings represented a true cross-section of Nova Scotian views on uranium mining in 1982, and revealed a movement divided between those who sought expert status – the citizen-scientists – and the disciples of political ecology, who relied upon various non-modern normative arguments to make their case. Among the latter, a small number of activists (like native Haligonian and filmmaker Ian Ball) were still willing to base their presentations on the limits to growth thesis or appropriate technology movement, a larger number were intent on denouncing the environmental injustice of the distribution of risks and benefits between rural and urban areas, and numerous witnesses presented anti-uranium activism as a defence of home place.[82] The most common presentations by far, however, were those assailing the modern politics of expertise.

From the very first hearing, in New Ross, Lunenburg County, where Michael Keddy warned the audience that "it is only after exploration has taken place that the Landowner sees the folly of putting his trust in someone whose interests lay not in the land but in the provincial deficit,"[83] presenters returned again and again to a claim of authority based on a close relationship with the land. The home place theme was neither parochial nor always confined to the countryside (though it did predominate there): the small activist group RESCUE explained the plural "communities" in its name as an acknowledgment of the duty to stand up for others in the same situation, and Dr. C.J. Byrne in Halifax complained about "some bloody economist or systems analyst talking about costs as if he or she were talking about buying jellybeans down at the corner store or Woolies [when] they never talk about the other and more serious cost, the heartache and sorrow brought about because people have to leave an area they have learned to live with and love."[84] By implicit, and occasionally explicit, contrast, a mining company could make all sorts of legal and economic commitments to a community, yet, as Ron Leitold of New Germany said, "the one it can't make is a personal commitment – concern, devotion, loyalty, love (call it what you will) for a particular area and its way of life – only the people who live there permanently can."[85] Though it is sometimes common to attempt a distinction between environmental defence of a home place and economic defence of the same, it is clear from the testimony of those who made such claims that the idea of pristine nature and the division between human and environment held little sway over their minds. The most articulate statement of their indivisibility came when Muriel Maybe and the Lunenburg County Women's Group drew upon Aldo Leopold's land ethic to describe how "we are obligated to respect and cooperate with the land if we hope to ensure our continued existence ... we are, in fact, members of a community of interdependent parts. We need the soil, the water, the plants, the animals."[86]

Presenters who made place-defence arguments were aware of their vulnerability to attack, particularly regarding two distinctions that clearly did prey on the minds of many witnesses, between native Nova Scotian and recent arrival from outside the province, and between reason and emotion. Frequent accusations that environmentalists were "foreign troublemakers" were well remembered. In a significant way, the claims of authority based on close relationships with a given place served to circumvent the discount on come-from-away opinion, or even turn it back against the critics. As Erin Gore said at the Chester hearing on April 30, "we are not outside agitators; we have lived here eleven years and are permanent residents. Our would-be corporate neighbours, the mining companies, are only temporary residents; what responsibility are they willing to accept?"[87] More than a few presenters also took time out of their presentations to defend the role of emotion in normal decision-making processes, decrying in some form or other the fallacy of rational choice (the idea that any decision can be made without reference to first principles based on emotion) and reminding the audience of the positive role of emotion in the history of the anti-slavery movement, struggles for economic justice, or democratic political reform.[88]

The more explicitly political presentations to the inquiry frequently included environmental justice arguments. In this, the presenters drew upon a durable tradition in Nova Scotian environmentalism going back to the SSEPA's defence of south shore fishermen's interests against the nuclear industry in 1973, but they also reflected a national trend. In fact, the discontents of metropolitanism formed a shared language of environmental activism across Canada. In British Columbia and especially in Saskatchewan, anti-uranium activists had vigorously challenged the imposition of environmental risks on western Canadian hinterland areas in order to produce benefits that would accrue mainly to urban centres and to the national capital.[89] Much the same arguments also appeared in the chemical forestry controversy, still in full swing in Cape Breton and the northern mainland even as the uranium inquiry progressed. The economic politics of the province since the 1920s have consisted largely of a series of rebellions against an exploit-and-export economy imposed by a Central Canadian–dominated system, and it had not escaped notice that since Aquitaine's withdrawal from Nova Scotia, the Millet Brook project had been pursued by the federal Canada Development Corporation, with the support and encouragement of AECL.[90] "They are here in Nova Scotia," argued CAUM's Brian McVeigh, "because this province acts as a hinterland for exploiting cheap resources to feed the manufacturing mecca of the central region of Canada, where one in three light bulbs are powered by nuclear power."[91] Worse yet for several of the rural presenters was the compounded imperial pressure from the provincial capital; as an angry Robert Finck complained to McCleave in Bridgewater, "it's just another example of second-class citizens getting the dirt while the Halifax gentry get the gravy."[92]

Nearly two hundred witnesses can cover a broad and diverse rhetorical territory. More than even the defence of home places and environmental justice, however, what united the makers of normative arguments at the inquiry was a rejection of the politics of expertise. When McCleave held his first hearing at New Ross on April 2, one of the first complaints was from a sawmiller at the Ross Farm Museum, Kenneth Seaboyer, one of scores of non-expert laymen self-educated in the nuclear industry's track record. "The experts on the safety of uranium seem to me to be experts only when everything functions properly," he said, inaugurating a seven-month-long litany of criticism of the technological optimism – the hubris – of scientists who presumed themselves able to contain mine wastes that would remain a radioactive hazard for half a million years or more.[93] "We must face the simple reality of Murphy's Law," insisted Muriel Maybe shortly after Seaboyer. "What can go wrong, sooner or later will. The Titanic did sink, Three Mile Island did leak, the Ocean Ranger did go down. Uranium tailings cannot be 100 percent contained."[94] With unintentional irony, the Atomic Energy Control Board's Kenneth Bragg appeared at the inquiry in June to claim that "it is reasonable to assume that future generations are going to be at least as smart as we are," and to detail the new methods in state-of-the-art tailings disposal at Key Lake, Saskatchewan, and Quirke Lake, Ontario (really doing himself no favours by describing Quirke Lake's deep-lake disposal scheme as being "completely out of the environment"). He was answered two weeks later by Valerie Wilson of People for Environmental Protection (PEP) in Annapolis Royal, quoting Hannes Alfven: "If a problem is too difficult to solve, one cannot claim that it is solved by pointing out all the efforts made to solve it."[95]

The practice of science, in addition to its capabilities, came in for occasional critique by the most politically savvy activists. FACT's Robert Bays made reference to the "gambling game" of scientifically derived but constantly revised safe exposure standards (once more conveniently supported by a remarkably candid AECB brief describing the "as low as reasonably achievable" (ALARA) principle of occupational radiation exposure as having its basis in "social and economic factors."[96] But Dr. Linda Christansen-Ruffman and Dr. Karen Flikeid for the Nova Scotian branch of the Canadian Research Institute for the Advancement of Women offered the most comprehensive deconstruction of the sociology of science: the unequal distribution of resources, the assumption of safety until proven otherwise, and the use of politically convenient assumptions in the design of studies.[97] Science and politics could scarcely be distinguished in Nova Scotia's uranium debate, the commissioner heard, when the Department of Health dispatched its Radiological Health Officer to discredit Dr. Rosalie Bertell in the press.[98]

Unusual as it may seem for environmentalists with a non-modern perspective to have come in such numbers to present their views to Judge McCleave, who had in one of his rare clear statements of intention promised to base his report on scientific fact and consult heavily with experts of many sorts during the second phase of his inquiry, those were the presenters with strategic considerations foremost in mind, less intent on convincing the judge and more on reaching and politicizing other Nova Scotians. Judy Davis expressed the attitude in *Rural Delivery* during the early days of the inquiry: "CARE does question whether or not our group – or any other environmental group – can receive a fair hearing from the Inquiry," but it would participate nonetheless.[99] A veteran of the Anil Hardboard campaign (see Chapter 1), Robert Whiting, demonstrated much the same sentiment at the inquiry, virtually ignoring McCleave to lecture the audience on industrial pollution as "political murder" and warn, "don't trust the government!"[100] Similarly skeptical presenters took the opportunity presented by McCleave's invitation to "be creative" and performed music impugning the honesty of supposedly neutral leaders in Halifax and Ottawa,[101] and Susan Hower took the creative pursuit of publicity several steps further than anyone else, at the Liverpool hearing on April 27, when she delivered a short speech on the need for a "responsible, informed electorate" who would care enough to notice events like McCleave's inquiry, then pulled a black hood over her head and performed a mock suicide with a loaded starter's pistol.[102]

Hower's shocking performance was only the most spectacular of many attempts to make people take note of what was going on at the inquiry hearings. As most activists were painfully aware, apart from Donna Smyth's regular reports in *Rural Delivery*, there was precious little press coverage of the inquiry's progress in the press or on the air. It was rare that a reporter would even show up to witness a hearing. Partly this was because Royal Commissions are known for their dry and technical hearings; partly it was due to the press's conditioning to view them as fact-finding exercises, newsworthy only when making their final recommendations to government; but it was also a failure on the part of environmentalists to make full use of their allies in the press. In Nova Scotia, the major news media are concentrated in the city of Halifax and emerge only reluctantly to cover events in the rest of the province when those appear most newsworthy. Few of McCleave's hearings were held in the city, and the environmental groups based there, with the best contacts in the press, sought the legitimacy of expertise and dedicated themselves more to influencing the commissioner than the public, failing to create the kind of spectacle that might have drawn greater attention.

The command of nuclear and geological science brought to bear on the McCleave Inquiry by Nova Scotian environmentalists was impressive. Following the lead of activists in British Columbia during the Bates Inquiry (Royal Commission of Inquiry, Health and Environmental Protection, Uranium Mining), they recruited from the medical field as much as possible,[103] and a battery of professional and citizen-scientists from the EAC, CAPE, KASE, and CAUM seldom let up on their barrage of facts related to the failures of tailings management and the viability of alternative energy systems. Ralph Torrie, Friends of the Earth Canada's expert on uranium, travelled from Ottawa to present for CAUM in Chester and gave an impeccably thorough description of all the many failures of the industry-standard "modified dumping" process. (He refused to use the terms "management" or "disposal" because, he said, "they aren't.") Again, the influence of a continent-wide anti-uranium movement came to the fore, as Torrie described the enormity of the Church Rock tailings spill in the United States, which only three years earlier had spread thousands of tons and tens of millions of gallons of radioactive mine waste across parts of New Mexico and Arizona. He drew special attention to acid leaching at tailings dams in Ontario. His presentation concluded with a list of potential avenues of technical development, including vitrification and the removal of radium and thorium from tailings – none of which were being pursued by the mining industry, nor were likely to be as long as the industry was permitted to dump radioactive elements in tailings heaps.[104] The EAC's own presentation included a similarly technical dissertation on the soft energy path, the failure of tailings containment, and the dangers of exposure to low-level radiation. The EAC also detailed an acceptable regulatory regime, in case the province decided to ignore the rest of its advice.

The presentations of the EAC, CAUM, KASE, and, to some extent, CAPE suggested nothing so much as a spirit of technological optimism equal to the best the AECB could offer. For them, it seemed as though there was nothing mankind could not do, only that which for reasons of ignorance and greed it chose not to do; and the remedy, far from being a renewed commitment to the local place or a re-evaluation of economic or ecological justice, was the recognition of scientific truth. Their moral arguments – for they were far from amoral technologists – revolved around the contribution of uranium to the nuclear arms race and carefully skirted the political ethics of dissent and public participation that so enraged McCleave. As for publicity, they seemed not to connect the inquiry hearings with the wider campaign for political pressure.[105] While the technically inclined presenters prided themselves on their ability to "make the uranium companies squirm," in Donna Smyth's words, their willingness to burnish the legitimacy of McCleave and his inquiry caused at least as much discomfort among the province's environmentalists, many of whom saw uranium companies as nearly irrelevant distractions from the real work of making government squirm.[106]

A Parting of the Ways

Their attitudes towards McCleave and his inquiry remained the single most salient dividing line among anti-uranium activists in Nova Scotia, reflecting all of the unresolved tensions from the previous two years about the politicization of environmental issues, the propriety of government funding, and the participation of environmental groups in non-public processes of consultation like the one concerning Jack Garnett's exploration guidelines. Unlike the courtroom setting of the herbicide trial, McCleave's inquiry presented no barriers to alternative modes of argument, and despite his continued and ever more desperate attempts to impose courtroom conventions on the inquiry and discourage debate in the press, McCleave could do nothing to prevent activists from using the question of the inquiry's legitimacy as a political weapon, as Paula Scott in Chester did in a letter to mines and energy minister Ron Barkhouse. Barkhouse had inaugurated a trend among MLAs by refusing to attend the initial hearing in New Ross, his own hometown and constituency, and Scott chided him for his absence:

I'd like to believe that we – the people – have the last word, but our cynicism tells us that you – the government – have already made up your collective mind under pressure from the mining industry and that the inquiry is merely a formality undertaken to quiet public opinion by going through the motions of asking our opinion. We've been disappointed too often to believe you easily.[107]

Her letter was copied to the three political party leaders, seven newspapers, and Judge McCleave.

Few if any environmentalists were willing to suspend their anti-uranium activities while McCleave made his wandering way around the province. Lectures continued, as did letters to politicians, the collection of petitions, and attempts to win the support of municipal councils.[108] But even among those who did not openly challenge the inquiry's legitimacy there was suspicion that the mining industry hoped to "wear out the energies of the industry's opponents" in the forum of a public inquiry.[109] McCleave's declared intention to have three phases of increasing scientific and legalistic complexity exacerbated such fears, though some persisted in distinguishing between the judge's intentions and those of the industry. For those who did not, the obvious solution was not only to use the inquiry to generate publicity but also to publicize the very fact that the inquiry served to stifle meaningful dissent. Those already on the fringes of political life had no great difficulty challenging McCleave. Sitting in front of the judge in Chester, Robert Whiting barked, "If I sound like a radical, then I damn well am!"[110] But it was the self-identifying leftists who were most successful in irritating him, correctly judging that "every time McCleave opens his mouth, more people are mobilized against him,"[111] and gambling that a collapse of the inquiry, or at least its transformation from easily ignored sideshow to source of political embarrassment, would be of greater benefit to the anti-uranium effort than seeing it through.

Conveniently, McCleave's conservative political history and his continued position as head of the Labour Relations Board gave him a strong contempt for persons of the left, who made excellent use of his prejudice. After a presentation in Amherst by Don Rushton of Concerned Citizens of Cumberland County (CCCC), on government collusion with corporate capitalism, a presentation that included material critical of the judge's conduct during the inquiry, McCleave traced the ownership of a slideshow used there back to another group in Halifax (specifically to Charles Lapp), noting in his records that "the groups did not seem to be related except in political philosophy, which was violently opposed to the Commissioner and the Inquiry, and the beliefs of more than 99% of the people living in Nova Scotia."[112] The Nova Scotia Federation of Labour took subtle aim at McCleave's labour board, and obliquely at his inquiry, when it insisted that government alone could not be trusted to keep uranium miners safe; "over the years any progress that has been made has been more as a result of confrontation than by any rational process of scientific enlightenment."[113] And Tony Seed, freelance journalist and candidate of the Marxist-Leninist Party, who had twice run against Robert Stanfield in Halifax when McCleave stood down for him, managed to get under the skin of both the commissioner and his supporters at the EAC by speaking well over his allotted time at a Halifax hearing in September, forcing the EAC presenters to reschedule for another day.[114] The irritants accumulated, and McCleave wrote to a friend that he felt under attack by "an unholy crowd of Marxist-Leninists, some fellow travellers in the Legislature [the NDP], some pretty irresponsible journalists ... and a few paranoid parties."[115]

If the voices from the left saw reason to question groups that refused to take issue with McCleave's behaviour or with his inquiry's legitimacy – if they wondered, with David Orton, "what kind of openness is it when the inquiry process is seen as one of educating Judge McCleave and not the public"[116] – their frustration was matched on the other side by those who questioned why anyone would hand more ammunition to a pro-uranium effort that habitually painted all environmentalists with the same red brush. What began early in 1982 with the Department of Health's Ted Dalgleish telling a *Digby Mirror* reporter how, in his view, environmental activism led to "mob violence," grew in intensity and frequency later in the year, after the overwhelmingly negative response to uranium mining at the inquiry became clear.[117] Kidd Creek's Roy John tried to be subtle at an Air Pollution Control Association meeting in Saint John, New Brunswick, in September, referring to "a segment of the population who have different social and political aspirations," but he was able to rely on his ally Greg Isenor, president of the Chamber of Mineral Resources, to offer more explicit context by borrowing lands and forests minister George Henley's term "subversive" to describe an "anti-development lobby" opposing uranium mining.[118] Instead of allying with environmentalists, the deputy minister of environment joined their critics early in 1983 with the most elegant innuendo yet against protesters of "particular political persuasions" and their association with "violence and civil disobedience."[119] None of this was unique to Nova Scotia; similarly targeted activists in the United States eventually named the anti-environmentalist campaign of the 1980s a "green scare," referencing the "red scare" persecution of anyone with left-leaning politics.[120] For environmentalists hoping to gain a place at the regulatory table, any taint of "subversion" could only be seen as counterproductive.

The easiest way to counter the charge of disreputable political leanings (aside from directly, as was often done[121]) was to redouble one's visible support for the legitimate consultative process, that is, for McCleave. Statements of support during the hearings were usually sent to McCleave himself. A good example is Ginny Point (formerly of the EAC) and her expression of gratitude for the inquiry's "enabling [of] people to publicly voice their thoughts through a legitimate process."[122] After the first stage came to a close and Kidd Creek Mines announced its withdrawal from the process in November, however (to be replaced at the head of the pro-uranium side by the Chamber of Mineral Resources, and soon by the chamber's new president, Don Pollock), the EAC had to resort to a press release in order to be seen to "reaffirm its commitment" to the process.[123]

Unfortunately for those seeking a reputable route to influence, Judge McCleave was not one to forget his enemies. Despite having been requested to make time for the north shore CARE group to speak about inquiry procedure and organization at the Pictou hearing on June 15, McCleave reacted with hostility to the brief read by CARE's Dean Whalen. Among other critiques, Whalen objected to the fact that inquiry transcripts were available only in Halifax and could be obtained only by being photocopied in the presence of either McCleave or the inquiry coordinator, Stanley Forgeron, or by being borrowed to be copied elsewhere.[124] He also asked McCleave to explain, at last, just what he had found "false, misleading, or unprincipled" about CARE's February press release.[125] Claiming to have been blindsided by this "attack on the inquiry," the judge wrote a letter the following day barring Whalen from any further role in the hearings until he apologized for the "slur" of implying that McCleave and Forgeron had been absent from the office without reason during Whalen's visit.[126]

The personal animosity between CARE and Judge McCleave coloured the remainder of the inquiry. Despite confusion on the part of the public as to how Whalen's story could be construed a slur, or why McCleave refused to consider his request to place hearing transcripts in regional libraries outside of Halifax (leaving, as one writer pointed out, the supposedly villainous CARE as the only ready source of information on the north shore[127]), the judge refused to back down or explain himself. He even went so far as to suppress the CARE brief when he finally relented and sent copies to the libraries.[128] The facilities for photocopying also became more accessible after the Pictou episode, yet CARE members received no credit for inspiring the reforms. Instead they remained personae non gratae at the inquiry (officially, only Whalen, "for his deceit," and Sherri Cline, "for her insolence" during the Pictou hearing, remained barred from participation[129]), and enjoyed scant support from their peers in the more southerly groups. Donna Smyth largely avoided publicizing the drama in *Rural Delivery*, which, publicity being the whole point of challenging the judge, prompted an angry Don Rushton to ask why she would "rather identify herself with McCleave."[130] In addition, the EAC's board of directors voted not to "publicly affiliate itself with the positions of other groups addressing the inquiry," when asked by CARE to speak out.[131]

Rather than attempt to repair the gulf between McCleave and CARE, or join the latter in further foredoomed criticism of the former's conduct, anti-uranium groups in the city and in the nearby mainland instead moved to formalize their own coalition ("publicly affiliate," one might say) ahead of stage two, isolating the north shore groups and the leftists. As the first stage of the inquiry closed, a committee of four activists from the central mainland area circulated (in their areas only) a proposal for the second stage, primarily requesting funding for their own expert witnesses, and suggesting that McCleave take over all questioning of witnesses himself.[132] Reaction from the north was predictably negative. Sherri Cline (CARE), Gail Fresia and Don Rushton (CCCC), and David Orton (SEPOHG) (who was on the verge of moving from Halifax to the North Shore with his family) wrote together to *Rural Delivery* to decry "a group putting itself forward as *the* representatives of the environmental movement – even though there was no open invitation to all interested parties," and a proposal that "reeks of faith in the system and authority."[133] The coalition's reply, that the northerners ought to make their own proposals and "embrace a range of divergent attitudes," left the distinct impression that the move to unify attitudes in and around the city had little to do with diversity and much to do with delegitimizing those who would not accept the premises of the supposedly impartially scientific second stage.[134]

IMPLOSION

Events do sometimes overtake expectations, and though no small number of activists in the province clearly expected a long and bitterly controversial second stage of Nova Scotia's uranium inquiry in 1983, in the end the battle was decided inside a single Halifax courtroom – perhaps even inside one man's mind – over the course of two angry days in March, a catharsis that forced both sides of the environmentalist conflict to finally acknowledge the deep roots of their differences. The departure from the script began with a letter sent by McCleave in February to the participants in the first stage of the inquiry. Apparently feeling pressured by activists eager for word on the planning of the second stage (much as he had felt ahead of the first), the judge asked everyone to "please hold your fire," and enjoined all parties to avoid "speculation" on the inquiry's future.[135] No one really did. Faced with a bitter exchange in the press between deputy environment minister E.L.L. Rowe and Elizabeth May, and wishing to pass judgment on some of the presenters whose "perverse ideology" had so bothered him in recent months,[136] McCleave convened a "special session" of the inquiry at the Halifax Law Courts on March 4, and requested the presence there of May, Rowe, Dean Whalen, Sherri Cline, Don Rushton, and a *Toronto Star* reporter (and yet another former Marxist-Leninist Party member) named Alan Story. Charles Lapp, creator of the slide/tape show presented by Rushton in Amherst, titled *Uranium: The Nova Scotia Experience,* was subpoenaed and arrived with a lawyer.[137] Story was there because he had interviewed McCleave by telephone the day before, acting on a tip from Lapp, and McCleave felt it "most improper that a witness should be talking to the press."[138] That the member of the press in question was a man who had only recently embarrassed the entire Nova Scotian judicial system by exposing the now-infamous wrongful conviction of Donald Marshall Jr. surely added to the impropriety. Over the course of the afternoon, McCleave made a remarkable series of rulings: that Lapp should return to court in two weeks to show and be judged for his slide/tape show (for reasons the judge refused to specify); that Story was forbidden to use two words from his interview with McCleave, "censorship" and "kangaroo," on pain of a contempt charge; that Rushton apologize to the inquiry for alleging slander by McCleave; and that Alan Ruffman, Howard Epstein, Tony Seed, and Gail Fresia were in "default of the inquiry" for not having produced written copies of presentations, and also had two weeks to comply or face the unspecified consequences of an unheard-of ruling.[139] It was by even the most generous assessment a bizarre performance, made only more so when McCleave addressed the Rowe/May controversy by asking the assembled audience to deliberate on "free speech [and] ... the limits that should be made on anything that is being said" in or out of the inquiry, presumably without using the word "censorship."[140]

Reactions to McCleave's extraordinary performance certainly exceeded any environmentalist's grandest hopes for publicity from the inquiry. His treatment of Alan Story in particular inflamed other members of the press, who repeated and reprinted the story of an "arrogant" and "paternalistic" judge making a reporter sit on the prisoner's bench and refusing him access to a lawyer when requested.[141] Inside the Legislature, McCleave came under repeated attack, and both the Liberal and NDP leaders requested his removal from the inquiry.[142] Perhaps worst of all for McCleave's immutable sense of the respect due his position was the ridicule; posters soon appeared around Halifax advertising a public screening of Charles Lapp's slide/tape show, and they were decorated with a picture of a kangaroo.[143] By the time of his second "special session," McCleave had clearly reached the end of his ability to handle the pressure.[144] "Beset by a radical group and beset by certain members of the press," the judge treated the assembly on March 18 to a long diatribe on the responsibilities of the press, while a crowd of protesters from CARE, CCCC, SEPOHG, and some Haligonian allies milled about the entrance to the Law Courts, holding signs reading "BOYCOTT THE INQUIRY" and "STOP ATTACKS ON ENVIRON-MENTALISTS."[145] Those inside had been made to pass through a metal detector and a line of police to reach the courtroom. After Lapp's slide/tape show was played to loud applause, McCleave simply ended the session. "We leave the courtroom wondering what has happened," Donna Smyth wrote. "Nobody has been charged with anything."[146]

No one was ever charged with anything, nor could they have been. Although the attorney general's examination (pointedly not an "investigation") of the situation ended publicly with a reaffirmation of McCleave's power to conduct the inquiry as he wished, privately the deputy attorney general informed the judge that his powers as a commissioner allowed him to compel only testimony and evidence under threat of contempt, and not to dictate anyone's speech or conduct inside or outside of the inquiry. In other words, while acting as a commissioner, he had not the powers of a judge.[147] McCleave, disappointed with "so little protection" offered him, retreated from public view, leaving environmentalists to debate the future of the inquiry among themselves without any hint of whether or when stage two would begin, or who might be called to account in court before then.[148]

Debate they did. Rather than unite against a commissioner whose own suspect political neutrality offered a clear chance to avoid the objectively scientific phase of the inquiry and return the debate fully to the political realm, the anti-uranium forces' internal divisions only deepened in response to McCleave's apparent breakdown. Susan Holtz did write to the attorney general on behalf of the EAC to lodge a formal complaint over the "curtailment of freedom of expression," but refused to join the chorus of calls for McCleave's ouster and replacement, advising instead a six-month recess.[149] Beyond the single complaint, in fact, those who had previously supported McCleave continued to do so. CAPE's Ralph Loomer even wrote again to the judge to "share [his] repugnance and distaste for the sleazy fanatics who would replace the institutions of democracy and jurisprudence of British tradition with a one-party dictatorship. Their strategy is intentionally disruptive and destructive of any worthy cause in which they participate."[150] Only slightly less divisive, but much more public, the EAC's Uranium Committee wrote in an open letter to the premier that "the recent series of criticisms and allegations in the House, the Courtroom, the media, and on the streets would seem to place the future of the Uranium Inquiry in jeopardy. Ecology Action Centre has not participated in these activities, nor does it intend to, because they are damaging to the process."[151] As their choice of language suggests, being seen to support the process meant more to the EAC than merely availing of another chance to reiterate its position on the science of radiation exposure. Its board of directors made it clear that the "Centre has been careful not to be entangled in any of the 'side shows' around the Inquiry and hopes to maintain its image as a responsible, credible organization whose attention remains focussed on the issues which generated the Inquiry."[152]

Those whose participation in any future inquiry stages was dubious at best maintained that a process without potential for organizing the public into a political pressure group could be no more than an "expensive pressure valve for public opinion" and a force for delegitimization of non-scientific positions.[153] They objected to any support given McCleave as a distortion of the truth – that he had attempted for a full year to stifle any dissent against his right to make the final decision – and they resented the refusal of CAPE and the EAC, as the best-known groups in the fight, to speak out against McCleave's attacks on leftist environmentalists. Only a little more than a year earlier, Susan Holtz had been bitterly upset at Sherri Cline and Hattie Perry for promoting "disunity" within the movement, and protested her toleration of political philosophies other than her own. Yet after McCleave's letter and his first special session made it clear to all that his anger was directed at "perverse" leftists, her concern was to protect the inquiry and the judge from criticism. A CAPE spokesperson explained to a *Mail-Star* reporter at the time that McCleave was upset at

FIGURE 5   A drawing of Judge McCleave and the EAC.
Source: Dawna Gallagher, in Don Rushton, "So Long, Stage II:
McCleave Gets Set to Ride Off into the Sunset," *New
Maritimes* 2, no. 6 (March 1984): 11.

"a few people isolated from the environmental movement [who] had it
out for him" from the beginning.[154] The northern groups, for their part,
renewed their criticisms of the EAC as "establishment bound," dependent
on government and corporate funding, and (along with CAPE) simply
politically naive.[155] (See Figure 5.)

Rather than the strength in diversity that some optimists had hoped
for, the stalemate between participants and non-participants in the inquiry
reflected the most basic differences in political assumptions. While Donna
Smyth lamented that personalizing the conflict with McCleave meant that
the "moderate, reasonable people who put such a lot of hard work into
Stage 1 are being ignored and forgotten,"[156] Don Rushton insisted that "the
discussion can never be completely separated from the forum in which it
takes place." And if the two camps willingly debated in the press, they

could not manage to discuss strategy in person. Rushton's attempt on behalf of the CCCC to convene a provincial meeting to discuss whether or not to pursue a boycott of the inquiry by all groups was not so much rebuffed as ignored by all of the leading groups other than SEPOHG, CARE (whose boycott could just as well be called a ban), and SSEPA.[157] Both sides had come to recognize that their positions were not only incompatible but at cross-purposes. One held that participation in and endorsement of the official system of decision making was the only effective way to convince political and industrial leaders of the truth of environmentalists' claims, the other that only political pressure from large numbers of citizens could overcome the established unity of political and industrial interests that the official system was designed to uphold, and that "playing the game" made it look like the problems were well in hand and therefore hindered the growth of political consciousness. Each side attacked the legitimacy of the other.

Conclusion

Nova Scotia's anti-uranium campaign was brief but spectacular, and was crucially significant to the formation of the provincial environmental movement. While it confirmed the lasting trends – the power of political pressure, the influence of global environmentalism on the provincial movement, the defence of home places as a motive, and the relevance of the modern/non-modern distinction in environmental arguments – it also impelled the makers of modernist arguments into closer cooperation with government and industry, and their non-modernist peers into their firmest yet rejection of the state's power to set the parameters of acceptable discussion.

The Royal Commission on Uranium Mining ended not with a bang or even a whimper but with a long shrug of confusion. Judge McCleave's disappearance from the public eye as an inquiry commissioner lasted a full year, and his reappearance in the winter of 1984 consisted of nothing more than the announcement of the premature end of the inquiry. The final report of the McCleave Inquiry was submitted to the Lieutenant Governor on January 30, 1985, nearly three years after it began (and about three and a half years after the moratorium began).[158] It was released to the public on March 19, and the reception among environmentalists revealed just how little had changed since March 1983. Despite McCleave's complete repudiation of most environmentalist claims about the dangers of radiation exposure, the impossibility of safe tailings disposal, and the inseparability of nuclear energy and nuclear weapons, the response was remarkably positive in Halifax and in Hants County.[159] The Buchanan government had accepted McCleave's recommendations to extend the moratorium until 1990 (on uranium only, releasing prospectors from the legal bind that prevented them from seeking other minerals in uranium claims) and erect a regulatory system based on "scientific assessment of risk and level of risk considered acceptable."[160] McCleave had also expressed a "personal preference" that any uranium mined in Nova Scotia be used in the province for electricity generation only, at least unless the tailings disposal could be proven safe, in which case export might be acceptable. Elizabeth May, Susan Holtz, and Donna Smyth insisted that the recommendations meant that uranium mining would never take place in Nova Scotia, despite warnings from others that the report could as easily be interpreted as an instruction to wait five years before declaring the tailings "safe" and carrying on with mining.[161] A regulatory system that made an honest assessment of the science, they believed, would find a safe nuclear energy system "of course impossible."[162]

Optimism aside, the close relationship between politics, the law, and the nuclear industry in Canada had certainly not changed during McCleave's absence from the scene. Even some of the faces remained the same. After a disastrous spill of radioactive water from a tailings pond at Key Lake, Saskatchewan, in the winter of 1984 – the same tailings system presented to McCleave as state-of-the-art – there was considerable work to be done to rehabilitate the image of uranium mining in Canada, work contributed to by the federal government's National Uranium Tailings Program and its newest recruit, Roy John.[163] And Kidd Creek's Nova Scotian lawyer, Ronald Pugsley, had filled his time since the hearings representing Dr. Leo Yaffe in the latter's libel suit against Donna Smyth. Smyth's trial was a nationally infamous bit of litigation in which only the author of the supposedly libellous text faced trial. The fact that the publisher was left out of the suit convinced activists that the accusation was meant only to silence and punish a particularly outspoken anti-uranium leader. Having waited two years for a trial (the same two years spent waiting on Judge McCleave), Dr. Smyth took two days to convince a jury that there was no risk to Dr. Yaffe's reputation in being described as "one of many 'experts' that the nuclear industry will parade in front of us."[164] A good deal of her success can be attributed to her own foresight in electing a jury trial, apparently having decided that, in her own case, faith in a judge's impartiality was probably not the wisest course. Indeed, the presiding judge seemed inclined to convict.[165] In any event, during the two years of waiting, Smyth was partially handicapped as a campaigner, unable to comment publicly on the case.

From the earliest days of the struggle in 1979 and 1980, Nova Scotian environmentalists had been fighting against not only (not even mainly) the mining industry but also the provincial and federal governments and their commitment to nuclear technology. They had been fighting, without always recognizing it, against a modern conception of economic development and risk assessment that devalued the personal experience of place and of home, discounted the inequity of metropolitanism, and preserved the privilege of expertise. In this, Nova Scotia serves as a lens through which to view the trends in Canadian (and worldwide) environmentalism. The Royal Commission on Uranium Mining, like the herbicide trial and the Royal Commission on Forestry, exemplified the use of judicial or quasi-judicial forums to dissipate activist energy and to enforce that same modernist discourse. McCleave may have been an unusually enthusiastic guardian of the traditional prerogatives of experts and duly constituted authority, adding to the value of the provincial story as an exemplar of environmentalist history, but it was not an accident of personality (his, or Orton's, or Holtz's) that the process of the inquiry completed the schism of the environmental movement in the province. None of them (not even McCleave) was a singular force of nature; they represented groups of like-minded activists or officials, who represent in turn the strands of environmental thought common throughout the international movement.

That the McCleave Inquiry ended when and how it did was almost entirely the work of the "radical" environmentalists he so despised, and given McCleave's repudiation of most of the arguments put to him by environmentalists, it would have been reasonable, in 1985 as it is now, to conclude that they were responsible for the moratorium that prevented uranium mining in Nova Scotia for years thereafter. They were responsible for turning the inquiry into a source of political pressure rather than a brake on it. But the mainstream groups in the province did not say so in 1985, because they could not accept the idea that conciliatory environmentalism had failed to do what it promised – namely, to gain a seat at the regulatory table. (So complete was his disdain for the activists' position that the government committee McCleave recommended be created to dictate safe standards had no room for even token public representation.[166]) The cost of victory against the uranium industry was an admission that such differences could no longer be bridged by blithe assurances of unity in the movement. Given the opportunity that McCleave provided, Nova

Scotian environmentalists first drifted then raced in opposite directions entirely under their own power, to the point that formal coalition was no longer possible. There was no longer an environmental movement in Nova Scotia; there were two. The rupture had been exhausting and embittering, and for most of the anti-uranium activists, McCleave's moratorium was victory enough.

5

Watermelons and Market Greens
Legacies of Early Activism

THE NUCLEAR ENERGY AND uranium mining industries profoundly disappointed the hopes of development promoters in the 1980s, as cost
overruns and soft international markets achieved what environmental
activists alone could not. Industrial development in general did not slow
appreciably, however, and Nova Scotia remained an exemplar of the activist trend, with new independent small groups springing up wherever great
enough numbers felt sufficiently threatened by new development, and with
small rural groups ready to offer more radical solutions to the downsides
of modernity.[1] By the middle of the 1980s, the division of Nova Scotian
environmentalism into mainstream and fringe was complete, but the
two streams were in frequent contact with each other. In Cumberland
County in 1981, for example, residents opposed to construction of a central
Maritime hazardous waste dump sought aid from the Ecology Action
Centre (EAC), but staked out their own, more radical position: if the experts could not be trusted, they said, the only morally consistent response
was to "learn to live without the products these companies manufacture."[2]
Nearly a decade later, Cape Bretoners belonging to the Save Boulanderie
Island Society attempted a court challenge with EAC support, after the
federal government cancelled an environmental assessment process for the
new Point Aconi coal-fired power station, but they were just as prepared
with threats and civil disobedience when the court ruled against them.[3]

Activist cooperation did not exist in a political vacuum or rely solely
on the capacity of mainstream and fringe to abide each other. The state
also maintained a role in their relationship, for better and for worse. Official

processes were often fruitless and divisive avenues of dissent, as those activists learned who tried to force the federal and provincial governments to clean up the residue of Sydney's steel industry, in Sydney Harbour and the tar ponds site. Despite air pollution up to 6,000 percent higher than the national standard, pollution in industrial Cape Breton had attracted little attention in the 1970s. As one resident recalled, "we just didn't want to hear, 'cause we wanted those damn jobs so bad."[4] It was, oddly, the Department of Fisheries that spurred the first major organized activist effort in 1982 when it closed the lobster fishery on the south side of the harbour, citing as justification the 735 pounds of phenol, 10,447 pounds of ammonium, 919 pounds of cyanide, and 2,058 pounds of thiocyanate dumped there every day by the Sydney Steel Corporation.[5] Once initiated, however, public activism accelerated rapidly. Faced with contamination of such magnitude – approximately 250 acres of tar ponds and former coke ovens sites, to a depth of 24 metres, arguably the largest concentration of toxic chemicals on the continent – there was no question that activists would target government directly, and groups of every sort contributed. Even the most determined opponent of government consultation could support direct lobbying for the relocation of residents away from the edges of the site (combined with a public pressure campaign, of course). The lessons of the 1970s were not lost on the environmental agencies of Ottawa and Halifax, however, and in order to contain the threat of radical/mainstream cooperation, the federal, provincial, and municipal governments created a Joint Action Group (JAG) to decide on clean-up methods. Ostensibly meant to include members of the public in the decision-making process, the JAG served equally well to focus discussion on the very questions – scientific versus experiential knowledge and the legitimacy of political positions – that put environmentalist participants at odds with each other. In the long run, the JAG proved little more than a tool to leach activist strength and a cover behind which politicians might hide.[6]

The tar ponds were the best known of the decade's controversies, but possibly the best illustration of the capacity for collaboration among environmental activists was the fight to prevent the demolition of Kelly's Mountain in Victoria County at the very end of the 1980s. In 1989, site preparation began for a "mega quarry" on the edge of St. Anne's Bay, a project that would have blasted out and crushed six million tons of granite annually to make gravel for sale to the United States. Local reaction was swift and typical: the Save Kelly's Mountain Society (SKMS) immediately began a campaign to have environmental assessment hearings, and sought help from the EAC resource centre.[7] It also reached out to allies among

Mi'kmaq traditionalists on Cape Breton Island, who objected to the presence of a quarry next to a sacred site at Gluscap's Cave. In the fight for Kelly's Mountain, Mi'kmaq activists made use of racial rhetoric in a way that Chief Ryan Googoo and the other Native participants in the herbicide trial had only begun to explore in the early 1980s. Traditionalists stood to gain both in their own communities and in the wider Canadian culture by promoting the image of an "Ecological Indian," whose unique cultural traditions gave him or her a privileged relationship with the natural world.[8] The SKMS, meanwhile, recognized the political power of the Native leaders' argument and did all they could to support it. Differences among environmentalists certainly did not disappear during the controversy, but they were contained by an agreement on rhetorical tactics and a refusal by all parties to be drawn into discussion of amelioration or remediation of environmental damage. As so often before, the provincial government was not very interested in permitting any obstruction to industrial development, but it was very difficult to ignore the combination of scientific arguments about noise, dust, toxicity, and fisheries from the SKMS, vandalism along the road to the quarry site by unknown agents, and promises by the Mi'kmaq Warrior Society to "wage war" in defence of the mountain. Though the "Battle of Kelly's Mountain" nearly became a genuine battle in the process, the activists got their hearings, preserved their common front, and, unlike the tar ponds clean-up advocates or the Point Aconi power plant opponents, emerged victorious.[9]

Successful collaboration on new issues was dependent on the ability of mainstream activists to refrain from attacking their allies in order to buttress their own legitimacy. That in turn relied on a measure of goodwill and a perception of common goals between the two groups. Stopping a certain project, be it power plant or quarry, was an appropriate common goal. Changing a broad policy was not. Activists concerned with industrial forestry in the 1980s had great difficulty agreeing on which policy to endorse and how to go about achieving it in the aftermath of their great failure at the herbicide trial.

While undeniably discouraged by the ignominious denouement of the trial settlement process in 1983, those activists who had led the legal effort remained determined to fight the use of herbicides in forestry (as well as the resurgent use of insecticides) by direct appeal and through the courts. In 1984, the EAC proposed a mandamus lawsuit against the provincial government, essentially an attempt to convince the Supreme Court of Nova Scotia that the Buchanan government, by refusing public hearings, had been negligent in its duty as a regulator. This "plan B," as it was called, carried less personal risk than a nuisance suit against the pulp companies, though everyone admitted that it might at best be considered a "stalling tactic."[10] Meanwhile, the EAC pursued a more cooperative relationship with the federal bureaucracy, applying for funding to produce reports on the risks of various pesticides or run conferences on "the ethical dimensions of risk."[11] Not surprisingly, Ottawa was reluctant to underwrite anything so public, but it was happy to offer support for the Atlantic Environmental Network or for Environment Canada's "stakeholder meetings," where the same issues would be discussed out of the public eye. Environment Canada was even happier to recruit Elizabeth May to work for the minister of environment as a liaison with the environmental movement and a distributor of funds. Minister Thomas McMillan's trust was well placed; May directed the bulk of funding to the Canadian Environmental Network (CEN), and not to any public advocacy efforts. Some activists even complained that she tried to stifle criticism of McMillan's Environmental Protection Act.[12]

A great many activists in Nova Scotia felt differently from May and the EAC about the efficacy of direct appeals to politicians' better natures, and very differently about the usefulness of further legal efforts. Premier John Buchanan and his ministers, who held the true decision-making power, had made it clear that they did not wish to listen to environmental activists and would do whatever possible to discourage them and quash dissent. As journalist Ralph Surette put it, "we may safely assume that if the government of Nova Scotia knew of hazards [from chemical spraying] it would not admit it."[13] The environmentalists' next step could only be to force them to listen by raising the number of voices, and voters, in the game. Even before the decision had been made to abandon an appeal of the herbicide trial verdict and settle with Stora Kopparberg, several groups in the province had soured on the search for legitimacy. A large coalition of groups planned a major publicity campaign in 1983, designed to embarrass the government at the peak of the tourist season with a coordinated, province-wide distribution of information on the "Dioxin Trail," a play on the province's promotional names for its coastal highways, such as the Sunrise Trail and Cabot Trail. Only days before the campaign was to begin, however, the EAC, Herbicide Fund Society, and Women's Health Education Network abandoned the plan and sent the printer's plates for the campaign leaflet, unused, to the Queens/Lunenburg South Shore Environmental Protection Association (Q/L SSEPA). "The only explanation we could get," wrote the latter, "was that those concerned in the pull-out thought the pamphlet was an unwise tactic." Activists on the south shore and north shore felt "anger and betrayal" but could only begin planning for the following year's campaign.[14] (Someone in Cumberland County did go ahead and erect the Dioxin Trail signs, but without the accompanying literature, it must have been a puzzling protest to the few who noticed it.[15])

The strongholds of radical activism were located on the province's north and southwest shores, as well as on Cape Breton Island. There, at some remove from Halifax, activists continued with their more populist style of advocacy without much assistance from the city. The Q/L SSEPA joined its north shore peers in the Scott Boycott Committee in late 1983, and the Concerned Residents of Clare and South West Environmental Protection Association organized public meetings on the continued use of 2,4-D in the province's roadside spray program. One prominent speaker for the two groups was a familiar face in anti-chemical circles: Robyn Warren.[16] On forestry matters, the rhetoric and action of the rural populists in the mid and late 1980s began to move towards overt challenges to government consultation. Charlie Restino and the Cape Breton Coalition Against Pesticides (CAP) declared that "more headway can be made by mobilizing public opinion" than by visiting politicians, and worked with David Orton and the North Shore Environmental Web (NSEW) on public appeals.[17] NSEW was even more forceful, uniting as it did many of the activists targeted by Judge Robert McCleave and many who had been dropped from Justice C. Denne Burchell's anti-herbicide injunction, who were overlooked in favour of the Cape Breton trial while the Scott Paper Company remained free to spray chemicals at will. Its members insisted that "it is usually foolish to look to governments (or the court system) to significantly redress environmental damage."[18] Diana Cole of the Concerned Residents of Clare group took such feelings one rhetorical step further. She had participated in one of Environment Canada's national environmental non-governmental organization (ENGO) conferences in 1983 and returned home describing the experience as an exercise in "totalitarian democracy," "empty rituals for the sake of the prestige and perpetuation of those in power," and an "exhausting and debilitating" distraction from real change. What was more, she suggested, this was exactly the intention of the agency, which "exists to preserve a repressive society and neutralize opposition." Activists from the north shore and Cape Breton applauded her statement.[19] Later in the decade, when the provincial government began drafting a new Pest Control Products Act and seeking responses from environmental groups, the harshest critics of consultation recalled the Department of Environment's testimony at the Royal Commission on Forestry in 1983, when its representatives spoke of the need for new legislation solely to raise "public confidence [and] reduce the level of public opposition to forest management programs involving the use of pest control products."[20] Rather than go along with the province's consultation as the mainstream groups did, unwilling groups followed the NSEW's lead in promoting an "informed consent or informed rejection" policy among municipal councils, which would have given local residents living within one kilometre of a woodlot a veto over chemical applications.[21] Ironically, despite refusing to participate in the formal consultation process, their challenge may have helped block Lands and Forests' attempt to exempt woodlands from regulation under the new act.[22]

Radical/mainstream conflict was never entirely limited to forestry debates. Since the early days of the uranium mining controversy, government funding of environmental groups had been a bone of contention, and the issue had not faded with the end of the Royal Commission on Uranium Mining hearings in 1983. When funding or political philosophy came up for discussion, sparks flew. When activists associated with the Sprayers of Dioxin Association in New Brunswick, a group representing victims of Agent Orange testing at CFB Gagetown, attempted to organize a Maritime Environmental Coalition in 1984 as a lobby group successor to the Maritime Energy Coalition, the response among more radical groups indicated just how far apart the factions had drifted. "We would not be involved in any group that accepts funds from those we're fighting," wrote Sherri Cline and Judy Davis of CARE, NSEW, and the Scott Boycott Committee, and "we prefer to put our energy and any funds into raising public awareness rather than into establishing a coalition oriented toward lobbying politicians or trying to match the opposition financially to fight on their terms through their channels."[23] Their critique targeted the EAC and its taste for legal tactics as much as any possible regional coalition. Unwilling to admit that the herbicide trial had been a disaster (Elizabeth May insisted it had prevented 2,4,5-T spraying), and dependent on government funding for its survival, the EAC drew heavy criticism from the radicals.[24] Diana Cole's reaction to the trial was as sharp as her critique of Environment Canada: it was, she said, a case of "poor leadership ... worse than no leadership at all because it lures the people to defeat in a dead end, making the failure appear as victory."[25] Rather than raise money for use in futile lawsuits, the radical activists preferred action, and in 1988 they blockaded roads to a woodlot in Colchester County for six weeks to prevent herbicide spraying there.[26]

Differences over funding were bound up with more profound disagreements about the relationship with government. The NSEW developed a written policy that "environmentalists should not work with corporations or government" on principle, and the more often instances of disagreement arose, the more clearly the radical activists articulated a philosophy that rejected economic growth, the primacy of scientific expertise, and homo-centric ethics. These were the articles of faith for modern society that they believed had led the mainstream down its mistaken path. Nor were they alone in staking out an ideological territory (though the radicals were usually more explicit in their rationale); the mainstream redoubled its own commitment to cooperation with government and industry throughout the 1980s. Nova Scotian environmentalists played as great a role in the promotion of the soft energy path (SEP) in the late 1980s as they had at the end of the 1970s. A 1988 update to the Friends of the Earth SEP study maintained the original document's insistence on the viability of the alternative energy scenario "under conditions of strong economic growth and substantial increases in material standards of living."[27] In 1984, and again in 1986, the province's flag bearer for SEP promotion, Susan Holtz, repeated the study's premise, that lifestyle change was a "primarily research question" and should be secondary (in environmentalist and government minds) to "a clear focus on environmental quality as the overriding value."[28] In doing so, she articulated what had become the basic assumption of the mainstream position: that it was possible to address environmental problems without significant or even noticeable change in the nature of society.

The logical extension of mainstream thinking about the minimal importance of social change was the expectation of government action. If environmental problems were "primarily the unintended effects of human activities and technical capabilities," then all that was required to solve them was to muster the political will needed to overrule those profiting from the status quo and simply "change our institutions to avoid or correct" the problems.[29] Pursuing such a goal meant launching ever more appeals to those in power, and sparing ever less attention for the public – not a new condition for the EAC and its national allies in the 1980s, but pursued with greater self-awareness than ever before. The EAC's board of directors openly described the centre's role in the early 1980s as changing "from one simply of creating environmental awareness to one of identifying the issues and proposing alternatives."[30] In 1988, they further specified the audience for their proposals: "We'll be expected by business, industry, labour, and government to act co-operatively in new forms for multistakeholder consultation on environmental and economic development."[31]

Governments, especially the federal government, did what they could to encourage mainstream environmentalists to cooperate, offering more opportunities for consultation with every passing year. The Canadian Council of Resource and Environment Ministers' National Task Force on Environment and Economy; the federal Energy Options Review; the Canadian Environmental Protection Act; the Environmental Assessment and Review Process; the Organisation for Economic Co-operation and Development (OECD) review of Canada's environment and energy policy; the Canadian Environmental Advisory Council; and of course the federally funded CEN and its regional parts – the list of round tables would have surpassed the dreams of twenty King Arthurs, though with rather less than a Camelot atmosphere of equality.[32] The flurry of consultations produced disappointing results: neither the federal gatherings nor their provincial counterparts (Voluntary Planning, the Nature Reserves Liaison Committee, and consultation on new legislation) ever diminished the speed with which new problems produced controversy at the local level. The EAC celebrated "politicians finally [realizing] ... that their constituents really care about the state of the environment," but at the same time found itself further away than ever from those constituents, who could not match the financial contributions on offer from Ottawa and Halifax.[33]

The not-so-secret ingredient that sweetened government consultation was participant funding. In 1981, the EAC admitted that it could not "exist financially solely on its membership," and it became progressively less self-conscious about reaching for other sources of money as the decade wore on.[34] Publication grants from Environment Canada supplemented payments for travel and for contribution to various consultative panels, encouraging further applications for carefully screened "public awareness lecture" grants, "job strategy" grants, student employment grants, and more.[35] Funding applications took up activist time (another reason to avoid topics likely to be rejected), but they represented a safer source of money in one significant way. After an audit by Revenue Canada following the herbicide trial, the EAC's directors became concerned with the security of the centre's tax-exempt charitable status and debated the legality of its actions: under the law, they found, charities were enjoined from attempting "to influence the policy making process [or] promote a change in the law," from lobbying politicians, and from demonstrating with the intent to embarrass or pressure a government.[36] Officially sanctioned processes were legitimate under their charitable status, offering an apparent route to influence without the legal liability. Thus, both the carrot and the stick encouraged mainstream environmentalists to pursue consultation and funding from government in the 1980s. The same activity saw more Nova Scotian environmentalists step onto a wider world stage as well, more often. Early participation by May and Holtz in a Canada-USA environmental council brought the EAC into global campaigns against whaling, ozone depletion, and transboundary pollution, especially acid rain. By the end of the 1980s, it was fully integrated into global networks with the World Wildlife Fund, Friends of the Earth, the Canada–United States Acid Rain Project, the New England Environmental Network, and the Fate of the Earth conferences, all of which increased its profile and reputation as the main Nova Scotian activist group, but none of which brought them any closer to those who remained in the province.[37]

In 1988, EAC coordinator Lois Corbett claimed that the environmental movement had matured from its antagonistic youth in the 1970s and now addressed itself to "two publics": the "grassroots supporters who may sometimes challenge the status quo, and bureaucrats and professionals whose interest may lie in a more mainstream image for the group."[38] Its radical critics, however, recognized that the EAC leaned far more towards the latter group, and deliberately so. "There are various tendencies which are essentially in contention for the soul of the movement," wrote Orton, and for those who would indeed challenge the status quo, "government lobbying or appeals to industry are seen as a waste of time" and a loss of power.[39] How could anyone celebrate concessions from above when they had so often turned out to be hollow? Activists who kept a close eye on the provincial government recognized the weakness of the "two publics" strategy at work in the EAC's cheerful announcement in 1986 that the environment minister had finally agreed to cease issuing permits for 2,4-D in the province, followed by a quiet resumption of its use on the power corporation's rights-of-way.[40] As Ian Sherman, one of the earliest opponents of Cape Breton insecticide spraying in the 1970s, wrote in the early years of the following decade:

> The last battle to stop chemical spraying back in 1976 was won *not* by environmental group pressure or media events but because all kinds of people

in communities throughout Cape Breton stood up to voice their opposition to chemical spraying. It was a door to door, face to face, grassroots petition campaign ... ultimately, only a similar grassroots movement of local people, especially those whose livelihoods depend directly on the forestry industry, will prevail to stop the spraying and regain control of local forest management.[41]

The myth of the neutral regulator held least sway over self-described "Greens," who insisted on the need for wholesale social change, beginning with the demise of industrial capitalism. A socialist strand in environmental thought was certainly nothing new. It had been around long enough to earn its proponents a nickname: "watermelons," green on the outside and red on the inside. In contrast to the mainstream's commitment to co-operation with government and industry, however, the fringe's antagonism towards both grew more intense in the 1980s. In 1981, Michael Clow argued that "it is very unlikely that an attempt to fuse a liberal understanding of capitalist society with an ecological awareness will be able to grapple with the dynamics of capitalism as a social system [that] requires growth to continue."[42] Six years later, after the herbicide trial, the uranium inquiry, and the associated activist schism, CAP's Charlie Restino observed in less academic terms that "when it comes to corporate profits, more is always better, even if it should mean the destruction of a forest ecosystem developed over thousands of years."[43] And there was nothing unintended about the destructive consequences of economic growth, he added.

Though the network of Greens centred on the north shore was determined to remain primarily concerned with local bioregional politics, they were equally committed internationalists and coalition builders, "part of a worldwide ecological movement."[44] The "informed consent or informed rejection" concept originating with the NSEW spread quickly into the rest of the world in 1987 and 1988, and the members of the NSEW and its successor organization, the Greenweb, participated in a truly global discussion around the theory and politics of "socialist biocentrism" or "left biocentrism."[45] In 1990, they joined as well in forming a new provincial coalition, the Nova Scotia Environment Alliance (NSEA), which made its philosophical foundations (and distinction from the mainstream) clear: the NSEA, they wrote,

> believes in the fundamental right of all living beings to clean air, clean water, and clean soil. The natural world has a right to exist as an entity unto itself,

independent of its utility to humans. We believe it is not possible to have infinite growth in a finite world. In questions of economic development versus the environment, the earth comes first.[46]

The NSEA statement was indeed a far cry from Holtz's complaint that talk of nature's rights "is a distraction from the main human issues facing us," but then it was meant to be.[47] The Greens were contemptuous of the "belief that the existing political and economic system can be reformed – if only the 'political will' can be summoned."[48] In their view, such attempts led to the kind of craven compromise that turned Elizabeth May and the EAC into advocates of *Btk* spraying and, by extension, defenders of the industrial forestry system.[49]

This conflict between the environmental mainstream and fringe was not, like so much else examined in previous chapters of this work, a factor in the shaping of the provincial movement. It was itself the shape of the movement, and would remain so for long after the events described herein. It should be noted that the factions did not persist in isolation. Cooperation was the norm, if not always the rule.[50] Ironically, the EAC and NSEW together comprised the entire Nova Scotian component of the International Uranium Congress in 1988, for example.[51] Theoretical differences might be left off the agenda in mixed company, yet the differences between the two sides were so profound that never could anyone speak honestly about a single movement, and never could any attempt at collaboration pass without debate over consultation versus confrontation – the tactics of the expert or of the citizen – as the proper stance towards those in power.

CONCLUSION

The initial motive for investigating the history of environmentalism in Nova Scotia – one province among ten possible subjects and a great many more possible combinations thereof – was unabashedly grounded in present-day observation of an environmental movement seemingly uncomfortable with acknowledging the clear divisions within itself: between, for example, those who would resist the further industrialization of the rural countryside and those who believe that the stabilization of the global climate depends somehow on the construction of wind turbine arrays. It was obvious early in the process that the variety and geographic distribution of environmental organizations in Nova Scotia warranted the focus on a single province; the wider the scope of investigation, the more small groups would necessarily be overlooked or omitted in the final account, and those small groups clearly showed the greatest vitality in the movement. They also presented the greatest contrast with the peripatetic, professional global environmentalism that forms such an insistent public face of environmental concern in the early twenty-first century. That contrast provided the central fact of the story and came to explain most clearly both the present-day divisions within environmentalism and the history of their development: environmental activism in Nova Scotia has always been essentially local in its inception, in its organization, and in its reason for being. The modernization of one segment of the movement marks the greatest change in its history. Each issue examined in this book has become the account of a different phase or aspect of the struggle between the place-loving soul of the movement and the temptations of a placeless modernity.

The essentially local nature of environmental activism is concealed not only by its rejection by some present-day modernist environmentalists but also by the direction of earlier activism towards the appropriate jurisdiction, which was not often the local municipal level. The Canadian Constitution places responsibility for natural resources at the provincial level, and activists attempting to influence the course of environmental policy most often cooperated on that level, forming issue-based coalitions or common fronts like CAP or the Uranium Committee of the EAC. Extraordinary cooperation among provinces can and did pull activist cooperation towards the same level, as when the Maritime Energy Corporation spawned the Maritime Energy Coalition. Once the immediate exigency had passed, however, the activist MEC dispersed back into mainly provincial networks. Federal-level coalitions proved more lasting, but those too simply followed jurisdiction; responsibility for atomic energy regulation or for the negotiation of international agreements dealing with acid rain belonged to Ottawa, for example, and the Canadian Coalition for Nuclear Responsibility or Canadian Coalition on Acid Rain formed to address that responsibility.

The nature of state power in action also goes some way towards concealing the local essence of environmentalism. One of the implicit arguments of the preceding chapters is that in situations where novelty or lack of firm direction from above allows bureaucrats to act on their differences of opinion, state actors may come to appear as issue-based activists first and guardians of bureaucratic prerogative second. This occurred with some professional foresters and federal energy bureaucrats in the late 1970s. But those liberties can be swiftly swept away by determined leadership, as Premier Buchanan demonstrated upon taking power in Nova Scotia in 1978.

In the final analysis, state power and state modernism are constant formative factors in environmental activism. The model of state/industry/activist relations presented in this book, including the centrality of the local experience, is entirely consistent with the history of other Canadian environmentalisms and might be profitably applied to more of them.[52]

Environmentalist cooperation at the higher levels of geographic integration – the provincial and federal levels – was sustained by the political structure of the Canadian state, but it never diminished the centrality or energy of the local. Every Nova Scotian group discussed in this book had its roots in a local reaction to environmental harm, either actual or prospective. Even the EAC, eventually the staunchest champion of modernist environmentalism, began as a small group at Dalhousie University concerned with recycling, urban air quality and sanitation, and municipal planning, local issues that earned it enough support among Haligonians to survive the end of full government funding in 1974. It is worth noting as well that localism is not a quality exclusive to environmental activist organizations; in contrast to the US experience, where unitary national groups have existed from the early days of many social movements, Canadian social movement organizations as a rule remained subnational in the 1970s. Logistical difficulties that hobbled communication – "no electronic mail and a costly long-distance telephone service" – have been singled out as an explanation, along with unequal access to financial resources across regions, leading to imbalance and resentment against the heartland of activism in Ontario. This is not obviously false, but it does take national organization as the norm, its absence to be explained.[53] As suggested above, Nova Scotia's environmentalist history demonstrates the importance of local, lived reality as a driver of activist recruitment. It is likely that social movement activism of any type that questions systems of modernity is inherently and normally local, and other explanations (from political culture and legal history, for instance) should be provided to account for national or provincial unity, rather than for its absence.

More than other kinds of activists, environmentalists have a particular attachment to the local level, due to their movement's origin in a reaction to industrial modernity's destructive effects. The modernist ethos is "a strategy of conceptual encompassment" of identities defined in relation to specific places and communities.[54] That is, it forces people to speak and think in the language of a universally applied scientific and technological idiom that makes any place interchangeable with any other. Those standing against it and defending specific places intuitively understand that an assertion of the irreducible local experience must necessarily end up in "some kind of boundary-drawing between the local and the global," implying a challenge to the combined forces of modernity: global capitalism, industrialism, normal science, and the rules of governmentality.[55] The story of Nova Scotian environmentalism set out in the previous chapters offers clear proof of the continuity of non-modern argument: from the south shore fishermen who in opposing the Stoddard Island nuclear plant defined their community as a fishing community and defied the language of cost-benefit analysis that suggested they might more profitably become a high-technology community, to the stubborn insistence of woodlot owners that local industry was preferable to multinational pulp and paper forestry and not readily compatible with it, to the many arguments presented against uranium mining in 1982 (and aggregate quarrying a decade later) that relied not on science alone but also on the meaningfulness of the provincial environment and the idea that a project might be legitimately resisted in order that a region or a mountain might continue to "be seen to be beautiful" or sacred.[56] The argument only gains from the move towards explicitly formulated localism by the most radical groups in the 1980s, such as the Greenweb's "informed consent or informed rejection" proposal, a profoundly subversive doctrine of extreme local democracy that effectively attacked every one of the modernist systems listed above and repudiated the image of environmentalist as expert manager.

Initially, in the early 1970s, activists with more and less radical views worked side by side in the same groups, defending harbours from destructive development, using rhetorical ammunition based on the language of environmental justice, biological science, traditional moral economies, precautionary principles, economic theory, and violent resistance. The non-modernist strand within early activism could not be eradicated, and finding a way to separate its most committed advocates from their more modernist peers was therefore essential for governments wishing to defuse the challenge. Eventually, they did just that: by the end of the story related in these chapters, the movement had failed decisively as a challenge to modernity. Activists in general never achieved the power to subvert modernity itself. Rather, large segments of the movement, large enough to define themselves as its mainstream, were subsumed into the modern, seeking expert status and accepting the sanctity of economic growth, capitalism, industry, science, and political liberalism. Government and industry cooperated in supporting the activists most likely to form a mainstream, and vilifying those who resisted. Consultation and red-baiting were the two most common tactics used to achieve those ends in the 1970s, but they were not alone, and the process continues in the twenty-first century

with wilderness areas and "green energy" megaprojects to tempt the bid-
dable activists and charges of "ecoterrorism" to marginalize the rest. Since
the end of the 1970s, the radical remainder relegated to fringe status has
enjoyed only enough power to mount occasionally successful challenges
under the most favourable circumstances, as the Mi'kmaq Warrior Society did by capitalizing on the popular notion of Native people's ecological
spirituality in order to advance a non-modern sanctity argument in defence
of Kelly's Mountain in 1990.

To explain this difference between the mainstream and the radical, the
balance of evidence from the history of Nova Scotian activism suggests that
urban activists, closer to the physical seat of bureaucratic power as well as
to the sources of funding in government and major industry, more readily
adopted modernist assumptions (or more often came with such assumptions already in place). Possibly as well, activists whose first campaigns
deal with urban planning and sanitation, areas of interest in which some
form of managerial control is unavoidable and debate centres on which
planning regime is best, are on track towards a universal, modernist sort
of activism from the beginning. That the same groups tend to be populated
by middle-class professionals, especially academics, could be a piece of
demographic coincidence owing to their concentration in urban environ-
ments, but for the fact that the same class of activists has displayed at every
location a greater level of comfort with the jargon of expertise as well as
with the bureaucratic system in which facility with that language functions
as a passport to greater access. For example, the relationship between the
academic Acadia Nuclear Power Study Group and the much more working-
class South Shore Environmental Protection Association – the former
abstractly scientific and the latter viscerally political (and successful) – finds
a parallel three chapters and nearly twenty years later in the tension between
the Save Kelly's Mountain Society and the Mi'kmaq Warrior Society.
Out-of-province origins account for no apparent difference in this regard
(though there does appear to have been a larger proportion of immigrants
among activists than in the general population), and neither does gender
(with no clear overrepresentation of either men or women in the movement): for every Susan Holtz, there was a Ralph Loomer, immigrant and
native modernists, respectively, and for every David Orton a Judy Davis,
their radical equivalents.[57] The geography and demography of dissent by
which those closer to the city and higher on the economic scale are faster
to abandon antagonism for conciliation is a tendency, not a rule, but as an
exemplar few can match Elizabeth May, whose move from Cape Breton

to Halifax and enrolment in law school in 1980 coincided with her shift in tactical choices.

The solidification of the mainstream's dominance required a rewriting of the history of the movement in the late 1980s – thus May's insistence that the herbicide trial had, by forestalling aerial spraying, prevented the use of Agent Orange in Nova Scotia, or that the better natures of Gerald Regan and Vince MacLean had been awakened in time to prevent insecticide spraying in the 1970s;[58] thus, as well, the suggestion that Judge McCleave's recommendations in 1985 constituted a rejection of uranium mining on terms articulated by environmentalists. In truth, however, the victories of environmental activists in Nova Scotia were invariably built on pressure politics and appeals to the interests of the local area, quite contrary to modernist claims to represent a wider society or expert perspective. The fishermen of the Yarmouth and Shelburne areas who cowed Premier Regan into shelving the Stoddard Island plan explicitly rejected the blithe assurances of the federal fisheries scientists and atomic energy experts who cooperated in the Canadian government's nuclear boosterism. The budworm battlers of Cape Breton Island knew enough to present themselves as the defenders of small lowland woodlot owners against the depredations of a foreign multinational, Stora. And McCleave's antagonists during the Royal Commission on Uranium Mining saw clearly the fallacy of public consultation and the utility of political embarrassment. Every one of them sought to mobilize the public en masse as well, and made frequent reference to the failure of the government in Halifax to uphold the interests of the hinterland.

The radicals (for that is what an insistence on non-modern localism made them) did not operate alone at any phase in their development; the organizational resources of the mainstream or those who would later form it created the conditions in which pressure politics could function, and armed all segments of the movement with the most compelling scientific arguments. Indeed, the radical approach did not ever function without a more rhetorically respectable counterpart, often voiced by the same people. The two strands of the movement diverged but remained intertwined, even within individuals. Every anti-nuclear campaigner in the province leaned gratefully on the research of the Acadia NPSG for instance, every anti-uranium campaigner on reports of the British Columbian experience circulated by the EAC. The Mi'kmaq Warrior Society readily acknowledged the value of its partnership with the research-heavy advocacy of the SKMS. May herself, who helped lead the greatest political pressure campaign in this story during the budworm spray controversy, frequently returned to praising the virtues of populist activism in later years, even as she pursued the lobbyist style. It was when the modernist tendency within the movement completely lost sight of its topophilic roots and contrived to act alone, excluding the radicals or softening the impact of their mass movement politics, that environmental activism failed. The herbicide trial, which compelled submission to the rules and procedures and evidentiary standards of the court, demonstrated the impotence of the method. So did the later Sydney tar ponds battle with its duly constituted but ineffective consultative bodies, and the fight over Point Aconi's power station, which brought yet another doomed legal challenge to court.

A lasting environmental movement in Nova Scotia began as an undifferentiated mixture of modern and non-modern activists. Conservationists, planners, and scientists worked alongside back-to-the-landers, student radicals, and the defenders of traditional economies. An eventual organizational distinction was predictable and probably innocuous, but the imbalance of political influence was not; it required a combination of encouragement for the modernists from two levels of government that wished to marginalize the radicals, willing participation by the modernists in the marginalization of their allies, and a global modernist environmentalism for the chosen few to participate in. What the creation of a mainstream could not do, however, was abolish the entirely local experience of environmental harm. Few environmentalists in the 1980s came to their activism out of concern for issues in the abstract; rather, most environmentalism in Nova Scotia remained an expression of love of place and a determination to defend it.

Notes

FOREWORD: ENVIRONMENTAL ACTION AND THE QUESTION OF SCALE

1 For Dubos's use of the term in 1972, at the United Nations Conference on the Human Environment in Stockholm, see Ralph Keyes, *The Quote Verifier: Who Said What, Where, and When* (New York: St Martin's Griffin, 2006), 79.

2 Denis E. Cosgrove, *Apollo's Eye: A Cartographic Genealogy of the Earth in the Western Imagination* (Baltimore and London: Johns Hopkins University Press, 2001); R. Buckminster Fuller, *An Operating Manual for Spaceship Earth* (Washington, DC: n.p., 1967); Kenneth E. Boulding, "The Economics of the Coming Spaceship Earth," in H. Jarrett, ed., *Environmental Quality in a Growing Economy: Essays from the Sixth RFF Forum* (Baltimore: Johns Hopkins University Press, 1966), 3–14.

3 John Muir, *My First Summer in the Sierra* (Boston: Houghton Mifflin, 1911; San Francisco: Sierra Club Books, 1988), 110 (citations are to the 1988 edition); Aldo Leopold, *A Sand County Almanac, and Sketches Here and There* (New York: Oxford University Press, 1949); Rachel Carson, *Silent Spring* (Boston: Houghton Mifflin, 1962); Paul R. Ehrlich, *The Population Bomb* (New York: Ballantine Books, 1968); Donnella H. Meadows, Dennis L. Meadows, Jørgen Randers, and William W. Behrens III, *The Limits to Growth* (New York: Universe Books, 1972); Andrew Kirk, *Counterculture Green* (Lawrence: University Press of Kansas, 2007). Ursula K. Heise offers a smart, wide-ranging, and important discussion of the relations between ways of imagining the global and ethical commitments to the local (upon which these opening paragraphs broadly rest) from the perspective of literary ecocriticism in *Sense of Place and Sense of Planet: The Environmental Imagination of the Global* (New York: Oxford University Press, 2008), where she pays particular attention to global-scale representations of ecological crisis.

4 Scott Russell Sanders, *Staying Put: Making a Home in a Restless World* (Boston: Beacon Press, 1993), xvi (cited by Heise, *Sense of Place*, 38).

5 Yi-fu Tuan, *Topophilia: A Study of Environmental Perception, Attitudes and Values* (Englewood Cliffs, NJ: Prentice Hall, 1974), 4. At the University of Toronto, Edward C. (Ted) Relph

submitted a PhD dissertation titled "The Phenomenon of Place" in 1973. Revised and published as *Place and Placelessness* (London: Pion, 1976), it became a "classic of human geography" (see *Progress in Human Geography* 24, 4 [2000]). Relph recalls that his search of library card catalogues ca. 1970 revealed that almost nothing had been written about place (see his website, Placeness, Place, Placelessness, http://www.placeness.com). In his discussion of Tuan's influential book on his website, Relph points out that both W.H. Auden and Gaston Bachelard had earlier used variants of the term.

6 Paul Shepard, "Place in American Culture," *North American Review* 262, 3 (Fall 1977): 32; Wendell Berry, *A Continuous Harmony: Essays Cultural and Agricultural* (New York: Harcourt Brace Jovanovich, 1972), 68–69; Sanders, *Staying Put*, xiii–xv (Shepard, Berry, and Sanders are cited in Heise, *A Sense of Place*, 29–31); Robert Fulford and John Sewell, *A Sense of Time and Place* (Toronto: City Pamphlets, [1971]); Northrop Frye, "Conclusion," in *Literary History of Canada: Canadian Literature in English*, ed. Carl F. Klinck (Toronto: University of Toronto Press, 1965); Neil Evernden, "Beyond Ecology: Self, Place and the Pathetic Fallacy," in *The Ecocriticism Reader: Landmarks in Literary Ecology*, ed. Cheryll Glotfelty and Harold Fromm (Athens: University of Georgia Press, 1996), 92–104; Philip Buckner, "'Limited Identities' Revisited: Regionalism and Nationalism in Canadian History," *Acadiensis* 30, 1 (Autumn 2000): 4–15, and J.M.S. Careless, "'Limited Identities' in Canada," *Canadian Historical Review* 50, 1 (March 1969): 1–10.

7 This paragraph distills much of the substance of Heise, *Sense of Place*, ch. 1, "From the Blue Planet to Google Earth: Environmentalism, Ecocriticism, and the Imagination of the Global." See also Arne Naess, "Identification as a Source of Deep Ecological Attitudes," in *Deep Ecology*, ed. Michael Tobias (San Diego: Avant Books, 1985), 268 (and in Heise, *Sense of Place*, 34).

8 T.W. Acheson, "The National Policy and the Industrialization of the Maritimes, 1880–1910," *Acadiensis* 1, 2 (1972): 3–28.

9 *Evidences of the Industrial Ascendancy of Nova Scotia* (Halifax: Canadian Manufacturers' Association, Publicity Committee of the Nova Scotia Branch, 1913).

10 E.R. Forbes, "Misguided Symmetry: The Destruction of a Regional Transportation Policy for the Maritimes," in *Canada and the Burden of Unity*, ed. D.J. Bercuson (Toronto: Macmillan of Canada, 1977), 60–86; and E.R. Forbes, *The Maritime Rights Movement, 1919–1927* (Montreal and Kingston: McGill-Queen's University Press, 1979). But see K. Cruikshank, "The Intercolonial Railway, Freight Rates and the Maritime Economy," *Acadiensis* 22, 1 (Autumn 1992): 87–110 for a different view.

11 Roy E. George, *A Leader and a Laggard: Manufacturing Industry in Nova Scotia, Quebec and Ontario* (Toronto: University of Toronto Press, 1970). In 1957, the federal government of Canada introduced a fiscal equalization program to achieve a national standard in public services. Tax dollars collected by the federal government were redistributed according to need, to enable poorer provinces to provide a minimum level of health care, social programs, and educational services. This did little to support regional growth beyond the service sectors; to this end, beginning in 1960, firms and regions were offered a sometimes bewildering array of tax incentives and assistance from special programs identified by an alphabet soup of acronyms.

12 These developments are well summarized in Donald J. Savoie, "Reviewing Canada's Regional Development Efforts," in [Newfoundland] *Royal Commission on Renewing and Strengthening Our Place in Canada* (St. John's: Queen's Printer, 2003), 147–83. For more details, see Donald J. Savoie, *Regional Economic Expansion: Canada's Search for Solutions* (Toronto:

University of Toronto Press, 1992), and Ralph Matthews, *The Creation of Regional Dependency* (Toronto: University of Toronto Press, 1983). In 1981, federal transfers (of all types, including unemployment insurance), counted for more than half of provincial revenues in all four Atlantic provinces. For later reflection on the political economy of regional development, see Donald J. Savoie, *Visiting Grandchildren: Economic Development in the Maritimes* (Toronto: University of Toronto Press, 2006).

13 Jenny Higgins and Melanie Martin, "The Come by Chance Oil Refinery," http://www.heritage.nf.ca/articles/politics/come-by-chance.php; H.A. Fredericks and Allan Chambers, *Bricklin* (Fredericton: New Brunswick Books, 1977). See also the equally notorious but earlier failure of Clairtone in Stellarton in 1967, attributed in an internal company report to the failings of the workforce, but the consequence of many influences, including poor infrastructure, overreach, and hubris: Nina Munk and Rachel Gotlieb, *The Art of Clairtone: The Making of a Design Icon, 1958–1971* (Toronto: McClelland and Stewart, 2008).

14 For resettlement schemes, see George Withers, "Resettlement of Newfoundland Inshore Fishing Communities, 1954–1972: A High Modernist Project" (MA thesis, Memorial University of Newfoundland, 2009); for the Comprehensive Development Plan, see Wade MacLauchlan, *Alex B. Campbell: The Prince Edward Island Premier Who Rocked the Cradle* (Charlottetown: Prince Edward Island Museum and Heritage Foundation, 2014).

15 Carson, *Silent Spring;* Ryan O'Connor, *The First Green Wave: Pollution Probe and the Origins of Environmental Activism in Ontario* (Vancouver: UBC Press, 2014); Mark J. MacLaughlin, "Rise of the Eco-Comics: The State, Environmental Education, and Canadian Comic Books, 1971–1975," *Material Culture Review* 77/78 (Spring/Fall 2013 [August 2014]): 9–20. Alan A. MacEachern, *The Institute of Man and Resources: An Environmental Fable* (Charlottetown: Island Studies Press, 2003); E.F. Schumacher, *Small Is Beautiful: A Study of Economics as if People Mattered* (New York: Harper and Row, 1973).

16 The phrase is adapted from the opening lines of William Wordsworth's poem called forth in opposition to the proposed Kendal and Windermere Railway in the English Lake District in 1844; in the original, it reads: "Is then no nook of English ground secure / From rash assault?" It was used in the title of James Winter, *Secure from Rash Assault: Sustaining the Victorian Environment* (Berkeley and London: University of California Press, 1999).

17 Garth Jowett and Victoria O'Donnell, *Propaganda and Persuasion,* 4th ed. (Thousand Oaks, CA: Sage Publications, 2006), 7.

18 See James C. Scott, *Weapons of the Weak: Everyday Forms of Peasant Resistance* (New Haven, CT: Yale University Press, 1985).

19 These facets of Royal Commission proceedings are well discussed in Molly Clarkson, "Speaking for Sockeye, Speaking for Themselves: First Nations Engagement in the Cohen Commission (2009–2012)" (MA thesis, University of British Columbia, 2016), ch. 2.

20 Frank S. Zelko, *Make It a Green Peace! The Rise of Countercultural Environmentalism* (New York: Oxford University Press, 2013); Robert Wilson, "Making Tracks," Seeing the Woods: A Blog by the Rachel Carson Center, August 25, 2016, https://seeingthewoods.org/2016/08/25/making-tracks-robert-wilson/#more-4668; Justin Page, *Tracking the Great Bear: How Environmentalists Recreated British Columbia's Coastal Rainforest* (Vancouver: UBC Press, 2014); Juliet Eilperin and Steven Mufson, "Activists Arrested at White House Protesting Keystone Pipeline," *Washington Post,* February 13, 2013.

21 "Judith Davis," http://www.inmemoriam.ca/view-announcement-198741-judith-davis.html.

22 David Orton, "Judy Davis – Portrait of an Activist," http://home.ca.inter.net/%7Egreenweb/Judy_Davis.pdf.

23 This discussion draws from Orton's own accounts of his life, available at Deep Green Web, http://deepgreenweb.blogspot.com.

24 David Orton, "The McCleave Uranium Inquiry in Nova Scotia," *Atlantic Socialist* 1, 2 (December 1982), http://home.ca.inter.net/%7Egreenweb/Uranium_Inquiry_NS.html, where items on other topics mentioned may also be found.

25 Mike Allen, "Portsmouth-Born Canadian Activist Reflects on Life," *The News* (Portsmouth), November 25, 1994, http://home.ca.inter.net/~greenweb/David_Orton_reflects_on_life.pdf.

26 David Orton, "Left Biocentrism," in *Encyclopedia of Religion and Nature*, vol. 2, *K–Z* (Bristol, UK: Thoemmes Continuum, 2005), 1003–5.

27 Orton's writings are collected at Green Web, http://home.ca.inter.net/~greenweb.

28 David Orton, "Climate Change Pollyannas," a review essay on Elizabeth May and Zoë Caron, *Global Warming for Dummies* (Mississauga, ON: John Wiley and Sons Canada, 2009); see in a similar vein David Orton's review of L. Anders Sandberg, *Trouble in the Woods: Forest Policy and Social Conflict in Nova Scotia and New Brunswick* (Fredericton: Acadiensis Press, 1992), both available at Green Web, http://home.ca.inter.net/~greenweb.

29 Orton, "Climate Change Pollyannas."

30 *Between the Issues* (October 1980), https://dalspace.library.dal.ca/xmlui/bitstream/handle/10222/14780/MS-11-13_31264026410238.pdf?sequence=1&isAllowed=y.

31 Amory B. Lovins, "Energy Strategy: The Road Not Taken?" *Foreign Affairs* (October 1976): 65–96. Many years later, Holtz's brief resume noted that she was "the founding Vice Chair of both the Nova Scotia and the now-disbanded National Round Table on the Environment and the Economy," and that she had served on numerous other advisory bodies and panels. In a 2013 article titled "Redirecting Anti-Wind Energy," the fundamental philosophical difference between her position and those of more radical environmentalists committed to oppositional politics (between Leeming's ecomodernist and non-modernist groups) was starkly exposed in her assertion that "some public stands are simply characterized as broad – and predictable – knee-jerk negativity: CAVE (Citizens Against Virtually Everything); BANANA (Build Absolutely Nothing Anywhere Near Anything); and NOPE (Not On Planet Earth)." In Holtz's mature view, adversarial processes, including Canada's legal and political systems, served to block progress on contentious public issues: Susan Holtz, "Redirecting Anti-Wind Energy: Individuals, Communities and Politicians Can Turn a Debate Stalemate into an Opportunity for Collaboration," *Alternatives Journal* 39, 5 (September/October 2013): 44–47.

32 Richard Grant, "Paul Watson: Sea Shepherd Eco-Warrior Fighting to Stop Whaling and Seal Hunts," *Daily Telegraph* (London), April 17, 2009; e.g., Mark S. Winfield, *Blue-Green Province: The Environment and the Political Economy of Ontario* (Vancouver: UBC Press, 2011).

33 Harald Welzer, *Climate Wars: What People Will Be Killed for in the 21st Century* (Cambridge: Polity Press, 2012); Michael F. Maniates, "Individualization: Plant a Tree, Buy a Bike, Save the World?" *Global Environmental Politics* 1, 3 (August 2001): 31–52.

34 Heise, *Sense of Place*, 21.

35 David Harvey, *The Condition of Post Modernity: An Enquiry into the Origins of Cultural Change* (Oxford, UK: Blackwell, 1989), 240.

36 Doreen Massey, *Space, Place and Gender* (Minneapolis: University of Minnesota Press, 1994), 147. For Orton, see Allen, "Portsmouth-Born."

37 Doreen Massey, "A Global Sense of Place," *Marxism Today* 38 (1991): 24–29, quote from p. 28; Tim Cresswell, *Geographic Thought: A Critical Introduction* (Chichester: Wiley

Blackwell, 2013), Chapter 11; Tim Cresswell, "Place," in John A. Agnew and James S. Duncan, eds., *The Wiley Blackwell Companion to Human Geography* (Chichester, Sussex: John Wiley and Sons, 2016), 238. See also Jamie Peck and Adam Tickell, "Neoliberalizing Space," *Antipode* 34 (2002): 380–404 and "Doreen Massey on Space," Social Science Bites at: http://www.socialsciencespace.com/2013/02/podcastdoreen-massey-on-space/.

INTRODUCTION

1 Ronald Inglehart, *The Silent Revolution: Changing Values and Political Styles among Western Publics* (Princeton, NJ: Princeton University Press, 1977). This is not actually a very new idea of environmentalism, especially if one adheres to a broad definition including the Romantic movement; it echoes what Aldous Huxley wrote in the essay "Wordsworth in the Tropics," in *Do What You Will* (London: Chatto and Windus, 1956 [1929]). See also Christopher Rootes, ed., *Environmental Movements: Local, National, and Global* (Portland, OR: Frank Cass, 1999); and Samuel Hays, *Beauty, Health, and Permanence: Environmental Politics in the United States, 1955–1985* (Cambridge: Cambridge University Press, 1987), 3.

2 Ramachandra Guha, *How Much Should a Person Consume? Environmentalism in India and the United States* (Los Angeles: University of California Press, 2006), 8. Guha credits the definition to G.M. Trevelyan in the 1931 Rickman Godlee Lecture, titled "The Calls and Claims of Natural Beauty." Similar ideas appear in Mahesh Rangarajan, *Fencing the Forest: Conservation and Ecological Change in India's Central Provinces 1860–1914* (Delhi: Oxford University Press, 1996); E.P. Thompson, *Customs in Common* (London: Merlin Press, 1991); Ramachandra Guha, *The Unquiet Woods: Ecological Change and Peasant Resistance in the Himalaya*, 2nd ed. (Oxford: Oxford University Press, 1989); Juan Martinez-Alier, *Ecological Economics: Energy, Environment, and Society* (Oxford: Basil Blackwell, 1990); Juan Martinez-Alier, *The Environmentalism of the Poor: A Study of Ecological Conflicts and Valuation* (Northhampton, UK: Edward Elgar, 2002); Guha and Martinez-Alier, *Varieties of Environmentalism: Essays North and South* (London: Earthscan, 1997); Jürgen Habermas, "New Social Movements," *Telos* 1981, no. 49 (1981): 33–37. Also, Nick Crossley and Jürgen Habermas, *Making Sense of Social Movements* (Buckingham, UK: Open University Press, 2002).

   Unfortunately, much of the international history remains trapped in the post-colonialists' jaundiced view of the global North: with the exception of the environmental struggles of a racial underclass, post-materialist notions of privileged "amenity" or "full-stomach" environmentalism dominate the view of activists in Europe and North America. See, for example, Ramachandra Guha, "Radical American Environmentalism and Wilderness Preservation: A Third World Critique," *Environmental Ethics* 11, no. 1 (1989): 71–83.

3 Andrew Dobson, *Green Political Thought* (London: Routledge, 2000 [1990]), 8; Alf Hornborg, "Environmentalism, Ethnicity, and Sacred Places: Reflections on Modernity, Discourse and Power," *Canadian Review of Sociology and Anthropology* 31, no. 3 (1994): 258; Alf Hornborg, "Undermining Modernity: Protecting Landscapes and Meanings among the Mi'kmaq of Nova Scotia," in *Political Ecology across Spaces, Scales, and Social Groups*, ed. Susan Paulson and Lisa Gezon (New Brunswick, NJ: Rutgers University Press, 2004), 196–216; Anthony Giddens, *The Consequences of Modernity* (Stanford, CA: Stanford University Press, 1990), 4–6; Bruno Latour, *We Have Never Been Modern* (Cambridge, MA: Harvard University Press, 1993 [1991]). The idea owes much to an earlier theorist of similar ideas: Thomas Kuhn, *The Structure of Scientific Revolutions* (Chicago: University of

Chicago Press, 1962). Similar ideas can be found in the work of Arne Naess and David Pepper: Naess, *The Ecology of Wisdom: Writings by Arne Naess* (Berkeley: Counterpoint, 2008); Pepper, *Modern Environmentalism* (London: Routledge, 1996).

4 This is especially true of neighbouring jurisdictions in New Brunswick and Prince Edward Island, where similar debates over nuclear power or chemical forestry, for example, were conducted contemporaneously with Nova Scotia's controversies. The intra-regional differences of political economy and political culture that justify treating Nova Scotia apart from its Maritime peers, even when their environmental movements joined forces as they did over nuclear power, are readily apparent in comparing the present study with those presenting histories of nuclear power, other energy politics, or forestry in New Brunswick and Prince Edward Island. See, for example, Adrian Egbers, "Going Nuclear: The Origins of New Brunswick's Nuclear Industry, 1950–1983" (MA thesis, Dalhousie University, 2008); James Kenny and Andrew Secord, "Engineering Modernity: Hydroelectric Development in New Brunswick, 1945–1970," *Acadiensis* 39, no. 1 (Winter/Spring 2010): 3–26; Alan MacEachern, *The Institute of Man and Resources: An Environmental Fable* (Charlottetown: Island Studies Press, 2003); and Mark McLaughlin, "Green Shoots: Aerial Insecticide Spraying and the Growth of Environmental Consciousness in New Brunswick, 1952–1973," *Acadiensis* 40, no. 1 (Winter/Spring 2011).

5 This book makes relatively little attempt to explain federal policy making in all of its unquestionably fractious detail. Readers keen to examine the federal relationship are encouraged to pay close attention to the role of the Canadian state in Chapters 2 and 4, to get the flavour of federal power in action. Federal development policy and environmental reactions to it are a well-studied component of Canada's political history, and interested readers would do well to look to James C. Scott, *Seeing Like a State: How Certain Schemes to Improve the Human Condition Have Failed* (New Haven, CT: Yale University Press, 1998). Canadian examples include Miriam Wright, *A Fishery for Modern Times: The State and Industrialization of the Newfoundland Fishery, 1934–1968* (Don Mills, ON: Oxford, 2001); Donald Savoie and John Chenier, "The State and Development: The Politics of Regional Development Policy," in *Still Living Together: Recent Trends and Future Directions in Canadian Regional Development,* ed. William Coffey and Mario Polese (Montreal: Institute for Research on Public Policy, 1998); Jennifer Nelson, *Razing Africville: A Geography of Racism* (Toronto: University of Toronto Press, 2008). Especially useful for tying together policy and theory is Hornborg, "Environmentalism, Ethnicity, and Sacred Places," 245–67; Hornborg, "Undermining Modernity"; Matthew Evenden, *Fish versus Power: An Environmental History of the Fraser River* (Cambridge: Cambridge University Press, 2004); Tina Loo, "People in the Way: Modernity, Society, and Environment on the Arrow Lakes," *BC Studies* 142/143 (Summer/Autumn 2004): 161–96; Philip Van Huizen, "Flooding the Border: Development, Politics, and Environmental Controversy in the Canadian-U.S. Skagit Valley" (PhD dissertation, University of British Columbia, 2013).

CHAPTER 1: AT HOME AND ABROAD

1 Mersey Tobeatic Research Institute, "Mersey Messages – Old Lake Rossignol," March 10, 2009, http://www.merseytobeatic.ca/pdfs/Mersey%20Messages/Mersey%20Messages%20March%2010%202009.pdf.

2 William Herrington and George Rounsefell, "Restoration of the Atlantic Salmon in New England," *Transactions of the American Fisheries Society* 70, no. 1 (1941): 123–27.

3   Frank Sobey to Russell McInnes, July 18, 1961, Public Archives of Nova Scotia, McInnes fonds, RG44, vol. 22, no. 4.

4   Russell McInnes, *Report of the Royal Commission into the Nova Scotia Light and Power Company Ltd. Gold River Hydro Development* (Halifax: Government of Nova Scotia, 1961).

5   Ibid.

6   Ibid.

7   "NSRC" document, 1981, Dalhousie University Archives and Special Collections, Ecology Action Centre fonds, MS-11-13 [hereafter DAL-EAC], box 33.2.

8   McInnes, *Report of the Royal Commission.*

9   The national scene shared much the same development in the mid-twentieth century, according to Tina Loo, whose book on Canadian wildlife conservation traced the rise of groups using both elite status and popular pressure to exert influence on the state, especially in the west. Tina Loo, *States of Nature: Conserving Canada's Wildlife in the Twentieth Century* (Vancouver: UBC Press, 2006).

10 "Wildlife Conservation," n.d. [c. 1950s], Acadia University Esther Clark Wright Archives, Harrison Flint Lewis fonds, accession 2006.020-LEW/36.

11 "Organization of the Nova Scotia Bird Society," 1955, Acadia University Esther Clark Wright Archives, Harrison Flint Lewis fonds, accession 2006.020-LEW/24.

12 Martin Rudy Haase, interview with the author, October 9, 2011; Sterling Evans, *The Green Republic: A Conservation History of Costa Rica* (Austin: University of Texas Press, 1999), 62.

13 Martin Rudy Haase, interview with the author, October 9, 2011; "FON newsletter," n.d. [c. 1970s], Rudy Haase papers.

.14 "DDT: Who It's Killing and Why," *Mysterious East,* December 1969, 19–26. See also Chapter 4. The story of New Brunswick's rivers and the effect of DDT on them appeared famously in Rachel Carson, *Silent Spring* (New York: Fawcett Crest, 1962).

15 Rust Associates, *A Review of the Boat Harbour Waste Treatment Facilities for Nova Scotia Water Resources Commission* (Montreal: Rust Associates, 1970).

16 "Special Report: The Death of Boat Harbour," *Mysterious East,* September 1970, 20–27.

17 Ibid., 22. The Pictou Landing Band did eventually win in court a recognition of the deception perpetrated in part by the federal government, and a settlement in 1993 that paid $35 million: Settlement agreement, July 20, 1993, http://boatharbour.kingsjournalism.com/wordpress/wp-content/uploads/pdfs/15.199335millionagreement.pdf, copy on file with the author.

18 Henry Ferguson to Rust Associates, March 24, 1970, http://boatharbour.kingsjournalism.com/wordpress/documents/, copy on file with the author.

19 "Special Report: The Death of Boat Harbour," 22.

20 Ibid., 26; Reverend D. Glass, Sharon-Saint John United Church Stellarton, to Premier G.I. Smith, August 16, 1970, http://boatharbour.kingsjournalism.com/wordpress/wp-content/uploads/pdfs/09.1970smithfromchurch.pdf, copy on file with the author; Dr. J.B. MacDonald to Rust Associates Consulting Engineers, March 22, 1970, http://boatharbour.kingsjournalism.com/wordpress/wp-content/uploads/pdfs/07.1970macDonaldLetterComplete.pdf, copy on file with the author.

21 "Special Report: The Death of Boat Harbour," 23; Reverend D. Glass, Sharon-Saint John United Church Stellarton, to Premier G.I. Smith, August 16, 1970, http://boatharbour.kingsjournalism.com/wordpress/wp-content/uploads/pdfs/09.1970smithfromchurch.pdf, copy on file with the author.

22 Tom Murphy, "Why Don't You Talk about the East River and Shut-Up about Boat Har-bour," *Mysterious East,* October 1970, 29–30.

23 "Report by J.A. Delaney and Associates on Pollution in Boat Harbour," Dalhousie University Archives and Special Collections, Dalhousie University Institute of Public Affairs/Henson College fonds, UA-26, box 116.6.

24 "Special Report: The Death of Boat Harbour," 22, 26.

25 Ibid., 22.

26 J.B. MacDonald to Premier Stanfield, March 19, 1966, http://boatharbour.kingsjournalism. com/wordpress/wp-content/uploads/pdfs/03.1966Macdonaldhammletter.pdf, copy on file with the author.

27 "Ecology Supplement," *4th Estate,* September 1972, 19.

28 "Special Report: The Death of Boat Harbour," 23.

29 A pipe.

30 Robert Martin, "Purcell's Cove: Look Quick a Sewage Treatment Plant," *Mysterious East,* June 1970), 11–13; J.J. Betlam, Bedford United Church Social Action Chairman, and R.G. McClung, United Church Women President, to Premier Regan, January 13, 1971, Public Archives of Nova Scotia, Gerald Regan fonds, RG100 [hereafter REGAN], vol. 78, no. 36–6b.

31 Bedford Basin Pollution Committee, *Report of the Bedford Basin Pollution Committee of the Council of the Bedford Service Commission* (Bedford: Bedford Service Commission, 1970).

32 S.R. Kerr, "The McLaren Report: A Treatment Plant at Purcell's Cove," *Mysterious East,* December 1970, 19–20; E.L.L. Rowe to D.R. MacDonald, Minister under the Water Act, August 14, 1970, REGAN, vol. 116, no. 1.

33 Major site proposals during the following years included the tip of Point Pleasant Park, McNab's Island, the area around York Redoubt, and the locations of the plants existing at the time of writing.

34 "Brief to Canadian Preparatory Committee for the UN Conference on the Human Environment," April 28, 1972, Dalhousie University Archives and Special Collections, Movement for Citizens Voice and Action fonds, MS-11-1, [hereafter MOVE] box 6.35.

35 "Pictou Area Successfully Combating Water Pollution at Boat Harbour," *Chronicle-Herald,* January 2, 1974, 25.

36 Donald Savoie, *Regional Economic Development: Canada's Search for Solutions* (Toronto: University of Toronto, 1992), 6–7.

37 Silver Donald Cameron, interview with the author, November 1, 2011.

38 *Report of the Task Force Operation Oil (Clean-up of the Arrow Oil Spill in Chedabucto Bay) to the Minister of Transport* (Ottawa: Information Canada, 1970), 9.

39 Ibid.

40 Ironically, the *Irving Whale* also sank with a load of Bunker C oil, north of Prince Edward Island on September 7, 1970, causing another major spill and necessitating a laborious recovery. Environment Canada, Emergencies Science and Technology Division, "Bunker C Fuel Oil (Irving Whale)," http://www.etc-cte.ec.gc.ca/databases/Oilproperties/pdf/ WEB_Bunker_C_Fuel_Oil_(Irving_Whale).pdf.

41 Silver Donald Cameron, interview with the author, November 1, 2011; *Report of the Task Force Operation Oil;* "Chedabucto Bay: Waiting for the End," *Mysterious East,* July 1971, 34.

42 "Chedabucto Bay: Waiting for the End," 34. The answer took eight more years to arrive, with the wreck of the *Kurdistan* on March 15, 1979. "Oil Clean-Up Continues on Cape Breton Coast," *Mail-Star,* April 18, 1979, 2; J.H. Vandermeulen and D.E. Buckley, "The

Kurdistan Oil Spill of March 16–17, 1979: Activities and Observations of the Bedford Institute of Oceanography Response Team," Canadian Technical Report of Hydrography and Ocean Sciences No. 35 (Ottawa, January 1985), http://www.dfo-mpo.gc.ca/Library/90618.pdf. However, spills from the Gulf refinery itself were frequent enough that the management called a 12,000-gallon spill in 1972 "one of the mildest ones we've had." David Bentley, "Bunker C Spill: No Danger," *Chronicle-Herald,* June 22, 1972, 23.

43 Shelburne Harbour, Sydney Harbour, and Cole Harbour were notable others.

44 Silver Donald Cameron, interview with the author, November 1, 2011; Alan Ruffman, interview with the author, February 21, 2012. The Bedford Basin group listed ninety-one members in its report, about half to each gender. Bedford Basin Pollution Committee, *Report of the Bedford Basin Pollution Committee.*

45 Silver Donald Cameron, interview with the author, November 1, 2011.

46 Donald Worster argues that the "age of ecology" began with the explosion of the first atomic bomb in 1945. Donald Worster, *Nature's Economy* (Cambridge: Cambridge University Press, 1998), 342.

47 Robert Gottlieb, *Forcing the Spring: The Transformation of the American Environmental Movement* (Washington, DC: Island Press, 1993); Martinez-Alier, *Environmentalism of the Poor;* Guha, *How Much Should a Person Consume?* Hays, *Beauty, Health, and Permanence.*

48 Carson, *Silent Spring;* Garret Hardin, "The Tragedy of the Commons," *Science* 162, no. 3859 (December 13, 1968): 1243–48; Donella Meadows et al., *The Limits to Growth* (New York: Universe, 1972); E.F. Schumacher, *Small Is Beautiful* (London: Blond and Briggs, 1973). For an overview, see Gary Haq and Alistair Paul, *Environmentalism since 1945* (London: Routledge, 2012).

49 Ken Hartnett, "Encounter on the Urban Environment: Historian's Report," MOVE, box 25.13.

50 "Encounter on Metro Area," MOVE, box 25.10.

51 "Kentville Meeting minutes," MOVE, box 25.8.

52 "Kentville Meeting minutes," MOVE, box 25.8; Brenda Large, "Kentville Huddle Huge Success," *4th Estate,* March 4, 1971, 1.

53 Africville was a Black community in Halifax demolished in the 1960s, leaving a lasting feeling of anger among many of its former residents. The Heritage Trust group simply wished to curb the city's habit of demolishing many other venerable neighbourhoods and buildings. See Nelson, *Razing Africville;* Heritage Trust of Nova Scotia, *Annual Report 1970–1971* (Halifax: The Trust, 1971).

54 Carla Laufer, "Information Must Be Taken to the People: Where Is the Real Power behind Metro Planning?" *4th Estate,* January 14, 1971, 3.

55 "Encounter on Metro Area," MOVE, box 25.10.

56 "ECO Newsletter," March 1970, Public Archives of Nova Scotia, Ecology Action Centre fonds, MG20 [hereafter PANS-EAC], vol. 3434, file 28.

57 "Kentville Meeting minutes," MOVE, box 25.8; Large, "Kentville Huddle Huge Success," 1; *MOVE Bulletin* 1, no. 1 (September 1971), MOVE, box 6.31.

58 Brian Gifford, interview with the author, February 22, 2013; "Experimental Ecology 300000: A Resume of the First Term," Brian Gifford papers.

59 Brian Gifford to Hal Blackadar, CHNS Radio, April 17, 1974, DAL-EAC, box 42.3.

60 *Fine Print,* 1974, DAL-EAC, box 38.1.

61 Brian Gifford, interview with the author, February 22, 2013; "Ecology Action Centre – L.I.P. Project & MOVE," October 1972, Brian Gifford papers.

62 Brian Gifford to "Ian [MacDougall], Linda [Ruffman], Bernie [Hart], and Dennis [Patterson]," October 31, 1973, PANS-EAC, vol. 3420, no. 17.

63 The limits theory held uncommon currency in 1972 and 1973, attracting support from a visiting Roman Catholic archbishop and a Universalist Unitarian minister in the city, for example, in addition to the more expected enthusiasts like Gifford at the EAC. "Society Must Change or Collapse – Bishop," *Chronicle-Herald*, June 8, 1972, 21; "Universalist Unitarian Church newsletter," 7, no. 23 (March 27, 1972), Public Archives of Nova Scotia, Universalist Unitarian Church fonds, MG4, vol. 347, no. 2. For more on limits thinking, see Chapter 3.

64 "EAC Brief on Environmental Protection Act," March 3, 1973, PANS-EAC, vol. 3432, no. 18.

65 Alan MacEachern, *Natural Selections: National Parks in Atlantic Canada, 1935–1970* (Montreal and Kingston: McGill-Queen's University Press, 2001).

66 Claire Campbell, ed., *A Century of Parks Canada, 1911–2011* (Calgary: University of Calgary, 2011).

67 Alison Froese-Stoddard, "Ship Harbour National Park: The Breakdown between National Agendas and Local Interests," unpublished manuscript (2011–12). Adapted to Alison Froese-Stoddard, "A Conservation 'Could Have Been': Ship Harbour National Park," NiCHE Canada, *The Otter* blog, http://niche-canada.org/node/10446.

68 Danielle Robinson, "Modernism at a Crossroad: The Spadina Expressway Controversy in Toronto, Ontario ca. 1960–1971," *Canadian Historical Review* 92, no. 2 (June 2011): 295–322.

69 Editorial, *Dartmouth Free Press,* July 12, 1973, clipping file, REGAN, vol. 84, no. 38; "March of the 'Kooks,'" *Mail-Star,* May 15, 1973, 6.

70 APES to Jean Chrétien, July 9, 1973, REGAN, vol. 84, no. 38.

71 "Rear Door Regan Ducks Park Protestors," *Dartmouth Free Press,* May 16, 1973, 1; APES to Jean Chrétien, July 9, 1973, REGAN, vol. 84, no. 38.

72 "Kooks Win! Regan Dumps Federal Park Plan," *Dartmouth Free Press,* December 26, 1973, clipping file, REGAN, vol. 84, no. 38; *Evening News,* May 11, 1973, clipping file, REGAN, vol. 84, no. 38; Gordon Hammond, "Government Should Withdraw Its Proposals," *Chronicle-Herald,* March 12, 1973, 7.

73 *MOVE Bulletin,* December 13, 1972, MOVE, box 6.23; "Kooks Win! Regan Dumps Federal Park Plan," ibid.; Hammond, ibid.

74 P. Thomas, "The Kouchibouguac National Park Controversy: Over a Decade Strong," *Park News* 17, no. 1 (1981): 11–13.

75 Wilfrid Creighton, and Kenneth Donovan, "Wilfrid Creighton & the Expropriations: Clearing Land for the National Park, 1936," *Cape Breton's Magazine* 69 (August 1995): 1–20; "Regan Government Regenerates Interest in Wreck Cove Hydro-Electric Development," *Chronicle-Herald,* April 9, 1974, 25.

76 "Park Controversy Ended?" *Mail-Star,* December 21, 1973, clipping file, REGAN, vol. 84, no. 38; "Kooks Win! Regan Dumps Federal Park Plan," *Dartmouth Free Press,* December 26, 1973, clipping file, REGAN, vol. 84, no. 38.

77 Irene Edwards to Premier Regan, July 6, 1973, REGAN, vol. 84, no. 38.

78 Carson, *Silent Spring.*

79 W.P. Kerr, Deputy Minister of Transportation, to Alan Steel, coordinator of the Royal Commission on Forestry, August 3, 1982, Public Archives of Nova Scotia, Nova Scotia Royal Commission on Forestry fonds RG44, [hereafter RCOF], vol. 186, no. 5.

80 Bill Templeman, "The Soft Spray Job," *Mysterious East,* October 1971, 7–14; "The Warren Story," *Queens/Lunenburg SSEPA News* [hereafter *Q/L SSEPA News*] 2, no. 3 (Summer 1983): 24–26; Brenda Large, "2,4,5-T Leaves Path of Destruction," *4th Estate,* June 17, 1971, 14–16.

81 Templeman, "The Soft Spray Job," 8.

82 Ibid., 8.

83 Ibid., 12.

84 Ibid., 8; Nick Fillmore, "A Long Nightmare Follows Defoliant," *Globe and Mail,* April 10, 1971, 8; Brenda Large, "Chemical Kills N.S. Animals: No Action on Warren Farm Case," *4th Estate,* June 17, 1971, 1, 18; Large, "2,4,5-T Leaves Path of Destruction," 14–16.

85 Templeman, "The Soft Spray Job," 12.

86 Ibid., 13.

87 Letter from Rosemary Eaton, Cole Harbour Environment Committee, to J.E. Hutchinson, NSRF, April 6, 1972, DAL-EAC, MS-11-13, box 24.5; *MOVE Bulletin,* January 11, 1973, MOVE, DAL, MS-11-1, box 6.21; Roger Bacon, MLA Cumberland East, to Premier Regan, regarding Amherst Pollution Committee, n.d., REGAN, vol. 78, no. 36–6b; Province of Nova Scotia, *Annual Report: Environmental Control Council: For the Year Ending Dec. 31 1974,* Department of Environment Library, reference number 354.33, 14–15.

88 Letter from Glen Bagnell to Rosemary Eaton, December 6, 1972, DAL-EAC, MS-11-13, box 24.5; *Chronicle-Herald,* October 19, 1972, clipping file, REGAN, vol. 94, no. 3.

89 Templeman, "The Soft Spray Job," 11.

90 Ibid., 13.

91 J.L. Kirby to David Watts, US Department of the Interior, January 15, 1971, REGAN, vol. 78, no. 36–6b.

92 Memo on "Possibilities of co-ordination and integration of effort in the field of environmental affairs for discussion at the forthcoming meeting of the Premiers of the three Maritime Provinces," May 1971, REGAN, vol. 78, no. 36–6b.

93 Kirby to Rowe, March 26, 1973, REGAN, vol. 83, no. 10.

94 E.L.L. Rowe, Chairman of Water Resources Commission, to Kirby, April 17, 1973, REGAN, vol. 84, no. 37; Rowe to Bagnell, March 22, 1973, REGAN, vol. 83, no. 10.

95 J.L. Kirby to David Watts, US Department of the Interior, January 15, 1971, REGAN, vol. 78, no. 36–6b.

96 Summary of points on citizen input, Glen Bagnell, Minister of Environment, 1973, Public Archives of Nova Scotia, Man and Resources Program fonds, MG20, vol. 891; E.L.L. Rowe, "Brief on Pollution for Premier," January 19, 1970, REGAN, vol. 78, no. 36–6b.

97 Memo, "Environmental Conservation Council," REGAN, vol. 78, no. 36–6b.

98 Joan Fraser, "New Hardboard Plant Opens in Nova Scotia," *Montreal Gazette,* July 5, 1967, 23; William Chanclor and Pat Murphy, "Robert Whiting: What One Man Can Do," *Axiom* 2, no. 4 (May 1976): 9.

99 Brief to ECC [Environmental Control Council] on "Potential Effects of Anil Effluent That Could Not Have Been Detected in Prior Surveys," by Norman Dale, Dalhousie University Institute for Environment Studies, 1974, PANS-EAC, vol. 3432, no. 19.

100 Environmental Control Council (C.A. Campbell, Malcolm Moores, and Dr. J.G. Ogden) to G.M. Bagnell, Minister of Environment, January 31, 1975, Department of Environment Library, reference no. L778 75/01.

101 Chanclor and Murphy, "Robert Whiting," 9.

102 Martin Hunt, "The Brave New World of the Canada Water Act," *Mysterious East,* November 1969, 10.

103 "Anil Complies with Cleanup," *Chronicle-Herald,* June 1, 1972, 4.

104 Chanclor and Murphy, "Robert Whiting," 6–9, 21; Martin Haase, interview with the author, October 9, 2011.

105 Brief to ECC on "Potential Effects of Anil Effluent That Could Not Have Been Detected in Prior Surveys," by Norman Dale, Dalhousie University Institute for Environment Studies, PANS-EAC, vol. 3432, no. 19; Martin Haase, interview with the author, October 9, 2011; Alan Ruffman, interview with the author, February 21, 2012; Chanclor and Murphy, "Robert Whiting," 6–9, 21.

106 Chanclor and Murphy, ibid.; ECC (C.A. Campbell, Malcolm Moores, and Dr. J.G. Ogden) to G.M. Bagnell, Minister of Environment, January 31, 1975, Department of Environment Library, reference no. L778 75/01.

107 Province of Nova Scotia, *Annual Report – Environmental Control Council: For the Year Ending Dec. 31 1975,* DAL-EAC, box 41.20; Alan Ruffman, interview with the author, February 21, 2012.

108 Chanclor and Murphy, "Robert Whiting," 9. The Nova Scotia government also participated in a national consultative process called the "Man and Resources" program in 1971–73, which once again left activists disappointed with the results. *Bulletin of the Man and Resources Conference Program* 2, no. 2 (November 1972), Public Archives of Nova Scotia, Man and Resources Program fonds, MG20, vol. 891; *Bulletin of the Man and Resources Conference Program* 2, no. 6 (April 1973), Public Archives of Nova Scotia, Man and Resources Program fonds, MG20, vol. 891; "Environmental Improvement Committee – notes of first meeting," February 2, 1973, MOVE, box 11.12.

109 There was one other notable public hearing by the ECC, into a landfill siting process in Halifax County, but the minister refused to release the council's report for publication. Alan Ruffman, interview with the author, February 21, 2012; Gordon Black, "Jack Lake – 'Interference at the Highest Level,'" *Chronicle-Herald,* March 10, 1976, 7.

110 Irene Edwards to Premier Regan, July 6, 1973, REGAN, vol. 84, no. 38.

CHAPTER 2: THE TWO MECs

1 "Air of Secrecy Surrounds Talks," *Chronicle-Herald,* June 6, 1972, 1.

2 Alan Ruffman, interview with the author, February 21, 2012.

3 Silver Donald Cameron, *The Education of Everett Richardson: The Nova Scotia Fishermen's Strike 1970–71* (Toronto: McLelland and Stewart, 1977).

4 Sales slowed even further in the mid-1970s, after India's first nuclear weapons test in 1974 using plutonium created in its Canadian reactor; stricter safeguards against the diversion of weapons-grade material blunted the appeal of a reactor that, civilian or not, had always been advertised as the world's most efficient plutonium maker.

5 Gordon Edwards, "Canada's Nuclear Industry and the Myth of the Peaceful Atom," in *Canada and the Nuclear Arms Race,* ed. Ernie Regehr and Simon Rosenblum (Toronto: James Lorimer, 1983), 122–70; Robert Bothwell, *Nucleus: The History of Atomic Energy of Canada Limited* (Toronto: University of Toronto Press, 1988).

6 "Regan Denies Talks," *Chronicle-Herald,* August 8, 1972, 3.

7 "Nuclear Danger Feared," *Chronicle-Herald,* June 13, 1972, 17.

8 "Federal Assistance for Proposed Nuclear Plant 'Not Likely Unless ...,'" *Chronicle-Herald,* June 10, 1972, 2; Ralph Surette, *Montreal Star,* September 22, 1973, clipping file, PANS-EAC, vol. 3434, no. 2.

9   Ralph Surette, ibid.; "N.S. Residents Should Demand Answers – Doane," *Chronicle-Herald,*
    June 7, 1972, 4; "Former Resources Official Concerned by Nuclear Plan," *Chronicle-Herald,*
    June 7, 1972, 4; Ralph Surette, "What Is Public, What Is Secret?" *4th Estate,* October 15,
    1975, 7.
10 Nuclear Power Plant Sub-Committee Meeting minutes, MOVE, box 11.16; Letter from
    S. Baskwill, MOVE, to the Editor, *Chronicle-Herald,* n.d., MOVE, box 11.16.
11 Perhaps unsurprising, given that physics professor Roy Bishop, who was largely respon-
    sible for its lengthy explanations of the science of nuclear fission, later went on to produce
    pro-nuclear material for AECL. Dr. Roy Bishop, interview with the author, January 16,
    2012.
12 Information packet on Stoddard Island, n.d., DAL-EAC, box 37.3.
13 "Some Facts on the Problems and Dangers of Atomic Energy," December 1972, PANS-
    EAC, vol. 3421, no. 20.
14 "Policy Statement on the Proposed Stoddard Island Nuclear Power Plant," March 20,
    1973, DAL-EAC, box 37.3; "Some Facts on the Problems and Dangers of Atomic Energy,"
    December 1972, PANS-EAC, vol. 3421, no. 20.
15 "Participate in Energy Policy," EAC application to OFY, 1973, DAL-EAC, box 39.6;
    Brian Gifford to Ian MacLaren, CNF Halifax, November 16, 1973, DAL-EAC, box 42.3.
16 "Lobster Fishermen Would Fight Plant," *Chronicle-Herald,* June 17, 1972, 4; "Some Facts
    on the Problems and Dangers of Atomic Energy," December 1972, PANS-EAC, vol. 3421,
    no. 20; "Necessary to Get All the Facts, Says Regan," *Chronicle-Herald,* October 25, 1973,
    10; "Participate in Energy Policy," EAC application to OFY, 1973, DAL-EAC, box 39.6.
17 "Necessary to Get All the Facts, Says Regan," *Chronicle-Herald,* 10.
18 Bruce Little, "Glace Bay Plant: Trying to Fix Costly Blunder," *Montreal Gazette,* May 16,
    1973, 33; AECL, *Canada Enters the Nuclear Age* (Montreal and Kingston: McGill-Queen's
    University Press, 1997), 337; Robert Campbell, "Heavy Water: Jewel to Millstone," *Mysterious
    East,* August 1970, 11–13. The Glace Bay plant was officially opened in 1967 but did not
    function until 1975, while the Port Hawkesbury plant began operations in 1970 but oper-
    ated at far less than full capacity until refurbished in 1974.
19 The risks posed by exposure to low-level radiation are a well-known and bitterly contested
    point of debate among activists, nuclear scientists, and medical researchers, as they have
    been since the famous Russell-Einstein Manifesto on radioactive fallout in 1955.
20 ECC meeting report, November 26, 1973, DAL-EAC, box 41.20.
21 Egbers, "Going Nuclear."
22 "Nuclear Route Likely for N.S.," *Mail-Star,* October 2, 1974, 1.
23 EAC Board of Directors Meeting minutes, October 7, 1974, PANS-EAC, vol. 3420, no.
    18; Brian Gifford to Bill Zimmerman, November 13, 1974, PANS-EAC, vol. 3420, no. 17.
24 Francis Early, "'A Grandly Subversive Time': The Halifax Branch of the Voice of Women
    in the 1960s," in *Mothers of the Municipality,* ed. Judith Fingard and Janet Guildford
    (Toronto: University of Toronto Press, 2005), 258.
25 *VOW National Newsletter* 9, no. 3, November 1972, PANS-EAC, vol. 3434, no. 28; "MOVE
    bulletin," December 13, 1972, MOVE, box 6.23.
26 Early, "'A Grandly Subversive Time,'" 263.
27 Susan Holtz to Doris Calder, November 1, 1974, DAL-EAC, box 41.7.
28 Ibid.
29 Alan Ruffman, interview with the author, February 21, 2012.
30 Susan Holtz to Anne [Wickens], February 14, 1975, DAL-EAC, box 41.7.

31 Dorothy Norvell to Donna Elliott, VOW Ontario, n.d., DAL-EAC, box 42.3; Dorothy
    Norvell to Orville Erickson, CWF, December 11, 1974, DAL-EAC, box 42.3; Lille d'Easum
    to Dorothy Norvell, December 28, 1974, DAL-EAC, box 42.3; Susan Holtz to Dick Lurie,
    December 9, 1974, DAL-EAC, box 42.3; Susan Holtz to Karen Alcock, Energy Probe, May
    12, 1975, DAL-EAC, box 38.16; EAC Board of Directors Meeting minutes, January 9, 1975,
    PANS-EAC, vol. 3420, no. 19.

32 Anne Wickens to Susan Holtz, October 4, 1974, October 25, 1974, PANS-EAC, vol. 3423,
    no. 30.

33 SSEPA press release, October 7, 1974, Public Archives of Nova Scotia, South Shore En-
    vironmental Protection Association fonds, MG20 [hereafter PANS-SSEPA], vol. 1016, no.
    27–27j; "Nuclear Route Likely for N.S.," *Mail-Star,* October 2, 1974, 1.

34 "Nuclear Power in Eastern Canada," Hattie Perry, n.d., PANS-SSEPA, vol. 1016, no.
    27–27j.

35 "Group Formed to Oppose N.B. Plant," *Chronicle-Herald,* April 8, 1974, 2.

36 "Maritime Coalition of Environmental Protection Associations," document, n.d. [1974],
    PANS-SSEPA, vol. 1016, no. 27–27j. Participation from Prince Edward Island was small,
    rather like its participation in the proposed energy corporation. Though the Institute of
    Man and Resources was widely admired, it did not join in activist activity and was gener-
    ally seen as a government agency. Only one group from PEI in the late 1970s, Help Our
    Provincial Environment (HOPE), is recorded to have participated in the MEC.

37 "CCNR newsletter," August 1976, PANS-EAC, vol. 3422, no. 1; EAC Board of Directors
    Meeting minutes, April 1, 1976, PANS-EAC, vol. 3420, no. 20; EAC Board of Directors
    Meeting minutes, March 4, 1976, PANS-EAC, vol. 3420, no. 20.

38 EAC Board of Directors Meeting minutes, June 25, 1975, PANS-EAC, vol. 3420, no. 19;
    Susan Holtz to Joe and Helen Bongiovanni, February 5, 1975, DAL-EAC, box 41.7. The
    CSG presence within the EAC lasted for many years, long enough to remake the organiza-
    tion entirely, especially during the "two Susans" era, from 1975 to 1980, when Susan Mayo
    as coordinator and Susan Holtz as energy researcher put their stamp on every aspect of
    the centre's activity. For some examples, see Board of Directors membership lists, PANS-
    EAC, vol. 3420, no. 23–24.

39 Donald Hamilton, NORDA, to Susan Mayo, July 29, 1975, DAL-EAC, box 42.4.

40 Susan Holtz to Liberty Pease, AECL, May 5, 1975, PANS-EAC, vol. 3421, no. 20; Susan
    Holtz to J. Blair Seaborn, Deputy Minister Environment Canada, June 8, 1975, PANS-
    EAC, vol. 3423, no. 7.

41 Susan Holtz to Anne Wickens, November 25, 1974, DAL-EAC, box 42.3.

42 "Dorothy" to "Carla," November 1974, DAL-EAC, box 41.7; Susan Holtz to Anne Wickens,
    November 25, 1974, DAL-EAC, box 42.3.

43 World oil prices had tripled by the end of that year, and would continue to rise through
    the rest of the decade, to the point that a barrel of oil that had cost $2 before the crisis cost
    an unprecedented $35 in 1981. Francisco Parra, *Oil Politics: A Modern History of Petroleum*
    (London: I.B. Tauris, 2004); MacEachern, *The Institute of Man and Resources,* 12–13.

44 *Energy and People Conference Proceedings,* October 1975, MOVE, box 22.1.

45 The term achieved unexpected popularity following the SCC report, and the council issued
    three follow-up reports over the next ten years, including one in 1977 written under the
    chairmanship of Ursula Franklin, then a renowned scientist, anti-nuclear activist, Voice
    of Women member, and Quaker. Science Council of Canada, *Natural Resource Policy Issues
    in Canada,* Report No. 19 (Ottawa: SCC, 1973); Science Council of Canada, *Canada as a*

Conserver Society: Resource Uncertainties and the Need for New Technologies, Report No. 27 (Ottawa: SCC, 1977).

46 *Energy and People Conference Proceedings,* October, 1975, MOVE, box 22.1.

47 See for example, Doris McMullan, Probe Ottawa, to Susan Holtz, August 11, 1975, PANS-EAC, vol. 3422, no. 1. McMullan was on CCNR's steering committee, and Holtz was on its policy committee. Also, describing the group's "rather conservative approach" as a positive feature: Susan Holtz to CCNR, March 14, 1979, DAL-EAC, box 44.1.

48 "Keynote – Dr. Gordon Edwards," *Energy and People Conference Proceedings,* October 1975, MOVE, box 22.1.

49 "Energy Policy Panel," *Energy and People Conference Proceedings,* October 1975, MOVE, box 22.1.

50 *Energy and People Conference Proceedings,* October 1975, MOVE, box 22.1. An account of the resolutions debate was also printed in the activist paper *Nuclear Reaction,* Winter 1975–76, DAL-EAC, box 41.7.

51 For example, "Some Energy Proposals 'Irresponsible,'" *Mail-Star,* September 22, 1975, 5.

52 "Follow-up Notes from Energy and People Conference," October 30, 1975, PANS-EAC, vol. 3422, no. 21.

53 J.R. Helliwell to Susan Holtz, March 3, 1976, DAL-EAC, box 42.4.

54 EAC Board of Directors Meeting minutes, May 6, 1976, PANS-EAC, vol. 3420, no. 20.

55 Notes from APEC energy seminar, January 21, 1976, PANS-EAC, vol. 3421, no. 18; Energy Options Committee Meeting minutes, January 1976, DAL-EAC, box 32.32.

56 Energy Options Committee minutes, November 29, 1975, DAL-EAC, box 32.32; Surette, "What Is Public, What Is Secret?" 7.

57 Susan Holtz to CCNR, March 14, 1979, DAL-EAC, box 44.1.

58 "Discussion on the Nuclear Option – Some Impressions," EMR planning and evaluation note, November 21, 1975, DAL-EAC, box 42.4; Gordon Edwards, "Nuclear Wastes: What, Me Worry?" 1978, http://www.ccnr.org/me_worry.html.

59 Call transcript, Susan Mayo and Ray Bouchard, August 12, 1977, PANS-EAC, vol. 3424, no. 5.

60 In New Brunswick, activists complained that "the federal government is buying off opposition to its policies" with empty consultation: *Nuclear Reaction,* June 9–11, 1978. The same journal had noted the previous year that "renewable energy is receiving attention mainly through the political expedient of keeping a vocal (eg: anti-nuclear) group quiet": *Nuclear Reaction,* April 30, 1977. Even one of Susan Holtz's friends at EMR told her that "Alastair [Gillespie of EMR] may be trading off renewables and conservation": John [MacEwan] to Susan [Holtz], n.d., PANS-EAC, vol. 3422, no. 38.

61 Notes from CEAC meeting, March 15, 1977, DAL-EAC, box 39.1.

62 Notes from CEAC meeting, November 21/22, 1977, DAL-EAC, box 39.1.

63 EAC Board of Directors Meeting minutes, June 1, 1978, PANS-EAC, vol. 3420, no. 22.

64 Robert Bott, David Brooks, and John Robinson, *Life after Oil* (Edmonton: Hurtig, 1983), 9; "FOE Workshop, October 8–10, 1978," DAL-EAC, box 28.15.

65 "A Soft Energy Path for Nova Scotia," EAC press release, November 24, 1977, DAL-EAC, box 28.14. Even the press noted the elite status of the audience for the soft path effort. Michael Shea at the CBC called the Halifax event a "bauble of the rich" and pointed out that nearly all of the attendees earned $20,000 or more annually: Sylvia Mangalam to Michael Shea, December 1977, DAL-EAC, box 42.9.

66   Amory Lovins and Hunter Lovins, "Introduction," in Bott, Brooks, and Robinson, *Life after Oil*, 4 [emphasis in original].

67   Ibid., 15–21.

68   "Special Report: A Soft Energy Path for Canada," *Alternatives* 12, no. 1 (Fall 1984).

69   Susan Holtz, *2025: Soft Energy Futures for Canada*, vol. 3, *A Soft Energy Path for Nova Scotia* (Ottawa: Energy, Mines and Resources, 1983). The population of the province in fact grew approximately 8.7 percent from 1981 to 2011. Statistics Canada, *Census Divisions and Subdivisions, Population, Occupied Private Dwellings, Private Households and Census and Economic Families in Private Households, Selected Social and Economic Characteristics, Nova Scotia* (Toronto: Statistics Canada, 1983), http://archive.org/details/1981959851983engfra.

70   "Interview with Susan Mayo and Susan Holtz," *Dartmouth Free Press*, January 31, 1979, DAL-EAC, box 39.7.

71   "Energy Corporation Urged for Maritimes," *Chronicle-Herald*, October 20, 1976, 1.

72   EAC Board of Directors Meeting minutes, April 14, 1977, PANS-EAC, vol. 3420, no. 21; EAC press release, June 11, 1976, PANS-EAC, vol. 3432, no. 19.

73   *Jusun* 5, no. 5 (November 1977), DAL-EAC, box 28.14.

74   *Nuclear Reaction*, June 9–11, 1978. Many directed the same critique at EMR's Committee on Nuclear Issues in the Community, created in 1978 "to ease fears" among the public. An ecologist among its founding members promptly resigned from the panel of nuclear apologists, but FOE's David Brooks did later join, along with Greenpeace's Patrick Moore: "Committee on Nuclear Issues in the Community," March 1978, PANS-EAC, vol. 3422, no. 1.

75   Paul Armstrong to unknown recipient, May 5, 1978, DAL-EAC, box 28.1.

76   Tricia O'Brien, AVEC, to "Susan(s)," EAC, April 17, 1978, PANS-EAC, vol. 3421, no. 19; Hattie Perry to Tony Reddin, HOPE (PEI), December 21, 1977, PANS-EAC, vol. 3424, no. 1. There was also by 1978 a group in Antigonish, the "alternative energy group," that followed a similar path without creating a "centre" of any sort, and a Cape Breton Alternate Energy Society, a Fundy Solar Research group, and other smaller congregations: Sister Donna Brady to Susan Holtz, January 12, 1978, DAL-EAC, box 42.9; Notes from EMR Office of Energy Conservation Atlantic Group Workshop, February 23–25, 1978, PANS-EAC, vol. 3422, no. 27.

77   Barb Taylor to Susan Holtz, March 31, 1979, DAL-EAC, box 41.3; Barb Taylor to Ian MacPherson (federal candidate), April 10, 1979, DAL-EAC, box 41.3; FACT committee to Premier Buchanan and all MLAs, April 1979, DAL-EAC, box 41.3; FACT (David Simon) to all Councilors, Commissioners, and planning advisors in Annapolis County and the towns of Annapolis Royal, Bridgetown, Digby, Lawrencetown, Kingston, and Middleton, n.d. [April-May 1979], DAL-EAC, box 41.3; George deAlth to Susan Holtz, May/June 1979, DAL-EAC, box 41.3. The group reported sixteen core members and strong public support in 1979, enough that they could later split into eastern and western branches: "FACT Newsletter," October 1979, DAL-EAC, box 41.3.

78   "MEC (NS)," PANS-EAC, vol. 3423, no. 29; "HOPE of Colchester County brochure," 1979, PANS-EAC, vol. 3424, no. 21.

79   John Morrison, "NS Energy Awards Application for CBAES," 1978/79, DAL-EAC, box 42.9.

80   Dr. Ron Loucks to Premier Buchanan, April 7, 1979, PANS-EAC, vol. 3421, no. 14.

81   *It Can Happen Here*, leaflet, May 1979, PANS-EAC, vol. 3421, no. 16; Hattie Perry to the premier and all MLAs, April 11, 1979, DAL-EAC, box 41.3.

82   Multiple newspaper clippings, PANS-EAC, vol. 3421, no. 16.

83 Dana Silk represented New Brunswick and Tom Reddin Prince Edward Island. *The Eastern Graphic,* clipping file, PANS-EAC, vol. 3421, no. 16.
84 EAC Board of Directors Meeting minutes, June 13, 1979, PANS-EAC, vol. 3420, no. 23.
85 J. Wilson Fitt to Susan Holtz, April 27, 1979, PANS-EAC, vol. 3421, no. 14.
86 Susan Holtz to Ursula Franklin, SCC, January 9, 1978, DAL-EAC, box 32.7; Susan Holtz to Premier Buchanan, June 18, 1979, DAL-EAC, box 44.1; Susan Holtz to Premier Buchanan, May 24, 1979, PANS-EAC, vol. 3421, no. 16.
87 Brief to the Premiers at Brudenell, June 4, 1979, PANS-EAC, vol. 3421, no. 16. The petition they offered accumulated fifteen thousand names by early 1980: EAC Board of Directors Meeting minutes, March 12, 1980, PANS-EAC, vol. 3420, no. 4.
88 David Folster, "Double Trouble – Power Bubble," *Maclean's,* October 6, 1980, 38–39.
89 EAC Board of Directors Meeting minutes, June 13, 1979, PANS-EAC, vol. 3420, no. 23.
90 Egbers, "Going Nuclear."
91 The warning came from Walt Patterson, in a letter to Susan Mayo, November 12, 1975, PANS-EAC, vol. 3422, no. 29.
92 Amy Zierler, "Laws to Protect Our Environment Are Feeble. They Just Don't Work," *Atlantic Insight* 1, no. 10 (December 1979): 44–49.
93 Notes from interview with minister Bill Gillis, by Susan Holtz, October 29, 1977, PANS-EAC, vol. 3424, no. 11.
94 Personal Report, Susan Holtz, January 6, 1977, PANS-EAC, vol. 3422, no. 34. The board of directors explicitly validated the strategy of "affecting policy decisions" by building the centre's "credible and respectable image and develop[ing] a sound basis for the policy options it presents." EAC, "Ecology Action Centre: A Public Interest Group Using Conservation and Renewable Energy," n.d. [c. 1979], PANS-EAC, vol. 3420, no. 4.
95 Solar Energy Research Institute symposium, September 12–13, 1978, DAL-EAC, box 28.13.
96 "Workshop on Energy Conservation and Community Economic Development," May 24–26, 1979, DAL-EAC, box 32.14.

CHAPTER 3: POWER FROM THE PEOPLE

1 Elizabeth May, *Budworm Battles* (Halifax: Four East, 1982), 8.
2 *Halifax Field Naturalists Newsletter* 3 (January-February 1976): 11–14.
3 Scott Cunningham, "Aerial Spraying Plan Dangerous," *Chronicle-Herald,* March 1, 1976, 7; Susan Mayo, April 12, 1976, DAL-EAC, box 42.5; "The Spruce Budworm Problem," April 1976, DAL-EAC, box 24.5; EAC Board of Directors Meeting minutes, December 17, 1975, PANS-EAC, vol. 3420, no. 19.
4 "Cabinet Approves Budworm Spraying Request," *Chronicle-Herald,* March 27, 1976, 1; "The Spruce Budworm Problem," April 1976, DAL-EAC, box 24.5; Lloyd Hawboldt to Burgess, December 29, 1975, Public Archives of Nova Scotia, Department of Lands and Forests fonds, MG20 [hereafter DLF], vol. 890, no. 1.
5 May, *Budworm Battles,* 16, 28–29.
6 Parker B. Donham, "Fatal to Children: Disease May Be Linked to Budworm Spray," *Cape Breton Post,* March 31, 1976, 1; May, *Budworm Battles,* 19.
7 May, *Budworm Battles,* 12.
8 Ian McLaren, NSRC, "Spruce Budworm Spraying – Human Health Hazard," *Mail-Star,* April 2, 1976, 7; David Folster, "The Agonizing Fight over Budworm Spray," *Atlantic Insight* 1, no. 3 (June 1979): 34–36.

9   Proceedings of Regional Meeting "Forest Protection Against Spruce Budworm in Maine, Quebec, and New Brunswick," November 2, 1976, DLF, vol. 890, no. 2.

10 Lloyd Hawboldt to Gary Saunders, May 14, 1976, DLF, vol. 890, no. 1; R.M. Bulmer to E.S. Atkins, November 16, 1976, DLF, vol. 890, no. 2. The anti-spray group included Gary Saunders and N.A. Wiksten; the pro-spray group included Ed Cloney and E.S. Atkins, Planning Services Coordinator.

11 L. Anders Sandberg, "Forest Policy in Nova Scotia: The Big Lease, Cape Breton Island, 1899–1960," *Acadiensis* 20, no. 2 (Spring 1991): 124, n. 86; Memo, N.A. Wiksten to E.S. Atkins, November 16, 1976, DLF, vol. 890, no. 2. Wiksten was blunt, insisting that "the company knew what they bought." Department of Lands and Forests, *Nova Scotia Forest Inventory: Cape Breton Island Subdivision 1970* (Halifax: Lands and Forests, 1970), 17, 25.

The values underpinning chemical forestry were as well established as chemical forestry, for example, in New Brunswick. See McLaughlin, "Green Shoots." The wider story of Nova Scotia's forest policy can be found in Nova Scotia, *House of Assembly Debates* (1960), 73–74. Cited in Sandberg, "Forest Policy in Nova Scotia," 105–28. *Statutes of Nova Scotia,* c.6, 1942; *Statutes of Nova Scotia,* c.5, 1962; Harry Thurston, "Prest's Last Stand," in Harry Thurston, *The Sea among the Rocks* (East Lawrencetown, NS: Pottersfield Press, 2002), 184–220; Sandberg, "Forest Policy in Nova Scotia"; Memo, Lloyd Hawboldt to Burgess, December 29, 1975, DLF, vol. 890, no. 1; Memo, N.A. Wiksten to E.S. Atkins, November 16, 1976, DLF, vol. 890, no. 2; Susan Mayo to Lloyd Hawboldt, October 15, 1976, DAL-EAC, box 42.5; L.S. Hawboldt, "Toward 'Budworm-Proofing' the Forests of Nova Scotia," 1976, DLF, vol. 890, no. 2.

12 Lloyd Hawboldt, "Toward 'Budworm-Proofing' the Forests of Nova Scotia," ibid.; R.M. Bulmer to Hollis Routledge, NSFI, November 16, 1976, DLF, vol. 890, no. 2; Proceedings of Regional Meeting "Forest Protection Against Spruce Budworm in Maine, Quebec, and New Brunswick," November 2, 1976, DLF, vol. 890, no. 2. The other jurisdictions' spray programs were established since 1954 (Maine), 1952 (New Brunswick), and 1970 (Quebec).

13 Lloyd Hawboldt to Gary Saunders, May 14, 1976, DLF, vol. 890, no. 1.

14 May, *Budworm Battles,* 27. Budworm forecasts are accomplished by counting the number of egg masses on conifer branches in the fall; the eggs will provide the following year's budworm population.

15 Editorial, *Evening News,* n.d., clipping file, Public Archives of Nova Scotia, Betty Peterson fonds, MG1 [hereafter PANS-BP], vol. 3469, no. 8.

16 Proceedings of Regional Meeting "Forest Protection Against Spruce Budworm in Maine, Quebec, and New Brunswick," November 2, 1976, DLF, vol. 890, no. 2.

17 Scott Paper, Bowater Mersey, Minas Basin Pulp and Power Company Limited, McLellan Lumber Company, Cobequid Lumber Company, to Vincent MacLean, Minister of Environment, December 8, 1976, DLF, vol. 890, no. 3.

18 "Budworm Politics: While Larvae Sleep, Political Forces Clash over Spraying," *4th Estate,* October 13, 1976, 4.

19 Miss Lettice Edwards, Shelburne, to Lands and Forests, December 6, 1976, DLF, vol. 890, no. 3; Susan Mayo to Norma Mosher, Women's Institutes of Nova Scotia, December 20, 1976, DAL-EAC, box 24.5; Susan Mayo to David Steadman, Association of Outdoor Nova Scotians, December 23, 1976, DAL-EAC, box 24.5; Anne Wickens, SSEPA, to all Cabinet members, December 2, 1976, DAL-EAC, box 24.5; Martin Haase to Ministers of Lands and Forests, Recreation, and Public Health, and the Premier, December 3, 1976, DAL-EAC, box 24.5.

At the intended date of decision, the Department of Lands and Forests had logged 899 letters and signatures received in opposition to the 1977 spray request, including the above as well as the Inverness/Guysborough Presbytery of the United Church, the Cape Breton Metropolitan Alliance for Development, and DEVCO's Marine Farming Research Branch at Baddeck. There were only 7 in favour: "Spruce Budworm Spray Program – 1976," December 15, 1976, DLF, vol. 890, no. 1.

20 May, *Budworm Battles*, 33, 45–46.

21 CBLAS poster, n.d., DAL-EAC, box 28.14.

22 Memo on CBLAS handbills, DLF, vol. 890, no. 4.

23 Jon Everett, "More Scary Questions in the Spray Debate," *Atlantic Insight* 4, no. 7 (July 1982): 12; May, *Budworm Battles*, 57–59.

24 Betty Peterson to Dorothy Norvell, February 7, 1977, PANS-BP, vol. 3469, no. 8; May, *Budworm Battles*, 74. Betty Peterson described the audience, which she put at over five hundred and decidedly anti-spray, as "organic food growers, beekeepers, oystermen, owners of lobster pounds, mothers with children, small woodlot owners, [and] rugged oldsters."

25 May, *Budworm Battles*, 76–77; "Spray Program Turned Down," *Chronicle-Herald*, February 4, 1977, 1.

26 May reported that information from Vince MacLean put the final number of supporting names at more than seven thousand (versus eighteen in favour of spraying): *Budworm Battles*, 78.

27 Ibid., 107–20.

28 Ibid., 116.

29 Ralph Surette, "Effort Made to End Abuse of Forests," *Globe and Mail*, August 5, 1978, 8. There was pressure within Lands and Forests to access even more federal money available to provinces with spray programs, such as the $7 million in additional funds granted to New Brunswick in 1976: Richard Butler, Administrative Assistant to the Deputy Minister of Environment, to Lloyd Hawboldt, April 6, 1976, DLF, vol. 890, no. 1.

30 Bruce Doern and Thomas Conway, *The Greening of Canada: Federal Institutions and Decisions* (Toronto: University of Toronto Press, 1994), 104–6. The Department of Energy, Mines and Resources did likewise in 1975, but soon lost interest: "Discussion on the Nuclear Option – Some Impressions," EMR planning and evaluation note, November 21, 1975, DAL-EAC, box 42.4.

31 The EAC met with the Atlantic Regional Board in 1976 and 1977: EAC Board of Directors Meeting minutes, July 8, 1976, PANS-EAC, vol. 3420, no. 20; EAC to Environment Canada Atlantic Regional Board, April 13, 1977, DAL-EAC, box 41.11.

32 EAC Board of Directors Meeting minutes, June 1, 1978, PANS-EAC, vol. 3420, no. 22.

33 Environment Canada, "Announcement: Public Consultation Transportation Expenses Assistance," 1984, DAL-EAC, box 43.5.

34 Joanne Lamey, Community Planning Association of Canada, "Notice of CEAC Meeting, Ottawa, 25–29 May, 1980," April 18, 1980, DAL-EAC, box 44.2. The threat of "co-optation" was consistently misunderstood by those who accepted such funding, even warily; there is little or no evidence that the money ever changed anyone's opinions, only that those already holding the desired opinions were selected (and self-selected) to receive the money. This is another case where blindness to the diversity of the environmental movement has created the illusion of novelty out of simple changes in the internal balance of power.

35 EAC Board of Directors Meeting minutes, November 14, 1979, PANS-EAC, vol. 3420, no. 23; Gilbert Savard, Citizens' Group Liaison, Department of Fisheries and Environment, to Susan Mayo, June 14, 1979, DAL-EAC, box 43.1.

36 Brian Gifford to Vince MacLean, April 14, 1977, DAL-EAC, box 24.5.

37 Memo, "Pete Ogden: Saturday March 9," DAL-EAC, box 41.20.

38 Vincent MacLean, n.d./n.s., clipping file, PANS-BP, vol. 3469, no. 8.

39 As an example of the use of concepts borrowed from industry, few can surpass the brief to the Task Force on Wood Allocation by the NSRC (and endorsed by the NSBS, HFN, and CNF). In it, these groups agreed on the need to "manage the threat of budworm," ideally with biological controls, to maximize the use of the forest resource, and to practise the most "intensive silviculture ... for the benefit of the total resources of Nova Scotia." DAL-EAC, box 24.5.

40 May, *Budworm Battles,* 132; Elizabeth May, "Qualified Approval Given Spray Test," n.d., PANS-BP, vol. 3469, no. 8.

41 EAC Board of Directors Meeting minutes, May 4, 1978, PANS-EAC, vol. 3420, no. 22; Vince MacLean to Brian Gifford, April 20, 1977, DAL-EAC, box 24.5; Vince MacLean to Gary Harris, Halifax, December 6, 1976, DLF, vol. 890, no. 3.

    It was also a false claim, as the ECC had requested, directly to the previous minister himself, permission to hold hearings into the budworm spray issue ten months earlier: "Points for information discussion about spruce budworm control in Nova Scotia from technical advisory sub-committee, Nova Scotia Environmental Control Council," January 27, 1976, DLF, vol. 890, no. 1; E.L.L. Rowe, ECC, to Bagnell, Minister of Environment, February 4, 1976, DLF, vol. 890, no. 1.

42 Eve Smith, BC, to the Editor, *Jusun* (EAC), n.d. [c. 1978], DAL-EAC, box 42.8. The Queens/Lunenburg SSEPA also warned Nova Scotian environmentalists to "approach with caution" the Atlantic ENGO meeting arranged by Environment Canada, quoting David Brooks on the "threat of co-optation ... any time you take money." *Q/L SSEPA News* 1, no. 9 (January 1983), PANS-BP, vol. 3470, no. 6.

43 Memo, "Pete Ogden: Saturday March 9," DAL-EAC, box 41.20. (The EAC staff member who spoke with Ogden was quite defensive about it and preferred to see Ogden's position as an "anti-action-group stance.")

44 Elizabeth May, "A Case History of Citizen Action Victory," *PEI Environeer* 6, no. 3 (1978): 23.

45 "Harmless Spray?" *Rural Delivery* 4, no. 7 (December 1979); "EAC Brief to Public Utilities Board," April 3, 1979, PANS-EAC, vol. 3424, no. 21.

46 *Rural Delivery* 5, no. 10 (March 1981); "Waste Farm Chemicals," *Rural Delivery* 6, no. 11 (April 1982); Chris Wood, "Crop Sprays: Sickness and Death Down on the Farm," *Atlantic Insight* 4, no. 2 (February 1982): 38–43.

47 The Royal Commission on Forestry eventually collected records from various departments and corporations, showing herbicide use (often 2,4-D/2,4,5-T) by the Department of Transportation back to 1964, Department of Agriculture back to 1960, and Nova Scotia Power Corporation back to 1970 (an incomplete record). "Memos on chemicals," Public Archives of Nova Scotia, Royal Commission on Forestry fonds [hereafter RCOF], vol. 186, no. 5.

    For the wider story, see David Kinkela, *DDT and the American Century* (Chapel Hill: University of North Carolina Press, 2011).

48 Gale Peterson, "The Discovery and Development of 2,4-D," *Agricultural History* 41, no. 3 (July 1967): 253.

49 Chris Arsenault, *Blowback: A Canadian History of Agent Orange and the War at Home* (Halifax: Fernwood, 2009), 60–64. The total use of Agent Orange alone in Vietnam is estimated at about 45 million litres. Edwin Martini, *Agent Orange: History, Science, and the Politics of Uncertainty* (Boston: University of Massachusetts Press, 2012).

50 Elizabeth Blum, *Love Canal Revisited: Race, Class, and Gender in Environmental Activism* (Lawrence: University Press of Kansas, 2008); Martini, *Agent Orange*, 126–37.

Others included the 1976 dioxin explosion at Seveso, Italy, the ongoing revelations of mercury poisoning and cover-ups at Minamata, Japan, and among the Grassy Narrows First Nation in Ontario, for example. Bruna De Marchi, "Seveso: From Pollution to Regulation," *International Journal of Environment and Pollution* 7, no. 4 (September 1997): 526–37; Paul Almeida and Linda Brewster Stearns, "Political Opportunities and Local Grassroots Environmental Movements: The Case of Minamata," *Social Problems* 45, no. 1 (February 1998): 37–60; George Hutchison and Dick Wallace, *Grassy Narrows* (Scarborough, ON: Van Nostrand Reinhold, 1977).

51 Arsenault, *Blowback*, 91.

52 Herbicides were used in forestry in New Brunswick as early as 1970: "Monitor," *Mysterious East*, November 1971, 24.

53 "ECC Annual Report, 1978," DAL-EAC, box 41.20; Alan Ruffman to Roger Bacon, Minister of Environment, July 16, 1979, DAL-EAC, box 41.20.

54 Government of Nova Scotia, *Supplement to the Public Accounts of the Province of Nova Scotia for the Year Ended March 31, 1977* (Halifax: Queen's Printer, 1978), 67; N.A. Wiksten, forestry economist, to Earl Atkins, Lands and Forests, January 9, 1978, DLF, vol. 890, no. 6; Memo from N.A. Wiksten to R.H. Burgess, October 20, 1977, DLF, vol. 890, no. 5; Parker Donham, "Who's Winning the War Over Chemical Spraying?" *Atlantic Insight* 4, no. 1 (January 1982): 24.

55 Gary Baudoux, New Glasgow, to George Henley, August 30, 1979, DLF, vol. 887, no. 3; George Henley to Gary Baudoux, n.d., DLF, vol. 887, no. 3; A.A. Pearson, Stellarton Town Clerk, to Doug Carter, Deputy Minister of Environment, September 20, 1979, DLF, vol. 887, no. 3; George Henley to Frank Craig, Trenton, September 10, 1979, DLF, vol. 887, no. 3.

Eldridge himself represented the Nova Scotia Department of Lands and Forests at the Eastern Spruce Budworm Council (with members from Ontario, Quebec, Nova Scotia, New Brunswick, Newfoundland, Maine, and the Canadian Forestry Service [CFS]), where he worked with spray advocates like CFS's I.W. Varty, who in 1980 offered a novel rationalization of the safety of insecticide sprays: no spray onto uninhabited land could be pollutant by definition, since only upon reaching people could any substance be considered pollution. *CANUSA Newsletter* 8 (January 1980), DLF, vol. 604, no. 5.

56 "Forest Industry 'Needs' Informed Public," *Forest Times* 3, no. 2 (March 1981).

57 *CANUSA Newsletter* 8 (January 1980), DLF, vol. 604, no. 5; Nova Scotia Government press release, 1982, DAL-EAC, box 24.5.

58 Environment Canada press release, August 23, 1978, DAL-EAC, box 24.5.

59 Quoted from Dirk VanLoon, "Roadside Spraying," *Rural Delivery* 7, no. 2 (July 1982). On the Crowdis Coalition, see Elizabeth May to "Susans," n.d., DAL-EAC, box 44.1.

60 Val Blofeld, Tantallon, to Susan Holtz, February 21, 1980, DAL-EAC, box 43.1.

61 "80/81 Board of Directors," *Jusun* 8, no. 3 (December 1980), PANS-EAC, vol. 3420, no. 3.

62 Vincent MacLean to Susan Mayo, September 21, 1977, DAL-EAC, box 42.6.

63 George Butters, "Spray Wars, Part Two," *Atlantic Insight* 4, no. 10 (October 1982): 10–11. The government also issued permits for the 1982 season on the day after the Legislature finished its session: "After the Herbicide Trial" (draft), DAL-EAC, box 24.9.

64 The injunction and suit were settled the following January without going to court. EAC Board of Directors Meeting minutes, March 25, 1981, PANS-EAC, vol. 3420, no. 25; J.D. Smith, Lands and Forests, to D.L. Eldridge (marginal notes), August 18, 1980, DLF, vol. 608, no. 13; Sheila Jones, "Big Pond Goes After a Big Fish," *Atlantic Insight* 2, no. 11 (November 1980): 26–27.

65 Vicki Palmer, "A Report on the Lochaber Spray Situation, Summer 1981," DAL-EAC, box 29.10; Petition, 1981, DAL-EAC, box 29.10.

66 Palmer, ibid.

67 "N.B. Looking at Same Defoliants Planned for Lochaber Spray Area," *Evening News,* July 16, 1981, 3.

68 Ginny Point, EAC, to "friends," January 14, 1982, DAL-EAC, box 29.1; Forestry Management Committee Strategy Meeting minutes, April 20, 1982, DAL-EAC, box 28.25. The April 20 meeting had sixteen attendees, all from the mainland; within about two months, the committee had representatives from several Cape Breton groups. The committee also went through several name changes, from "Forestry Management Committee" initially, to "Forestry (Toxic Substances) Committee" while still mainly a mainland group, to simply the "Forestry Committee." EAC Board of Directors Meeting minutes, March 23, 1982, PANS-EAC, vol. 3421, no. 1.

69 J.D. Smith, Lands and Forests, to D.L. Eldridge (marginal notes), August 18, 1980, DLF, vol. 608, no. 13.

70 EAC Board of Directors Meeting minutes, May 24, 1982, PANS-EAC, vol. 3421, no. 2; Q/L SSEPA press release, April 30, 1982, DAL-EAC, box 29.9; Notice of Public Lecture "You and Me and 2,4-D" by Murray Prest, Vicki Palmer, and Dr. Cameron McQueen, April 20, 1982, DAL-EAC, box 29.1.

71 Adam Ashforth, "Reckoning Schemes of Legitimation: On Commissions of Inquiry as Power/Knowledge Forms," *Journal of Historical Sociology* 3, no. 1 (March 1990): 1–22.

72 Alexa McDonough to EAC, April 27, 1982, DAL-EAC, box 29.1. There was also the co-incidence of a resolution by the Canadian Institute of Foresters, asking Nova Scotia to repeal the Forest Improvement Act. EAC press release, March 24, 1982, DAL-EAC, box 29.9.

73 EAC Board of Directors Meeting minutes, May 24, 1982, PANS-EAC, vol. 3421, no. 2; Ginny Point, EAC, to Greg Kerr, Minister of Environment, June 2, 1982, DAL-EAC, box 29.1; Greg Kerr to Ginny Point, June 17, 1982, DAL-EAC, box 29.1.

74 EAC Memo on FACT meeting with John Sansom, Department of Environment, April 19, 1982, DAL- EAC, box 29.7. Laura Colpitts joined the Women's Health Education Network presentation at the Royal Commission on Forestry to tell of her experience with public announcements coming after spraying had already been done: Presentation by WHEN, March 24, 1983, RCOF, vol. 159, no. 4.

75 Alternative modes of argument were best represented, incongruously, among the rural Cape Breton activists who made up the strongest supporters of the seventeen plaintiffs in the herbicide trial. Presentation by FALASH, October 8, 1982, RCOF, vol. 158a, no. 2.

76 "SSEPA Fact Sheet," May 1982, DAL-EAC, box 28.28; EAC press release, July 2, 1982, DAL-EAC, box 28.28; "Forest Herbicide Trial – Background," PANS-BP, vol. 3469, no. 8; Peter Cumming, "The Herbicide Fight," *Rural Delivery* 7, no. 12 (May 1983): 32.

Disputes over terminology were numerous. Perhaps the best example came in the course of Murray Prest's testimony before the Royal Commission on Forestry, when challenged by Commissioner Rev. Greg MacKinnon:

MacKinnon: I don't think it's fair. I think it's using scare tactics to talk about Agent Orange. You know, that conjures up all kinds of things from Vietnam, when in fact the solution they were using was in fact laced with dioxin. I've seen the figures, I can't remember but –

Prest: Up to 70 parts per million, as compared to our levels now is 0.1 parts per million. But the thing is, it's still poison at parts per trillion so, you know –

MacKinnon: Well, of course.

Prest: You can only kill a fellow so dead, you know. The overkill doesn't matter."

Presentation by Murray Prest, Mooseland, April 22, 1983, RCOF, vol. 159, no. 9.

77   "MLA Supports Band's Efforts to Stop Spraying," *Chronicle-Herald,* July 8, 1982, 21; EAC Forestry Committee Meeting minutes, July 14, 1982, DAL-EAC, box 29.7.

78   "Forest Herbicide Trial – Background," PANS-BP, vol. 3469, no. 8; Peter Cumming, "The Herbicide Fight," *Rural Delivery* 7, no. 12 (May 1983): 32; EAC press release, July 8, 1982, DAL-EAC, box 29.9. It is important to note that May and the EAC were not criticizing Googoo or the people of Whycocomagh for their choice, only the fact that they should have had to make it.

79   George Butters, "Spray Wars, Part Two," *Atlantic Insight* 4, no. 10 (October 1982): 10–11.

80   Peter Cumming, "The Herbicide Case: Part I: Plaintiffs against the Spray," *Rural Delivery* 8, no. 1 (June 1983): 24.

81   "Publicity Sheet," [late] 1982, DAL-EAC, box 28.28; *Q/L SSEPA News Special Edition* 1, no. 5 (August 1982), PANS-BP, vol. 3469, no. 8. In fact, only fifteen of the "plaintiffs" were true plaintiffs. Elizabeth May had been rejected by Justice Burchell as a representative of all Nova Scotians, but she and Ruth Schneider of North River Bridge, Cape Breton County, kept their financial undertakings before the court in solidarity with the fifteen class-action representatives, making seventeen in total.

82   The total in the end was $14,465.17, and it was paid by Stephanie and John May after selling property near Baddeck. Peter Cumming, "The Herbicide Fight," *Rural Delivery* 7, no. 12 (May 1983); *Mail-Star,* April 27, 1983, clipping file, PANS-BP, vol. 3469, no. 8; Bruce Wildsmith to Peter Bessen, United States, December 8, 1983, DAL-EAC, box 29.6.

83   Elizabeth May to the plaintiffs, February 1, 1983, DAL-EAC, box 24.7.

84   Presentation by People for Environmental Protection (PEP), March 22, 1983, RCOF, vol. 159, no. 3; Rhonda Ryan, ed. "PEP TALK newsletter" 1, February 1983, DAL-EAC, box 40.13; Rhonda Ryan to EAC, n.d., DAL-EAC, box 28.25. New groups, small and local, continued to arrive throughout the approximately year and a half of high controversy, including those like PEP which jumped into the job of province-wide organization, and those like the Concerned Citizens Against 2,4-D in Yarmouth and Argyle, which spent most of their efforts on local proselytizing and left a very scant documentary record despite, in its case, successfully halting a roadside spray program in Yarmouth County: Veralyn Rogers to Roger Bacon, Minister of Agriculture and Marketing, June 24, 1983, DLF, vol. 882, no. 3.

85 Heather Laski, "A Costly Fight in Cape Breton," *Maclean's*, May 9, 1983, 20; "Sackville Environmental Protection Association" document, 1979, DAL-EAC, box 43.2.

86 Mailing List, July 1982, DAL-EAC, box 29.1; "Nova Scotia Environmental Network, 82-83" document, DAL-EAC, box 28.25.

87 The EAC, mired in perennial fundraising difficulties of its own, was happy to direct support toward the HFS and devote its own fundraising energies to finding its own funds. EAC Board of Directors Meeting minutes, February 16, 1983, PANS-EAC, vol. 3421, no. 3; EAC Communications Resource Group Meeting minutes, January 27, 1983, PANS-EAC, vol. 3421, no. 11.

88 Rhonda Ryan, ed., "PEP TALK newsletter" 1, February 1983, DAL-EAC, box 40.13.

89 McLaughlin, "Green Shoots," 22.

90 Pesticide Action Network, *Pesticide Digest* 2 (1983), PANS-EAC, vol. 3434, no. 3. The worldwide PAN was created in Malaysia on May 28, 1982, at an NGO workshop on the global trade in pesticides.

91 EAC Board of Directors Meeting minutes, March 23, 1983, PANS-EAC, vol. 3421, no. 3.

92 "HFS Public Relations Committee Note: Fund-Raising Tours," DLF, vol. 882, no. 3; Stephanie May, "Getting into PAN," *Between the Issues* 4, no. 1 (May-June 1984), on her trip to see the PAN–North America in Washington, DC, in November 1983. Support, financial and otherwise, did come from all over: Farley Mowat, Ralph Nader, the Ontario Federation of Labour, and similarly diverse sources. Elizabeth May to the plaintiffs, February 1, 1983, DAL-EAC, box 24.7; "Draft Motion from the Ontario Federation of Labour Health and Safety Committee, re: 2,4,5-T / 2,4-D Use in Nova Scotia," DAL-EAC, box 24.7.

93 Peter Cumming, "The Stink of Defeat (draft)," 5, DAL-EAC, box 24.8; Liz Calder, EAC, to Lars Moberg, Sweden, January 25, 1984, DAL-EAC, box 44.7.

94 David Folster, "The Agonizing Fight over Budworm Spray," *Atlantic Insight* 1, no. 3 (June 1979): 34–36.

95 *Between the Issues* 3, no. 2 (May-June 1983). Amusingly, NSFI's lawyer, George Cooper, said almost the same thing in May: *Chronicle-Herald*, May 6, 1983, clipping file, PANS-BP, vol. 3469, no. 8.

96 The nearest to open discussion was Peter Cumming's comment on the decision to go to court that "the issue seems to have taken on even more of a 'technical' and 'scientific' nature and less of a 'political' and 'social' and 'philosophical' and 'values' nature ... the change from a 'people's case' to an 'expert's case' may prove to be the most expensive few minutes of these 'ordinary people's' lives." *Between the Issues* 3, no. 2 (May-June 1983).

97 Bruce Wildsmith and Richard Murtha to plaintiffs, December 17, 1982, DAL-EAC, box 24.7. Nunn decided against a jury trial on account of the complexity of the evidence expected to be presented.

98 *Between the Issues* 3, no. 2 (May-June 1983).

99 Logan Norris, Oregon State University Forest Science Department, to the President, Dean, and Directors of the same, October 20, 1983, DAL-EAC, box 24.9.

100 Cumming, "The Herbicide Case: Part I," 26.

101 Ibid., 24–26; Peter Cumming, "Herbicide Case: Part II: Expert Dispute," *Rural Delivery* 8, no. 2 (July 1983): 24–27.

102 Fred McMahon, "Residents 'Guinea Pigs' in Spray Plan," *Chronicle-Herald*, June 7, 1983, 4.

103 Cumming, "Herbicide Case: Part II," 24–25.

104 Ibid., 22.

105 *Between the Issues* 3, no. 2 (May-June 1983).
106 Members of the EAC board of directors were quite worried that "the Judge does not appear to understand some of the very basic scientific principles involved." EAC Board of Directors Meeting minutes, May 18, 1983, PANS-EAC, vol. 3421, no. 3. Nunn himself asked the assembled lawyers and witnesses at trial how "can I possibly absorb all this stuff?" Cumming, "Herbicide Case: Part II," 26.
107 McMahon, "Residents 'Guinea Pigs' in Spray Plan," 4.
108 One witness for the plaintiffs, Theodore Sterling, statistician and epidemiologist at Simon Fraser University, was incensed at Nunn's attitude, asking why only Nova Scotia, of all jurisdictions in the world, still allowed 2,4,5-T aerial spraying: "Even the Russians don't use it now." Cumming, "Herbicide Case: Part II," 26–27.
109 McMahon, "Residents 'Guinea Pigs' in Spray Plan," 4.
110 Cumming, "Herbicide Case: Part II," 27.
111 Glen Wannamaker, "Spray Opponents Lick Their Wounds and Get Ready for Another Fight," *Atlantic Insight* 5, no. 11 (November 1983): 7. Anger among the witnesses as well: the Swedish researcher Mikael Eriksson complained that "the judge himself admitted that he had difficulties comprehending, yet he acquits the company and accuses me and other independent researchers of being partial." *Dala-Demokraten,* November 10, 1983, enclosed with Letter from Hollis Routledge, NSFI, to D.L. Eldridge, November 21, 1983, DLF, vol. 882, no. 3.
112 *Between the Issues* 3, no. 4 (October 1983).
113 Peter Cumming, "The Herbicide Fight," *Rural Delivery* 7, no. 12 (May 1983).
114 *Between the Issues* 3, no. 5 (December 1983).
115 Cumming, "The Stink of Defeat (draft)," 3.
116 Susan Murray, "Trouble in the Anti-Spray House," *Atlantic Insight* 6, no. 6 (June 1984): 7. At least one plaintiff, Bob Sampson, also told the CBC that he would have given up much sooner if he could have: "Transcript of CBC Inquiry," March 16, 1983, DLF, vol. 882, no. 3.
117 Cumming, "The Stink of Defeat (draft)," 4.
118 United Press Agency of the Swedish Press, December 13, 1983, clipping file, DAL-EAC, box 29.9. May had no right to appeal on her own, as she was not a true plaintiff. Stephanie May, "Getting into PAN," *Between the Issues* 4, no. 1 (May-June 1984); "Interview with Ryan Googoo," *Between the Issues* 3, no. 6 (February-March 1984).
119 Cumming, "The Stink of Defeat (draft)," 8–9. In the end, only $3,993.91 went to NSFI: "Details of Finalization of Settlement," DAL-EAC, box 24.8. The HFS formed a new Citizens Against Pesticides (CAP) group on December 14, 1983, initially under the leadership of John Shaw, Jeff Brownstein, Aaron Schneider, Peter Cumming, and John Roberts, with support from Margaree Environmental Association, FALASH, Victoria County Landowners, Mabou Support Group, Glendale Against Herbicide Spraying, Whycocomagh Against Herbicide Spraying, Antigonish Support Group, Baddeck–Middle River Support Group, Sydney–Big Pond Support Group: "Announcement: 1st AGM of CAP – 11 February 1984, Baddeck," n.d., DAL-EAC, box 24.9.
120 Cumming, "The Stink of Defeat (draft)," 6.
121 EAC Communications Resource Group Meeting minutes, January 27, 1983, PANS-EAC, vol. 3421, no. 11.
122 Elizabeth May to the Editor, *Chronicle-Herald,* February 25, 1983, clipping file, DAL-EAC, box 39.7; *Q/L SSEPA News* 1, no. 12 (May 1983).

123 Speech by George Henley, n.d., DAL-EAC, box 24.5. Connie Schnell in South Haven, Victoria County, soon began making "I'm an ENVIRONMENTAL SUBVERSIVE" buttons to sell for the HFS: Peter Cumming to EAC, DAL-EAC, box 24.5.

124 John Goodall, Stellarton, to the Editor, *Chronicle-Herald,* October 12, 1982, clipping file, RCOF, vol. 192, no. 2.

125 EAC Board of Directors Meeting minutes, May 24, 1982, PANS-EAC, vol. 3421, no. 2; EAC Board of Directors Meeting minutes, February 13, 1980, PANS-EAC, vol. 3420, no. 24. The EAC's leaders discussed the change themselves in 1978, noting the organization's "lack of contact with membership," "lack of outreach," and poor "relationship of centre to hinterlands." (The fact that they viewed the Centre as "the centre" is also quite telling, given the great vitality of environmentalism outside of Halifax.) Special Meeting of Board of Directors minutes, December 9, 1978, PANS-EAC, vol. 3420, no. 22.

126 Quoted in *Pesticide Action Network Newsletter* 2 (1983), PANS-EAC, vol. 3434, no. 3. It is worth noting that Nunn's ruling was against the plaintiffs, but not necessarily for the defence – the failure of one to make the case does not necessarily mean that the other's position is therefore correct – but that kind of nice distinction rarely comes across.

127 Department of Environment spokesperson Linda Laffin, quoted in *Between the Issues* 3, no. 4 (October 1983).

128 "Minutes of Joint Meeting, Deputy Ministers of Health or Representatives with ESBC," May 1, 1980, DLF, vol. 604, no. 5.

129 Christine Blair, "Tour Provides Sobering Look at Budworm-Damaged Forest," *Forest Times* 5, no. 3 (August 1983): 8; T.D. Smith, Lands and Forests, to R.E. Bailey, Lands and Forests, June 10, 1983, DLF, vol. 882, no. 3; "Misconceptions about Forest Industry Must Be Corrected," *Chronicle-Herald,* January 19, 1984, 40; "Funding Approved for N.S. Forestry Exhibition," *Chronicle-Herald,* January 21, 1984, 18. (The amount was $4,270, from Lands and Forests.)

130 Environmental Control Council, "Discussion Paper on Herbicide Use in Nova Scotia," n.d. [c. late 1983], DAL-EAC, box 24.9.

131 Alan Jeffers, "Streatch Will Take Over Forestry Board," *Chronicle-Herald,* March 30, 1984, 17; EAC press release, n.d., DAL-EAC, box 29.9.

132 Blair, "Tour Provides Sobering Look at Budworm-Damaged Forest," 8.

133 EAC to Charles Caccia, federal Minister of Environment, February 20, 1984, DAL-EAC, box 29.9; "Forestry Taught in Schools," *Chronicle-Herald,* December 17, 1983, 23.

134 Peter Cumming, "CAP RAP," May 2, 1984, DAL-EAC, box 24.9. "Truth in Forestry" was a pro-spray pressure group originating in Cape Breton but registered to NSFI's lawyers' firm in Halifax: *Between the Issues* 3, no. 6 (February-March 1984).

135 On ground sprays: D.L. Eldridge to G. Kerr, Minister of Environment, August 12, 1983, DLF, vol. 606, no. 8; Gerard MacLellan, Department of Environment, to D.L. Eldridge, August 19, 1983, DLF, vol. 882, no. 3; Greg Kerr to Ake Thor, NSFI, August 4, 1983, RCOF, vol. 186, no. 10. (Environment and Lands and Forests also encouraged Scott Paper to use ground spray on Christmas tree farms when aerial permits were denied by the Department of Public Health due to the nearby source of water for the Town of Wolfville: D.L. Eldridge to G. Henley, April 28, 1983, DLF, vol. 606, no. 8.) On the minister's vow: M. Haase to Premier Buchanan, July 15, 1983, DAL-EAC, box 43.5. On the end of public notice: George Moody, Minister of Environment, to Ryan Googoo, November 6, 1985, DAL-EAC, box 43.7; Letter discussing end of newspaper notification, from E.L.L. Rowe, Department of Environment, to file, January 5, 1983, DLF, vol. 606, no. 8.

136 EAC press release, August 17, 1983, DAL-EAC, box 24.7; "NDP Protest Spray Plans," *Chronicle-Herald,* August 26, 1983, 15; "NSFI Target of Anti-Spray Protestors," *Chronicle-Herald,* August 16, 1983, 21. Government had of course not stopped listening to the pulp and paper industry, and it is in that correspondence that Kerr revealed the total acreage sprayed with chemical herbicides, mostly phenoxy, in 1982 (fifteen thousand acres, both aerial and ground), and pushed for more "public education" from industry. Greg Kerr, Minister of Environment, to Ake Thor, NSFI, August 4, 1983, DLF, vol. 186, no. 10.

137 Quote from Agricultural Association of Canada president Jack Elliott, at a Department of Agriculture workshop, "Pesticides in Perspective – An Educational Seminar," April 9, 1984, quoted in *Between the Issues* 4, no. 1 (May-June 1984).

138 "Infamous Gypsy Moth Found in Alarming Numbers in Western Half of Nova Scotia," *Chronicle-Herald,* February 7, 1984, 25. The pressure to spray chemical insecticides to kill spruce budworm never really ceased either: "Cape Breton Landowners Information Update," January 1982, DAL-EAC, box 24.9.

139 Most, possibly all, of the sites at issue in the trial (among others) were sprayed with glyphosate and 2,4-D in 1984: *Between the Issues* 4, no. 1 (May-June 1984); George Moody, Minister of Environment, to Ryan Googoo, November 6, 1985, DAL-EAC, box 43.7; Paul Schneidereit, "Varying Views on Herbicide Spraying," *Chronicle-Herald,* July 21, 1984, 21.

140 Cumming, "The Stink of Defeat (draft)," 7.

141 Ibid., 10. Much the same sentiment came from Hester Lessard of the SSEPA: Murray, "Trouble in the Anti-Spray House," 7.

142 Peter Cumming, "Out of the Courts and Back to the Issue," *Between the Issues* (October 1983).

143 Cape Breton Island Wildlife Association, "Spraying: Removing Nature's Food Supply," *Chronicle-Herald,* December 15, 1983, 7.

144 David Orton's opinion was that success against biocides would require civil disobedience, and activists ought to know that "the provincial government will be on the other side of the barricades." David Orton to the Editor, *Rural Delivery* 8, no. 1 (June 1983).

145 *Q/L SSEPA News* 2, no. 2 (July-August 1983).

CHAPTER 4: TWO ENVIRONMENTALISMS

1 Lawrence Welsh, "Uranium Industry Faces Low Sales, Poor Prices as Power Demand Falls," *Globe and Mail,* March 29, 1982, B3;Bruce Little, "Shades of the Klondike in Atlantic Canada," *Atlantic Insight* 1, no. 2 (May 1979): 25; Province of Nova Scotia, Department of Mines and Energy, *Uranium in Nova Scotia: A Background Summary for the Uranium Inquiry – Nova Scotia, Report 82-7* (Halifax: Government of Nova Scotia, 1982), 3–9. Companies exploring for uranium in the province included Lacana, Gulf, and Noranda on the north shore; and Esso, Aquitaine, Shell, Eldorado, Norcen, and Saarberg on the southern mainland.

2 Ralph Torrie, "BC's Inquiry and Moratorium," *CCNR's Transitions* 3, no. 1 (May 1980); Jim Harding, *Canada's Deadly Secret: Saskatchewan Uranium and the Global Nuclear System* (Halifax: Fernwood, 2007), 28; Document on Inuit opposition to uranium mining, DAL-EAC, box 42.9; Amy Zierler, "Labrador's Great Debate: Who Wants a Uranium Mine?" *Atlantic Insight* 2, no. 4 (May 1980): 18; Saskatchewan Environmental Society and Regina Group for a Non-Nuclear Society to Susan Holtz, 1977, PANS-EAC, vol. 3434, no. 14.

Government support went as far as participation in an international price-fixing cartel involving all of Canada's uranium producers, the United Kingdom, France, South Africa, and Australia. "Uranium Cartel Probe Reports Laws Broken," *Globe and Mail*, May 27, 1981, quoted in Donna Smyth, *Subversive Elements* (Toronto: Women's Press, 1986), 116–17.

3 Ken Kelley, Sable River, to the Editor, *Rural Delivery* 4, no. 4 (September 1979); Ken Kelley, "Mining for Death," *Rural Delivery* 4, no. 6 (November 1979); EAC Board of Directors Meeting minutes, August 8, 1979, PANS-EAC, vol. 3420, no. 23; Ron and Ruth Loucks to the Editor, *Rural Delivery* 4, no. 8 (January 1980); FACT Newsletter, March 1980, DAL-EAC, box 41.3.

4 Joanne Lamey, CPAC, to Ginny Point, EAC, January 14, 1980, DAL-EAC, box 30.8.

5 Joanne Lamey, CPAC, to Ginny Point, EAC, February 13, 1980, DAL-EAC, box 30.8; *Jusun* 8, no. 2 (Summer 1980), PANS-EAC, vol. 3420, no. 24.

6 Early, "'A Grandly Subversive Time,'" 28, 36.

7 Jocelyn Rhodenizer, South Berwick Women's Institute, to EAC, September 17, 1980, DAL-EAC, box 43.2; Document on Nova Scotia Women's Institutes, January 21, 1980, PANS-EAC, vol. 3433, no. 38.

8 "Farm Women's Conference," *Rural Delivery* 5, no. 8 (January 1981); Burlington and Summerville Women's Institutes, *Hants Journal*, February 25, 1981, 5.

9 "Conference on Health Effects of Radiation, Moncton," February 20–22, 1981, DAL-EAC, box 30.10.

10 Al Kingsbury, "Uranium Moratorium Urged," *Mail-Star*, March 13, 1981, 52; Gwen Davies and Bill Johnston, "Aquitaine Mines a Rich Vein of Controversy," *Atlantic Insight* 3, no. 8 (October 1981): 16; EAC Board of Directors Meeting minutes, January 28, 1981, PANS-EAC, vol. 3420, no. 25; Burlington Women's Institute, "Brief Presented to the West Hants Municipal Council on the Subject of Uranium Mining," March 12, 1981, PANS, Royal Commission on Uranium Mining fonds, RG 44 [hereafter RCU], vol. 206, no. 28.

11 SSEPA Resolution, March 26, 1981, DAL-EAC, box 30.8; EAC Resolution, March 25, 1981, DAL- EAC, box 41.21; VOW Meeting minutes, May 4, 1981, PANS-BP, vol. 3470, no. 6.

12 VOW Atlantic Newsletter, May 1, 1981, PANS-BP, vol. 3472, no. 1; Donna Smyth, "Selling Radon Daughters: Uranium in Nova Scotia," *Canadian Forum* 61, no. 714 (December/ January 1981): 6–8.

13 Al Kingsbury, "Hants Uranium Prospects Good," *Mail-Star*, April 13, 1981, 19.

14 Ralph Loomer, "Nova Scotia Can Do without Uranium Invasion," *Chronicle-Herald*, June 29, 1981, 7; VOW Atlantic Newsletter, May 1, 1981, PANS-BP, vol. 3472, no. 1.

15 Nova Scotia, *Hansard*, April 2, 1981; Davies and Johnston, "Aquitaine Mines a Rich Vein of controversy," 16; Hal Mills and Ron Loucks, "Uranium Mining Meeting Notes," April 6, 1981, DAL-EAC, box 30.10; MLA David Nantes, Select Committee on Uranium Mining, to Susan Holtz, September 3, 1981, DAL-EAC, box 44.4.

16 "New Group Joins Anti-Nuclear Drive," *Chronicle-Herald*, June 2, 1981, 22; CAPE press release, July 1, 1981, DAL-EAC, box 30.8.

17 Gillian Thomas, CAPE, to Daphne Taylor, EAC (marginal notations), May 25, 1981, DAL-EAC, box 30.8.

18 J. Brian Harley, "Maps, Knowledge, and Power," in *The Iconography of Landscape*, ed. D. Cosgrove and S. Daniels (Cambridge: Cambridge University Press, 1988), 277.

19 Clayton Campbell, "Pressure Groups Threaten Uranium Mining Operation, Conference Told," *Chronicle-Herald*, June 25, 1981, quoted in Smyth, *Subversive Elements*, 63.

20 Alexa McDonough to Ginny Point, EAC, December 18, 1981, DAL-EAC, box 30.8.

21 Muriel Siemers, "The Uranium Issue: The Harmful Effect on Agriculture," *Chronicle-Herald,* September 24, 1981, 7.

22 Press release, September 21, 1981, DAL-EAC, box 33.4; Alan Jeffers, "Issuing of Drilling Permits 'Madness,'" *Chronicle-Herald,* September 22, 1981, 2.

23 Jeffers, ibid.

24 Department of Mines and Energy, *Uranium in Nova Scotia,* 13.

25 "Saarberg Drops Plans for Uranium Exploration," *Chronicle-Herald,* September 24, 1981, 5.

26 *Chronicle-Herald,* November 6, 1981, clipping file, RCU, vol. 206, no. 28; "Medical Society of Nova Scotia General Council, Community Health Committee Report," November 20–21, 1981, RCU, vol. 201, no. 10.

27 *Chronicle-Herald,* October 26, 1981, clipping file, RCU, vol. 206, no. 28.

28 Donna Smyth, "The Public Debate Begins," *Rural Delivery* 6, no. 7 (December 1981).

29 Daphne Taylor, EAC, to Betty Peterson, December 17, 1981, DAL-EAC, box 30.8; Smyth, ibid.

30 Roy John to Fred Barrett, FACT, December 9, 1981, DAL-EAC, box 30.8.

31 Unknown to the Editor, *The Mirror,* Digby, n.d., clipping file, DAL-EAC, box 30.10.

32 Roy John to Fred Barrett, FACT, December 9, 1981, DAL-EAC, box 30.8.

33 "Hazardous Wastes in Atlantic Canada," Acadia University Conference notes, DAL-EAC, box 35.12.

34 Notice of Bertell's press conference, November 12, 1981, DAL-EAC, box 33.4; Donna Smyth, "Uranium Update," *Rural Delivery* 6, no. 8 (January 1982). At the request of some Nova Scotian activists, Bertell penned a withering response to Dalgleish in a letter to Joan Brown-Hicks at the Halifax library, December 16, 1981, DAL-EAC, box 30.10.

35 Rod Bantjes and Tanya Trussler, "Feminism and the Grass Roots: Women and Environmentalism in Nova Scotia, 1980–1983," *Canadian Review of Sociology and Anthropology* 36, no. 2 (1999): 194.

36 EAC press release, May 10, 1981, DAL-EAC, box 30.10.

37 "Health Hazards of Mineral Mining," EAC lecture (John Hartlen, Jack Garnett, Ralph Torrie), February 17, 1981, DAL-EAC, box 30.10.

38 EAC press release, May 10, 1981, DAL-EAC, box 30.10. CAPE's Gillian Thomas also objected in a letter to MLA Ron Russell, May 17, 1981, DAL-EAC, box 30.8.

39 Susan Holtz to Jack Garnett, December 20, 1980, DAL-EAC, box 30.11.

40 EAC Energy Committee Meeting minutes, March 18, 1981, DAL-EAC, box 30.11; EAC Energy Committee Meeting minutes, May 13, 1981, PANS-EAC, vol. 3420, no. 25.

41 "(EAC) Environmental Action Task Force Report," March 30, 1982, PANS-EAC, vol. 3421, no. 1; EAC Board of Directors Meeting minutes, June 24, 1981, PANS-EAC, vol. 3420, no. 25.

42 EAC Uranium Committee Meeting minutes, February 8, June 9, June 23, and July 7, 1982, DAL-EAC, box 30.11; David Orton, "Draft Resolution Regarding the Terms of Reference of the Uranium Committee, EAC," May 27, 1981, DAL-EAC, box 30.11.

43 David Orton, "What Makes an Activist? (Part 1)," *Deep Green Web,* April 2, 2011, http://deepgreenweb.blogspot.ca/2011/04/what-makes-activist-part-1.html.

44 David Orton, "Uranium Mining: Hearings into Dangers Needed," *Chronicle-Herald,* August 24, 1981, 7.

45 EAC Uranium Subcommittee minutes, May 27 and June 9, 1981, DAL-EAC, box 30.11. Orton was far from the only one to press a critique of the EAC's methods during this time.

See also Michael Clow, "A Left-Environmentalist Perspective on Canadian Industrial Strategy," paper presented at the Ecology Energy and Resources section of the Canadian Political Science Association 1981 Annual Meeting, Dalhousie University, Halifax, Nova Scotia, May 27–29, 1981; Peter Rowbottom, Cape Breton, to Susan Hotlz, March 11, 1981, DAL-EAC, box 44.3; "Moncton Conference Notes: 'Experts in Society' by Dr. Woollard," *Southern New Brunswick Nuclear News* 15 (February-March 1981); "Conference on Health Effects of Radiation, Moncton," February 20–22, 1981, DAL-EAC, box 30.10; David Underwood, Sable River, to the Editor, *Rural Delivery* 5, no. 11 (April 1981); Linda Christiansen-Ruffman, "The Ecology Action Center in the Seventies," February 1979, PANS-EAC, vol. 3420, no. 23.

46 *Chronicle-Herald,* November 6, 1981, clipping file, RCU, vol. 206, no. 28.

47 Sherri Cline to EAC, October 1, 1981, DAL-EAC, box 44.4.

48 David Orton to the head of the Halifax City Regional Library, October 1, 1981, DAL-EAC, box 30.8.

49 SSEPA press release, n.d., DAL-EAC, box 44.6 [emphasis in original].

50 Smyth, "The Public Debate Begins."

51 Susan Holtz to Hattie Perry, October 29, 1981, PANS-EAC, vol. 3421, no. 1.

52 Susan Holtz, "EAC's Position on Uranium Development and the N.S. Uranium Exploration Guidelines/Regulations," November 18, 1981, PANS-EAC, vol. 3431, no. 7, and PANS-EAC, vol. 3424, no. 12.

53 Anne Wickens to Susan Holtz, n.d., PANS-EAC, vol. 3421, no. 1.

54 Sherri Cline to Susan Holtz, January 20, 1982, PANS-EAC, vol. 3421, no. 1.

55 Anne Wickens to Susan Holtz, n.d., PANS-EAC, vol. 3421, no. 1.

56 Dorien Freve to Hattie Perry, December 8, 1981, PANS-EAC, vol. 3421, no. 1.

57 Linda Christiansen-Ruffman, "The Ecology Action Center in the Seventies," February 1979, PANS- EAC, vol. 3420, no. 23.

58 Susan Holtz to Bob Paehlke, January 14, 1981, DAL-EAC, box 44.4.

59 Bessa Ruiz to Sherri Cline, February 11, 1982, PANS-EAC, vol. 3421, no. 1. The EAC also denied its reliance on government funding, an effort undermined by its explanation in the press of the need for more of it: Sandy Smith, "Government Funds Needed to Continue EAC's Work," *Chronicle-Herald,* September 20, 1981, 5.

60 EAC news release, January 13, 1982, DAL-EAC, box 33.4; "Nuclear Technology in Nova Scotia and Canada: An Unacceptable Risk," 1982, DAL-EAC, box 37.4. Groups represented were VOW, EAC, CAPE, FACT, KASE, and Project Ploughshares.

61 Gwen Davies and Ginny Point, "Looking Back over the Past Two Years," draft, EAC Annual Report 1980–82, PANS-EAC, vol. 3420, no. 4.

62 *Toronto Star,* March 12, 1983, enclosed with letter from G.C. Eglington, Ottawa, to McCleave, May 25, 1983, RCU, vol. 204, no. 14. McCleave had inquired about possible libel action against the newspaper.

63 Order in Council #82-200, February 9, 1982, in *Report of the Nova Scotia Royal Commission of Inquiry on Uranium* (Halifax: Queen's Printer, 1985).

64 EAC Board of Directors Meeting minutes, February 24, 1982, PANS-EAC, vol. 3421, no. 1. The nine groups were: CARE, CCCC, EAC, SEPOHG, Fish or Cut Bait Collective, Recreation Directors of Nova Scotia, Taxpayers for a Safe Environment, WHEN, and the United Church – Church in Society Committee.

65 CARE press release, February 14, 1982, DAL-EAC, box 30.8.

66 EAC Brief to McCleave, January 29, 1982, marginal notations, RCU, vol. 201, no. 10.

67 EAC Board of Directors Meeting minutes, February 24, 1982, PANS-EAC, vol. 3421, no. 1.

68 Hattie Dyck, "McCleave Asks Groups' Cooperation in Inquiry," *Chronicle-Herald*, February 27, 1982; "Susan Holtz Time Allocation Sample," PANS-EAC, vol. 3421, no. 1; Martin Haase to McCleave, March 6, 1982, RCU, vol. 202, no. 5.

69 McCleave to Glenys Livingstone and Stanley Forgeron, Inquiry staff, March 18, 1982, RCU, vol. 203, no. 6.

70 Editorial, *Truro Daily News*, March 9, 1982, 4; Robert McCleave to John Conrad, Managing Editor, *Amherst Daily News*, March 9, 1982, RCU, vol. 201, no. 7; Unknown to Gillian Thomas, March 24, 1982, DAL-EAC, box 30.9.

71 Robert Whiting to McCleave, March 2, 1982, RCU, vol. 202, no. 5.

72 Judy Davis, Tatamagouche, to McCleave, March 4, 1982, RCU, vol. 201, no. 2.

73 Winston Settle to Robert McCleave, February 27, 1982, RCU, vol. 203, no. 6.

74 "Liberal MLAs Believe Uranium Inquiry Chairman Judge Robert McCleave Has Displayed a Contempt for Free Speech and They Want the Government to Remove Him from the Post," unknown publication, March 5, 1982, clipping file, DAL-EAC, box 33.3.

75 Jim Lotz to McCleave, March 8, 1982, RCU, vol. 201, no. 1.

76 McCleave to Alexa McDonough, March 17, 1982, RCU, vol. 201, no. 10.

77 McCleave to Gillian Thomas, April 10, 1982, DAL-EAC, box 30.9.

78 McCleave to Russell Logan, President, Maritime Diamond Drilling Co. Ltd., n.d., RCU, vol. 201, no. 2.

79 Ralph Loomer to Robert McCleave, February 27, 1982, RCU, vol. 201, no. 2. The Women's Health Education Network and EAC also distanced themselves from CARE: Janet Campbell (WHEN) to Robert McCleave, March 11, 1982, RCU, vol. 201, file 8; Dr. J.E. Baker (EAC) to Robert McCleave, March 4, 1982, PANS-EAC, vol. 30, no. 8.

80 For example, a new branch of the CCCC in Parrsboro: Environmental Protection in Cumberland South (EPICS) press release, n.d., DAL-EAC, box 33.4.

81 McCleave to Archbishop Hayes (Catholic), Halifax, March 23, 1982, RCU, vol. 201, no. 10; Betty Peterson, Halifax Monthly Meeting of Friends (Quaker), to McCleave, February 5, 1983, RCU, vol. 202, no. 1; McCleave to Reverend Archdeacon C.R. Elliott (Anglican), Halifax, April 10, 1982, RCU, vol. 202, no. 1; W.R. MacDonald, Clerk of Session, Saint Andrew's United Church, Wolfville, to McCleave, February 22, 1982, RCU, vol. 201, no. 2; Reverends Clint Mooney, Donald McLeod, and K.O. Robinson, to McCleave, March 8, 1982, RCU, vol. 201, no. 8; Joan Cunningham, Nova Scotia Provincial Council of the Catholic Women's League of Canada, to McCleave, n.d., RCU, vol. 201, no. 10.

82 Anne Bishop, Lismore Community Concern Committee, and Umbrella Co-operative Ltd., transcripts of Pictou hearing, hearing #16, June 15, 1982, RCU, vol. 197, nos. 5 and 6; Robert Whiting, transcripts of Chester hearing, hearing #21, July 7, 1982, RCU, vol. 198, no. 3; Jeffrey Gold, for Beverly Brett, transcripts of Baddeck hearing, hearing #31, September 8, 1982, RCU, vol. 198, no. 23; Ian Ball, transcripts of Halifax hearing, hearing #34, September 17, 1982, RCU, vol. 199, nos. 5 and 6.

83 Michael Keddy, transcripts of New Ross hearing, hearing #1, April 2, 1982, RCU, vol. 195, no. 12.

84 Norma Flynn, RESCUE, transcripts of Vaughan hearing, hearing #19, June 25, 1982, RCU, vol. 197, no. 12; Dr. C.J. Byrne, transcripts of Halifax hearing, hearing #10, May 21, 1982, RCU, vol. 196, no. 6.

85 Ron Leitold, New Germany, transcripts of Bridgewater hearing, hearing #3, April 20, 1982, RCU, vol. 195, no. 16.

86   Muriel Maybee, Lunenburg County Women's Group, transcripts of Bridgewater hearing, hearing #3, April 20, 1982, RCU, vol. 195, no. 16.

87   Erin Gore, transcripts of Chester hearing, hearing #5, April 30, 1982, RCU, vol. 195, no. 21. Bantjes and Trussler surveyed women involved in the anti-uranium movement in Nova Scotia and found a majority – 78 percent of those who could be placed – had come "from away," that is, had been born out of the province: Bantjes and Trussler, "Feminism and the Grass Roots," 192.

88   The clearest such argument came from Ross Baker, transcripts of Truro hearing, hearing #24, July 13, 1982, RCU, vol. 198, no. 9. Several female presenters also explicitly tied the legitimacy of their moral arguments to their status as mothers; for example, Cathleen Kneen, Pictou County VOW, transcripts of Pictou hearing, hearing #16, June 15, 1982, RCU, vol. 197, no. 6.

89   Torrie, "BC's Inquiry and Moratorium"; Harding, *Canada's Deadly Secret.*

90   Ernest Forbes, *The Maritime Rights Movement 1919–1927: A Study in Canadian Regionalism* (Montreal and Kingston: McGill-Queen's University Press, 1979); Margaret Conrad, "The Atlantic Revolution of the 1950s," in *Beyond Anger and Longing: Community and Development in Atlantic Canada,* ed. Berkeley Fleming (Fredericton: Acadiensis, 1988), 55–98; Jennifer Smith, "Intergovernmental Relations, Legitimacy, and the Atlantic Accords," *Constitutional Forum* 17, no. 3 (2008): 81–98.

91   Brian McVeigh, transcripts of Chester hearing, hearing #5, April 30, 1982, RCU, vol. 195, no. 21. Michael Marshall pointed out that mining companies accustomed to operating in the Canadian north, where they experienced few constraints on their activities, failed to see the difference between Nova Scotia and Canada: transcripts of Halifax hearing, hearing #10, May 21, 1982, RCU, vol. 196, no. 6.

92   Robert Finck, transcripts of Bridgewater hearing, hearing #3, April 20, 1982, RCU, vol. 195, no. 16.

93   Kenneth Seaboyer, transcripts of New Ross hearing, hearing #1, April 2, 1982, RCU, vol. 195, no. 12.

94   Muriel Maybee, Lunenburg County Women's Group, transcripts of Bridgewater hearing, hearing #3, April 20, 1982, RCU, vol. 195, no. 16.

95   AECB, transcripts of Halifax hearing, hearing #17, June 22, 1982, RCU, vol. 197, no. 7; Valerie Wilson, transcripts of Annapolis Royal hearing, hearing #20, July 6, 1982, RCU, vol. 198, no. 1.

96   Robert Bays, "Uranium Exploration and Mining: Fact and Fiction," RCU, vol. 198, no. 2; AECB, transcripts of Halifax hearing, hearing #17, June 22, 1982, RCU, vol. 197, no. 7.

97   Drs. Linda Christansen-Ruffman and Karin Flikeid, Canadian Research Institute for the Advancement of Women – Nova Scotia, transcripts of Halifax hearing, hearing #15, June 11, 1982, RCU, vol. 197, no. 4.

98   Martin Haase, transcripts of Chester hearing, hearing #5, April 30, 1982, RCU, vol. 195, no. 21.

99   Judy Davis, CARE, to the Editor, *Rural Delivery* 6, no. 12 (May 1982).

100   Robert Whiting, transcripts of Chester hearing, hearing #21, July 7, 1982, RCU, vol. 198, no. 3; tape recording of the same, RCU, Ac 673. Whiting's speech also included one of several open threats of violent civil disobedience received by the inquiry.

101   Gordon Campbell, "Our Ways Never Failed Us," music and lyrics, performed at Parrsboro, October 7, 1982, RCU, vol. 199, no. 23.

102 Susan Hower, transcripts of Liverpool hearing, hearing #4, April 27, 1982, RCU, vol. 195, no. 18.

103 Dr. William Thurlow, Digby, transcripts of Bear River hearing, hearing #6, May 6, 1982, RCU, vol. 195, no. 22; Dr. A.R. Robertson, for staff group at St Martha's Hospital and Antigonish Committee Concerned with Uranium Exploration, Antigonish, and Dr. Robert Sers, Antigonish, transcripts of Antigonish hearing, hearing #37, September 28, 1982, RCU, vol. 199, no. 11; Drs. W.L. Phillips and J.D.A. Henshaw, Valley Medical Society, transcripts of Wolfville hearing, hearing #9, May 19, 1982, RCU, vol. 196, no. 4.

104 Ralph Torrie, Ottawa, transcripts of Chester hearing, hearing #21, July 7, 1982, RCU, vol. 198, no. 3; Judy Pasternak, *Yellow Dirt: A Poisoned Land and a People Betrayed* (New York: Free Press, 2010).

105 One very good example is seen in the May family presentations in Margaree Harbour on September 7, 1982, hearing #30, which included (from Elizabeth May) criticism of the mining industry for attempting to politically influence government during the inquiry: RCU, vol. 198, no. 21.

106 Smyth, *Subversive Elements,* 233.

107 Paula Scott to MLA Ron Barkhouse, April 3, 1982, RCU, vol. 201, no. 3.

108 Donna Smyth, "Uranium Update: Trouble in the Barnyard," *Rural Delivery* 6, no. 11 (April 1982); EAC Board of Directors Meeting minutes, May 24, 1982, PANS-EAC, vol. 3421, no. 2.

109 Gillian Thomas, transcripts of Vaughan hearing, hearing #19, June 25, 1982, RCU, vol. 197, no. 12.

110 Tape recording of Chester hearing, hearing #21, July 7, 1982, RCU, Ac 673.

111 David Orton, quoted in Susan Murray, "Sour Notes in the 'Judge McCleave Waltz,'" *Atlantic Insight* 5, no. 6 (June 1983): 12–13.

112 Transcripts of Amherst hearing, hearing #42, October 7, 1982, RCU, vol. 199, no. 20.

113 Nova Scotia Federation of Labour, transcripts of Halifax hearing, hearing #22, July 9, 1982, RCU, vol. 198, no. 5.

114 Tony Seed, transcripts of Halifax hearing, hearing #38, September 29, 1982, RCU, vol. 199, no. 13.

115 McCleave to Vivian Wittenburg, May 3, 1983, RCU, vol. 203, no. 7.

116 David Orton to the Editor, *Rural Delivery* 8, no. 3 (August 1982).

117 Unknown to the Editor, *Digby Mirror,* DAL-EAC, box 30.10.

118 Atlantic Canada Chapter, Northeast Atlantic International Section, Air Pollution Control Association, *Proceedings of the Sixth Annual Technical Meeting,* September 29–30, 1982; Don MacDonald, "'Subversives Holding Up Uranium Development,'" *Chronicle-Herald,* September 3, 1982, with brief by Elizabeth May, RCU, vol. 198, no. 21.

119 Fred McMahon, "Environmental Activists' Motives Questioned," *Chronicle-Herald,* February 23, 1983, with brief by Elizabeth May, RCU, vol. 198, no. 21.

120 David Helvarg, *The War against the Greens: The 'Wise Use' Movement, the New Right, and Anti-Environmental Violence* (San Francisco: Sierra Club Books, 1994).

121 Elizabeth May complained about "a kind of McCarthyism" in a letter to Robert McCleave, February 23, 1983, RCU, vol. 203, no. 6. Marilyn Manzer did similarly in a letter to McCleave, October 1, 1982, RCU, vol. 202, no. 2; Donna Smyth, "Uranium Update: Can It!" *Rural Delivery* 7, no. 5 (October 1982); Dirk van Loon, Editorial, *Rural Delivery* 7, no. 5 (October 1982).

122 Ginny Point, transcripts of Halifax hearing, hearing #15, June 11, 1982, RCU, vol. 197, no. 4.

123 EAC press release, November 18, 1982, DAL-EAC, box 33.4.

124 Dean Whalen (CARE) to Robert McCleave, March 4, 1982, RCU, vol. 203, no. 6; CARE brief, June 15, 1982, RCU, vol. 203, no. 6.

125 CARE brief, June 15, 1982, RCU, vol. 203, no. 6; "CARE presentation to the NS Uranium Inquiry," June 15, 1982, DAL-EAC, box 30.8.

126 "Dean Whalen – Sherri Cline Incident at Pictou," RCU, vol. 203, no. 6; McCleave to Dean Whalen, June 16, 1982, DAL-EAC, box 30.9.

127 Tony Law, Scotsburn, to Robert McCleave, June 22, 1982, RCU, vol. 202, no. 7.

128 Stanley Forgeron, inquiry secretary, to Fred Popowich, assistant librarian, Pictou-Antigonish Regional Library, November 17, 1982, RCU, vol. 203, no. 3. The librarian was told that CARE had never submitted a written brief.

129 "Dean Whalen – Sherri Cline Incident at Pictou," RCU, vol. 203, no. 6.

130 Don Rushton to the Editor, *Rural Delivery* 7, no. 2 (July 1982).

131 EAC Board of Directors Meeting minutes, June 23, 1982, PANS-EAC, vol. 3421, no. 2.

132 Multiple letters, RCU, vol. 200, no. 9; Gwenyth Phillips (KASE) to Robert McCleave, February 26, 1983, RCU, vol. 202, no. 4; Gillian Thomas, Bill Zimmerman, and George Gore, the Western Group Drafting Committee, to the Editor, *Rural Delivery* 7, no. 8 (January 1983). The four were Ruth Conrad, Ron Leitold, Elizabeth May, and Martin Haase.

133 Sherri Cline (CARE), Gail Fresia (CCCC), Don Rushton (CCCC), and David Orton (SEPOHG), to the Editor, *Rural Delivery* 7, no. 7 (December 1982) [emphasis in original].

134 Gillian Thomas (CAPE), Bill Zimmerman (SSEPA), and George Gore (CAUM), the "Western Group Drafting Committee," to the Editor, *Rural Delivery* 7, no. 8 (January 1983).

135 McCleave to participants, February 18, 1983, DAL-EAC, box 30.8.

136 Uranium Commission press release, n.d., RCU, vol. 203, no. 6; McCleave to all stage 1 participants, February 18, 1983, DAL-EAC, box 30.8.

137 Subpoena to Charles Lapp, RCU, vol. 203, no. 6. Lapp later sent a bill to McCleave's office for $905.30 to cover his lawyer's fee: RCU, vol. 203, no. 6.

138 Transcript, March 4, 1983, RCU, vol. 200, no. 10; McCleave to David Chipman, President of the Nova Scotia Barristers' Society, April 6, 1983, RCU, vol. 210, no. 10.

139 Transcript, March 4, 1983, RCU, vol. 200, no. 10; McCleave to Don Rushton, March 7, 1983, RCU, vol. 203, no. 6.

140 Transcript, March 4, 1983, RCU, vol. 200, no. 10.

141 *Truro Daily News*, March 14, 1983, 4; Transcript of CBC Radio Noon, March 7, 1983, RCU, vol. 203, no. 6.

142 *Truro Daily News*, March 14, 1983, 4; Nova Scotia House of Assembly Debates and Proceedings, March 7, 1983, RCU, vol. 204, no. 13.

143 "Public Screening of: Uranium the Nova Scotia Experience – the 30 minute slide-tape show subpoenaed by Judge McCleave," poster, RCU, vol. 203, no. 6. Donna Smyth's account of the whole episode in *Rural Delivery* also came with a small drawing of a redacted kangaroo, added by the publisher: Donna Smyth, "Uranium Update: Beyond Uranium," *Rural Delivery* 7, no. 11 (April 1983).

144 Transcript, March 18, 1983, RCU, vol. 200, no. 11. McCleave admitted to requiring medical advice to handle the pressure, in a letter to E.L.L. Rowe, Deputy Minister of Environment, May 3, 1983, RCU, vol. 201, no. 10.

145 Transcript, March 18, 1983, RCU, vol. 200, no. 11; Don Rushton, "So Long, Stage II: McCleave Gets Set to Ride Off into the Sunset," *New Maritimes* 2, no. 6 (March 1984): 10–11. Smyth estimated twenty protesters, Rushton thirty.

146 Donna Smyth, "Uranium Update: To Be Continued...," *Rural Delivery* 7, no. 12 (May 1983).

147 Jim Vibert, "How Will 'Examine' Opposition Charges," *Chronicle-Herald*, n.d., clipping file, RCU, vol. 202, no. 10; Attorney General press statement, March 22, 1983, DAL-EAC, box 30.8; Deputy Attorney General to McCleave, March 22, 1983, RCU, vol. 204, no. 4.

148 Robert McCleave to Clive Schaefer, May 6, 1983, RCU, vol. 203, file 7.

149 Susan Holtz to Harry How, Attorney General, March 9, 1983, DAL-EAC, box 43.6; Smyth, "Uranium Update: To Be Continued."

150 Ralph Loomer to Robert McCleave, March 10, 1983, vol. 202, no. 2. Stephanie May also wrote to reassure McCleave that the May family was still "singing your praises": Letter from Stephanie May to McCleave, July 7, 1983, RCU, vol. 203, no. 7.

151 Hal Mills, co-chair, EAC Uranium Committee, to Robert McCleave (copy open letter to Premier Buchanan), c.c. to the Attorney General, March 21, 1983, RCU, vol. 203, no. 4. Also printed in *Between the Issues* 3, no. 2 (May-June 1983).

152 EAC Board of Directors Meeting minutes, March 23, 1983, PANS-EAC, vol. 3421, no. 3.

153 Sherri Cline to Susan Holtz, March 28, 1983, DAL-EAC, box 30.8.

154 McCleave to participants, February 18, 1983, DAL-EAC, box 30.8; Rushton, "So Long Stage II," 11.

155 Rushton, ibid.; Sherri Cline to the Editor, *Rural Delivery* 7, no. 12 (May 1983).

156 Smyth, "Uranium Update: Beyond Uranium."

157 *Q/L SSEPA News* 1, no. 12 (May 1983).

158 *Report of the Nova Scotia Royal Commission on Uranium Mining.*

159 EAC Uranium Committee minutes, April 10, 1985, PANS-EAC, vol. 3421, no. 5; Lyndon Watkins, "Anti-Uranium Winners Plan Picnic," *Mail-Star,* March 20, 1985, quoted in Smyth, *Subversive Elements,* 262–63.

160 *Report of the Nova Scotia Royal Commission on Uranium Mining;* "Position Paper," RCU, vol. 195, no. 3.

161 EAC Uranium Committee minutes, April 10, 1985, PANS-EAC, vol. 3421, no. 5; Watkins, "Anti-Uranium Winners Plan Picnic."

162 EAC Uranium Committee minutes, ibid.

163 McCleave to Roy John, National Uranium Tailings Program, April 12, 1984, RCU, vol. 203, no. 7; Martin Haase to McCleave, February 3, 1984, RCU, vol. 203, no. 3.

164 Michael Doyle, "Use of Word 'Parade' Is Defamatory: Dr. Yaffe," *Chronicle-Herald,* January 18, 1985, quoted in Smyth, *Subversive Elements,* 257; Gillian Thomas to the Editor, *Rural Delivery* 8, no. 4 (September 1983).

165 *Arts Magazine,* n.d., clipping file, PANS-BP, vol. 3470, no. 6; "Civil Jury Rules Dr. Smyth Did Not Defame Dr. Yaffe," quoted in Smyth, *Subversive Elements,* 258.

166 *Report of the Nova Scotia Royal Commission on Uranium Mining;* EAC Board of Directors Meeting minutes, July 17, 1985, PANS-EAC, vol. 3421, no. 5.

CHAPTER 5: WATERMELONS AND MARKET GREENS

1 The value of Nova Scotia as a lens through which to view Canadian and North American environmentalism in general is particularly great in view of the persistence of this rural

vitality: while much of the rest of the country continued in rapid urbanization throughout the 1980s, Nova Scotia's proportion of rural population remained large and consistent throughout the late decades of the twentieth century, a population from which new rural activists appeared. Nationally, the percentage of rural population fell from 24 to 19 percent from 1981 to 2011; in Nova Scotia it fell only slightly, from 45 to 43 percent. New Brunswick and Prince Edward Island shared similarly high proportions of rural residents. Statistics Canada, "Population, Urban and Rural, by Province and Territory," http://www.statcan.gc.ca/tables-tableaux/sum-som/l01/cst01/demo62d-eng.htm.

2 Barbara Clark, Wallace, Cumberland Resists Unhealthful Development (CRUD), to the Editor, *Rural Delivery* 5, no. 11 (April 1981); Harry Thurston, "The C.R.U.D. Hits the Fan in Cumberland County," *Atlantic Insight* 3, no. 6 (July 1981): 21–22; Chris Wood, "The Toxic Time Bomb," *Atlantic Insight* 3, no. 7 (August/September 1981): 84, 87–88, 91.

3 Civil disobedience in this case included dumping twenty kilograms of nails on the road to the site, among other actions. Kevin Cox, "Power but No Glory in Point Aconi," *Globe and Mail*, March 16, 1991, D2; Campbell Morrison, "Appeal Judges Reject Aconi Foes' Arguments," *Daily News*, June 7, 1991, 3; Susan LeBlanc, "Trial Begins for Greenpeace Activists," *Mail-Star*, December 14, 1990, D20; Alan Jeffers, "C.B.'s Future Black Like Coal," *Chronicle-Herald*, December 6, 1990, D8; Patricia Lynn Hutchinson, "Committee Concerned Over Effects of Proposed Point Aconi Project," *Chronicle-Herald*, June 26, 1990, D8.

4 Maude Barlow and Elizabeth May, *Frederick Street: Life and Death on Canada's Love Canal* (Toronto: HarperCollins, 2000), 55.

5 Ibid., 75; G.L. Trider and O.C. Vaidya, *An Assessment of Liquid Effluent Streams at Sydney Steel Corporation* (Ottawa: Environment Canada, Environmental Protection Service, Atlantic Region, 1980).

6 Bruno Marcocchio to the Editor, *Cape Breton Post*, February 7, 2004, http://www.safecleanup.com; Glenn Hanam to the Editor, *Cape Breton Post*, March 6, 2004, http://www.safecleanup.com. Much the same process was applied to the Point Aconi controversy: Randy Jones, "Citizen's Groups Want to Be Taken Seriously," *Mail-Star*, March 3, 1993, D5.

7 Dr. Beth MacCormick and Leon Dubinsky, to EAC, n.d., DAL-EAC, box 45.11.

8 Shepard Krech, *The Ecological Indian: Myth and History* (New York: W.W. Norton, 1999).

9 "The Battle of Kelly's Mountain," 1989, Acadia University Esther Clark Wright Archives, Kings Environment Group fonds [hereafter KEG], file 74; Hornborg, "Environmentalism, Ethnicity and Sacred Places."

    The business of quarrying gravel for the US market moved on to Digby County soon after the Kelly's Mountain failure, and spent many years attempting to find a foothold there: "Society for the Preservation of the Eastern Head," August 13, 1990, KEG, file 81.

10 EAC to "Friends," March 8, 1984, DAL-EAC, box 29.6; "Plan 'B': The Administrative Law Route," n.d., DAL-EAC, box 24.9.

11 "Grant Application to Public Awareness Program for Science and Technology" (rejected), May 14, 1984, DAL-EAC, box 37.10; "Grant Application to Public Awareness Program for Science and Technology" (rejected), October 15, 1985, DAL-EAC, box 37.10; "Report on FOE Pesticides Lobby," 1983, DAL-EAC, box 37.8. EAC also pursued more consultation with the province: memo, 1985, PANS-EAC, vol. 3433, no. 27.

12 Clipping file, PANS-BP, vol. 3469, no. 8; David Israelson, *Toronto Star*, January 10, 1987, clipping file, PANS-EAC, vol. 3433, no. 29.

13 Ralph Surette, "Fear of Spraying," *Atlantic Insight* 5, no. 7 (July 1983): 32. Forester and author Ralph Johnson, for example, had his history of the forests of Nova Scotia edited by government printers to remove anti-clearcutting sections. Johnson to T.C. De Fayer, February 7, 1985, Public Archives of Nova Scotia, Ralph S. Johnson fonds, vol. 2863, no. 40, cited in Paul Webster, "Pining for Trees: The History of Dissent against Forest Destruction in Nova Scotia 1749–1991" (MA thesis, Dalhousie University, 1991).

14 *Q/L SSEPA News* 2, no. 3 (September/October 1983).

15 *Chronicle-Herald,* August 31, 1983, 2.

16 *Q/L SSEPA News* 2, no. 3 (September/October 1983).

17 Duncan MacLellan, "A Study of Selected New Brunswick, Nova Scotia, Regional, and National Forestry Sector Interest Groups," report for the Council of Maritime Premiers, n.d., DAL-EAC, box 46.13.

18 David Orton, "Deep Ecology and the Green Movement," 1986–87, Greenweb, http://home.ca.inter.net/~greenweb/DE&Green_Movement.html.

19 Diana Cole, "National ENGO Meeting in Ottawa, 1983," *Q/L SSEPA News* 2, no. 1 (June 1983).

20 Presentation by Nova Scotia Department of Environment, Halifax hearing, January 27, 1983, RCOF, vol. 158b, no. 3; "EAC Response to the NS Pest Control Products Act," October 1985, DAL- EAC, box 41.15.

21 David Orton, "Informed Consent or Informed Rejection of Pesticide Use: A Concept for Environmental Action," *Philosophy and Social Action* 16, no. 4 (October/December 1990), KEG, file 61.

22 Charlie Restino, CAP, to Premier Buchanan, May 26, 1988, KEG, file 61.

23 Jerry White, Sprayers of Dioxin, to Neil Livingston, September 10, 1984, DAL-EAC, box 24.9; Sherri Cline and Judy Davis, to Neil Livingston, October 15, 1984, DAL-EAC, box 24.9.

24 Farley Mowat, "Grass Roots Crusader: Elizabeth May and the Budworm Battle," in *Rescue the Earth; Conversations with the Green Crusaders* (Toronto: McLelland and Stewart, 1990), 188–205; *Between the Issues* 4, no. 2 (July/August 1984).

25 Quoted in David Orton, "Problems Facing the Green Movement in Canada and Nova Scotia – Greenweb Bulletin #17," *Greenweb,* January 1990, http://home.ca.inter.net/~greenweb/%20Problems_Green_Movement.pdf, accessed Summer 2013.

26 Ibid.; Webster, "Pining for Trees," 209.

27 "II. Energy Conservation – Economic Aspects," CCNR briefs, June 3, 1980, PANS-EAC, vol. 3422, no. 2; "A Soft Energy Path for Canada," *Alternatives* 12, no. 1 (Fall 1984); "2025: Soft Energy Futures for Canada – 1988 Update," DAL-EAC, box 12.8.

28 Susan Holtz, report on Environment Canada's stakeholder group – State of the Environment Advisory Group, May 1986, DAL-EAC, box 32.16; Deborah Jones, "Energy Conservation Saving Money in Atlantic Canada," *Atlantic Insight* 6, no. 11 (November 1984): 19–21; Susan Holtz, "Energy Policy from an Environmental Perspective: The Core Elements," September 1987, DAL-EAC, box 15.15.

29 Susan Holtz, "The Philosophers Confer," *Between the Issues* 3, no. 6 (February/March 1984).

30 "Proposal for an Atlantic Environmental Newsletter" to Environment Canada by EAC, February 12, 1981, DAL-EAC, box 44.3.

31 EAC Board of Directors Meeting minutes, November 5, 1988, PANS-EAC, vol. 3421, no. 6.

32 Susan Holtz to "environmentalists," October 7, 1980, PANS-EAC, vol. 3434, no. 9; Susan Holtz, "Report on 1987-88 Activities of Ecology Action Center Senior Researcher Susan

Holtz," April 30, 1988, PANS-EAC, vol. 3421, no. 6; Annual Report 1987–88, April 30, 1988, PANS-EAC, vol. 3421, no. 6.

33 EAC Annual Report 1987–88, April 30, 1988, PANS-EAC, vol. 3421, no. 6.

34 EAC Board of Directors Meeting minutes, February 25, 1981, PANS-EAC, vol. 3420, no. 25. Membership supplied $10,000 to $11,000, less than a third of the centre's total funds, in 1986, for example, declining to only about $8,000 the following year. "EAC Unaudited Financial Statements for the Year Ended March 31, 1986," DAL-EAC, box 23.19; "EAC Unaudited Financial Statements for the Year Ended March 31, 1987," DAL-EAC, box 23.19.

35 EAC Board of Directors Meeting minutes, December 4, 1985, PANS-EAC, vol. 3421, no. 5.

36 EAC Board of Directors Meeting minutes, multiple dates, 1985, PANS-EAC, vol. 3421, no. 3.

37 EAC Board of Directors Meeting minutes, May 15, 1985, PANS-EAC, vol. 3421, no. 5; EAC Board of Directors Meeting minutes, March 25, 1981, PANS-EAC, vol. 3420, no. 25; Lois Corbett, "Co-Director's Report," April 21, 1990, PANS-EAC, vol. 3421, no. 8.

38 EAC Board of Directors Meeting minutes, June 13, 1988, PANS-EAC, vol. 3421, no. 6.

39 David Orton, "Problems Facing the Green Movement in Canada and Nova Scotia – Greenweb Bulletin #17," Greenweb, http://home.ca.inter.net/~greenweb/%20Problems_Green_Movement.pdf, accessed Summer 2013.

40 EAC press release, October 9, 1986, DAL-EAC, box 38.15; "Application for Approval – Resource Management and Pollution Control Division, NSPC," 1990, KEG, file 66.

41 Ian Sherman to the Editor, Scotia Sun, n.d., RCOF, vol. 191, file 1 [emphasis in original].

42 Michael Clow, "A Left-Environmentalist Perspective on Canadian Industrial Strategy," paper presented at the Canadian Political Science Association 1981 Annual Meeting, May 27–29, 1981, DAL-EAC, box 44.4. An earlier example of socialist ecological thought in Canada was the editorial position of the radical leftist journal Our Generation, which insisted on the common opposition to "a system of natural and human exploitation and domination." Our Generation 7, no. 1 (January-February 1970): 4.

43 Charlie Restino, article clipping, 1987, DAL-EAC, box 29.9.

44 David Orton, "Discussion Paper: The Green Movement and Our Place in It – Position of the North Shore Environmental Web," 1988, Greenweb, http://home.ca.inter.net/~greenweb/NSEW_Position_on_Green_Movement.pdf.

45 Orton, "Informed Consent or Informed Rejection of Pesticide Use"; Patrick Curry, Ecological Ethics: An Introduction (Cambridge: Polity, 2006), 84–88.

46 "Gathering of the Greens," August 3–5, 1990, PANS-BP, vol. 3476, no. 12; "NSEN," 1990, KEG, file 66.

47 Holtz, "The Philosophers Confer."

48 David Orton, "The Greens: An Introduction," in Toward a New Maritimes: A Selection from Ten Years of New Maritimes, ed. Scott Milsom and Ian McKay (Charlottetown: Ragweed Press, 1992 [1990]), 281–85.

49 David Orton, "The North Shore Environmental Web," n.d., PANS-EAC, vol. 3423, no. 14; Letter, Elizabeth May to "Susans," n.d., DAL-EAC, box 44.1; David Orton, "Forestry Herbicide Use: A Hazard to Our Environment," 1988, Greenweb, http://home.ca.inter.net/~greenweb/Forestry_Herbicide_Use.pdf; Q/L SSEPA News 2, no. 1 (June 1983).

50 "Building a Green Nova Scotia," NSEN submission to Nova Scotia Roundtable on Environment and Economy, 1991, DAL-EAC, box 19.4; "AEN Meeting: General Footage, May 11–13, 1990," video, DAL-EAC, box 5.53.

51 "International Uranium Congress," June 15, 1988, PANS-EAC, vol. 3424, no. 28.

52 The three-part relationship finds echoes in, for example, Jennifer Read, "'Let Us Heed the Voice of Youth': Laundry Detergents, Phosphates, and the Emergence of the Environmental Movement in Ontario," *Journal of the Canadian Historical Association* 7, no. 1 (1996): 227–50; Jane Barr, "The Origins and Emergence of Quebec's Environmental Movement: 1970–1985" (MA thesis, McGill University, 1995); Frank Zelko, "Making Greenpeace: The Development of Direct Action Environmentalism in British Columbia," *BC Studies* 142/143 (Summer 2004): 197–239; Ryan O'Connor, *The First Green Wave: Pollution Probe and the Origins of Environmental Activism in Ontario* (Vancouver: UBC Press, 2014); Van Huizen, "Flooding the Border"; and John-Henry Harter, "Environmental Justice for Whom? Class, New Social Movements, and the Environment: A Case Study of Greenpeace Canada, 1971–2000," *Labour/Le Travail* 54 (Fall 2004): 83–119.

53 Dominique Clément, *Canada's Rights Revolution* (Vancouver: UBC Press, 2008), 204. The US comparison may be made with Paul Sutter, *Driven Wild: How the Fight against Automobiles Launched the Modern Wilderness Movement* (Seattle: University of Washington Press, 2002).

54 Hornborg, "Environmentalism, Ethnicity, and Sacred Places," 259.

55 Ibid., 263.

56 Marie Vandergraaf, transcripts of New Ross hearing, hearing #1, April 2, 1982, RCU, vol. 195, no. 12.

57 As remarked by Rod Bantjes (and in Chapter 4), this is not the story often told of a female-dominated movement: Bantjes and Trussler, "Feminism and the Grass Roots," 183. For similar findings in places outside Nova Scotia, see S. Cable, "Women's Social Movement Involvement: The Role of Structural Availability in Recruitment and Participation Processes," *Sociological Quarterly* 33, no. 1 (1992): 35–50. The mythology, however, has had a strong grip on the movement itself, as seen in Elizabeth May's statement that "women are essentially *different* from men ... operate more from a left-brained intuitive thought process. We are biologically and spiritually connected to the cosmos ... by nature, much more *selfless* than men." May, "Gaia Women," in Farley Mowat, ed., *Rescue the Earth! Conversations with the Green Crusaders* (Toronto: McClelland and Stewart, 1990), 249.

58 May, *Budworm Battles*, 116; May, "Gaia Women."

Bibliography

PRIMARY SOURCES

Archives

Acadia University Esther Clark Wright Archives
Harrison Flint Lewis fonds, accession 2006.020-LEW
Kings Environment Group fonds, accession 1996–001-KEG

Dalhousie University Archives and Special Collections
Dalhousie University Institute of Public Affairs/Henson College fonds, UA-26
Ecology Action Centre fonds, MS-11-13
Movement for Citizens Voice and Action fonds, MS-11-1

Personal Papers
Brian Gifford papers, in the possession of Brian Gifford, Halifax, Nova Scotia
Rudy Haase papers, in the possession of Rudy Haase, Chester, Nova Scotia

Public Archives of Nova Scotia (PANS)
Department of Lands and Forests fonds, MG20
Ecology Action Centre fonds, MG20
Man and Resources Program fonds, MG20
McInnes fonds, RG44
Nova Scotia Royal Commission on Forestry fonds, RG44
Betty Peterson fonds, MG1
Gerald Regan fonds, RG100
Royal Commission on Uranium Mining fonds, RG44
South Shore Environmental Protection Association fonds, MG20
Universalist Unitarian Church fonds, MG4

Interviews

Bishop, Roy, January 16, 2012, via telephone
Cameron, Silver Donald, November 1, 2011, Halifax, Nova Scotia
Gifford, Brian, February 22, 2013, Halifax, Nova Scotia
Haase, Martin Rudy, October 9, 2011, Chester, Nova Scotia
Ruffman, Alan, February 21, 2012, Halifax, Nova Scotia

Periodicals

4th Estate (Halifax)
Atlantic Insight
Between the Issues (Halifax)
Cape Breton Post
Chronicle-Herald (Halifax)
Daily News (Halifax)
Dartmouth Free Press
Evening News (New Glasgow)
Forest Times (Halifax)
Globe and Mail (Toronto)
Halifax Field Naturalists Newsletter
Hants Journal
Mail-Star (Halifax)
Montreal Gazette
Montreal Star
Mysterious East (Fredericton)
Nuclear Reaction (New Brunswick)
Queens/Lunenburg SSEPA News
Rural Delivery (Liverpool)
Truro Daily News

Government Reports

Department of Lands and Forests. *Nova Scotia Forest Inventory: Cape Breton Island Subdivision 1970.* Halifax: Lands and Forests, 1970.
Government of Nova Scotia. *Supplement to the Public Accounts of the Province of Nova Scotia for the Year Ended March 31, 1977.* Halifax: Queen's Printer, 1978.
McInnes, Russell. *Report of the Royal Commission into the Nova Scotia Light and Power Company Ltd. Gold River Hydro Development.* Halifax: Government of Nova Scotia, 1961.
Province of Nova Scotia. *Annual Report: Environmental Control Council: For the Year Ending Dec. 31 1974.* Department of Environment Library, reference number 354.33.
Province of Nova Scotia, Department of Mines and Energy. *Uranium in Nova Scotia: A Background Summary for the Uranium Inquiry – Nova Scotia, Report 82–7.* Halifax: Government of Nova Scotia, 1982.
Report of the Nova Scotia Royal Commission on Uranium Mining. Halifax: Queen's Printer, 1985.

Report of the Task Force Operation Oil (Clean-up of the Arrow Oil Spill in Chedabucto Bay) to the Minister of Transport. Ottawa: Information Canada, 1970.

Rust Associates. *A Review of the Boat Harbour Waste Treatment Facilities for Nova Scotia Water Resources Commission*. Montreal: Rust Associates, 1970.

Science Council of Canada. *Canada as a Conserver Society: Resource Uncertainties and the Need for New Technologies*. Report No. 27. Ottawa: SCC, 1977.

–. *Natural Resource Policy Issues in Canada*. Report No.19. Ottawa: SCC, 1973.

Statistics Canada. *Census Divisions and Subdivisions, Population, Occupied Private Dwellings, Private Households and Census and Economic Families in Private Households, Selected Social and Economic Characteristics, Nova Scotia*. Toronto: Statistics Canada, 1983. http://archive.org/details/1981959851983engfra.

–. "Population, Urban and Rural, by Province and Territory." http://www.statcan.gc.ca/tables-tableaux/sum-som/l01/cst01/demo62d-eng.htm.

Trider, G.L., and O.C. Vaidya. *An Assessment of Liquid Effluent Streams at Sydney Steel Corporation*. Ottawa: Environment Canada, Environmental Protection Service, Atlantic Region, 1980.

Other Primary Sources

Atlantic Canada Chapter, Northeast Atlantic International Section, Air Pollution Control Association. *Proceedings of the Sixth Annual Technical Meeting*. September 29–30, 1982.

Bedford Basin Pollution Committee. *Report of the Bedford Basin Pollution Committee of the Council of the Bedford Service Commission*. Bedford: Bedford Service Commission, 1970.

Bott, Robert, David Brooks, and John Robinson. *Life after Oil*. Edmonton: Hurtig, 1983.

Chanclor, William, and Pat Murphy. "Robert Whiting: What One Man Can Do." *Axiom* 2, no. 4 (May 1976): 6–9, 21.

Clow, Michael. "A Left-Environmentalist Perspective on Canadian Industrial Strategy." Paper presented at the Ecology Energy and Resources section of the Canadian Political Science Association 1981 Annual Meeting, Dalhousie University, Halifax, Nova Scotia, May 27–29, 1981.

Editorial. *Our Generation* 7, no. 1 (January-February 1970): 4.

Edwards, Gordon. "Nuclear Wastes: What, Me Worry?" 1978. http://www.ccnr.org/me_worry.html.

Folster, David. "Double Trouble – Power Bubble." *Maclean's*, October 6, 1980, 38–39.

Heritage Trust of Nova Scotia. *Annual Report 1970–1971*. Halifax: The Trust, 1971.

Holtz, Susan. *2025: Soft Energy Futures for Canada*. Vol. 3, *A Soft Energy Path for Nova Scotia*. Ottawa: Energy, Mines and Resources, 1983.

Laski, Heather. "A Costly Fight in Cape Breton." *Maclean's*, May 9, 1983, 20.

Lovins, Amory, and Hunter Lovins. "Introduction." In *Life after Oil*, by Robert Bott, David Brooks, and John Robinson. Edmonton: Hurtig, 1983.

May, Elizabeth. *Budworm Battles*. Halifax: Four East, 1982.

–. "A Case History of Citizen Action Victory." *PEI Environeer* 6, no.3 (1978): 23.

"Moncton Conference Notes: 'Experts in Society' by Dr Woollard." *Southern New Brunswick Nuclear News* 15 (February-March 1981).

Rushton, Don. "So Long, Stage II: McCleave Gets Set to Ride Off into the Sunset." *New Maritimes* 2, no. 6 (March 1984): 10–11.

Smyth, Donna. "Selling Radon Daughters: Uranium in Nova Scotia." *Canadian Forum* 61, no. 714 (December/January 1981): 6–8.

–. *Subversive Elements.* Toronto: Women's Press, 1986.

"Special Report: A Soft Energy Path for Canada." *Alternatives* 12 (Fall 1984): 1.

Torrie, Ralph. "BC's Inquiry and Moratorium." *CCNR's Transitions* 3, no. 1 (May 1980): 1.

SECONDARY SOURCES

AECL. *Canada Enters the Nuclear Age.* Montreal: McGill-Queen's University Press, 1997.

Almeida, Paul, and Linda Brewster Stearns. "Political Opportunities and Local Grassroots Environmental Movements: The Case of Minamata." *Social Problems* 45, no. 1 (February 1998): 37–60. http://dx.doi.org/10.2307/3097142.

Arsenault, Chris. *Blowback: A Canadian History of Agent Orange and the War at Home.* Halifax: Fernwood, 2009.

Ashforth, Adam. "Reckoning Schemes of Legitimation: On Commissions of Inquiry as Power/Knowledge Forms." *Journal of Historical Sociology* 3, no. 1 (March 1990): 1–22. http://dx.doi.org/10.1111/j.1467-6443.1990.tb00143.x.

Bantjes, Rod, and Tanya Trussler. "Feminism and the Grass Roots: Women and Environmentalism in Nova Scotia, 1980–1983." *Canadian Review of Sociology and Anthropology* 36, no. 2 (1999): 179–98. http://dx.doi.org/10.1111/j.1755-618X.1999.tb01274.x.

Barlow, Maude, and Elizabeth May. *Frederick Street: Life and Death on Canada's Love Canal.* Toronto: HarperCollins, 2000.

Barr, Jane. "The Origins and Emergence of Quebec's Environmental Movement: 1970–1985." MA thesis, McGill University, 1995.

Blum, Elizabeth. *Love Canal Revisited: Race, Class, and Gender in Environmental Activism.* Lawrence: University Press of Kansas, 2008.

Bothwell, Robert. *Nucleus: The History of Atomic Energy of Canada Limited.* Toronto: University of Toronto Press, 1988.

Cable, S. "Women's Social Movement Involvement: The Role of Structural Availability in Recruitment and Participation Processes." *Sociological Quarterly* 33, no. 1 (1992): 35–50. http://dx.doi.org/10.1111/j.1533-8525.1992.tb00362.x.

Cameron, Silver Donald. *The Education of Everett Richardson: The Nova Scotia Fishermen's Strike 1970–71.* Toronto: McLelland and Stewart, 1977.

Campbell, Claire, ed. *A Century of Parks Canada, 1911–2011.* Calgary: University of Calgary, 2011.

Carson, Rachel. *Silent Spring.* New York: Fawcett Crest, 1962.

Clément, Dominique. *Canada's Rights Revolution.* Vancouver: UBC Press, 2008.

Conrad, Margaret. "The Atlantic Revolution of the 1950s." In *Beyond Anger and Longing: Community and Development in Atlantic Canada,* ed. Berkeley Fleming, 55–98. Fredericton: Acadiensis, 1988.

Crossley, Nick. *Making Sense of Social Movements.* Buckingham, UK: Open University Press, 2002.

Curry, Patrick. *Ecological Ethics: An Introduction.* Cambridge: Polity, 2006.

De Marchi, Bruna. "Seveso: From Pollution to Regulation." *International Journal of Environment and Pollution* 7, no. 4 (September 1997): 526–37.

Dobson, Andrew. *Green Political Thought.* London: Routledge, 2000 (original work published in 1990).

Doern, Bruce, and Thomas Conway. *The Greening of Canada: Federal Institutions and Decisions*. Toronto: University of Toronto Press, 1994.

Early, Francis. "'A Grandly Subversive Time': The Halifax Branch of the Voice of Women in the 1960s." In *Mothers of the Municipality*, ed. Judith Fingard and Janet Guildford, 253–80. Toronto: University of Toronto Press, 2005.

Edwards, Gordon. "Canada's Nuclear Industry and the Myth of the Peaceful Atom." In *Canada and the Nuclear Arms Race*, ed. Ernie Regehr and Simon Rosenblum, 122–70. Toronto: James Lorimer, 1983.

Egbers, Adrian. "Going Nuclear: The Origins of New Brunswick's Nuclear Industry, 1950–1983." MA thesis, Dalhousie University, 2008.

Environment Canada. Emergencies Science and Technology Division. "Bunker C Fuel Oil (Irving Whale)." Ottawa: n.d. http://www.etc-cte.ec.gc.ca/databases/Oilproperties.

Evans, Sterling. *The Green Republic: A Conservation History of Costa Rica*. Austin: University of Texas Press, 1999.

Evenden, Matthew. *Fish versus Power: An Environmental History of the Fraser River*. Cambridge: Cambridge University Press, 2004. http://dx.doi.org/10.1017/CBO9780511512032.

Forbes, Ernest. *The Maritime Rights Movement 1919–1927: A Study in Canadian Regionalism*. Montreal and Kingston: McGill-Queen's University Press, 1979.

Froese-Stoddard, Alison. "Ship Harbour National Park: The Breakdown between National Agendas and Local Interests." Unpublished manuscript (2011–12). Adapted to Alison Froese-Stoddard, "A Conservation 'Could Have Been': Ship Harbour National Park," NiCHE Canada, *The Otter* blog, http://niche-canada.org/2012/08/02/a-conservation-could-have-been-ship-harbour-national-park/.

Giddens, Anthony. *The Consequences of Modernity*. Stanford, CA: Stanford University Press, 1990.

Gottlieb, Robert. *Forcing the Spring: The Transformation of the American Environmental Movement*. Washington, DC: Island Press, 1993.

Guha, Ramachandra. *Environmentalism: A Global History*. New York: Longman, 2000.

–. *How Much Should a Person Consume? Environmentalism in India and the United States*. Los Angeles: University of California Press, 2006.

–. "Radical American Environmentalism and Wilderness Preservation: A Third World Critique." *Environmental Ethics* 11, no. 1 (1989): 71–83. http://dx.doi.org/10.5840/enviroethics198911123.

–. *The Unquiet Woods: Ecological Change and Peasant Resistance in the Himalaya*. 2nd ed. Oxford: Oxford University Press, 1989.

Guha, Ramachandra, and Juan Martinez-Alier. *Varieties of Environmentalism: Essays North and South*. London: Earthscan, 1997.

Habermas, Jürgen. "New Social Movements." *Telos* 1981, no. 49 (1981): 33–37. http://dx.doi.org/10.3817/0981049033.

Haq, Gary, and Alistair Paul. *Environmentalism since 1945*. London: Routledge, 2012.

Hardin, Garret. "The Tragedy of the Commons." *Science* 162, no. 3859 (December 13, 1968): 1243–48. http://dx.doi.org/10.1126/science.162.3859.1243.

Harding, Jim. *Canada's Deadly Secret: Saskatchewan Uranium and the Global Nuclear System*. Halifax: Fernwood, 2007.

Harley, J. Brian. "Maps, Knowledge, and Power." In *The Iconography of Landscape*, ed. D. Cosgrove and S. Daniels, 277–312. Cambridge: Cambridge University Press, 1988.

Harter, John-Henry. "Environmental Justice for Whom? Class, New Social Movements, and the Environment: A Case Study of Greenpeace Canada, 1971–2000." *Labour/Le Travail* 54 (Fall 2004): 83–119. http://dx.doi.org/10.2307/25149506.

Hays, Samuel. *Beauty, Health, and Permanence: Environmental Politics in the United States, 1955–1985.* Cambridge: Cambridge University Press, 1987. http://dx.doi.org/10.1017/CBO9780511664106.

Helvarg, David. *The War against the Greens: The 'Wise Use' Movement, the New Right, and Anti-Environmental Violence.* San Francisco: Sierra Club Books, 1994.

Herrington, William, and George Rounsefell. "Restoration of the Atlantic Salmon in New England." *Transactions of the American Fisheries Society* 70, no. 1 (1941): 123–27. http://dx.doi.org/10.1577/1548-8659(1940)70[123:ROTASI]2.0.CO;2.

Hornborg, Alf. "Environmentalism, Ethnicity and Sacred Places: Reflections on Modernity, Discourse and Power." *Canadian Review of Sociology and Anthropology* 31, no. 3 (1994): 245–67. http://dx.doi.org/10.1111/j.1755-618X.1994.tb00948.x.

–. "Undermining Modernity: Protecting Landscapes and Meanings among the Mi'kmaq of Nova Scotia." In *Political Ecology across Spaces, Scales, and Social Groups*, ed. Susan Paulson and Lisa Gezon, 196–216. New Brunswick, NJ: Rutgers University Press, 2004.

Hutchison, George, and Dick Wallace. *Grassy Narrows*. Scarborough, ON: Van Nostrand Reinhold, 1977.

Huxley, Aldous. *Do What You Will*. London: Chatto and Windus, 1956 (original work published in 1929).

Inglehart, Ronald. *The Silent Revolution: Changing Values and Political Styles among Western Publics*. Princeton, NJ: Princeton University Press, 1977.

Kenny, James, and Andrew Secord. "Engineering Modernity: Hydroelectric Development in New Brunswick, 1945–1970." *Acadiensis* (Fredericton) 39, no. 1 (Winter/Spring 2010): 3–26.

Kinkela, David. *DDT and the American Century*. Chapel Hill: University of North Carolina Press, 2011.

Krech, Shepard. *The Ecological Indian: Myth and History*. New York: W.W. Norton, 1999.

Kuhn, Thomas. *The Structure of Scientific Revolutions*. Chicago: University of Chicago Press, 1962.

Latour, Bruno. *We Have Never Been Modern*. Cambridge, MA: Harvard University Press, 1993 (original work published in 1991).

Loo, Tina. "People in the Way: Modernity, Society, and Environment on the Arrow Lakes." *BC Studies* 142/143 (Summer/Autumn 2004): 161–96.

–. *States of Nature: Conserving Canada's Wildlife in the Twentieth Century*. Vancouver: UBC Press, 2006.

MacEachern, Alan. *The Institute of Man and Resources: An Environmental Fable*. Charlottetown: Island Studies Press, 2003.

–. *Natural Selections: National Parks in Atlantic Canada, 1935–1970*. Montreal and Kingston: McGill-Queen's University Press, 2001.

Martinez-Alier, Juan. *Ecological Economics: Energy, Environment, and Society*. Oxford: Basil Blackwell, 1990.

–. *The Environmentalism of the Poor: A Study of Ecological Conflicts and Valuation*. Northhampton, UK: Edward Elgar, 2002. http://dx.doi.org/10.4337/9781843765486.

Martini, Edwin. *Agent Orange: History, Science, and the Politics of Uncertainty.* Boston: University of Massachusetts Press, 2012.

May, Elizabeth. "Gaia Women." In *Rescue the Earth! Conversations with the Green Crusaders,* ed. Farley Mowat, 247–65. Toronto: McClelland and Stewart, 1990.

McLaughlin, Mark. "Green Shoots: Aerial Insecticide Spraying and the Growth of Environmental Consciousness in New Brunswick, 1952–1973." *Acadiensis* (Fredericton) 40, no. 1 (Winter/Spring 2011): 3–23.

Meadows, Donella, Dennis Meadows, Jørgen Randers, and William Behrens III. *The Limits to Growth.* New York: Universe, 1972.

Mersey Tobeatic Research Institute. "Mersey Messages – Old Lake Rossignol." March 10, 2009. http://www.merseytobeatic.ca/pdfs/Mersey%20Messages/Mersey%20Messages %20March%2010%202009.pdf.

Mowat, Farley. *Rescue the Earth; Conversations with the Green Crusaders.* Toronto: McLelland and Stewart, 1990.

Naess, Arne. *The Ecology of Wisdom: Writings by Arne Naess.* Berkeley: Counterpoint, 2008.

Nelson, Jennifer. *Razing Africville: A Geography of Racism.* Toronto: University of Toronto Press, 2008.

O'Connor, Ryan. *The First Green Wave: Pollution Probe and the Origins of Environmental Activism in Ontario.* Vancouver: UBC Press, 2014.

Orton, David. "The Greens: An Introduction." In *Toward a New Maritimes: A Selection from Ten Years of New Maritimes,* ed. Scott Milsom and Ian McKay, 281–85. Charlottetown: Ragweed Press, 1992 (original work published 1990).

Parra, Francisco. *Oil Politics: A Modern History of Petroleum.* London: I.B. Tauris, 2004.

Pasternak, Judy. *Yellow Dirt: A Poisoned Land and a People Betrayed.* New York: Free Press, 2010.

Pepper, David. *Modern Environmentalism.* London: Routledge, 1996. http://dx.doi.org/ 10.4324/9780203412244.

Rangarajan, Mahesh. *Fencing the Forest: Conservation and Ecological Change in India's Central Provinces 1860–1914.* Delhi: Oxford University Press, 1996.

Read, Jennifer. "'Let Us Heed the Voice of Youth': Laundry Detergents, Phosphates, and the Emergence of the Environmental Movement in Ontario." *Journal of the Canadian Historical Association* 7, no. 1 (1996): 227–50. http://dx.doi.org/10.7202/031109ar.

Robinson, Danielle. "Modernism at a Crossroad: The Spadina Expressway Controversy in Toronto, Ontario ca. 1960–1971." *Canadian Historical Review* 92, no. 2 (June 2011): 295–322.

Rootes, Christopher, ed. *Environmental Movements: Local, National, and Global.* Portland, OR: Frank Cass, 1999.

Sandberg, L. Anders. "Forest Policy in Nova Scotia: The Big Lease, Cape Breton Island, 1899–1960." *Acadiensis* (Fredericton) 20, no. 2 (Spring 1991): 105–28.

Savoie, Donald. *Regional Economic Development: Canada's Search for Solutions.* Toronto: University of Toronto, 1992.

Savoie, Donald, and John Chenier. "The State and Development: The Politics of Regional Development Policy." In *Still Living Together: Recent Trends and Future Directions in Canadian Regional Development,* ed. William Coffey and Mario Polese, 407–22. Montreal: Institute for Research on Public Policy, 1998.

Schumacher, E.F. *Small Is Beautiful.* London: Blond and Briggs, 1973.

Scott, James C. *Seeing Like a State: How Certain Schemes to Improve the Human Condition Have Failed.* New Haven, CT: Yale University Press, 1998.

Smith, Jennifer. "Intergovernmental Relations, Legitimacy, and the Atlantic Accords." *Constitutional Forum* 17, no. 3 (2008): 81–98.

Sutter, Paul. *Driven Wild: How the Fight against Automobiles Launched the Modern Wilderness Movement.* Seattle: University of Washington Press, 2002.

Thomas, P. "The Kouchibouguac National Park Controversy: Over a Decade Strong." *Park News* 17, no. 1 (1981): 11–13.

Thompson, E.P. *Customs in Common.* London: Merlin Press, 1991.

Thurston, Harry. *The Sea among the Rocks.* East Lawrencetown, NS: Pottersfield Press, 2002.

Van Huizen, Philip. "Flooding the Border: Development, Politics, and Environmental Controversy in the Canadian-U.S. Skagit Valley." PhD dissertation, University of British Columbia, 2013.

Vandermeulen, J.H., and D.E. Buckley. "The Kurdistan Oil Spill of March 16–17, 1979: Activities and Observations of the Bedford Institute of Oceanography Response Team." Canadian Technical Report of Hydrography and Ocean Sciences No. 35. Ottawa, January 1985. http://www.dfo-mpo.gc.ca/Library/90618.pdf.

Webster, Paul. "Pining for Trees: The History of Dissent against Forest Destruction in Nova Scotia 1749–1991." MA thesis, Dalhousie University, 1991.

Worster, Donald. *Nature's Economy.* Cambridge: Cambridge University Press, 1998.

Wright, Miriam. *A Fishery for Modern Times: The State and Industrialization of the Newfoundland Fishery, 1934–1968.* Don Mills, ON: Oxford, 2001.

Zelko, Frank. "Making Greenpeace: The Development of Direct Action Environmentalism in British Columbia." *BC Studies* 142/143 (Summer 2004): 197–239.

Index

NATURE|HISTORY|SOCIETY

GENERAL EDITOR: GRAEME WYNN